60 9702540 4

AF616348

Computer Analysis of Structural Frameworks

Second Edition

James A. D. Balfour
BSc, PhD, CEng, MICE

OXFORD
BLACKWELL SCIENTIFIC PUBLICATIONS
LONDON EDINBURGH BOSTON
MELBOURNE PARIS BERLIN VIENNA

Blackwell Scientific Publications
Editorial Offices:
Osney Mead, Oxford OX2 0EL
25 John Street, London WC1N 2BL
23 Ainslie Place, Edinburgh EH3 6AJ
3 Cambridge Center, Cambridge,
Massachusetts 02142, USA
54 University Street, Carlton
Victoria 3053, Australia

Other Editorial Offices:
Librairie Arnette SA
2, rue Casimir-Delavigne
75006 Paris
France

Blackwell Wissenschafts-Verlag
Meinekestrasse 4
D-1000 Berlin 15
Germany

Blackwell MZV
Feldgasse 13
A-1238 Wien
Austria

First Edition published 1986 by
Collins Professional and Technical Books
Second Edition published 1992

Printed and bound in Great Britain by
Hartnolls Ltd, Bodmin, Cornwall

DISTRIBUTORS

Marston Book Services Ltd
PO Box 87
Oxford OX2 0DT
(Orders: Tel: 0865 791155
Fax: 0865 791927
Telex: 837515)

USA
Blackwell Scientific Publications, Inc.
3 Cambridge Center
Cambridge, MA 02142
(*Orders:* Tel: 800 759-6102
617 225-0401)

Canada
Oxford University Press
70 Wynford Drive
Don Mills
Ontario M3C 1J9
(*Orders:* Tel: 416 441-2941)

Australia
Blackwell Scientific Publications
(Australia) Pty Ltd
54 University Street
Carlton, Victoria 3053
(*Orders:* Tel: 03 347-0300)

A catalogue record for this book is available from the British Library.

ISBN 0–632–02912–9

A Library of Congress catalogue reference is available.

Contents

Preface

Practising engineers now use computers to solve all non-trivial problems of structural analysis. Almost certainly the computer program used will be based upon the stiffness method, and equally certainly it will not have been written by the engineer using it.

To obtain a solution to a structural analysis problem the engineer first selects a suitable computer program and then prepares data describing the problem in accordance with the user manual. Say, for instance, a static analysis is required, then the data will consist of member properties and locations, restraints, material properties, and details of the loading. Using the supplied data the computer then performs the structural analysis and outputs the displacements and stresses (or stress resultants). If buckling, plastic or dynamic analysis is required then the input and output will, of course, be different, but the basic procedure will be the same.

The twin facts that the stiffness method is extremely general and powerful, and computer programs are expensive to develop has divided the engineering profession into two groups: a small group of specialist program writers, and a much larger group of practising engineers who use the programs. Today's engineers have at their fingertips an analysis capability undreamt of by previous generations of engineers, and it is possible to wield that power without knowing anything of its source. In general, if the input is correct then so too is the output. This raises a thorny issue; if the practising engineer *need* know nothing about structural analysis programs in order to use them, then how much *should* he know? Opinions on this matter vary. I am convinced, however, that the practising engineer should know enough about structural analysis programs to enable him to:

- Idealise structures in a sensible manner.
- Prepare input data knowing, at least in general terms, how it will be used by the computer.

- Interpret the computer output.
- Know when, and why, a structure is beyond the capabilities of a particular computer program.
- Select, from a number of different computer programs the one best suited to the problem in hand.
- Devise cross checks when problems arise.

This book is intended to give engineering students and practising engineers the background knowledge necessary to make safe and efficient use of commercial frame analysis programs. If, after working through the text, the reader feels that he has mastered the six skills listed above then the author's efforts will have been worthwhile. This book is not intended for the expert structural analyst, but it should serve as an introductory text to more advanced studies as the methodology and notation have been chosen to make easy the transition to finite element theory.

In my experience applying newly learned concepts to practical problems is an essential part of the learning process. This makes the teaching of computer methods of structural analysis difficult because computer methods are intended for use on a computer. When computer methods are applied by hand the volume of arithmetic generated by even the simplest problem is likely to dull the enthusiasm of the keenest student. One way out of this difficulty is to quickly move from hand calculation to solving problems using a standard structural analysis program. While this will familiarise the student with the mechanics of using a standard program it is unlikely to give him much insight into how the program works. Unfortunately standard structural analysis programs tend to appear as 'black boxes' to their users with input going in one end and solution appearing magically at the other.

This book tries to avoid these difficulties by using short computer programs to perform each of the standard procedures used in commercial structural analysis programs. Each program is complete in itself, and all were written with the sole aim of clarity (efficiency, compactness and elegance were non-considerations). By transferring the burden of arithmetic calculation to the computer these programs allow the analysis of reasonably complex structures to be undertaken, yet leave control of the analysis procedure in the hands of the student. Also presented are automatic structural analysis programs for a number of different framework types. These programs have been constructed by splicing together the programs for individual procedures.

The programs are written in BASIC and are designed to run on any computer from a desk top microcomputer to a mainframe machine. Unfortunately there are many versions of BASIC in common use. The programs in this book are written in the popular Microsoft BASIC (GW-BASIC, QuickBASIC, QBasic), and the programs may need some minor conversion work before they will run under other versions of BASIC.

The reader of this book should have an elementary knowledge of statics, strength of materials, matrix algebra, and computer programming in BASIC. If deficient in any of these subjects then I suggest that the reader should make good these deficiencies before proceeding (references *5, 8, 9, 14, 23, 24* and *27* should prove particularly helpful with this task).

Second Edition

Early structural analysis computer programs dealt exclusively with linear elastic problems of statics. There were two principal reasons for this:

- Linear elastic analysis is of particular interest to engineers.
- Such problems did not make excessive demands on the small memory sizes and slow speeds offered by early computers.

The stiffness method has long been recognised as being capable of solving problems involving large displacements, dynamic effects, buckling and plasticity. Indeed current commercial structural analysis programs will solve many, if not all, of these types of problems.

The first edition of this book dealt solely with the static analysis of linearly elastic structures. The second edition is not limited to such problems: it illustrates the simplicity and flexibility of the stiffness method by also considering problems in the field of structural dynamics. This should be of particular interest to practising civil and structural engineers who may not have covered this material at undergraduate level, but are now required to check the dynamic behaviour of many of the structures that they design.

One surprising aspect of the first edition was the interest generated by the suite of structural analysis programs presented in the book. The analysis programs in the first edition were intended as a simple illustrations of the stiffness method at work and not as tools to be used in day-to-day situations. Their usefulness and popularity has prompted me to give them a complete rewrite in this edition. The text now contains listings of 24 BASIC programs, including the six automatic analysis programs and the two utility programs described below. These six programs make up an integrated structural analysis suite.

PRE.BAS	This is the biggest program in the book and is some 1750 lines long. Its function is to allow the user to easily build/modify data files for the six analysis programs described below.
DCHECK.BAS	This program conducts a range of checks on the data prepared by PRE.BAS.
PTRUSS.BAS	This program will solve static problems involving linearly elastic plane truss structures.

STRUSS.BAS	This program will solve static problems involving linearly elastic space truss structures (i.e. three-dimensional pin jointed structures).
PFRAME.BAS	This program will solve static problems involving linearly elastic plane frame structures (i.e. rigidly connected plane frame structures). PFRAME.BAS has facilities to handle element loads, local axes and elements containing pins.
PFDYNAM.BAS	This program will calculate the natural frequencies and mode shapes for a plane frame structures (i.e. dynamic analysis).
GRID.BAS	This program will solve static problems involving linearly elastic grillages. GRID.BAS has ability to handle element loads.
SFRAME.BAS	This program will solve static problems involving linearly elastic space frame structures (i.e. three-dimensional rigidly connected structures).

Any of the six analysis programs, the data pre-processor PRE.BAS and the data checking program DCHECK.BAS can CHAIN to each other passing the name of the data file in COMMON. They can all output results to the screen, a printer or a file. Error handling has been improved as has the data file structure. Data files can now be easily built/modified using a standard text editor (e.g. EDIT in DOS 5) rather than PRE.BAS should the user desire.

The programs are available in either source code form to be run under GW-BASIC, QBASIC, QuickBASIC, etc, or in compiled form so that they can be executed from the DOS command line (compiled programs cannot, of course, be modified by the user). The programs have been widely used as teaching tools and have been checked using output from a commercial finite element program and are believed to be reasonably bug-free. They are not, however, fully validated commercial programs and should not, for this reason, be used for commercial purposes. Readers who would like to obtain a copy of the programs will find details of how to order the programs at the back of the book.

I am indebted to my colleague, Professor A.D. Edwards, for many helpful suggestions made during the preparation of the first edition of this book.

I would like to conclude by recognising the debt owed by myself and the engineering profession to the many researchers and practising engineers who have brought the computer analysis of structures to its current level of sophistication. By their efforts the arithmetic tedium has been taken out of structural analysis, and the designer has been freed to ask the question - *what if?*

Chapter 1

Introduction

1.1 Structural Analysis in Perspective

Civil engineering structures are many and varied, including tall buildings, airports, factory buildings, roads, harbours, railways, water treatment works, sewage disposal works, offshore structures, dams, pipelines, and bridges. To design and build such structures is a long and complex process where the time from conception to completion is measured in years rather than in months. Say, for instance, a manufacturing company requires a new factory to expand its operations. Having acquired a suitable site the company then commissions a firm of consulting engineers to design the factory and supervise its construction. Once the consulting engineer has established the requirements of the client, he then sets about the process of designing the structure.

The first stage in the design process is to produce a preliminary scheme or schemes. This phase requires decisions to be made regarding the type of structure, its layout, and the materials to be used in its construction. In other words the preliminary scheme determines the structure's appearance, cost, functionality and to a large extent its ultimate success or failure. Here the designer uses a blend of art and science to produce a design that is close to the optimum. To do this he must integrate his understanding of engineering materials, building costs, structural behaviour, construction methods, and site conditions. If successful then the structure will be economical, functional, durable, safe, and pleasing to the eye. It must be remembered, however, that the preliminary design for some structures, such as tall buildings, is the responsibility of the architect, who may or may not involve a structural engineer at the preliminary design stage.

Structural analysis is conducted during the preliminary design to ensure that the proposed scheme is not structurally impractical. The calculations are usually kept short by oversimplifying the assumed structural behaviour. At this stage the designer is only interested in approximate values of member sizes, settlements, deflections, etc.

Once the outline design is complete the structure is analysed in detail to ensure that it has adequate strength and stiffness. The strength is checked to ensure that the maximum expected loads can be carried without collapse, and the stiffness is checked to avoid excessive deflection under everyday (service load) conditions. Other aspects of structural behaviour such as vibration, cracking in concrete, and foundation settlement are also assessed during this phase. From the detailed structural analysis come final member sizes, reinforcement layouts, connection details, and foundation dimensions. Codes of practice are extensively used during the detailed analysis and design of a structure. These codes relate to different structural types and construction materials. They give guidance on many aspects of analysis and design, including loadings, material specifications, stresses and deflections.

Structural analysis is, therefore, inextricably bound up with structural design. It is one of the tools that the designer uses to ensure the economy and safety of the final structure. It is important to remember, however, that structural analysis is a means to an end and not an end in itself. In essence it is a check, and it is salutary to remember that engineers of the past built highly successful structures without the aid of structural analysis.

Of course, the structural analysis must be competently executed. As structures are invariably idealised prior to analysis, care must be taken to ensure that the idealised structure incorporates the principal characteristics of the actual structure. The loads must be properly assessed and the calculations must be carried out accurately. However, given a preliminary design, any two competent engineers will compute deflections and stresses that are essentially the same. Hence structural analysis, while a necessary part of the design process, does not affect the ultimate success of the structure (gross errors excepted).

The creative part of the structural analysis phase occurs when the analysis shows that the preliminary design is unsatisfactory. Parts of the structure may be overstressed making the structure unsafe, while other parts may be grossly understressed making the structure uneconomical. In addition, vibration, deflection, ground pressures, etc., may be found to be unacceptable. Seldom will a preliminary design emerge as a final design without modification. If the detailed structural analysis shows the structure to be unacceptable then the designer must decide what changes are required in order to satisfy the structural requirements. It is here that the engineer can use his experience and understanding of structural behaviour to good effect. It takes a certain skill to select, from the infinite number of possible structural modifications, one that will change an unsatisfactory structure into a satisfactory one.

Once the design has been finalised the contract documents are prepared. These documents, which include working drawings, specifications, conditions of contract, and bills of quantity, are then issued to civil engineering contractors who tender competitively for the work. The contract is normally awarded to the contractor making the lowest bid. The construction is monitored by the resident engineer to ensure that the structure is built according to the contract documents, and that the contractor is fairly rewarded for the work done.

1.2 The Computer Revolution

Today, computers play an integral part in the analysis of civil engineering structures. Hand calculation is limited to simple structures, and initial member sizing (either during the preliminary design or prior to computer analysis). Prudent engineers also estimate, by hand, the expected deflections and forces at selected points of a structure about to undergo computer analysis. This can provide a useful check that there is nothing seriously wrong with the input data, and gives the engineer a feel for how the structure will work. When compared with the analysis procedures of only a few years ago it is evident that a revolution has taken place. There follows a brief account of the events leading up to that revolution. The background to a revolution is always interesting.

By the end of the nineteenth century the principles governing the elastic behaviour of structures had been laid down by men such as Mohr, Rankine, Euler, Maxwell, Castigliano and Muller-Breslau. These principles had the virtues of generality and mathematical elegance, but although they gave engineers a better understanding of structural behaviour, they proved to be of limited practical value due to the volume of arithmetic associated with their application. In many cases to find a solution entailed solving a system of simultaneous equations. Consequently the first half of this century saw the appearance of a myriad of structural analysis techniques designed to reduce the arithmetic to manageable proportions. Many of these techniques were problem orientated, being concerned with only one type of structure, and yet, in most cases, these techniques were based upon the same underlying principles. Of the value to practising engineers of techniques such as moment distribution there can be no doubt, yet it is apparent that engineers were required to grapple with a huge range of apparently different analysis techniques.

During the early 1940s a machine called ENIAC (Electronic Numerical Integrator and Calculator) was built at the University of Pennsylvania. It was housed in a room approximately 20 metres by 10 metres, it contained 18,000 valves and weighed about 30 tonnes. ENIAC is generally recognised as being the world's first digital computer. The use of a large number of valves, each having a relatively short lifespan, made keeping ENIAC operational a major problem. Typically the machine would run for only a few minutes before valve failure would halt operations.

The reliability problem was largely solved by the invention of the transistor in 1948, and by the mid-1960s a second generation of computers based upon solid state technology was being commercially produced. At about the same time manufacturing techniques was being developed that allowed an integrated circuit (IC) consisting of a number of transistors to be produced on a single silicon wafer. Since that major technological breakthrough computers have become steadily smaller, cheaper, and more powerful. Today's microcomputers are small enough to sit on a desk top. They offer computing power that is vastly superior to that of ENIAC and they are extremely reliable. The low cost of these machines, coupled with a growing computer awareness in society, has led to the widespread use of computers in schools, offices, homes, and factories.

Developments in computer software have tended to parallel those in computer hardware. New software has been required to take advantage of the improved facilities offered by each new generation of computers. Each year sees an improvement in the quality and scope of commercially available software. In contrast to the cost of computer hardware, which has been falling, the cost of software has shown no corresponding decrease. The net effect of these two factors is that in many computer installations the capital investment in software is similar to that in hardware.

Of the many different ways in which computer technology has affected the engineering profession, it is in the field of structural analysis that the impact has been most profound (although some engineers claim that word processing has had a greater impact on their daily work). Early computers were used primarily for numerical calculation, and engineers were quick to realise that the computational speed and accuracy offered by these machines might herald a new era in structural analysis. Much of the pioneering work was done by aircraft engineers because at that time aircraft design was severely hampered by a lack of sufficiently accurate analysis techniques. Aircraft structures are extremely complex and the type of analysis that can be undertaken by hand will give only a poor representation of the true structural behaviour. Some attempts were made to computerise hand methods of calculation, but it was soon realised that a more general approach stemming directly from the fundamental principles of structural analysis would be better suited to computer implementation. The computer's ability to handle vast amounts of arithmetic with speed and accuracy has made viable these computationally intensive methods.

By 1953 engineers were writing stiffness equations in matrix notation and solving the resulting simultaneous equations using digital computers. Matrix notation is used to describe modern structural analysis techniques because it is efficient, compact and is well suited to computer implementation. Of the different matrix methods that have been successfully translated into computer code, the stiffness method has proved to be by far the most generally useful. The principal advantage of the stiffness method is that it is easy to automate and hence a single program can be used to analyse a structure of any geometry. Another advantage of the stiffness method is that it is not limited to linearly elastic problems of statics. It has been successfully extended to problems involving buckling, plasticity, and dynamics.

The development cost of structural analysis programs is high. A simple single element, static analysis program consisting of about 1000 statements is likely to take a number of man-months to develop, test, and document. At the other end of the scale a large general purpose finite element program with pre- and post-processors may consist of 250,000 statements and will take hundreds of man-years to develop. This has led to the situation where a small number of establishments have made a large investment in the development of structural analysis software. These programs are marketed to the engineering community, thus dividing engineers into a small group of specialist program writers, and a large group of practising engineers who use the programs.

Many civil engineering structures are supported by frameworks and consequently frame analysis programs are extensively used by civil engineers. To the user a frame analysis program appears as a "black box" which will produce results at one end if fed with input data at the other. This book is concerned with the structural theory behind these programs and with how that theory is turned into useful computer programs.

Chapter 2

Fundamentals of Structural Behaviour

2.1 Introduction

The theory of structures is a wide and sometimes complex subject. Fortunately the structural theory used in this book is limited to simple beam theory, Castigliano's First Theorem, virtual work and the fundamental concepts of equilibrium, compatibility, and superposition. This chapter presents all of these topics with the exception of Castigliano's First Theorem which is developed in Chapter 3. Should the reader require a more general introduction to the theory of structures then he should refer to one of the many excellent textbooks on the subject. This chapter begins by discussing the way in which the highly complex behaviour of real structures is made tractable by the process of structural idealisation.

2.2 Definition of a Structure

The function of a structure is to transmit force. The trunk of a tree transmits wind and gravity forces from the upper parts of the tree to the roots and hence safely into the ground. Floor beams are used to transmit floor loadings to the building walls. The walls then carry the load from the beams to the foundations and the foundations transmit the load into the ground below. From these simple examples it is evident that the world is full of both man-made and natural structures, and that force can be transmitted in a variety of ways.

2.3 Stable and Unstable Structures

A structure is said to be stable if it can support an arbitary system of infinitesimal loads. No part of a stable structure is a mechanism, nor can the structure, as a whole, undergo rigid body motion. Figure 2.1 shows a structure that is a mechanism. As can be seen the structure can support the particular system of loads shown in the left-hand diagram, but not the system of loads shown in the right-hand diagram.

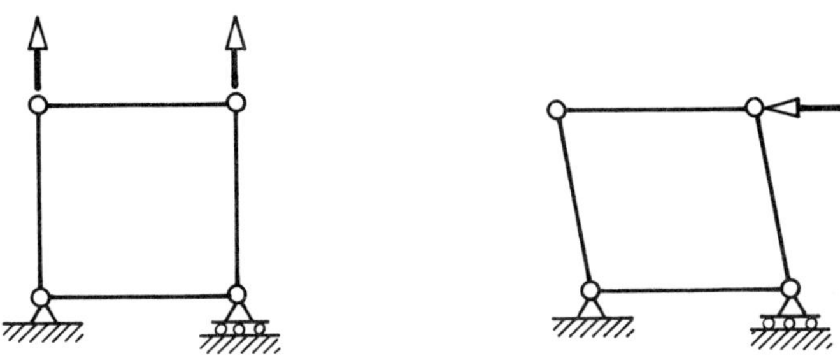

Figure 2.1 Example of a mechanism

Figure 2.2 shows a structure that is inadequately restrained and is therefore capable of rigid body motion. Again a particular system of loads can be supported, but not a general system.

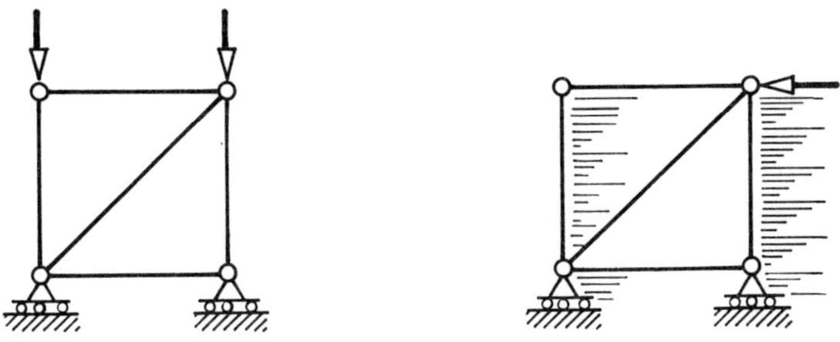

Figure 2.2 Example of an inadequately restrained structure

Civil engineers are concerned to produce stable structures. If a structure is unstable then the structural displacements caused by the loads become indeterminate, and no meaningful solution can be found.

2.4 The Need to Idealise

The actual behaviour of even a simple structure is extremely complex. Take for example the building shown in fig 2.3. Obviously the structure is three-dimensional. Of course the roof and cladding will act together with the frame of the structure to resist loads. We know that there will be a complex interaction between the foundations and the subsoil. The loading will be highly variable and impossible to exactly predict in advance. Yet more likely than not, for analysis purposes, the structure will be idealised as a two-dimensional series of interconnected line elements built-in at ground level as shown on the right of the diagram. Moreover, the complex stress-strain properties of the structural materials will be assumed to be something much simpler, and the loading may well be assumed to be uniformly distributed.

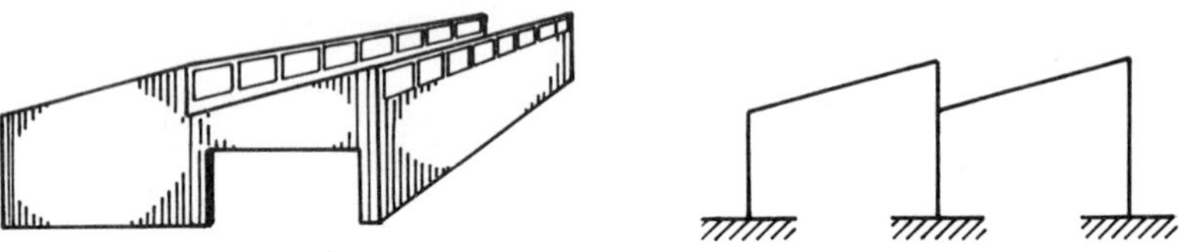

Figure 2.3 Structure idealisation

The structural engineer does not therefore analyse the real structure, but an idealised version of the structure. Any results obtained will not tell him about the behaviour of the actual structure, but about the behaviour of this mathematical model. It is obvious that unless the mathematical model is sufficiently realistic then the results will be of questionable value. Part of the skill of the structural engineer lies in being able to idealise a structure to an appropriate level of sophistication. If the idealisation is too detailed then the analysis will be excessively long and costly. If, on the other hand, the idealisation is too crude then some important aspect of the structural behaviour may not be properly modelled, making the structure either unsafe or uneconomical.

2.5 Idealisation of Structural Loading

The principal objective of structural analysis is to ensure that all anticipated loadings (design loads) are carried safely and economically. Loads are usually classified as dead or live. Dead loads are permanent loads. In practice dead load is the self weight of the structure, which can be subdivided into load carrying components (such as a building frame) and non-load carrying components (such as cladding and partition walls). Live loads are non-permanent loads such as snow, people, furniture, machinery, vehicles, and wind. While some loads, such as the weight of water in a water tower, can be assessed with some degree of confidence, most design loadings are but crude approximations to the actual loads the structure will carry. For instance, the calculation of the wind pressures experienced by a building during a once in 100-year storm is a calculation fraught with uncertainty. The following load types are usually used to simplify the actual structural loadings:

- Uniformly distributed load
- Uniformly varying load

- Point load
- Line load.

A high intensity uniformly distributed load that acts over a relatively small area is sometimes called a patch load.

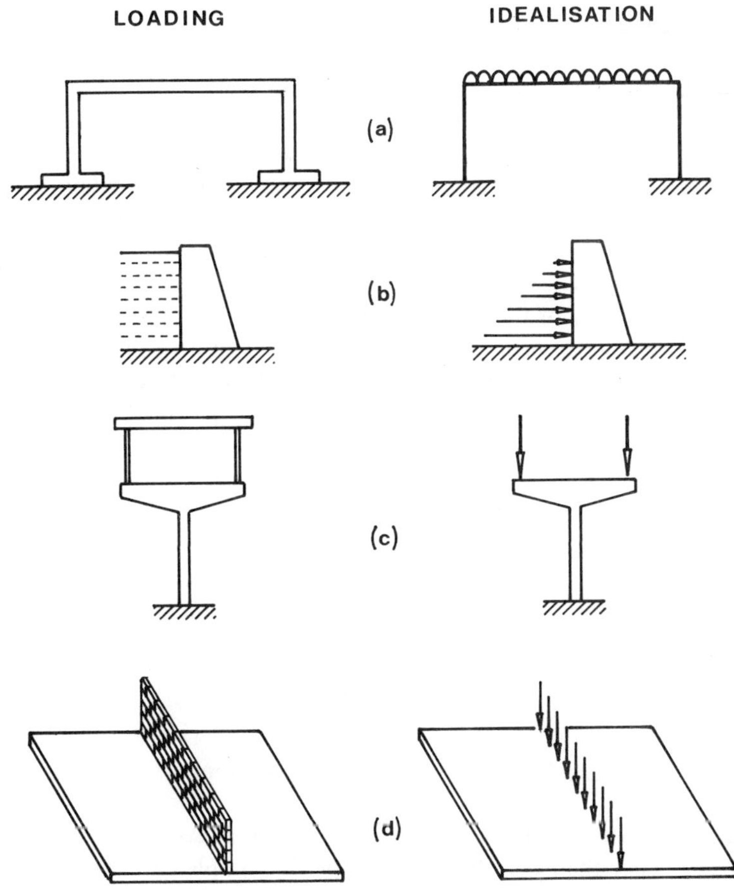

Figure 2.4 Load idealisations

Figure 2.4 gives examples of all these load types:

(a) Shows how the self weight of a concrete bridge deck might be considered as a uniformly distributed load.

(b) Shows how water pressure, which varies linearly with depth, can be regarded as a uniformly varying load.

(c) Shows how the loads from the upper level of a two level roadway might be idealised as point loads.

(d) Shows how a partition wall crossing a slab might be idealised as a line load.

2.6 Idealisation Within the Structure

Structural Elements

Modern structural analysis techniques divide the structure into a number of elements. Each element is relatively simple, making the calculation of the properties of individual elements straightforward. Assembling the elements in accordance with the geometry of the structure produces a model of the complete structure that can be used to investigate the effects of the applied loading. Real structures consist purely of three-dimensional elements. In practice, however, it is not always necessary to model the structure in three dimensions and engineers take advantage of this to simplify the analysis procedure. Depending upon the problem in hand, structures are idealised using *line, plate or brick elements* (illustrated in fig 2.5).

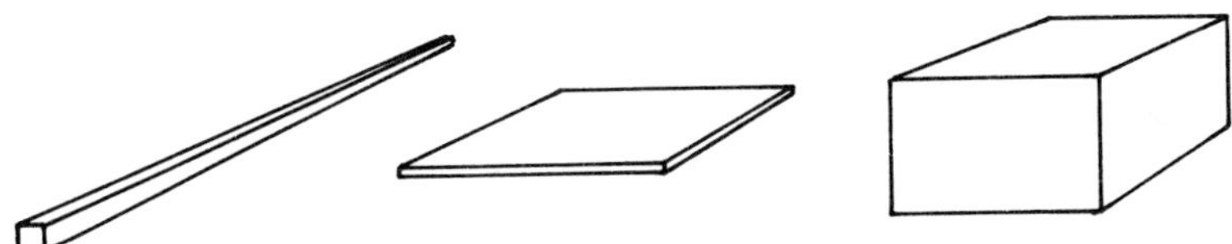

Figure 2.5 Line, plate and brick elements

Line elements are used to model structural components that have one dimension that is much greater than the other two. Typical examples are truss members, beams, columns and slender arches. Structures consisting entirely of line elements are often referred to as skeletal, or frame, structures. Plate elements are used to model structural components, such as slabs and shells, of which two dimensions are much greater than the third. Plate elements are also used where, although the structure is obviously three-dimensional, the structural action is essentially two-dimensional (e.g. cases of plane strain). Brick elements are used to model structures such as thick slabs and machine components where the structural action is truly three-dimensional. As this text deals exclusively with structural frameworks, only line elements are of concern.

Structural Materials

The science of engineering materials is a large and often complex subject. When designing structures, however, the engineer invariably adopts a simplified model of material behaviour.

The framed structures encountered in civil engineering are usually constructed from steel or reinforced concrete. Figure 2.6 shows typical *idealised* stress-strain diagrams for steel reinforcement and structural concrete.

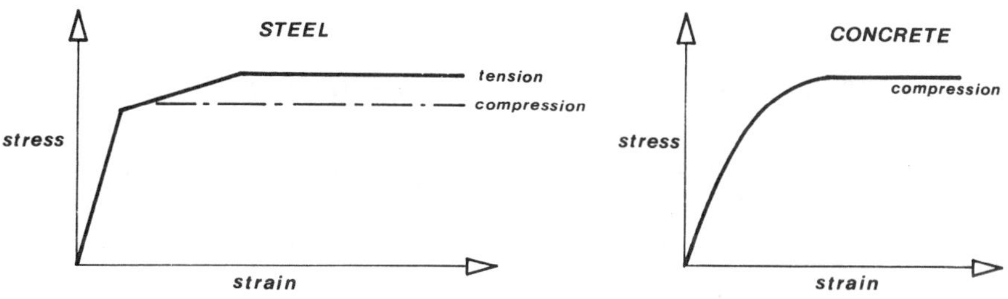

Figure 2.6 Typical idealised stress-strain diagrams

Both stress-strain curves have a section where strain increases yet the stress remains constant (i.e. the horizontal portion of the curve). Deformations associated with that portion of the curve are plastic. Plastic deformation is non-recoverable deformation that remains after the removal of the applied stress, and it should be noted that the onset of plastic deformation occurs before the stress reaches the level of the horizontal portion of the curve. Steel will sustain large plastic strains prior to failure and is therefore described as a *ductile material.* Concrete, by contrast, fails shortly after the onset of plasticity and is therefore described as a brittle material.

Engineers make extensive use of codes of practice when designing structures. Most of these codes are based upon a limit state design philosophy which demands that the structure never reaches the limit states of serviceability or collapse. The serviceability limit state is reached when the structure begins to lose utility or to cause public concern through causes such as excessive cracking in concrete, large deflections, or unacceptable levels of vibration. The engineer must ensure that the structure will remain serviceable under the service loads as laid down in the relevant codes of practice. Service loading represents a severe working load and the structure is invariably assumed to be linearly elastic under service conditions.

The ultimate limit state is reached when the structure is on the point of collapse. Ultimate loads represent the extreme loading for the structure and the structure must be able to carry these loads without collapse. Structures close to collapse are usually behaving inelastically. In spite of this the forces within the structure are often determined using linear elastic theory. The individual members are then designed on the basis of inelastic material behaviour. Linear elastic analysis under ultimate loads is used because, with the exception of yield line analysis of slabs and simple plastic analysis of steel frames, appropriate analysis techniques are not readily available. Limit state codes recognise this by allowing the member forces to be assessed on the basis of linear elastic behaviour. In recent years there has been a steady improvement in computer based techniques for the analysis of structures under ultimate conditions. Such techniques are, however, not within the scope of this book, which assumes linear elastic behaviour throughout.

2.7 Idealisation at the Structural Boundaries

In section 2.3 the need to restrain the structure against rigid body motion was discussed. To facilitate analysis the engineer makes assumptions about the displacement conditions pertaining at the points of restraint. As these points are at the boundary of the structure they are known as *boundary conditions*.

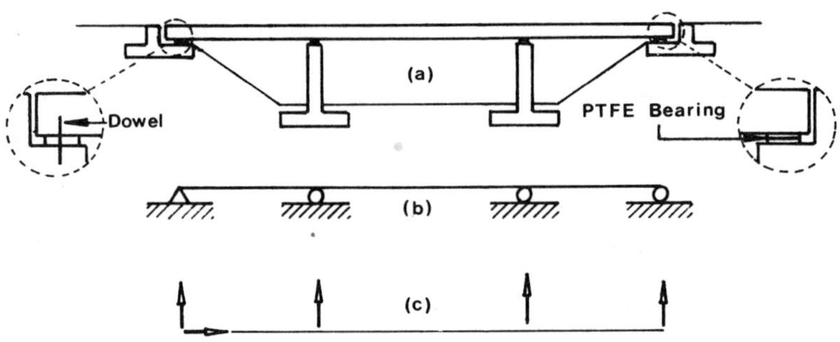

Figure 2.7 Non-monolithic bridge structure

Take for example the bridge shown in fig 2.7(a). PTFE, a very low friction material, has been incorporated in the bearings at the abutments and on the top of the columns. The bridge deck is dowelled to the left-hand abutment and will slide freely on the PTFE bearings. Consequently the boundary conditions assumed during the analysis of the bridge deck are likely to be those illustrated in fig 2.7(b). These boundary conditions imply zero displacement in the directions shown in fig 2.7(c).

The bridge deck might, however, be constructed as monolithic with the columns as shown in fig 2.8(a). In this case the bridge deck should not be analysed independently from the columns. Figure 2.8(b) shows the boundary conditions that might be assumed, and fig 2.8(c) indicates the locations and directions of zero displacement. Note that although rotation has been restrained at the base of the columns, the degree of restraint will depend upon the nature of the connection between the columns and the foundations.

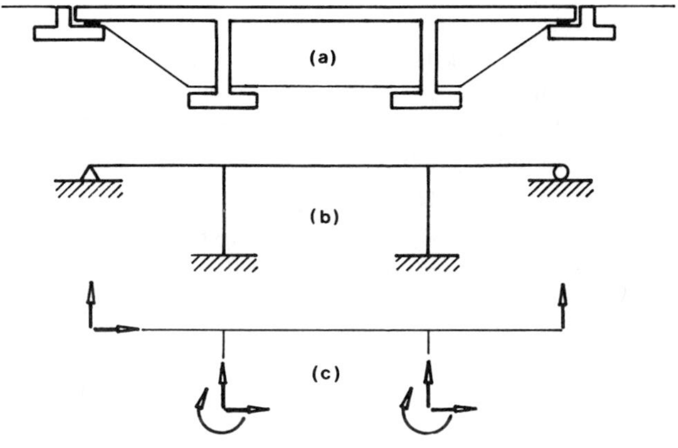

Figure 2.8 Monolithic bridge structure

To make assumptions about the boundary conditions is mathematically convenient, yet perfect boundary conditions never occur in practice. For instance, pinned connections always have some moment carrying capacity, and a fully fixed connection will always shed some moment due to the rotation of the connection itself.

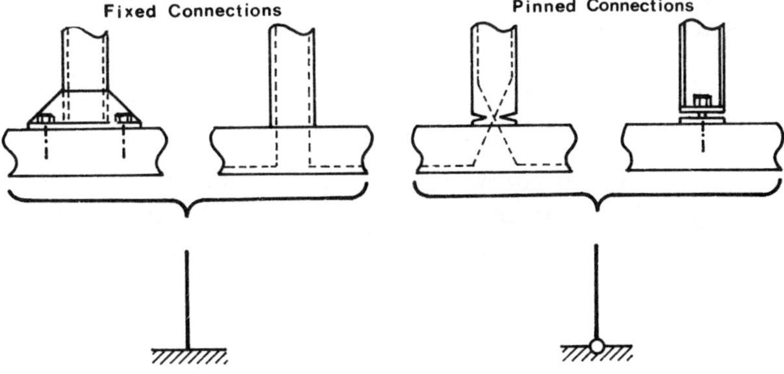

Figure 2.9 Fixed and pinned base connections

Consider the boundary conditions shown in fig 2.7(a). The PTFE bearings are assumed to offer no resistance to horizontal movement. In practice they will offer some resistance. Also the vertical displacement at the top of the columns and at the abutments is assumed to be zero, although it is evident that stress in the columns and the subsoil will result in some vertical displacement. As the boundary conditions have a significant effect on the structural behaviour it is important that the boundary conditions assumed during the analysis are realistic. Figure 2.9 gives examples of connection details that might be considered as fixed and pinned.

A further complication arises when more than one member is involved. Consider, for example, fig 2.10 which shows how three members of a framework might be connected to a foundation. Obviously the members are rigidly connected together, yet the connection to the foundation is effectively pinned. Hence, for the purposes of analysis, the boundary condition should be assumed to be that shown in the centre diagram and not that shown in the right-hand diagram. The boundary condition shown in the right-hand diagram allows the ends of the members to rotate by different amounts and does not, therefore, adequately represent the behaviour of the actual structure.

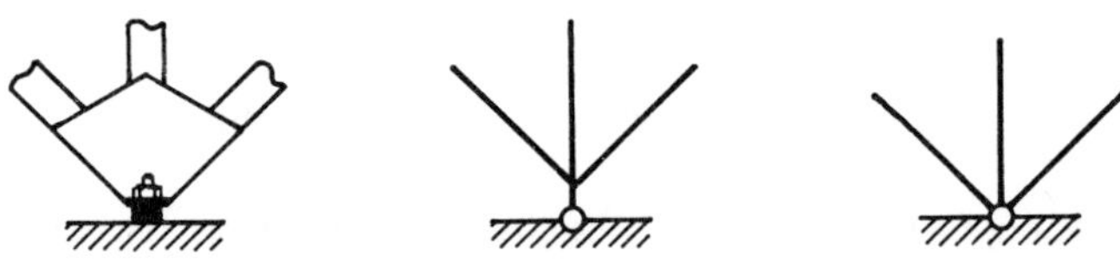

Figure 2.10 Boundary conditions

Other common boundary conditions include roller supports, elastic supports and semi-rigid supports. In addition it may be necessary to consider the effect of settlements at the foundations as part of the analysis.

2.8 Equilibrium

Equilibrium is the most important concept in structural analysis. Engineers talk about two types of equilibrium, static and dynamic, although it can be argued that static equilibrium is a special case of dynamic equilibrium. This section will consider only static equilibrium, dynamic equilibrium is discussed in Chapter 11.

Figure 2.11 Vector components in two and three dimensions

Static equilibrium exists if all parts of a structure can be considered static (i.e. motionless) under a particular system of loads. As all elements of the structure are stationary there is no net force acting on any element (otherwise acceleration would occur). In two dimensions force can be completely described by two mutually perpendicular linear components and one rotational component, as illustrated in fig 2.11.

Also shown in fig 2.11 are the three mutually perpendicular linear components and the three rotational components that completely describe force in three dimensions. This enables the well-known equations of static equilibrium to be written:

Two-dimensional case

$$\sum P_x = 0 \qquad \sum P_y = 0 \qquad \sum M_z = 0$$

Three-dimensional case

$$\sum P_x = 0 \qquad \sum M_x = 0$$
$$\sum P_y = 0 \qquad \sum M_y = 0$$
$$\sum P_z = 0 \qquad \sum M_z = 0$$

Internal and External Forces

Equilibrium is to do with forces, and forces can be regarded as being either internal or external. External forces consist of applied forces and reactions. The applied forces can be regarded as active forces, and the reactions as passive forces generated in response to the applied loads. The applied loads usually have preset values, whereas the reactions assume values that will maintain the equilibrium of the structure. The engineer can distinguish between applied loads and reactions, but the structure cannot, with reactions and applied loads appearing simply as external forces. Internal forces are the forces generated within the structure in response to the applied loads. In essence structural action is the transmission of the applied loads to the reactions via internal member forces.

Stress Resultants

A better name for the forces generated within a framework in response to the applied loads is stress resultants. In fact internal forces do not actually exist, as force is transmitted through the structure by stress within the members. Internal forces (stress resultants) are simply useful vector quantities obtained by integrating stress, or moment of stress, over the cross-section of the member.

Freebody Diagrams

Stemming directly from the principle of equilibrium is the concept of freebody diagrams. The basic argument is this:

- If a structure as a whole is in static equilibrium then every part (or element) of the structure must be in static equilibrium.
- Hence the internal and external forces (and moments) acting on every part of the structure must balance. If they did not, then the out-of-balance force would cause that part of the structure to accelerate.
- If a part of the structure is "cut" from the rest of the structure, then the stresses which were formerly present at the cuts would be released, causing a change in the structural behaviour. If, however, stresses (or stress resultants) identical to those which existed prior to cutting are applied to the cut faces, then the structure and the severed part will be unaware that they are no longer connected.

This reasoning holds because the only communication between the elements of a structure is the stress present at the inter-element boundaries. To illustrate this concept consider the simple structure shown in fig 2.12.

As can be seen two cuts have been made, dividing the original structure into three elements. Freebody diagrams are obtained by applying stress resultants to the cut faces as shown. Due to the nature of the loading the structure is subjected to only horizontal forces. Hence only one equation of static equilibrium is of value (i.e. $\sum P_x = 0$) as the other two equations are automatically satisfied.

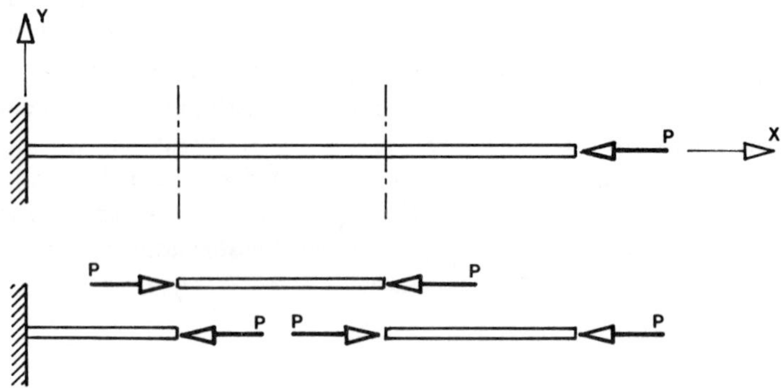

Figure 2.12 A simple example of freebody diagrams

In the case of the beam shown in fig 2.13 the three equations of static equilibrium, $\sum P_x = 0$, $\sum P_y = 0$ and $\sum M = 0$, can be used to find the shearing force and bending moment at any section along the beam. Say, for instance, the shearing force and bending moment at sections *A-A* and *B-B* are required to be found.

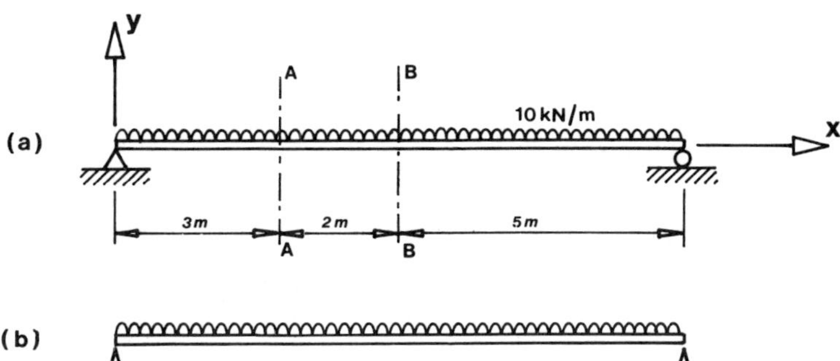

Figure 2.13 A simple beam structure

To find the reactions the freebody diagram shown in fig 2.13(b) is used. The right-hand support is a roller, therefore the reaction must be vertical, and consequently there can be no horizontal component of force at the left-hand reaction. Having established that both reactions are vertical, the equations of equilibrium $\sum P_y = 0$ and $\sum M = 0$ are used to find the value of the reactions. These equations can be applied in an infinite number of ways. The engineer's reasoning would be:

For vertical forces to balance $\quad R_1 + R_2 = 100$

by symmetry $\quad R_1 = R_2$

therefore $\quad R_1 = R_2 = 50$ kN

The student engineer, lacking experience, may decide to take moments about the centre of the beam (section *B-B*).

$\sum M = 0$ - taking anticlockwise turning effects as positive

$$-5\,R_1 + 5\,R_2 - 0 \times 100 = 0 \tag{2.1}$$

Vertical forces must sum to zero, hence (taking upwards as positive)

$$R_1 + R_2 - 100 = 0 \tag{2.2}$$

Equation (2.1) - 5 x equation (2.2) gives

$$-5\,R_1 + 5\,R_2 - 5\,R_1 - 5\,R_2 + 500 = 0$$

hence

$$R_1 = 50 \text{ kN}$$

Substitute for R_1 in equation (2.1)

$$-5 \times 50 + 5R_2 = 0$$

hence

$$R_2 = 50 \text{ kN}$$

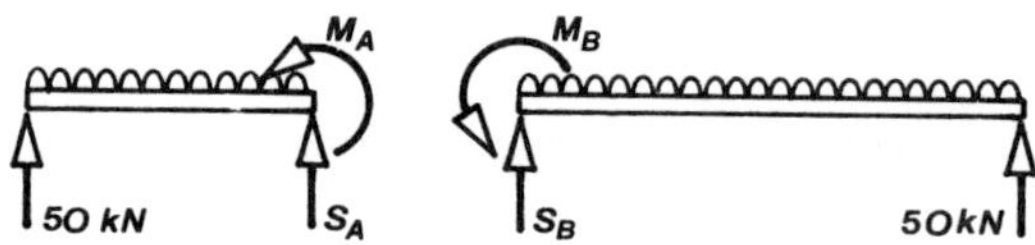

Figure 2.14 Freebody diagrams for the beam ends

Having found the reactions the freebody diagrams produced by cuts *A-A* and *B-B* can be sketched as shown in fig 2.14. To find the required bending moments and shearing forces consider first the left-hand portion of the beam.

$\sum P_y = 0$ - taking upwards forces as positive

$$50 - 3 \times 10 + S_A = 0$$

hence

$$S_A = -20 \text{ kN}$$

$\sum M = 0$ - taking moments about *A-A* and taking anticlockwise turning effects as positive.

$$-50 \times 3 + (3 \times 10) \times 1.5 + M_A = 0$$

hence

$$M_A = 105 \text{ kN m}$$

Now consider the right-hand portion of the beam.

$\sum P_y = 0$ - taking upwards forces as positive

$$50 - 5 \times 10 + S_B = 0$$

hence

$$S_B = 0 \text{ kN}$$

$\sum M = 0$ - taking moments about *B-B* and taking anticlockwise turning effects as positive.

$$50 \times 5 - (5 \times 10) \times 2.5 + M_B = 0$$

hence

$$M_B = -125 \text{ kN m}$$

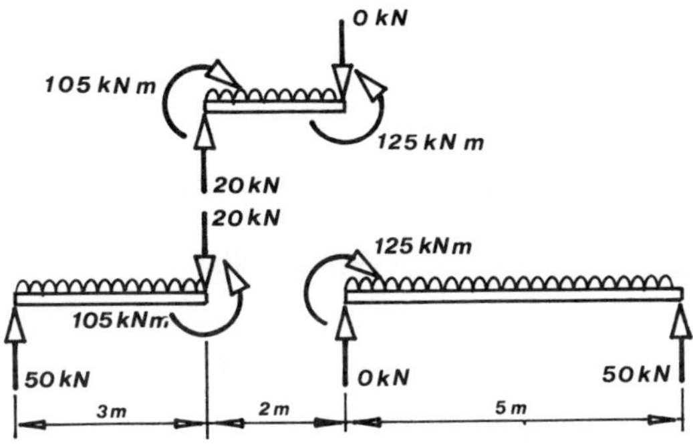

Figure 2.15 Final freebody diagrams

The final freebody diagrams are shown on fig 2.15 and the reader should verify that the centre portion of the beam is in equilibrium.

Now consider the use of freebody diagrams in the analysis of the frame shown in fig 2.16(a).

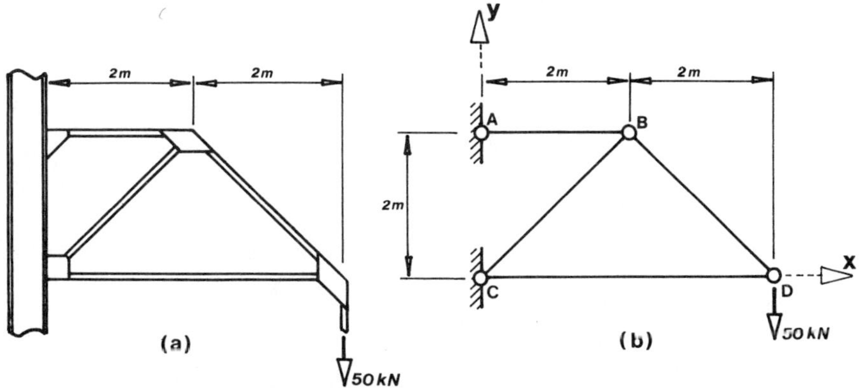

Figure 2.16 A simple framework

Triangulated frameworks such as that shown in fig 2.16(a) transmit load primarily by axial force in the members. Consequently to idealise the joints as perfect pins as shown in fig 2.16(b) is reasonable. In order to find the reactions the freebody diagram shown in fig 2.17(a) is used. There are four components of reaction (h_a, v_a, h_c, v_c): one of which can be found by consideration of the freebody diagram for joint "A" (shown in fig 2.17(b)).

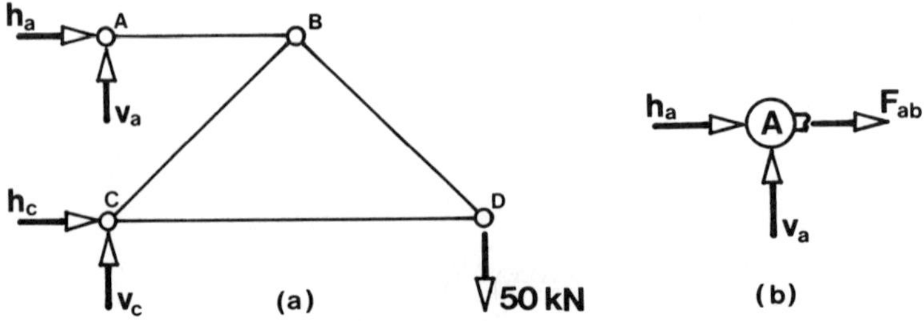

Figure 2.17 Freebody diagrams

Now consider the equilibrium of the whole frame.

$$\sum M = 0 \text{ - taking moments about "}c\text{"}$$

$$-2 \times h_a - 4 \times 50 = 0$$

hence

$$h_a = -100 \text{ kN}$$

$$\sum P_x = 0$$

$$h_a + h_c = 0$$

hence

$$h_c = -h_a = 100 \text{ kN}$$

$$\sum P_y = 0$$

$$v_a + v_c - 50 = 0$$

hence

$$v_c = 50 \text{ kN} \quad \text{as} \quad v_a = 0$$

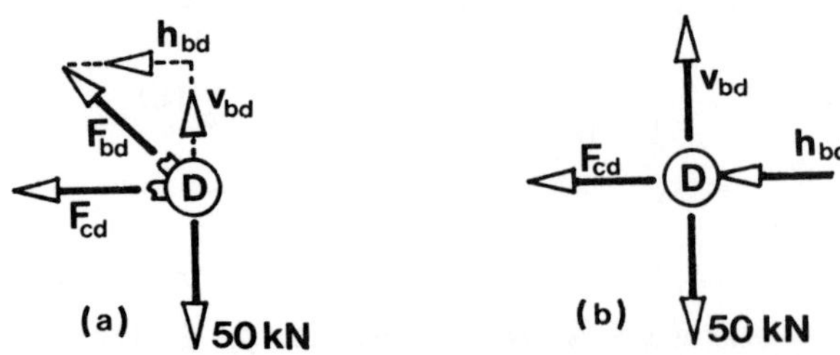

Figure 2.18 Freebody diagram for joint "D"

If the forces in members *BD* and *CD* are of interest, then the best way forward is to consider the freebody diagram for joint "*D*" as shown in fig 2.18(a).

Replace the force in member *BD* by its components as shown in fig 2.18(b) and then consider the equilibrium of the joint.

$$\sum P_y = 0,$$

$$v_{bd} - 50 = 0$$

hence

$$v_{bd} = 50 \text{ kN}$$

As *BD* is at 45 degrees,

$$h_{bd} = v_{bd} = 50 \text{ kN}$$

and

$$F_{bd} = \sqrt{v_{bd}^2 + h_{bd}^2} = 50\sqrt{2} \text{ kN}$$

$$\sum P_x = 0$$

$$-h_{bd} - F_{cd} = 0$$

hence

$$F_{cd} = -50 \text{ kN}$$

The technique that has been used here is, of course, the well-known method of joints.

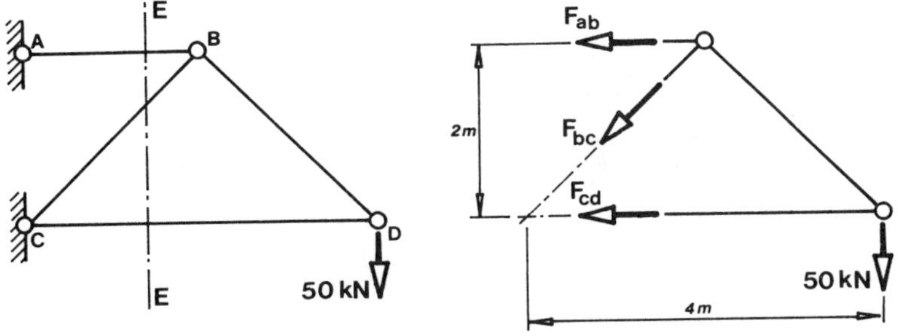

Figure 2.19 Freebody diagram to find the member forces

The force in the remaining members can be found by considering the equilibrium of the freebody diagram produced by cutting the frame on section *E-E*, as shown in fig 2.19.

$$\sum M = 0, \text{ taking moments about "C".}$$

$$(-50 \times 4) + (2 \times F_{ab}) = 0$$

hence

$$F_{ab} = 100 \text{ kN}$$

$$\sum P_y = 0$$

$$-50 - (F_{bc} / \sqrt{2}) = 0$$

hence

$$F_{bc} = -50\sqrt{2}$$

The foregoing calculations are in fact an example of the method of sections commonly used in the analysis of pin jointed frames.

2.9 Deformation of Structures

Structures support applied loads by developing internal stresses within their members. As stress cannot exist without strain, it follows that a structure cannot resist loads without deformation.

Compatibility

The principle of compatibility is assumed to apply to all of the structures encountered in this book. Compatibility is concerned with deformation. If compatibility is assumed then geometric fit is implied (i.e. if a joint of a structure moves, then the ends of the members connected to that joint move by the same amount, consistent with the nature of the connection).

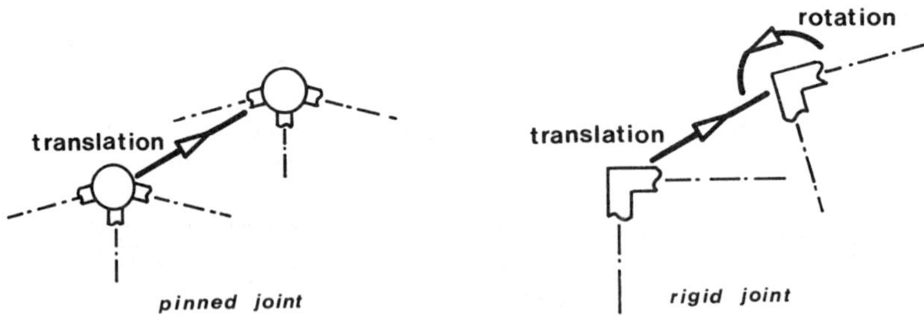

Figure 2.20 Compatibility and joint displacements

In the case of a pin jointed frame, compatibility means that the ends of the members meeting at a joint undergo equal translation. If the framework is rigidly jointed then, in addition to equal translation, the rotation of the ends of the members meeting at a joint must also be equal, as shown in fig 2.20.

Small Deflection Theory

As all structures deform when loaded it follows that the equations of equilibrium should be based upon the deformed shape of the structure. In most civil engineering structures, however, the deformations caused by the loading are extremely small in comparison to the size of the structure. Consequently the assumption of unchanged geometry can (usually) be safely made, thereby greatly simplifying the analysis procedure. Small deflection theory assumes that the equilibrium equations for the loaded structure can be safely based upon the unloaded geometry. Consider for example the simple frame shown in fig 2.21(a).

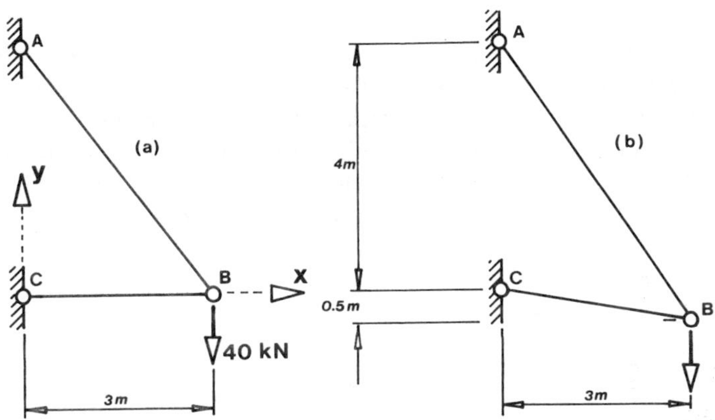

Figure 2.21 An example of large displacements

On the assumption of unchanged geometry the forces in members *AB* and *BC* can easily be shown to be 50 kN tension and 30 kN compression respectively.

If, however, the original member *AB* is replaced by a very flexible member which causes the frame to deform as shown in fig 2.21(b), then small deflection theory may not be sufficiently accurate. Obviously the geometry has undergone a significant change during loading, and the equilibrium equations for joint B are as follows:

$$\sum P_y = 0$$

$$(F_{ab} \times 4.5/5.408) - (F_{cb} \times 0.5/3.041) - 40 = 0$$

$$\sum P_x = 0$$

$$(F_{cb} \times 3.0/3.041) - (F_{ab} \times 3.0/5.408) = 0$$

Solution of these equations yields F_{ab} = 54.08 kN tension, and F_{bc} = 30.41 kN compression.

Fortunately in civil engineering structures the change in geometry under loading is a second order effect that can usually be safely ignored by the engineer. Only when the deformations are large does the use of small deflection theory become suspect. Large deformations sometimes occur in flexible structures such as suspension bridges, slender arches, and tall buildings. Large deformations are also associated with structures that are approaching collapse. In such cases the deformations are usually large enough to warrant basing the equations of equilibrium on the loaded geometry.

2.10 Statically Indeterminate Structures

All of the structures encountered so far have been *statically determinate*. In other words the forces in the members can be determined by the application of the equations of equilibrium alone. Structural deformation will, of course, be dependent on the materials and cross-sections used. However, provided that the structure is linearly elastic and small deflection theory remains valid, then the member forces will be independent of the materials and cross-sections used.

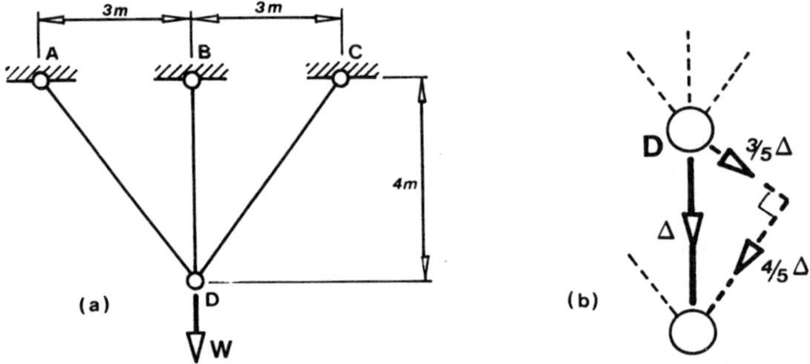

Figure 2.22 A statically indeterminate structure

Structures that cannot be analysed using only the equations of equilibrium are said to be *statically indeterminate*. The analysis of indeterminate structures demands a knowledge of the structural properties of the members, and in general the force carried by each member will depend upon its material and cross-section.

To illustrate, consider the frame shown in fig 2.22(a) which is constructed using a linearly elastic material. The cross-sectional area of member *BD* is A_1 and the cross-sectional area of members *AD* and *CD* is A_2. Due to the symmetry there will be no horizontal movement of joint "*D*" which will displace as shown in fig 2.22(b). If "*D*" displaces downwards by "Δ" then the change in length of members *AD* and *CD* will be 4Δ/5.

Strain in *BD*

$$\varepsilon_{bd} \;=\; \delta L/L \;=\; \Delta/4$$

Strain in *AD* and *CD*

$$\varepsilon_{ad} = \varepsilon_{cd} = \tfrac{4}{5}\Delta/5$$

Hence if the stress in *BD* is σ_{bd} then, as stress is proportional to strain, the stress in *AD* and *CD* will be $16\sigma_{bd}/25$.

Force = stress x cross-sectional area

therefore

$$F_{bd} = \sigma_{bd} A_1$$

and

$$F_{ad} = F_{cd} = \tfrac{16}{25}\sigma_{bd} A_2$$

Vertical equilibrium of joint "*D*" gives:

$$\tfrac{4}{5}(F_{ad} + F_{cd}) + F_{bd} - W = 0$$

hence

$$\frac{128}{125}\sigma_{bd} A_2 + \sigma_{bd} A_1 = W$$

therefore

$$\sigma_{bd} = \frac{W}{(\frac{128}{125}A_2 + A_1)}$$

We will now consider the following three cases.

Case 1 - $A_2 = A_1$

$$\sigma_{bd} = \frac{125\ W}{253\ A_1}$$

$$F_{bd} = \frac{125\ W}{253}$$

$$F_{ad} = F_{cd} = \frac{80\ W}{253}$$

Case 2 - $A_2 = A_1/10$

$$\sigma_{bd} = \frac{1250\ W}{1378\ A_1}$$

$$F_{bd} = \frac{1250\ W}{1378}$$

$$F_{ad} = F_{cd} = \frac{80\ W}{1378}$$

Case 3 - $A_2 = 10A_1$

$$\sigma_{bd} = \frac{125\ W}{1405\ A_1}$$

$$F_{bd} = \frac{125\ W}{1405}$$

$$F_{ad} = F_{cd} = \frac{800\ W}{1405}$$

These results are illustrated in fig 2.23, and as might be expected the stiffer the member the more load it tends to attract, showing clearly how the member properties influence the force distribution in a statically indeterminate structure.

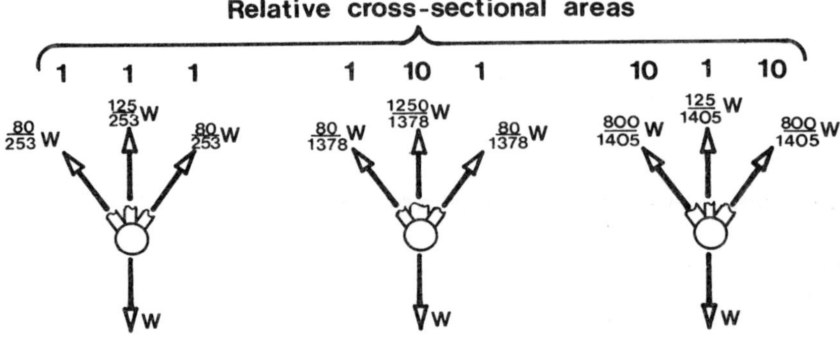

Figure 2.23 Member forces

Redundancy

If a structure is statically indeterminate then it has at least one reactive or member force that can be released without the structure becoming a mechanism. Such forces are known as redundants, and the total number of redundants is known as the degree of indeterminacy of the structure. Figure 2.24 shows three statically determinate structures. Such structures have zero degree of indeterminacy, implying that to remove any internal or reactive component of force will produce a mechanism.

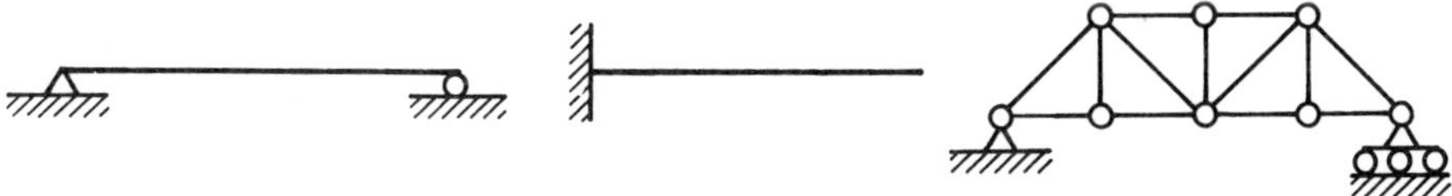

Figure 2.24 Examples of statically determinate structures

By introducing an additional support to each of the beams, and two additional members to the truss as shown in fig 2.25, the beams are made indeterminate to the first degree, and the truss is made indeterminate to the second degree.

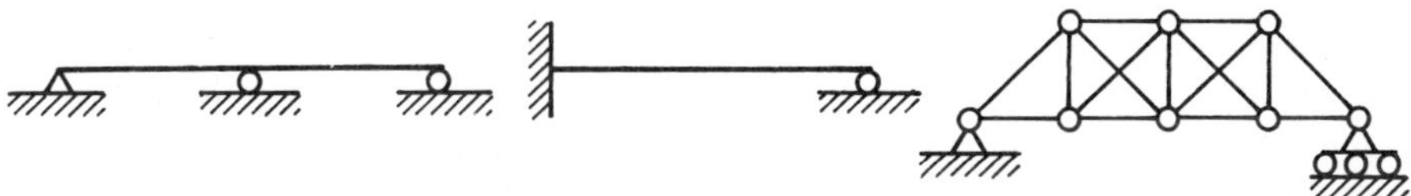

Figure 2.25 Examples of statically indeterminate structures

It should be noted that the beams can be rendered statically determinate by releasing either a support or an internal moment. The internal moment could be released by introducing a pin at any point along the beam (except at the free ends). In the case of the truss the reader should note that only the two centre panels are statically indeterminate which means that releasing a member force outside these panels would produce a mechanism.

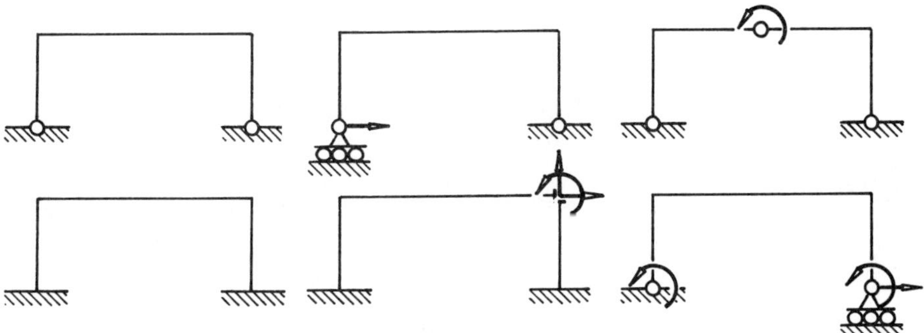

Figure 2.26 Portal frames with pinned and fixed feet

The left-hand diagrams in fig 2.26 show portals with pinned and fixed feet. The portal with pinned feet is indeterminate to the first degree as it can be rendered determinate by releasing a single component of force. The portal with fixed feet is indeterminate to the third degree. The other diagrams in fig 2.26 illustrate how components of force equal in number to the degree of indeterminacy can be released without the structure becoming unstable. The circular arrows indicate points where a moment has been released, and the linear arrows incidate where a linear force has been released.

2.11 Superposition

If a structure is constructed from a linearly elastic material, and small deflection is adopted then the structure behaves linearly. Hence deflections, reactions, and member forces are directly proportional to the applied loads. Consequently the effect of a given load system on a linear structure will be independent of the order in which the loads are applied, making valid the principle of superposition. Superposition is most commonly used to investigate the combined effect of two load systems. If, for instance, a structure has been independently analysed for imposed load and for wind load, then the combined effect of imposed and wind load can be found by simply superimposing the individual effects of the two load systems. The propped cantilever shown in fig 2.27 illustrates the principle of superposition.

The diagram shows that, if the effect of the propping force (i.e. the right-hand reaction) is superimposed on the effect of the applied load on the unpropped cantilever, then the true results are obtained. Here the propping force and the applied load have been treated as two independent load systems and then their effects have been superimposed.

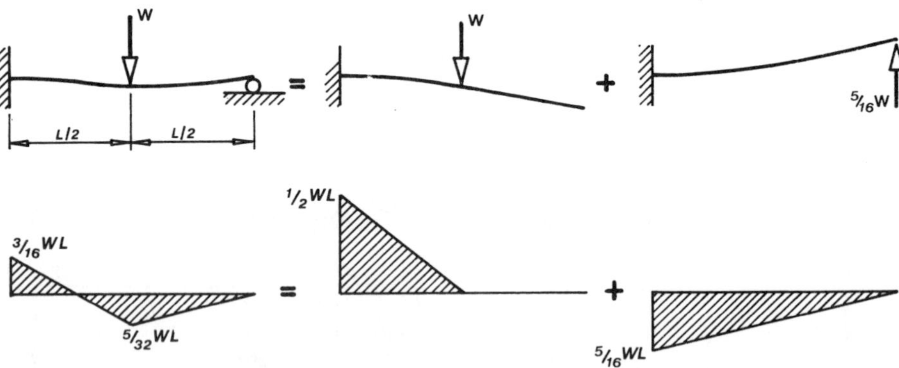

Figure 2.27 An example of superposition

Superposition is not valid when the structural behaviour is non-linear. There are two principal sources of non-linearity in structures. The first is when the displacements are large, making small deflection theory inapplicable. The second is when the stiffness of the members of the structure is non-linear due to non-linear material behaviour. Non-linear behaviour is beyond the scope of this book, with linear behaviour being assumed throughout.

2.12 Simple Beam Behaviour

Axial Deformation

Axial deformation is the lengthening or shortening of a line element caused by equal and opposite longitudinal forces as shown in fig 2.28.

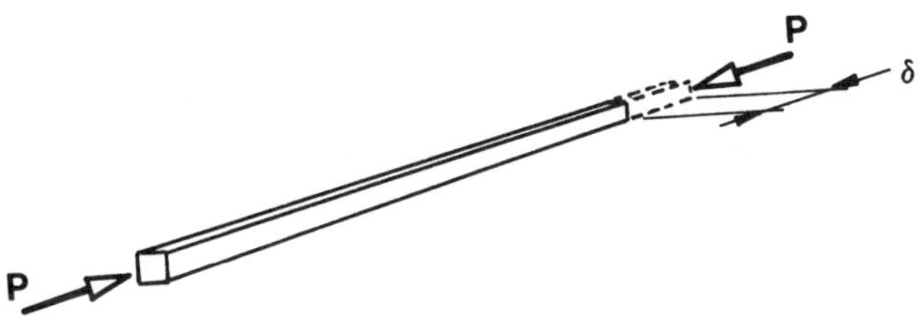

Figure 2.28 Axial deformation

The axial deformation of the element is governed by equation (2.3).

$$\delta = \frac{PL}{EA} \tag{2.3}$$

where

δ	is the axial deformation
P	is the axial load
L	is the element length
E	is the elastic constant (Young's modulus)
A	is the cross-sectional area.

Torsional Deformation

Torsion is the twisting of a line element caused by equal and opposite moments at its ends, as shown in fig 2.29.

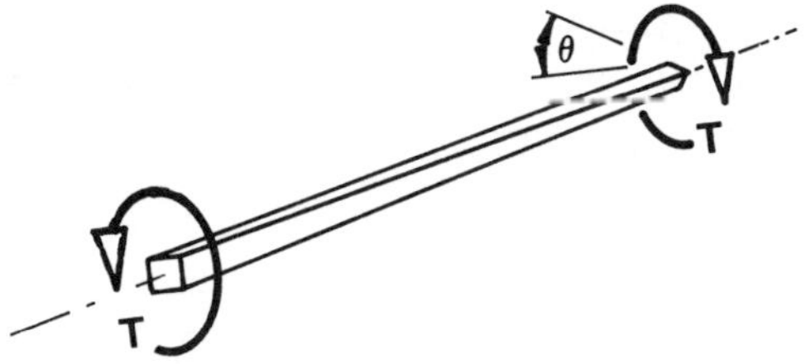

Figure 2.29 Torsional deformation

The angle of twist as shown in fig 2.29 is given by equation (2.4).

$$\theta = \frac{TL}{GJ} \tag{2.4}$$

where

θ is the relative rotation between the ends of the element
T is the applied torsional moment
L is the element length
G is the modulus of rigidity
J is the torsional constant.

The modulus of rigidity, "G", is sometimes referred to as the modulus of rigidity in shear, or simply the shear modulus. The modulus of rigidity, the elastic constant, and Poisson's ratio, "ψ", are related by the following formula:

$$G = \frac{E}{2(1 + \psi)}$$

The torsional constant, "J", is a property of the cross-section and has units of second moment of area. The torsional constant cannot be evaluated exactly for most cross-sections and must be calculated from approximate formulae.

Flexural Deformation

Flexural deformation (or bending) of a line element arises from transverse forces and moments as illustrated in fig 2.30. These forces cause translation and rotation of the ends of the members as shown. The remainder of this section is devoted to finding the relationship between the element end forces and the flexural deformation. Rotational and translational displacements at the member ends will be treated independently as any composite deformation system can be found by superposition.

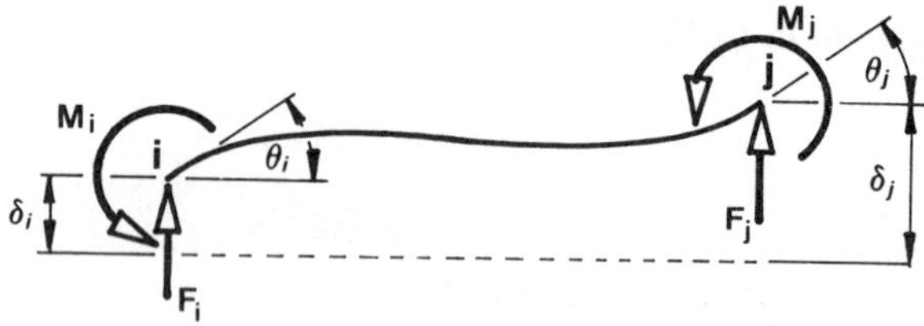

Figure 2.30 Flexural deformation

If a line element is loaded only at its ends then the shear force along the length of the element is constant.

$$SF = k$$

but

$$SF = \text{rate of change of bending moment}$$

i.e.

$$\frac{dM}{dx} = k$$

therefore, by integrating the above expression

$$M = kx + l$$

Where "l" is a constant of integration. Simple beam theory (sometimes known as the engineer's theory of bending) assumes that bending moment is proportional to curvature, with EI as the constant of proportionality.

$$M = EI\frac{d^2y}{dx^2} = kx + l$$

Incorporating EI into the constants on the right-hand side and integrating twice yields:

$$\frac{d^2y}{dx^2} = \frac{kx}{EI} + \frac{l}{EI}$$

$$\frac{dy}{dx} = \frac{kx^2}{2EI} + \frac{lx}{EI} + c$$

$$y = \frac{kx^3}{6EI} + \frac{lx^2}{2EI} + cx + d$$

or

$$y = ax^3 + bx^2 + cx + d \tag{2.5}$$

This proves that the flexural deformation of a prismatic line element loaded only at its ends conforms to a cubic polynomial. Knowing the rotation or slope (dy/dx) and the transverse displacement (y) at both ends of the element gives a total of four boundary conditions, and these enable the four coefficients (a, b, c, d) of the polynomial to be evaluated. An expression that algebraically describes the deformed shape of an element is known as a *shape function*.

Now consider the force system associated with rotation at end "i" of the line element "i,j" shown in fig 2.31.

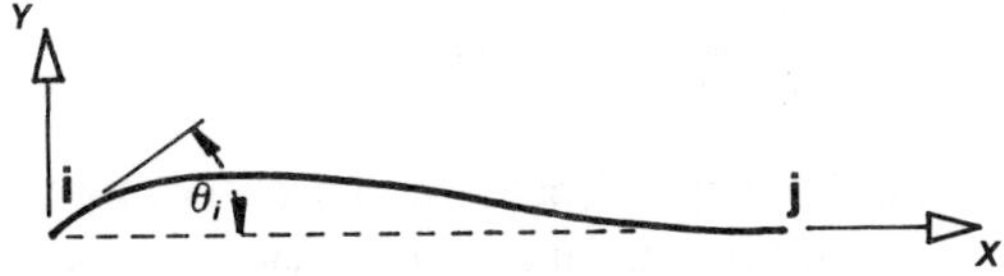

Figure 2.31 Rotation at end "i"

The shape function and its derivatives are

$$y = ax^3 + bx^2 + cx + d$$

$$\frac{dy}{dx} = 3ax^2 + 2bx + c$$

$$\frac{d^2y}{dx^2} = 6ax + 2b$$

The boundary conditions are

at $x = 0$

$$y = 0 \qquad \text{therefore} \qquad 0 = 0 + 0 + 0 + d$$

$$\frac{dy}{dx} = \theta_i \qquad \text{therefore} \qquad \theta_i = 0 + 0 + c$$

at $x = l$

$$y = 0 \qquad \text{therefore} \qquad 0 = aL^3 + bL^2 + cL + d$$

$$\frac{dy}{dx} = 0 \qquad \text{therefore} \qquad 0 = 3aL^2 + 2bL + c$$

hence

$$d = 0, \qquad c = \theta_i \qquad a = \theta_i/L^2, \qquad b = -2\theta_i/L$$

giving

$$y = \frac{x^3}{L^2}\theta_i - \frac{2x^2}{L}\theta_i + x\theta_i \tag{2.6}$$

but

$$M = EI\frac{d^2y}{dx^2} \qquad \text{(i.e. bending moment} = EI \text{ x curvature)}$$

Substitution of the second differential of equation (2.6) into the above equation yields:

$$M = EI\theta_i\left(\frac{6x}{L^2} - \frac{4}{L}\right)$$

hence

$$M = -\frac{4EI}{L}\theta_i \qquad \text{when} \qquad x = 0$$

$$M = \frac{2EI}{L}\theta_i \qquad \text{when} \qquad x = L$$

The signs mean that the applied bending moments causes negative curvature at the left-hand end of the beam, and positive curvature at the right-hand end of the beam. Hence the force system associated with rotation at "*i*" is as shown in fig 2.32. Note that the forces $6EI\theta_i/L^2$ are necessary to maintain the equilibrium of the element.

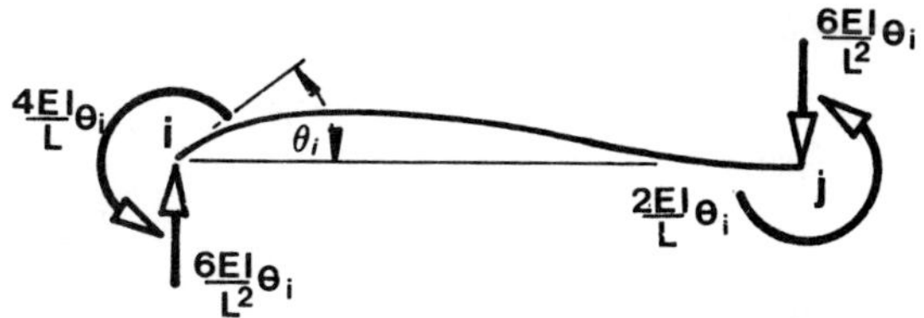

Figure 2.32 Force system associated with rotation at node "*i*"

Now consider the force system associated with transverse translation at end "*i*" as illustrated in fig 2.33.

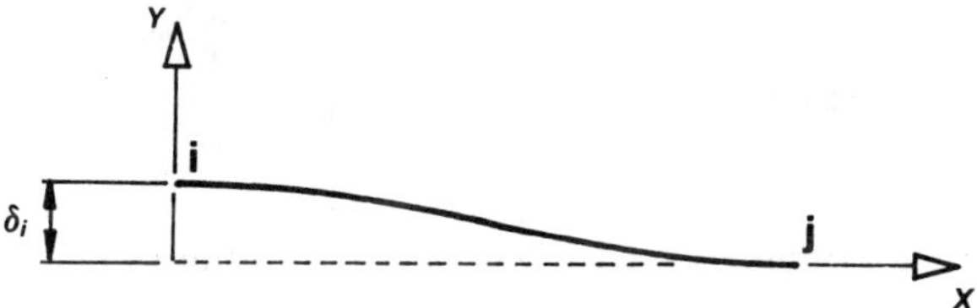

Figure 2.33 Transverse translation at node "*i*"

The boundary conditions are:

at $x = 0$

$$y = \delta_i \qquad \text{therefore} \qquad \delta_i = 0 + 0 + 0 + d$$

$$\frac{dy}{dx} = 0 \qquad \text{therefore} \qquad 0 = 0 + 0 + c$$

at $x = L$

$$y = 0 \qquad \text{therefore} \qquad 0 = aL^3 + bL^2 + cL + d$$

$$\frac{dy}{dx} = 0 \qquad \text{therefore} \qquad 0 = 3aL^2 + 2bL + c$$

hence

$$d = \delta_i, \qquad c = 0, \qquad b = -\frac{3}{L^2}\delta_i, \qquad a = \frac{2}{L^3}\delta_i$$

giving

$$y = \frac{2x^3}{L^3}\delta_i - \frac{3x^2}{L^2}\delta_i + \delta_i \tag{2.7}$$

but

$$M = EI\frac{d^2y}{dx^2} \qquad \text{(i.e. bending moment} = EI \text{ x curvature)}$$

Substitution of the second differential of equation (2.7) into the above equation yields

$$M = EI\delta_i \left[\frac{12x}{L^3} - \frac{6}{L^2} \right]$$

hence

$$M = -\frac{6EI}{L^2}\delta_i \qquad \text{when} \quad x = 0$$

$$M = \frac{6EI}{L^2}\delta_i \qquad \text{when} \quad x = L$$

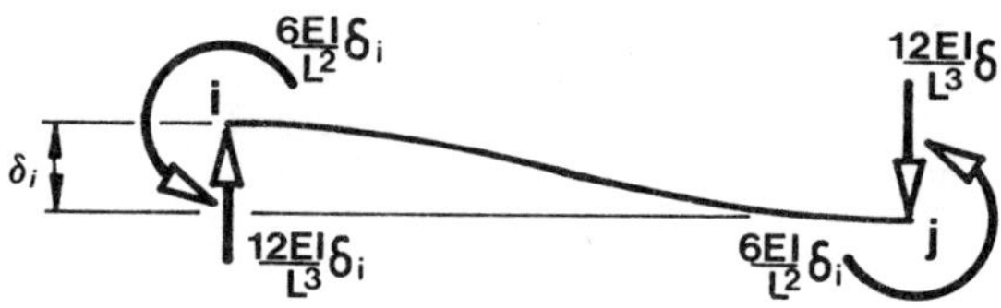

Figure 2.34 Force system associated with translation at "i"

The signs mean that the applied bending moments causes negative curvature at the left-hand end of the beam, and positive curvature at the right-hand end of the beam. Hence the force system associated with translation at "*i*" is as shown in fig 2.34.

2.13 Virtual Work

The technique of applying a small test displacement system to a structure that is in equilibrium, and then equating the work done by the existing forces to the strain energy gained by the structure, is known as the *method of virtual work.* The principle of virtual work states that if a structure that is in equilibrium under a given set of forces is subjected to some small test displacement system (a virtual displacement system) then the work done by the external forces is equal to the change in strain energy in the structure, i.e.

Work done by the forces acting on the structure = *Strain energy gained by the structure*

Note that the force system includes reactions and the displacement system is compatible, i.e. it must not violate the boundary conditions and the displacements must be continuous throughout the structure. Hence the application of the principle of virtual work requires an *equilibrium force system* and a *compatible displacement system.*

The principle of virtual work is great importance in the field of structural analysis and is widely used to determine displacements and forces in statically determinate and statically indeterminate structures. Only a very brief introduction can be given to the subject here and the reader is referred to the bibliography for more detail.

Consider, for example, using the principle of virtual work to calculate the value of the right-hand reaction and the bending moment under the load for the beam shown in fig 2.35.

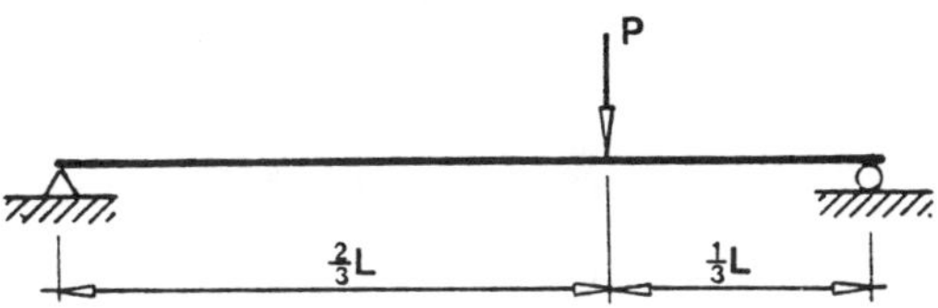

Figure 2.35 A simply supported beam

Figure 2.36(a) shows the freebody diagram for the beam. If the right-hand reaction is given a virtual displacement "δ" then work is done by the applied load and right-hand reaction (note that the work is negative if the displacement is in the opposite direction of the force). There is no change in the strain energy stored in the structure, hence the virtual work equation is

$$(R_R \times \delta) - (P \times \tfrac{2}{3}\delta) = 0$$

hence

$$R_R = \tfrac{2}{3}P$$

The bending moment under the load is obtained by applying a virtual rotation of "θ" under the load as shown in fig 2.36(b).

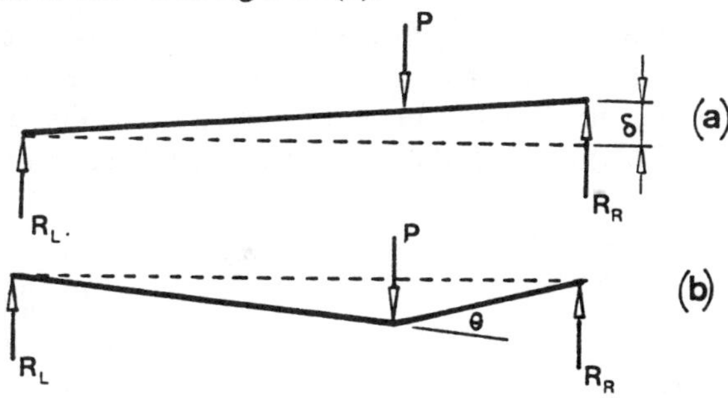

Figure 2.36 Virtual Displacements

The resulting work equation is

$$(P \times \tfrac{2}{3}L \times \tfrac{1}{3}\theta) - M\theta = 0$$

hence

$$M = \tfrac{2}{9} PL$$

Often the virtual displacement systems result in a change of the flexure of the elements of the structure. In such cases it is necessary to evaluate the change in the strain energy stored in the structure brought about by the application of the virtual displacement system.

Figure 2.37 shows a differential element of a member of a moment carrying structure.

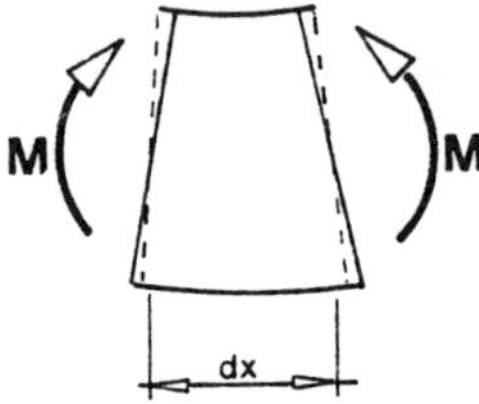

Figure 2.37 Differential element of a beam

If the moment acting on the element is "M" and the virtual displacement system causes a change in curvature ΔR then change in the strain energy stored in the element is

$$dU = M \, \Delta R \, dx$$

and the change in the strain energy of the whole member is

$$U = \int_0^L M \, \Delta R \, dx$$

The change in the strain energy of the whole structure is, of course, obtained by summing the change in strain energy of all of the members of the structure.

Often the virtual displacement system used is the displacement system due to some particular loading. If the moment due that loading is m then

$$\Delta R = \frac{m}{EI}$$

and

$$U = \int_0^L M \frac{m}{EI} dx$$

To illustrate how this theory is applied the principle of virtual work will be used to find the slope "θ" at the end of the uniform cantilever shown in fig 2.38(a).

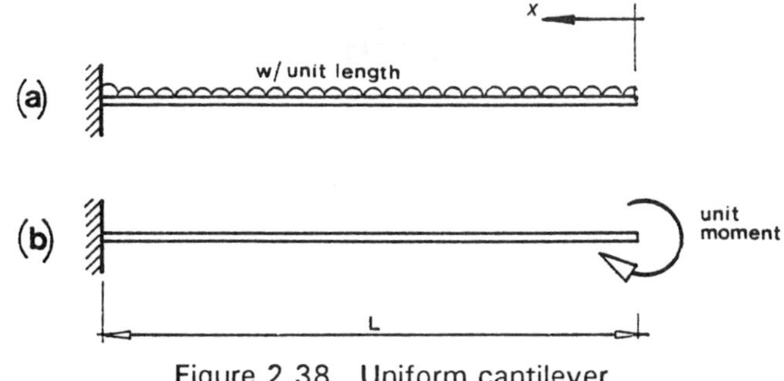

Figure 2.38 Uniform cantilever

As the actual displacements are of interest the actual displacements will be used as the *virtual displacement system* and the force system shown in fig 2.38(b) will be used as the *equilibrium force system*. Hence

$$M = 1 \qquad \text{and} \qquad m = \frac{w\,x^2}{2}$$

$$\textit{work done} = \textit{strain energy gained}$$

$$1 \times \theta = \int_0^L M \frac{m}{EI}\,dx$$

$$= \int_0^L 1 \frac{w\,x^2}{2EI}\,dx$$

which yields

$$\theta = \frac{w\,L^3}{6EI}$$

Chapter 3

Introduction to the Stiffness Method

3.1 Classification of Framed Structures

Framed structures are structures that can be satisfactorily idealised using line elements. Such structures are often referred to as skeletal structures. Usually the elements of the structure are assumed to be connected either by frictionless pins or by rigid joints. Framed structures are usually idealised as one of the five types of skeletal structures shown in fig 3.1.

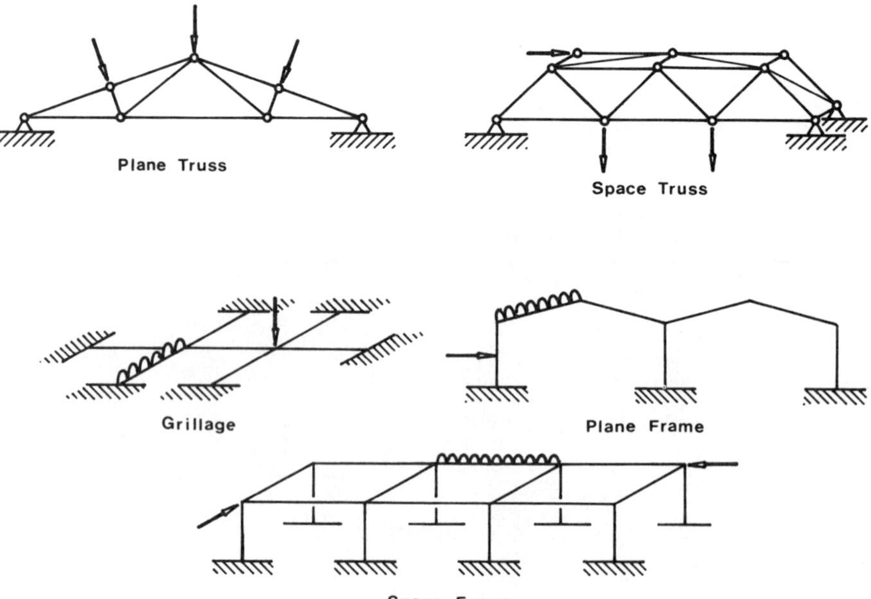

Figure 3.1 Structural frameworks

Trusses are usually loaded only at their joints and, because the joints cannot transmit bending moment, trusses must be triangulated to avoid mechanism formation. Plane trusses are loaded in their own plane, whereas the joints of a space truss can be loaded from any direction.

FRAMEWORK TYPE	JOINT TYPE	JOINT LOADS	ELEMENT LOADS	LOAD DIRECTIONS
Plane Trusses	Pinned	Yes	No	In Plane
Space Trusses	Pinned	Yes	No	Any Direction
Plane Frames	Rigid	Yes	Yes	In Plane
Grillages	Rigid	Yes	Yes	Normal to Plane
Space Frame	Rigid	Yes	Yes	Any Direction

Rigidly jointed frames are often loaded along their elements as well as at their joints. Plane frames, like plane trusses, are loaded only in their own plane. In contrast, grillages are always loaded normal to the structure. Space frames can, of course, be loaded in any plane.

3.2 Degree of Freedom

In the context of the stiffness method the degree of freedom of a structure is the number of displacement components to be found during the analysis. Finding these displacements by the stiffness method involves the solution of a system of linear equations relating the known applied forces to the unknown displacements, where the number of equations is equal to the degree of freedom of the structure.

In general, elastic structures have an infinite number of independent displacement components as displacements vary continuously throughout the structure. Obviously, to solve an infinite number of simultaneous equations will take an unacceptably long time. The stiffness method renders the analysis tractable by considering displacements and forces only at selected points, called *nodes*.

When the stiffness method is applied to problems in continuum mechanics it is commonly known as the finite element method. Node points are selected and imaginary lines between the nodes divide the structure into elements. To assess the properties of the elements an assumption is made about how the displacements vary within the elements. The finite element method is an approximate method because of this assumption. In general, using a finer element mesh (i.e. more elements) will produce a more accurate solution.

When the stiffness method is applied to structural frameworks a node is usually located at each joint of the structure. The analysis determines the nodal displacements caused by the applied loading. The number of possible displacement components at each node is known as the *nodal degree of freedom*, and the nodal degree of freedom for different framework types is as shown in the following table.

FRAMEWORK TYPE	NODAL DEGREE OF FREEDOM
Plane Truss	2
Space Truss	3
Plane Frame	3
Grillage	3
Space Frame	6

Figure 3.2 illustrates the nodal freedoms for different types of structural frameworks.

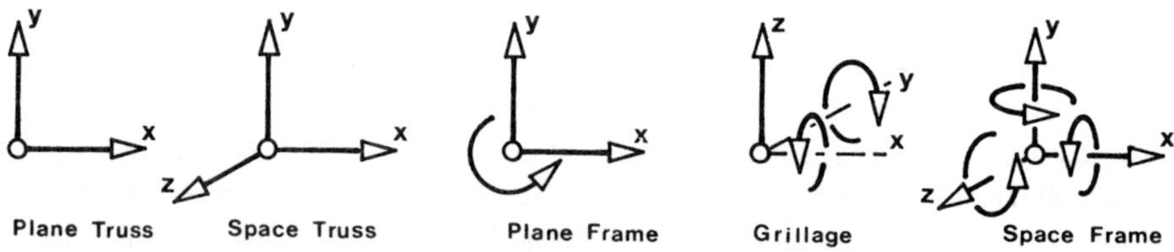

Figure 3.2 Nodal degrees of freedom

The stiffness method is concerned with the relationship between nodal (for the time being nodes equate to joints) forces and displacements. The components of force and displacement at a node are conveniently stored in one-dimensional arrays known as the nodal force and displacement vectors. The number of elements in each vector is equal to the nodal degree of freedom as shown in the following arrays. In general the displacement vector contains rotations as well as translations, and the force vector contains moments and forces. Consequently these vectors are sometimes referred to as generalised force and displacement vectors.

STRUCTURE TYPE	NODAL FORCE VECTOR	NODAL DISPLACEMENT VECTOR
Plane Truss	$\begin{bmatrix} P_x \\ P_y \end{bmatrix}$	$\begin{bmatrix} \Delta_x \\ \Delta_y \end{bmatrix}$
Space Truss	$\begin{bmatrix} P_x \\ P_y \\ P_z \end{bmatrix}$	$\begin{bmatrix} \Delta_x \\ \Delta_y \\ \Delta_z \end{bmatrix}$
Plane Frame	$\begin{bmatrix} P_x \\ P_y \\ M_z \end{bmatrix}$	$\begin{bmatrix} \Delta_x \\ \Delta_y \\ \theta_z \end{bmatrix}$
Grillage	$\begin{bmatrix} P_z \\ M_x \\ M_y \end{bmatrix}$	$\begin{bmatrix} \Delta_z \\ \theta_x \\ \theta_y \end{bmatrix}$
Space Frame	$\begin{bmatrix} P_x \\ P_y \\ P_z \\ M_x \\ M_y \\ M_z \end{bmatrix}$	$\begin{bmatrix} \Delta_x \\ \Delta_y \\ \Delta_z \\ \theta_x \\ \theta_y \\ \theta_z \end{bmatrix}$

where

P is a linear force component.
M is a rotational force component (i.e. a moment).
Δ is a linear displacement component.
θ is a rotational displacement component.

The space frame is the most complicated type of rigidly jointed framework. Each element of a space frame can undergo axial deformation, torsional deformation, and flexural deformation (in two planes). Axial and torsional deformations of a prismatic line element loaded only at its ends vary linearly along the length of the element. Hence it is a simple matter to determine the axial and torsional deformation at any point along the element from a knowledge of the end displacements. Section 2.12 showed that the cubic polynomial defining the flexural deformation of a line element loaded at its ends can be determined from the transverse displacements and the rotations at its ends. It follows that element end displacements define flexural deformation at all points along the element. Hence the deformed shape of a prismatic space frame element loaded only at its ends is uniquely defined by its end displacements.

All other types of rigidly jointed frameworks are special cases of space frame structures. Hence the deformed shape of any type of structural framework, loaded only at its joints, is completely defined by the nodal displacements. Consequently the stiffness method is exact when applied to frameworks loaded only at their joints. That only joint loads have been considered here is not, in fact, a restriction. Element loads, which have been excluded in the interests of clarity, can easily be included in the analysis through the use of equivalent joint forces as described in Chapter 8.

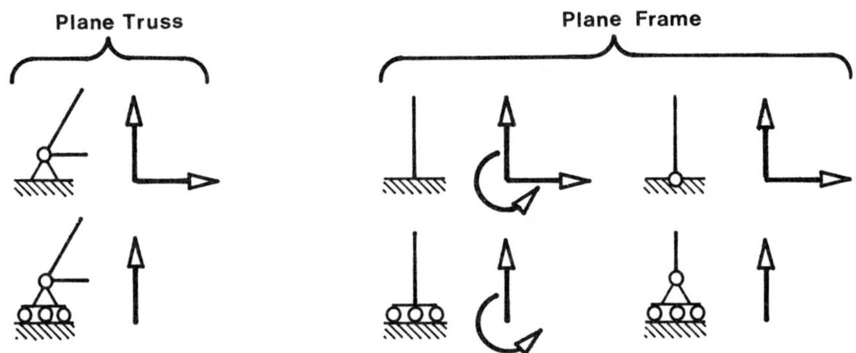

Figure 3.3 Boundary conditions for plane trusses and plane frames

Boundary Conditions

For stability a structure is always restrained at one or more nodes (see section 2.3). These restraints are known as boundary conditions and they have the effect of reducing the degree of freedom of the restrained node.

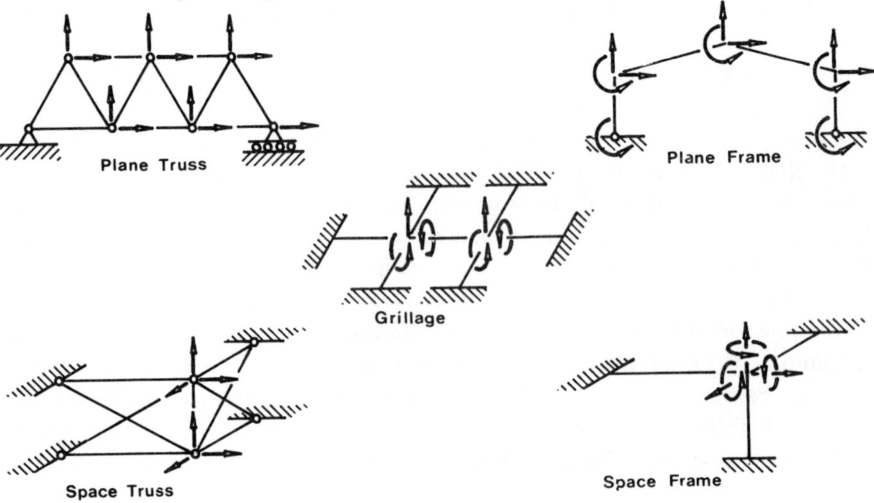

Figure 3.4 Unrestrained nodal displacements

Figure 3.3 shows typical boundary conditions and restraint directions for plane trusses and plane frames. The boundary conditions for other framework types are similar in concept.

The determination of the degree of freedom of a framed structure is a straightforward process, requiring only a knowledge of the nodal degree of freedom, and an understanding of the boundary conditions. The degree of freedom of the structure is simply the product of the number of nodes and the nodal degree of freedom, less the number of restrained displacement components. Figure 3.4 shows the unrestrained nodal displacement components for some simple structural frameworks. Henceforth the unrestrained nodal displacement components will be referred to as *freedoms*.

3.3 Matrix Methods of Structural Analysis

If a structure is in a state of stable equilibrium and small displacement theory is valid then there is a unique relationship between the deformation of the structure and the load system applied to it. In other words the structure will take up one, and only one, deformed shape under the action of a given set of loads. Consider, for example, the rectangular portal shown in fig 3.5.

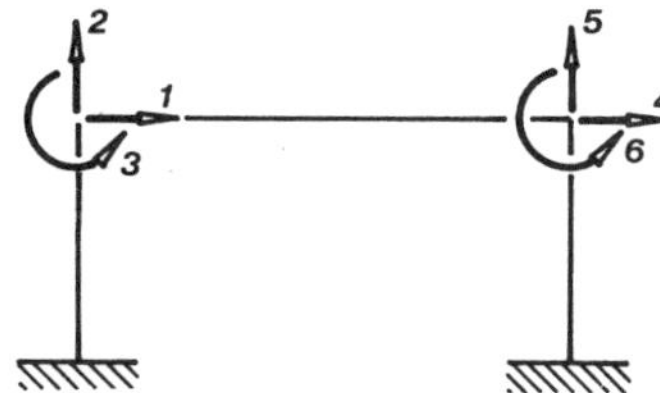

Figure 3.5 Unrestrained nodal displacements for a rectangular portal

The deformed shape is completely defined by the freedoms shown. If the loads in the directions of the freedoms are P_1, P_2, P_3, ... P_6 and the resulting displacements are Δ_1, Δ_2, Δ_3, ... Δ_6 as shown, then

$$\Delta_1 = G_1(P_1, P_2, \ldots\ldots P_6) \quad \text{or} \quad P_1 = H_1(\Delta_1, \Delta_2, \ldots\ldots \Delta_6)$$

$$\Delta_2 = G_2(P_1, P_2, \ldots\ldots P_6) \quad \text{or} \quad P_2 = H_2(\Delta_1, \Delta_2, \ldots\ldots \Delta_6)$$

$$\vdots \qquad\qquad\qquad\qquad \vdots$$

$$\Delta_6 = G_6(P_1, P_2, \ldots\ldots P_6) \quad \text{or} \quad P_6 = H_6(\Delta_1, \Delta_2, \ldots\ldots \Delta_6)$$

The equations on the left-hand side state that the unknown nodal displacements are a function of the applied loads. Of course, the expressions can be rewritten to give the applied loads as a function of the unknown nodal displacements, as shown by the equations on the right-hand side. If the structure is assumed to be linearly elastic then there is a linear relationship between the loads and the resulting displacements. That relationship can be expressed with either the loads or the displacements as the independent variables. If the displacements are chosen as the indepenent variables then

$$
\begin{aligned}
\Delta_1 &= F_{11}P_1 + F_{12}P_2 + \ldots\ldots + F_{16}P_6 \\
\Delta_2 &= F_{21}P_1 + F_{22}P_2 + \ldots\ldots + F_{26}P_6 \\
&\vdots \\
\Delta_6 &= F_{61}P_1 + F_{62}P_2 + \ldots\ldots + F_{66}P_6
\end{aligned}
$$

Rewriting these equations in matrix form gives

$$
\begin{bmatrix} \Delta_1 \\ \Delta_2 \\ \Delta_3 \\ \Delta_4 \\ \Delta_5 \\ \Delta_6 \end{bmatrix} = \begin{bmatrix} F_{11} & F_{12} & F_{13} & F_{14} & F_{15} & F_{16} \\ F_{21} & F_{22} & F_{23} & F_{24} & F_{25} & F_{26} \\ F_{31} & F_{32} & F_{33} & F_{34} & F_{35} & F_{36} \\ F_{41} & F_{42} & F_{43} & F_{44} & F_{45} & F_{46} \\ F_{51} & F_{52} & F_{53} & F_{54} & F_{55} & F_{56} \\ F_{61} & F_{62} & F_{63} & F_{64} & F_{65} & F_{66} \end{bmatrix} \begin{bmatrix} P_1 \\ P_2 \\ P_3 \\ P_4 \\ P_5 \\ P_6 \end{bmatrix} \tag{3.1}
$$

$$\boldsymbol{\Delta} = \boldsymbol{F}\ \boldsymbol{P}$$

If the loads are chosen as the independent variables then

$$
\begin{aligned}
P_1 &= K_{11}\Delta_1 + K_{12}\Delta_2 + \ldots\ldots + K_{16}\Delta_6 \\
P_2 &= K_{21}\Delta_1 + K_{22}\Delta_2 + \ldots\ldots + K_{26}\Delta_6 \\
&\vdots \\
P_6 &= K_{61}\Delta_1 + K_{62}\Delta_2 + \ldots\ldots + K_{66}\Delta_6
\end{aligned}
$$

Writing these equations in matrix form gives

$$
\begin{bmatrix} P_1 \\ P_2 \\ P_3 \\ P_4 \\ P_5 \\ P_6 \end{bmatrix} = \begin{bmatrix} K_{11} & K_{12} & K_{13} & K_{14} & K_{15} & K_{16} \\ K_{21} & K_{22} & K_{23} & K_{24} & K_{25} & K_{26} \\ K_{31} & K_{32} & K_{33} & K_{34} & K_{35} & K_{36} \\ K_{41} & K_{42} & K_{43} & K_{44} & K_{45} & K_{46} \\ K_{51} & K_{52} & K_{53} & K_{54} & K_{55} & K_{56} \\ K_{61} & K_{62} & K_{63} & K_{64} & K_{65} & K_{66} \end{bmatrix} \begin{bmatrix} \Delta_1 \\ \Delta_2 \\ \Delta_3 \\ \Delta_4 \\ \Delta_5 \\ \Delta_6 \end{bmatrix} \tag{3.2}
$$

$$\boldsymbol{P} = \boldsymbol{K}\ \boldsymbol{\Delta}$$

where

$\boldsymbol{P}$ is the loading vector

Δ is the displacement vector
F is the structure flexibility matrix
K is the structure stiffness matrix

Equations (3.1) and (3.2) form the basis of the flexibility and stiffness methods of structural analysis respectively. It should be noted, however, that equation (3.1) gives an oversimplified view of the flexibility method.

At first sight the flexibility method looks the more attractive because, once the flexibility matrix F has been found, the structural displacements Δ can be obtained by simple matrix multiplication. By contrast, to find the unknown displacements using the stiffness method involves solving a system of simultaneous equations. However, the stiffness matrix turns out to be much easier to develop than the flexibility matrix. Further, the stiffness method is better suited to computer implementation than the flexibility method. For these reasons the stiffness method now completely dominates structural analysis computer programs. As the displacements are the primary unknowns, the stiffness method is sometimes referred to as the displacement method.

The flexibility method will often be more economical than the stiffness method for hand calculation. It is the opinion of the author, however, that matrix methods should not be used for hand calculation because traditional methods of structural analysis will almost always be quicker and less error prone. Finally, as the flexibility method is less well suited to computer implementation than the stiffness method it will not be considered further and the remainder of this book will deal exclusively with the stiffness method.

3.4 Stiffness Coefficients

The analysis of structures using the stiffness equation $P = K \Delta$ has three distinct phases

1. Assembly of the stiffness matrix K. The elements of K are, in effect, the coefficients of a set of simultaneous equations. Note that K is a property of the structure and is independent of the loading.

2. Generation of the loading vector P.

3. Solution of the simultaneous equations. This yields the unknown structural displacements, Δ, caused by the known applied loads, P.

The remainder of this section takes a closer look at the structure stiffness matrix K. The solution of simultaneous equations will be considered in Chapter 4.

Consider a linearly elastic body acted upon by a system of forces P, which cause displacements Δ in the direction of the forces as shown in fig 3.6.

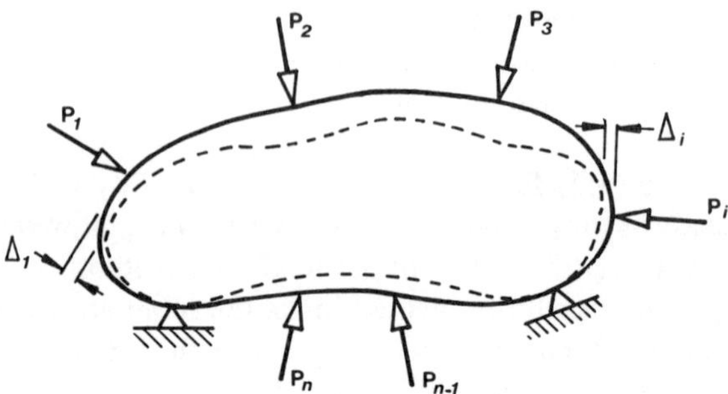

Figure 3.6 Deformation of an elastic body

The total work done, W, during the application of the forces is

$$W = \frac{1}{2}(P_1\Delta_1 + P_2\Delta_2 + \ldots\ldots + P_n\Delta_n) \tag{3.3}$$

As the body is elastic, energy is conserved, and the work done by the external forces is equal to the strain energy, U, gained by the body.

$$W = U$$

Now if one displacement, say Δ_1, is varied by an infinitesimal amount $d\Delta_1$ while all other displacements are held constant, then the change in the strain energy of the body is

$$dU = \frac{\partial U}{\partial \Delta_1} d\Delta_1 \tag{3.4}$$

and the corresponding change in P_1 is

$$dP_1 = \frac{\partial P_1}{\partial \Delta_1} d\Delta_1$$

Hence the change in the work done is equal to

$$dW = \left[\frac{P_1 + (P_1 + dP_1)}{2}\right] d\Delta_1$$

$$= P_1\, d\Delta_1 + \frac{1}{2}\frac{\partial P_1}{\partial \Delta_1}(d\Delta_1)^2$$

In the limit

$$dW = P_1\, d\Delta_1 \tag{3.5}$$

Energy is conserved during this variation of displacement and therefore the work done is equal to the change in the strain energy, hence

$$dW = dU$$

Substituting using equations (3.4) and (3.5) produces the following

$$P_1 \, d\Delta_1 = \frac{\partial U}{\partial \Delta_1} d\Delta_1$$

hence

$$P_1 = \frac{\partial U}{\partial \Delta_1} \tag{3.6}$$

The previous equation is *Part I of Castigliano's First Theorem.* Differentiation of equation (3.3) yields (noting that $U = W$)

$$\frac{\partial U}{\partial \Delta_1} = \frac{1}{2}\left[P_1 + \frac{\partial P_1}{\partial \Delta_1}\Delta_1 + \frac{\partial P_2}{\partial \Delta_1}\Delta_2 + \ldots\ldots + \frac{\partial P_n}{\partial \Delta_1}\Delta_n \right]$$

Using Castigliano's First Theorem to substitute for $\partial U/\partial \Delta_1$ in the above equation gives

$$P_1 = \frac{\partial P_1}{\partial \Delta_1}\Delta_1 + \frac{\partial P_2}{\partial \Delta_1}\Delta_2 + \ldots\ldots + \frac{\partial P_n}{\partial \Delta_1}\Delta_n$$

Varying all other displacements in turn yields

$$P_2 = \frac{\partial P_1}{\partial \Delta_2}\Delta_1 + \frac{\partial P_2}{\partial \Delta_2}\Delta_2 + \ldots\ldots + \frac{\partial P_n}{\partial \Delta_2}\Delta_n$$

$$P_3 = \frac{\partial P_1}{\partial \Delta_3}\Delta_1 + \frac{\partial P_2}{\partial \Delta_3}\Delta_2 + \ldots\ldots + \frac{\partial P_n}{\partial \Delta_3}\Delta_n$$

$$\vdots$$

$$P_n = \frac{\partial P_1}{\partial \Delta_n}\Delta_1 + \frac{\partial P_2}{\partial \Delta_n}\Delta_2 + \ldots\ldots + \frac{\partial P_n}{\partial \Delta_n}\Delta_n$$

Writing these equations in matrix form

$$\begin{bmatrix} P_1 \\ P_2 \\ \cdot \\ \cdot \\ P_n \end{bmatrix} = \begin{bmatrix} \frac{\partial P_1}{\partial \Delta_1} & \frac{\partial P_2}{\partial \Delta_1} & \ldots\ldots & \frac{\partial P_n}{\partial \Delta_1} \\ \frac{\partial P_1}{\partial \Delta_2} & \frac{\partial P_2}{\partial \Delta_2} & \ldots\ldots & \frac{\partial P_n}{\partial \Delta_2} \\ \cdot & \cdot & & \cdot \\ \cdot & \cdot & & \cdot \\ \frac{\partial P_1}{\partial \Delta_n} & \frac{\partial P_2}{\partial \Delta_n} & \ldots\ldots & \frac{\partial P_n}{\partial \Delta_n} \end{bmatrix} \begin{bmatrix} \Delta_1 \\ \Delta_2 \\ \cdot \\ \cdot \\ \Delta_n \end{bmatrix} \tag{3.7}$$

This is, of course, the stiffness equation

$$\boldsymbol{P} = \boldsymbol{K}\,\boldsymbol{\Delta}$$

Note that the elastic body is not subjected to applied moments. Introducing applied moments causes no difficulty, their omission was simply for the sake of clarity.

Inspection of the stiffness matrix shows that it consists of derivatives that represent the rate of change of force with displacement, i.e. they are *stiffnesses*. If the structure is linearly elastic then these terms are constant. The physical meaning of $\partial P_i/\partial \Delta_j$ is the rate of change of the force at "i" with variation of displacement at "j" ($\partial P_i/\partial \Delta_j$ is a partial derivative which implies that all other displacements are held constant during the variation of displacement at "j"). If Δ_j is varied by unity, then $\partial P_i/\partial \Delta_j$ becomes *the force at "i" associated with unit change in the displacement at "j"* (all other displacements held constant). Hence, column "j" of the stiffness matrix is the force system required to maintain unit displacement at "j". This means that the complete stiffness matrix can be generated by calculating the force system required to maintain unit displacement at each freedom in turn.

Physical insight into the stiffness method can be gained by considering unit displacement at each freedom in turn, and then scaling the force systems for the unit displacements by the actual displacements to allow the true structural behaviour to be obtained by superposition. Consider the rigidly jointed frame shown in fig 3.7.

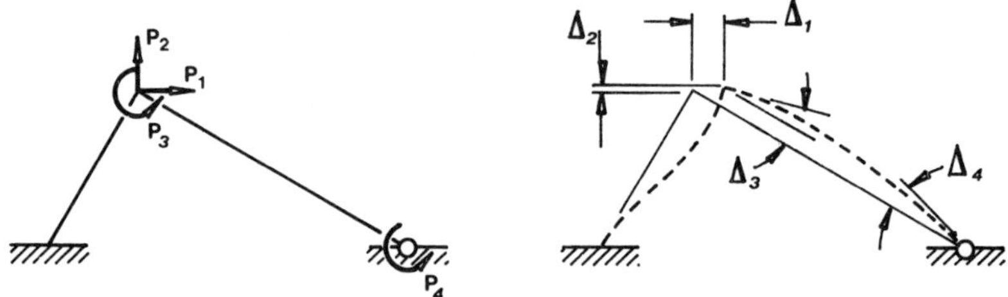

Figure 3.7 Forces and displacements

The frame has four freedoms and is acted upon by the loads P_1, P_2, P_3 and P_4 in the directions of the freedoms. These loads cause displacements Δ_1, Δ_2, Δ_3, and Δ_4 in the directions of the loads.

Figure 3.8 shows the force systems required to maintain unit displacement in the direction of each freedom in turn (note that the reactive forces have been omitted and that K_{ij} is the force developed in the direction of freedom "i" when there is unit displacement in the direction of freedom "j").

The force system associated with displacement Δ_i will be equal to Δ_i times the force system associated with unit displacement in the direction of freedom "i". If the individual force systems associated with each displacement are summed then, by the principle of superposition, the result must be equal to the system of applied loads that is shown in fig 3.7 and the equilibrium equation for each freedom in turn can be written as follows

$$\Delta_1 K_{11} + \Delta_2 K_{12} + \ldots\ldots + \Delta_4 K_{14} = P_1$$
$$\Delta_1 K_{21} + \Delta_2 K_{22} + \ldots\ldots + \Delta_4 K_{24} = P_2$$
$$\Delta_1 K_{31} + \Delta_2 K_{32} + \ldots\ldots + \Delta_4 K_{34} = P_3$$
$$\Delta_1 K_{41} + \Delta_2 K_{42} + \ldots\ldots + \Delta_4 K_{44} = P_4$$

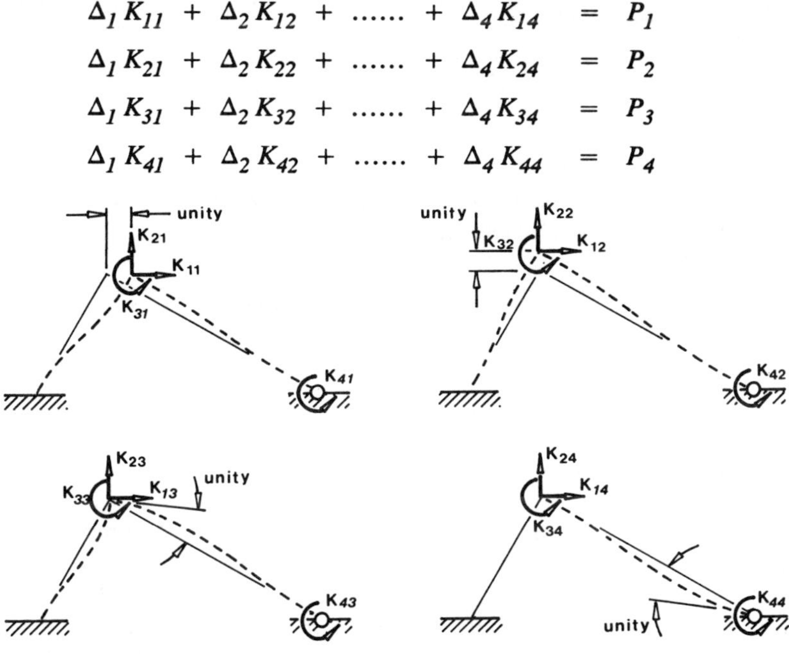

Figure 3.8 Evaluation of stiffness coefficients

expressing the above equations in matrix form

$$\begin{bmatrix} K_{11} & K_{12} & K_{13} & K_{14} \\ K_{21} & K_{22} & K_{23} & K_{24} \\ K_{31} & K_{32} & K_{33} & K_{34} \\ K_{41} & K_{42} & K_{43} & K_{44} \end{bmatrix} \begin{bmatrix} \Delta_1 \\ \Delta_2 \\ \Delta_3 \\ \Delta_4 \end{bmatrix} = \begin{bmatrix} P_1 \\ P_2 \\ P_3 \\ P_4 \end{bmatrix}$$

The above equation is, of course, the structure stiffness equation.

3.5 Direct Application of the Stiffness Method

The material in the previous three sections has shown that the stiffness method can be applied directly as a four stage procedure.

1. Identify the nodal freedoms.
2. Find the coefficients of the stiffness matrix by applying unit displacement at each freedom in turn.
3. Develop the loading vector from the applied loads.
4. Solve the simultaneous equations to find the structural displacements.

Once the structural displacements are known the element forces can be easily evaluated.

Example 3.1

Find the displacements and the forces at the top of the left-hand column of the portal shown in the left-hand diagram below (assume the elements to be axially rigid). EI is constant for all elements.

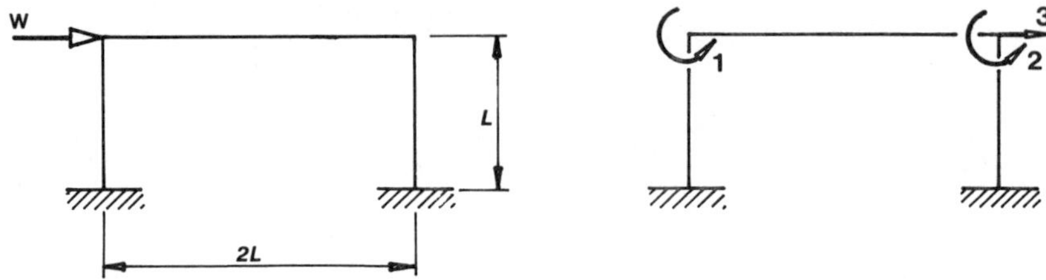

The frame shown is similar to the portal shown in fig 3.5. A computer analysis of such a frame would take the axial strain in the elements into account. Consequently the frame would be regarded as having the six freedoms shown in fig 3.5. Hand methods of calculation would, however, regard the elements as axially rigid, thus simplifying the problem by reducing the degree of freedom of the structure from six to three as shown in the right-hand diagram. The error introduced by this assumption is small. Axial rigidity is assumed here in order to limit the size of the problem.

The elements of the stiffness matrix are found by evaluation of the forces required to maintain unit displacement at each freedom in turn. The stiffness characteristics of line elements are derived in section 2.12.

Freedom 1 (impose unit rotation in the direction of freedom 1)

unity

K_{11} K_{21} K_{31}

$$K_{11} = \frac{4EI}{L} + \frac{4EI}{2L} = \frac{6EI}{L}$$

$$K_{21} = \frac{2EI}{2L} = \frac{EI}{L}$$

$$K_{31} = \frac{6EI}{L^2}$$

Freedom 2 (impose unit rotation in the direction of freedom 2)

unity

K_{12} K_{22} K_{32}

$$K_{22} = \frac{4EI}{L} + \frac{4EI}{2L} = \frac{6EI}{L}$$

$$K_{12} = \frac{2EI}{2L} = \frac{EI}{L}$$

$$K_{32} = \frac{6EI}{L^2}$$

Freedom 3 (impose unit translation in the direction of freedom 3)

$$K_{33} = 2 \times \frac{12EI}{L^3} = \frac{24EI}{L^3}$$

$$K_{13} = \frac{6EI}{L^2}$$

$$K_{23} = \frac{6EI}{L^2}$$

Hence the structure stiffness matrix and the loading vector are

$$\boldsymbol{K} = \frac{EI}{L^3}\begin{bmatrix} 6L^2 & L^2 & 6L \\ L^2 & 6L^2 & 6L \\ 6L & 6L & 24 \end{bmatrix} \qquad \boldsymbol{P} = \begin{bmatrix} 0 \\ 0 \\ W \end{bmatrix}$$

and the structure stiffness equation $\boldsymbol{K}\,\boldsymbol{\Delta} = \boldsymbol{P}$ is

$$\frac{EI}{L^3}\begin{bmatrix} 6L^2 & L^2 & 6L \\ L^2 & 6L^2 & 6L \\ 6L & 6L & 24 \end{bmatrix}\begin{bmatrix} \Delta_1 \\ \Delta_2 \\ \Delta_3 \end{bmatrix} = \begin{bmatrix} 0 \\ 0 \\ W \end{bmatrix}$$

and

$$\boldsymbol{K}^{-1} = \frac{L}{480EI}\begin{bmatrix} 108 & 12 & -30L \\ 12 & 108 & -30L \\ -30L & -30L & 35L^2 \end{bmatrix}$$

hence $\boldsymbol{\Delta} = \boldsymbol{K}^{-1}\boldsymbol{P}$ yields

$$\begin{bmatrix} \Delta_1 \\ \Delta_2 \\ \Delta_3 \end{bmatrix} = \begin{bmatrix} -WL^2/16EI \\ -WL^2/16EI \\ 7WL^3/96EI \end{bmatrix}$$

Note that the inverse of $\boldsymbol{K}$ is used to solve the structure stiffness equation as this technique is widely understood. The reader should note however that in practice a more computationally efficient method would be used.

The forces at the top of the left-hand column are found by superposition of the effects of the end displacements.

Forces due to Δ_1

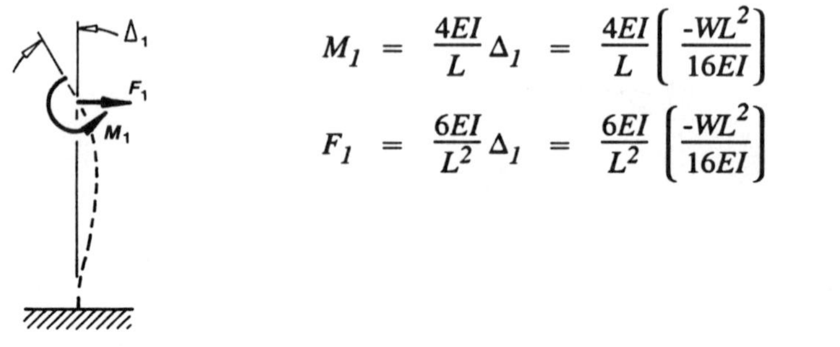

$$M_1 = \frac{4EI}{L}\Delta_1 = \frac{4EI}{L}\left(\frac{-WL^2}{16EI}\right) = -\frac{WL}{4}$$

$$F_1 = \frac{6EI}{L^2}\Delta_1 = \frac{6EI}{L^2}\left(\frac{-WL^2}{16EI}\right) = -\frac{3W}{8}$$

Forces due to Δ_3

$$M_3 = \frac{6EI}{L^2}\Delta_3 = \frac{6EI}{L^2}\left(\frac{7WL^3}{96EI}\right) = \frac{7WL}{16}$$

$$F_3 = \frac{12EI}{L^3}\Delta_3 = \frac{12EI}{L^3}\left(\frac{7WL^3}{96EI}\right) = \frac{7W}{8}$$

The final forces at the top of the left-hand column can now be found by superposition

$$M = M_1 + M_3 = \frac{-WL}{4} + \frac{7WL}{16} = \frac{3WL}{16}$$

$$F = F_1 + F_3 = \frac{-3W}{8} + \frac{7W}{16} = \frac{W}{2}$$

Example 3.2

Find the nodal displacements for the pin jointed frame shown in the following diagram. Use these displacements to find the force in elements *AC*, *BC*, and *CD*. Demonstrate the equilibrium of joint "C" and find the nodal displacements if element *AD* is removed. $EA = 4 \times 10^6$ N for all elements.

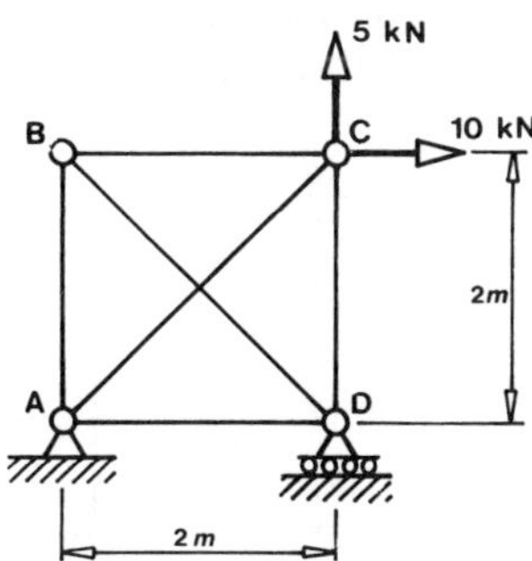

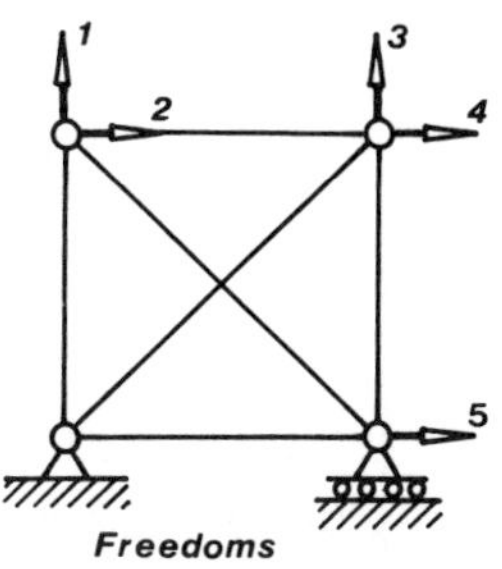

The structure has five freedoms as shown in the right-hand diagram. The structure stiffness matrix is generated from the force systems required to maintain unit displacement at each freedom in turn. The relationship between axial force and axial deformation is given by equation (2.3).

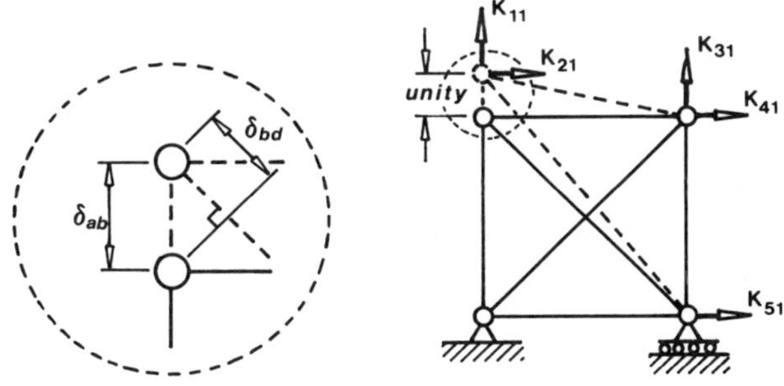

K_{11} is equal to the force in element *AB* plus the vertical component of the force in element *BD*. Hence

$$K_{11} = EA\left[\frac{\delta_{ab}}{L_{ab}} + \frac{\delta_{bd}}{L_{bd}}\frac{1}{\sqrt{2}}\right]$$

$$= 4 \times 10^6 \left[\frac{1}{2} + \frac{(\frac{1}{\sqrt{2}})}{2\sqrt{2}}\frac{1}{\sqrt{2}}\right] \qquad = \quad 2.707 \times 10^6 \text{ N/m}$$

$$K_{21} = -\frac{EA}{L_{bd}}\delta_{bd}\frac{1}{\sqrt{2}}$$

$$= \frac{-4 \times 10^6}{2\sqrt{2}}\frac{1}{\sqrt{2}}\frac{1}{\sqrt{2}} \qquad = \quad -0.707 \times 10^6 \text{ N/m}$$

$$K_{51} = -K_{21} = 0.707 \times 10^6 \text{ N/m}$$

$$K_{31} = K_{41} = 0$$

Applying unit displacement in a similar manner at the other four freedoms in turn yields the remaining stiffness coefficients, and the final stiffness equation is

$$10^6 \begin{bmatrix} 2.707 & -0.707 & 0 & 0 & 0.707 \\ -0.707 & 2.707 & 0 & -2.0 & -0.707 \\ 0 & 0 & 2.707 & 0.707 & 0 \\ 0 & -2.0 & 0.707 & 2.707 & 0 \\ 0.707 & -0.707 & 0 & 0 & 2.707 \end{bmatrix} \begin{bmatrix} \Delta_1 \\ \Delta_2 \\ \Delta_3 \\ \Delta_4 \\ \Delta_5 \end{bmatrix} = \begin{bmatrix} 0 \\ 0 \\ 5000 \\ 10000 \\ 0 \end{bmatrix}$$

Solving using any of the equation solving programs given in Chapter 4 yields

$$\begin{bmatrix} \Delta_1 \\ \Delta_2 \\ \Delta_3 \\ \Delta_4 \\ \Delta_5 \end{bmatrix} = \begin{bmatrix} 0.17 \times 10^{-2} \text{ m} \\ 0.83 \times 10^{-2} \text{ m} \\ -0.77 \times 10^{-3} \text{ m} \\ 0.10 \times 10^{-1} \text{ m} \\ 0.17 \times 10^{-2} \text{ m} \end{bmatrix}$$

Element forces

To find the forces in elements *AC*, *BC*, and *CD* first calculate extensions

$$\delta_{ac} = \frac{\Delta_4}{\sqrt{2}} + \frac{\Delta_3}{\sqrt{2}} = 0.00653 \text{ m}$$

$$\delta_{bc} = \Delta_4 - \Delta_2 = 0.0017 \text{ m}$$

$$\delta_{cd} = \Delta_3 = -0.00077 \text{ m}$$

and

$$F = \frac{EA}{L}\delta$$

hence

$$F_{ac} = \frac{4 \times 10^6}{2\sqrt{2}} 0.00653 = 9230 \text{ N}$$

$$F_{bc} = \frac{4 \times 10^6}{2} 0.0017 = 3400 \text{ N}$$

$$F_{cd} = \frac{4 \times 10^6}{2} (-0.00077) = -1540 \text{ N}$$

Equilibrium of joint C

The freebody diagram for joint "*C*" is as shown below, and its equilibrium is checked by summing horizontal and vertical forces.

$$\sum H = 3400 - 10000 + \frac{9230}{\sqrt{2}} \approx 0$$

$$\sum V = 5000 + 1540 - \frac{9230}{\sqrt{2}} \approx 0$$

Removal of element AD

If element AD is removed then K_{55} becomes 0.707 x 10^6, and the new solution vector is

$$\begin{bmatrix} \Delta_1 \\ \Delta_2 \\ \Delta_3 \\ \Delta_4 \\ \Delta_5 \end{bmatrix} = \begin{bmatrix} 0.0 \text{ m} \\ 1.66 \times 10^{-2} \text{ m} \\ -0.25 \times 10^{-2} \text{ m} \\ 1.66 \times 10^{-2} \text{ m} \\ 1.66 \times 10^{-2} \text{ m} \end{bmatrix}$$

As expected the displacements are much larger, and their relative magnitudes indicate that $F_{ab} = F_{bc} = F_{bd} = 0$. The reader should confirm that these forces must be zero (no calculation should be required).

Summary

These two examples have shown the stiffness method to be simple in concept and flexible in use. However, because each additional freedom produces another equation which must be solved simultaneously, hand application is limited to all but the simplest of structures. The criteria used to assess hand methods of analysis are very different from those used to assess computer-based methods. A hand technique stands or falls by the amount of arithmetic it generates. By contrast, the success of a computer-based method depends primarily upon its generality and how amenable it is to computer implementation. The generality and amenability to computer implementation offered by the stiffness method has led to its complete dominance of commercial structural analysis software.

The remainder of this text is concerned with the techniques required to computerise the stiffness method for the analysis of structural frameworks. The remainder of this chapter will consider axes systems and transformation of force and displacement between axes systems. Chapters 5 - 10 then consider the static analysis of different types of structural frameworks, and Chapter 11 shows how the stiffness method can be easily extended to the dynamic analysis of structural frameworks.

3.6 Axes Systems

To computerise the stiffness method requires formalisation of the solution procedure. In practice the use of a number of different axes systems in a single analysis is found to facilitate automation of the method. Three different axes systems are commonly used in the analysis of structural frameworks. These are

1. Coordinate axes.
2. Nodal axes.
3. Element axes.

Coordinate Axes

Coordinate axes are used to define the location of the nodes, and hence the structural geometry. The origin and orientation of the coordinate axes are totally arbitrary. Figure 3.9 shows a pitched roof portal and a coordinate axes system which might be used to describe the nodal coordinates.

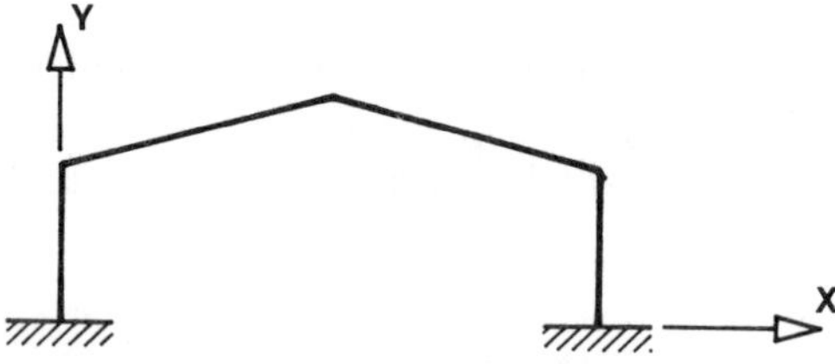

Figure 3.9 Coordinate axes

Nodal Axes

The forces and displacements at each node are described in terms of an axis system located at the node. If the nodal axes lie parallel to the coordinate axes then the nodal axes are referred to as *global axes*. If all nodal axes systems are global then the solution procedure is simplified. Figure 3.10 shows the global axes for the pitched roof portal shown in fig 3.9 (note that the nodal "x" axes are horizontal).

Figure 3.10 Global axes

The boundary conditions for most structures can be described in terms of the global axes system. Where the structure has inclined supports, however, a local axes system may have to be employed. Figure 3.11(a) shows the coordinate axes system adopted for a rectangular portal with an inclined support, and fig 3.11(b) shows how a local axes system has been used to allow the application of the boundary conditions.

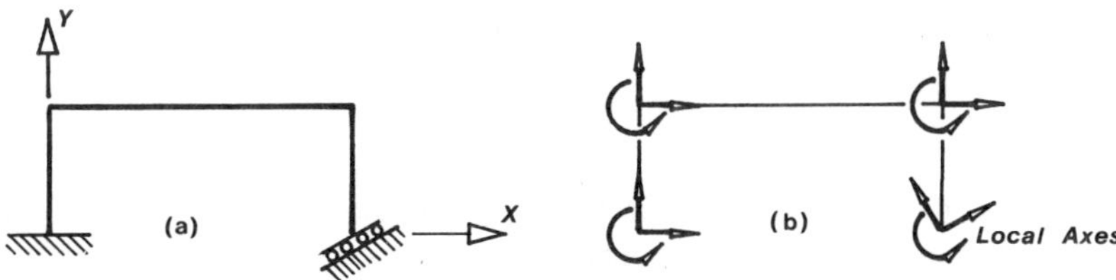

Figure 3.11 Local axes

Of course, if the coordinate axes system had been chosen to lie parallel and perpendicular to the inclined support then the use of a local axes system would, in this case, have been avoided.

Element Axes

Nodal axes systems (global and local) are used to describe the displacements and forces at the nodes. These axes are used when generating the stiffness equation $\boldsymbol{K}\,\boldsymbol{\Delta} = \boldsymbol{P}$ because, when considering nodal equilibrium, forces and displacements at each node must be expressed in a single axes system. The solution of the stiffness equation yields the unknown structural displacements, and from these displacements the element end forces are found.

In this text the element axes systems shown in fig 3.12 are used to describe the element stress resultants. These axes yield the stress resultants that are most useful to the engineer.

1. Plane truss - axial force.
2. Space truss - axial force.
3. Plane frame - axial force, shear force and bending moment

4. Grillage - bending moment, torsion and shear force.
5. Space frame - axial force, bending moment (in two planes), torsion, and shear force (in two planes).

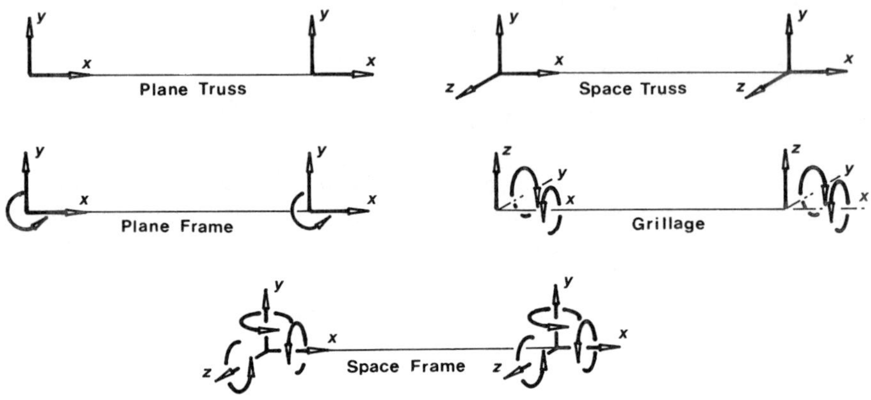

Figure 3.12 Element axes systems

Note that the foregoing assumes that for plane frames, grillages, and space frames the element *"y"* axis coincides with a principal axis of the section. The reader should also note the following

1. Right-hand axes systems are used throughout.
2. Plane structures always lie in the coordinate "*x,y*" plane.
3. In all cases *the "x"axis of element "i,j" runs along the element from end "i" to end "j"*

3.7 Properties of the Structure Stiffness Matrix

There now follows a description of some of the properties of the structure stiffness matrix.

It is square

The structure stiffness matrix is a square matrix of dimension "*n*" where "*n*" is the degree of freedom of the structure.

It is symmetric

The symmetry of the structure stiffness equation can be proved through the use of Castigliano's First Theorem (see section 3.4).

$$P_i = \frac{\partial U}{\partial \Delta_i} \qquad \text{hence} \qquad \frac{\partial P_i}{\partial \Delta_j} = \frac{\partial^2 U}{\partial \Delta_j \, \partial \Delta_i}$$

similarly

$$\frac{\partial P_j}{\partial \Delta_i} = \frac{\partial^2 U}{\partial \Delta_i \, \partial \Delta_j}$$

As the order of differentiation is immaterial

$$\frac{\partial^2 U}{\partial \Delta_i \, \partial \Delta_j} = \frac{\partial^2 U}{\partial \Delta_j \, \partial \Delta_i} \qquad \text{hence} \qquad \frac{\partial P_i}{\partial \Delta_j} = \frac{\partial P_j}{\partial \Delta_i}$$

i.e.

$$K_{ij} = K_{ji}$$

It is banded

Consider the multi-bay portal shown in fig 3.13. Figure 3.14 shows the non-zero forces required to maintain unit rotation at freedom 12.

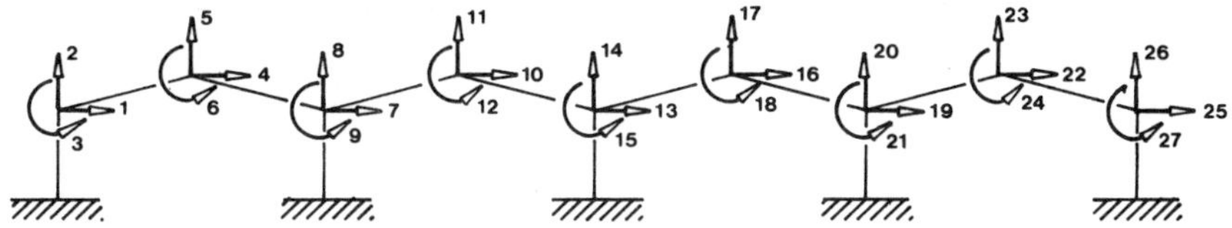

Figure 3.13 Multi-bay portal

As can be seen, only 9 of the possible 27 stiffness terms in column 12 of the stiffness matrix will be non-zero, and if the freedoms are well numbered the non-zero terms will be clustered around the diagonal (i.e. the matrix will be banded).

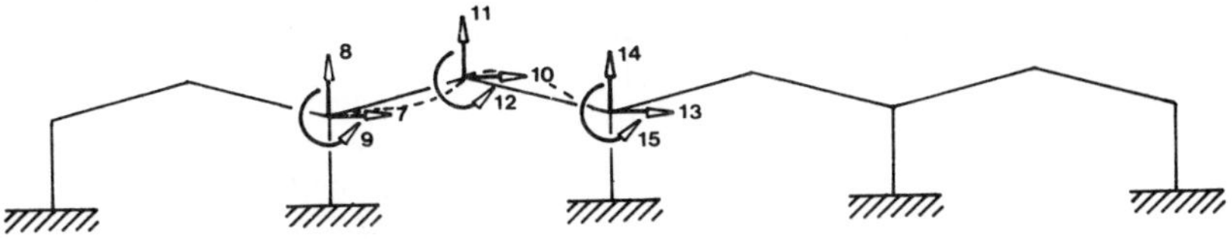

Figure 3.14 Forces for unit displacement at freedom 12

By taking advantage of the band in the structure stiffness matrix savings can be made of both computer storage and time. Indeed the problem size that can be handled by many structural analysis computer programs is limited, not by the degree of freedom, but by the bandwidth of the structure stiffness matrix.

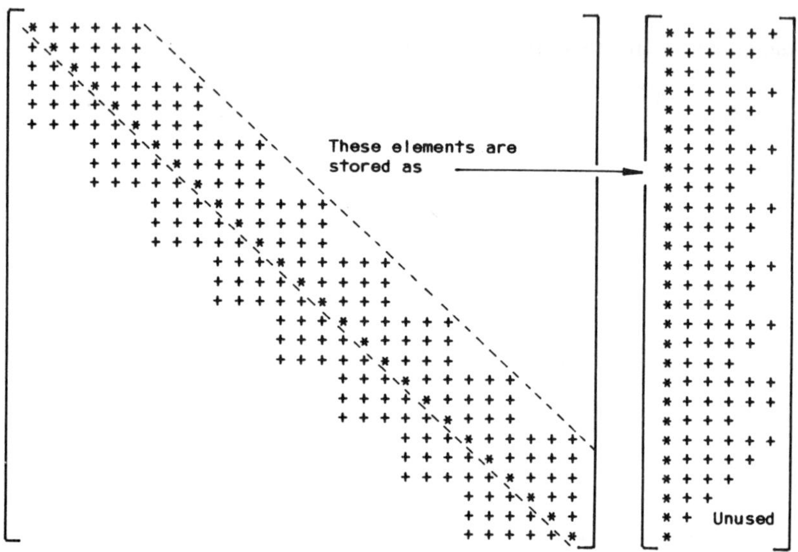

Figure 3.15 Non-zero elements in the structure stiffness matrix

Figure 3.15 shows the non-zero elements of the structure stiffness matrix for the multi-bay portal shown in fig 3.13, and also illustrates how computer storage can be reduced by storing only the upper semi-band of the matrix.

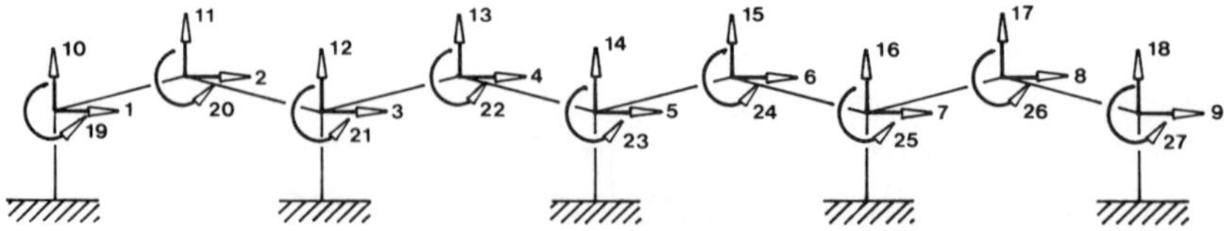

Figure 3.16 Poorly numbered freedoms

Had the freedoms been poorly numbered as shown in fig 3.16, then the band in the structure stiffness matrix would be lost, as can be seen in fig 3.17.

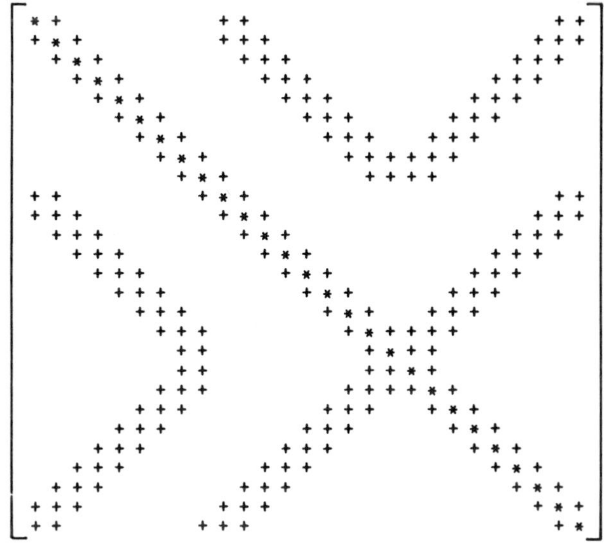

Figure 3.17 Effect of poor freedom numbering on the stiffness matrix

It is positive definite

The structure stiffness matrix is positive definite as pre- and post-multiplication by an arbitrary non-zero vector always results in a positive quantity since

$$\boldsymbol{P} = \boldsymbol{K}\,\boldsymbol{\Delta}$$

and the gain in strain energy due to work done by the loads is

$$U = \tfrac{1}{2}\boldsymbol{P}^T\,\boldsymbol{\Delta}$$

hence

$$U = \tfrac{1}{2}\,\boldsymbol{\Delta}^T\,\boldsymbol{K}^T\,\boldsymbol{\Delta}$$

but

$$\boldsymbol{K}^T = \boldsymbol{K}$$

therefore

$$U = \tfrac{1}{2}\,\boldsymbol{\Delta}^T\,\boldsymbol{K}\,\boldsymbol{\Delta}$$

If the structure is in a state of stable equilibrium then the structure gains strain energy on any arbitrary small deformation, proving that the structure stiffness matrix is positive definite. Having a positive definite coefficient matrix is a useful property when solving simultaneous equations.

3.8 Transformation of Force and Displacement

In general the element axes systems will not coincide with the nodal axes systems, making necessary the transformation of forces and displacements from one axes system to another.

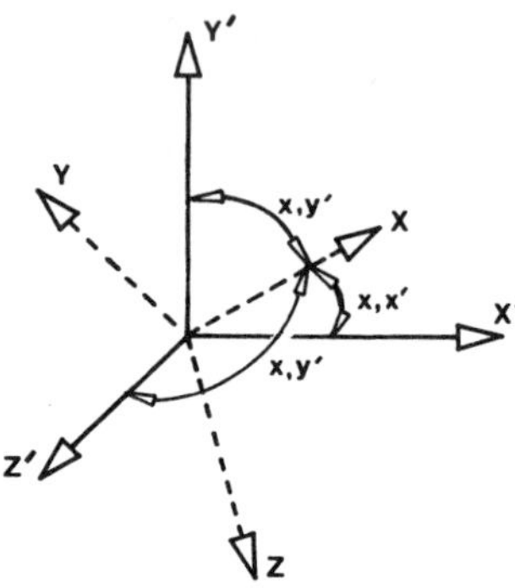

Figure 3.18 Rotation of axes in three-dimensional space

Figure 3.18 shows two, right-handed, three-dimensional cartesian coordinate axes systems with a common origin, but oriented at different angles. The problem in hand is to find a matrix that will transform a vector of force or displacement from the unprimed to the primed axes system. Consider first the x component which is, of course, a function of x', y', and z'

$$x = F(x', y', z')$$

hence

$$dx = \frac{\partial x}{\partial x'} dx' + \frac{\partial x}{\partial y'} dy' + \frac{\partial x}{\partial z'} dz' \tag{3.8}$$

and

$$\frac{\partial x}{\partial x'} = \cos(x, x') = l_x$$

$$\frac{\partial x}{\partial y'} = \cos(x, y') = m_x$$

$$\frac{\partial x}{\partial z'} = \cos(x, z') = n_x$$

where $\cos(x,x')$ is the cosine of the angle between the "x" and "x'" axes, $\cos(x,y')$ is the cosine of the angle between the "x"and "y'" axes, etc.

Note that the direction in which the angle is measured is immaterial as $\cos(360 - \alpha) = \cos(\alpha)$. "$l_x$", "$m_x$", and "$n_x$" are known as the *direction cosines* of the "x" axis with respect to the x',y',z' axes system. Equation (3.8) can be rewritten

$$dx = l_x\, dx' + m_x\, dy' + n_x\, dz'$$

Integrating

$$x = l_x x' + m_x y' + n_x z'$$

Treating the "y" and "z" components similarly and writing the equations in matrix form gives

$$\begin{bmatrix} x \\ y \\ z \end{bmatrix} = \begin{bmatrix} l_x & m_x & n_x \\ l_y & m_y & n_y \\ l_z & m_z & n_z \end{bmatrix} \begin{bmatrix} x' \\ y' \\ z' \end{bmatrix} \qquad (3.9)$$

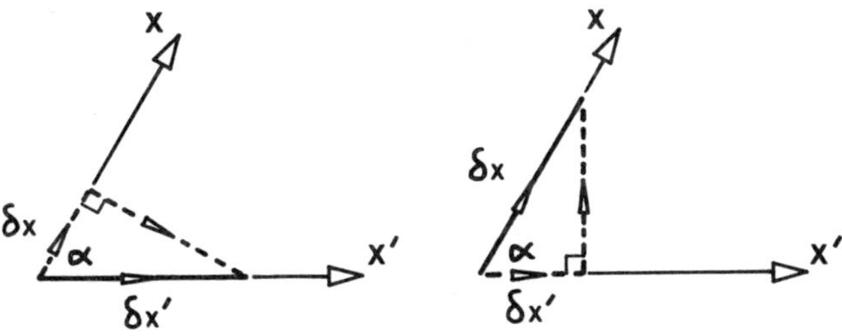

Figure 3.19 Evaluation of direction cosines

If the transformation is to be made from the unprimed to the primed axes system then

$$dx' = \frac{\partial x'}{\partial x} dx + \frac{\partial x'}{\partial y} dy + \frac{\partial x'}{\partial z} dz \qquad (3.10)$$

Figure 3.19 shows the "x" and "x'" axes from the axes systems shown in fig 3.18, only the paper represents the x,x' plane. Note that, in general, no other axis will lie in this plane.

The partial derivative $\partial x/\partial x'$ represents the rate of change of "x" with "x'", ("y'" and "z'" remaining constant). Figure 3.19(a) shows a change $\delta x'$ being made to "x'". As the change takes place along the "x'" axis, "y'" and "z'" are, by definition, constant. The corresponding change in "x" is δx as shown, hence

$$\frac{\partial x}{\partial x'} = \lim_{\delta x' \Rightarrow 0} \left[\frac{\delta x}{\delta x'}\right] = \cos\alpha \qquad (3.11)$$

Now consider the partial derivative $\partial x/\partial x'$ which represents the rate of change of "x'" with "x" ("y" and "z" remaining constant). Figure 3.19(b) shows "x" being varied by δx and the corresponding change in "x'". By inspection

$$\frac{\partial x'}{\partial x} = \lim_{\delta x \Rightarrow 0} \left[\frac{\delta x'}{\delta x}\right] = \cos\alpha \qquad (3.12)$$

Hence by comparison of equations (3.11) and (3.12)

$$\frac{\partial x'}{\partial x} = \frac{\partial x}{\partial x'} \qquad \text{(or } l_{x'} = l_x\text{)}$$

similarly

$$\frac{\partial x'}{\partial y} = \frac{\partial y}{\partial x'} \qquad \text{(or } m_{x'} = l_y\text{)}$$

$$\frac{\partial x'}{\partial z} = \frac{\partial z}{\partial x'} \qquad \text{(or } n_{x'} = l_z\text{)}$$

hence equation (3.10) reduces to

$$dx' = l_x\,dx + l_y\,dy + l_z\,dz$$

Integrating

$$x' = l_x x + l_y y + l_z z$$

Treating the "y'" and "z'" components similarly yields the following matrix equation

$$\begin{bmatrix} x' \\ y' \\ z' \end{bmatrix} = \begin{bmatrix} l_x & l_y & l_z \\ m_x & m_y & m_z \\ n_x & n_y & n_z \end{bmatrix} \begin{bmatrix} x \\ y \\ z \end{bmatrix} \qquad (3.13)$$

By definition the transformation matrices in equations (3.9) and (3.13) must be the inverse of each other, which means that

$$\begin{bmatrix} l_x & m_x & n_x \\ l_y & m_y & n_y \\ l_z & m_z & n_z \end{bmatrix} \begin{bmatrix} l_x & l_y & l_z \\ m_x & m_y & m_z \\ n_x & n_y & n_z \end{bmatrix} = \begin{bmatrix} 1 & 0 & 0 \\ 0 & 1 & 0 \\ 0 & 0 & 1 \end{bmatrix}$$

or

$$l_i l_j + m_i m_j + n_i n_j = 0,\ 1 \qquad i \neq j,\ i = j\ \ (i, j = x, y, z)$$

This is the *orthogonality property* of direction cosines. The six orthogonality equations mean that only three direction cosines out of the total of nine are required to define the position of one axes system relative to the other, provided that not all three known direction cosines lie in one row or one column. For instance if l_x, m_x, and n_x only were known (see fig 3.18) then the position of the "x" axis would be uniquely defined, but the positions of the "y" and "z" axes would not, as the "x,y,z" axes system could be rotated about the "x" axis.

Example 3.3

With reference to fig 3.18 calculate the coordinates of the point (1,1,1) in the primed axis in terms of the unprimed axes system given

$$l_x = m_x = n_x = \frac{1}{\sqrt{3}}$$

and

$$m_y = \frac{1}{\sqrt{2}}$$

Note that those readers who think well in three dimensions will be able to see, by inspection, that the solution is $(\sqrt{3}, 0, 0)$. This will be checked by developing the transformation matrix that will transform vectors from the primed axes system to the unprimed axes system. The required transformation requires calculation of the unknown six direction cosines which are found through the use of the orthogonality properties of direction cosines.

$$l_x l_y + m_x m_y + n_x n_y = 0$$

$$\frac{1}{\sqrt{3}} l_y + \frac{1}{\sqrt{3}} \frac{1}{\sqrt{2}} + \frac{1}{\sqrt{3}} n_y = 0$$

hence

$$n_y = -\frac{1}{\sqrt{2}} - l_y \qquad (3.14)$$

and

$$l_y^2 + m_y^2 + n_y^2 = 1$$

$$l_y^2 + \frac{1}{2} + \left(-\frac{1}{\sqrt{2}} - l_y\right)^2 = 1$$

hence

$$l_y = -\frac{1}{\sqrt{2}} \qquad (3.15)$$

and, from eqt (3.14)

$$n_y = 0 \qquad (3.16)$$

also

$$l_x l_z + m_x m_z + n_x n_z = 0$$

$$\frac{l_z}{\sqrt{3}} + \frac{m_z}{\sqrt{3}} + \frac{n_z}{\sqrt{3}} = 0$$

hence

$$n_z = - l_z - m_z \qquad (3.17)$$

and finally

$$l_y l_z + m_y m_z + n_y n_z = 0$$

Substituting for l_y and n_y using equations (3.15) and (3.16) gives

$$-\frac{l_z}{\sqrt{2}} + \frac{m_z}{\sqrt{2}} = 0$$

hence

$$m_z = l_z \tag{3.18}$$

$$l_z^2 + m_z^2 + n_z^2 = 1$$

Substituting for m_z and n_z using equations (3.17) and (3.18) gives

$$l_z^2 + l_z^2 + 4\,l_z^2 = 1$$

hence

$$l_z = \frac{1}{\sqrt{6}}, \qquad m_z = \frac{1}{\sqrt{6}}, \qquad n_z = -\frac{2}{\sqrt{6}}$$

Equation (3.9) gives the transformation from primed to unprimed axes.

$$\begin{bmatrix} x \\ y \\ z \end{bmatrix} = \begin{bmatrix} l_x & m_x & n_x \\ l_y & m_y & n_y \\ l_z & m_z & n_z \end{bmatrix} \begin{bmatrix} x' \\ y' \\ z' \end{bmatrix}$$

and using the direction cosines calculated above

$$\begin{bmatrix} x \\ y \\ z \end{bmatrix} = \begin{bmatrix} 1/\sqrt{3} & 1/\sqrt{3} & 1/\sqrt{3} \\ -1/\sqrt{2} & 1/\sqrt{2} & 0 \\ 1/\sqrt{6} & 1/\sqrt{6} & -2/\sqrt{6} \end{bmatrix} \begin{bmatrix} 1 \\ 1 \\ 1 \end{bmatrix}$$

which yields the expected solution

$$\begin{bmatrix} x \\ y \\ z \end{bmatrix} = \begin{bmatrix} \sqrt{3} \\ 0 \\ 0 \end{bmatrix}$$

Chapter 4

Simultaneous Linear Equations

4.1 Introduction

Linear simultaneous equations take the form shown below

$$
\begin{array}{lllllll}
a_{11}\,x_1 & + & a_{12}\,x_2 & + \; .... \; + & a_{16}\,x_n & = & b_1 \\
a_{21}\,x_1 & + & a_{22}\,x_2 & + \; .... \; + & a_{26}\,x_n & = & b_2 \\
\vdots & & & & & & \\
a_{61}\,x_1 & + & a_{62}\,x_2 & + \; .... \; + & a_{66}\,x_n & = & b_n
\end{array}
$$

where a_{11}, a_{12}, ... a_{nn} and b_1, b_2, ... b_n are known constants, and x_1, x_2, ... x_n are required to be found. The above system of equations can be expressed more elegantly in matrix notation as follows

$$
\begin{bmatrix}
a_{11} & a_{12} & a_{13} & .... & a_{1n} \\
a_{21} & a_{22} & a_{23} & .... & a_{2n} \\
\cdot & & & & \\
\cdot & & & & \\
a_{n1} & a_{n2} & a_{n3} & .... & a_{nn}
\end{bmatrix}
\begin{bmatrix} x_1 \\ x_2 \\ \cdot \\ \cdot \\ x_n \end{bmatrix}
=
\begin{bmatrix} b_1 \\ b_2 \\ \cdot \\ \cdot \\ b_n \end{bmatrix}
$$

or

$$\boldsymbol{A}\,\boldsymbol{X} = \boldsymbol{B}$$

where

- $\boldsymbol{A}$ is the coefficient matrix
- $\boldsymbol{X}$ is the vector of unknowns (the solution vector)
- $\boldsymbol{B}$ is the right hand side vector

The analysis of linearly elastic structures using the stiffness method produces the system of linear equations $\boldsymbol{K}\,\boldsymbol{\Delta} = \boldsymbol{P}$ which must be solved simultaneously. The methods used to solve systems of linear equations are classed either as direct methods or iterative methods. In section 3.7 the structure stiffness matrix was seen to have the following properties.

1. It is symmetric.
2. It is banded.
3. It is positive definite.

These properties make the use of direct methods preferable to iterative methods, and consequently only direct methods will be considered in this text.

4.2 Gauss-Jordan Elimination - Program GAUSSJ.BAS

Gauss-Jordan elimination is a direct method that is well suited to digital computation and is found in many stock codes for the solution of simultaneous linear equations. These codes usually employ row interchange to increase numerical accuracy by using the biggest available pivot in the elimination procedure. The stiffness method, however, produces large elements on the diagonal, and this renders row interchange unnecessary, which is fortunate since row interchange destroys banding. In essence Gauss-Jordan elimination solves the system of linear equations by reducing the coefficient matrix to the unit matrix. The method, without row interchange, is illustrated in the following example. Normalisation is used to avoid the pivots becoming uncomfortably large or small (i.e. the pivotal row is divided by the pivot prior to elimination).

Example 4.1

Use Gauss-Jordan elimination to solve this system of equations

$$\begin{bmatrix} 2 & 1 & 0 \\ 1 & 3 & -2 \\ 0 & -2 & 5 \end{bmatrix} \begin{bmatrix} x_1 \\ x_2 \\ x_3 \end{bmatrix} = \begin{bmatrix} 4 \\ 1 \\ 11 \end{bmatrix}$$

First pivot - Divide row 1 by the pivot (the diagonal element)

$$\begin{bmatrix} 1 & 0.5 & 0 \\ 1 & 3 & -2 \\ 0 & -2 & 5 \end{bmatrix} \begin{bmatrix} x_1 \\ x_2 \\ x_3 \end{bmatrix} = \begin{bmatrix} 2 \\ 1 \\ 11 \end{bmatrix}$$

Subtract - 1 x row 1 from row 2
0 x row 1 from row 3

$$\begin{bmatrix} 1 & 0.5 & 0 \\ 0 & 2.5 & -2 \\ 0 & -2 & 5 \end{bmatrix} \begin{bmatrix} x_1 \\ x_2 \\ x_3 \end{bmatrix} = \begin{bmatrix} 2 \\ -1 \\ 11 \end{bmatrix}$$

Second pivot - Divide row 2 by the pivot

$$\begin{bmatrix} 1 & 0.5 & 0 \\ 0 & 1 & -0.8 \\ 0 & -2 & 5 \end{bmatrix} \begin{bmatrix} x_1 \\ x_2 \\ x_3 \end{bmatrix} = \begin{bmatrix} 2 \\ -0.4 \\ 11.0 \end{bmatrix}$$

Subtract - 0.5 x row 2 from row 1
-2 x row 2 from row 3

$$\begin{bmatrix} 1 & 0 & 0.4 \\ 0 & 1 & -0.8 \\ 0 & 0 & 3.4 \end{bmatrix} \begin{bmatrix} x_1 \\ x_2 \\ x_3 \end{bmatrix} = \begin{bmatrix} 2.2 \\ -0.4 \\ 10.2 \end{bmatrix}$$

Third pivot - Divide row 3 by the pivot

$$\begin{bmatrix} 1 & 0 & 0.4 \\ 0 & 1 & -0.8 \\ 0 & 0 & 1 \end{bmatrix} \begin{bmatrix} x_1 \\ x_2 \\ x_3 \end{bmatrix} = \begin{bmatrix} 2.2 \\ -0.4 \\ 3.0 \end{bmatrix}$$

Subtract - 0.4 x row 3 from row 1
-0.8 x row 3 from row 2

$$\begin{bmatrix} 1 & 0 & 0 \\ 0 & 1 & 0 \\ 0 & 0 & 1 \end{bmatrix} \begin{bmatrix} x_1 \\ x_2 \\ x_3 \end{bmatrix} = \begin{bmatrix} 1 \\ 2 \\ 3 \end{bmatrix}$$

Which yields

$$\begin{bmatrix} x_1 \\ x_2 \\ x_3 \end{bmatrix} = \begin{bmatrix} 1 \\ 2 \\ 3 \end{bmatrix}$$

This example has shown how each diagonal element is used to eliminate the off-diagonal elements above and below it.

Program GAUSSJ.BAS

This program solves a system of linear simultaneous equations by Gauss-Jordan elimination with no row interchange. Solution is for a single right-hand side.

Input

Coding is simplified if the coefficient matrix and the right-hand side vector(s) are stored in one array, which is called the augmented matrix. The augmented matrix for the system of equations solved in the previous example is

$$\begin{bmatrix} 2 & 1 & 0 & 4 \\ 1 & 3 & -2 & 1 \\ 0 & -2 & 5 & 11 \end{bmatrix}$$

Comments on the Algorithm

Note that if a zero pivot is encountered then division by zero occurs and the algorithm breaks down. Singularity and ill-conditioning are discussed in section 4.7.

Output

The input data is echoed and after solution the solution vector is output.

Listing

```
1000 '========================    GAUSSJ.BAS    =============================
1010 '
1020 OPTION BASE 1
1030 CLS
1040 PRINT
1050 PRINT "=================================================================="
1060 PRINT
1070 PRINT "PROGRAM GAUSSJ.BAS                       Copyright (c) J.Balfour 1991"
1080 PRINT
1090 PRINT "       Program for the solution of simultaneous equations"
1100 PRINT "            by Gauss-Jordan with no row interchange"
1110 PRINT
1120 PRINT "               For further information contact"
1130 PRINT " James A.D.Balfour, Heriot-Watt University, Riccarton, Edinburgh"
1140 PRINT "                Tel 031-449-5111, Fax 031-451-3170"
1150 PRINT
1160 PRINT "=================================================================="
1170 '
1180 ' ... Note that variables are defined in Appendix A
1190 '
1200 MAXNEQ% = 50
1210 DIM A(MAXNEQ%,MAXNEQ%+1)
1220 '
1230 ' ... Maximum number of equations is set by the above statement
1240 '
1250 PRINT
1260 INPUT "Number of equations = ", NEQ%
1270 IF (NEQ%>0) AND (NEQ%<=MAXNEQ%) THEN GOTO 1290
1280 PRINT "Maximum number equations is currently set at"; MAXNEQ% : GOTO 1660
```

```
1290 PRINT
1300 PRINT "Input the augmented matrix" : PRINT
1310 FOR I% = 1 TO NEQ%
1320   FOR J% = 1 TO NEQ%+1
1330     PRINT "Element ("; I%; ","; J%; ") = "; : INPUT "", A(I%,J%)
1340   NEXT J%
1350 NEXT I%
1360 '
1370 ' ... Echo the augment matrix
1380 '
1390 PRINT : PRINT
1400 PRINT "+ + + + + + + + + + + + + +"
1410 PRINT "+   AUGMENTED MATRIX   +"
1420 PRINT "+ + + + + + + + + + + + + +"
1430 PRINT
1440 FOR I% = 1 TO NEQ%
1450   FOR J% = 1 TO NEQ%+1
1460     PRINT A(I%,J%),
1470   NEXT J%
1480   PRINT : PRINT
1490 NEXT I%
1500 '
1510 ' ... Call the subroutine to solve the equations
1520 '
1530 GOSUB 1720
1540 '
1550 ' ... Print solution
1560 '
1570 PRINT
1580 PRINT "* * * * * * * * * * * * * *"
1590 PRINT "*    SOLUTION VECTOR    *"
1600 PRINT "* * * * * * * * * * * * * *"
1610 PRINT
1620 FOR I% = 1 TO NEQ%
1630   PRINT A(I%,NEQ%+1) : PRINT
1640 NEXT I%
1650 PRINT
1660 END
1670 '
1680 '*************    SOLVE LINEAR SIMULTANEOUS EQUATIONS    ***************
1690 '
1700 ' ... Loop for all pivots
1710 '
1720 FOR I%  = 1 TO NEQ%
1730   TEMP = A(I%,I%)
1740   '
1750   ' ... Normalise
1760   '
1770   FOR J%  = 1 TO NEQ%+1
1780     A(I%,J%) = A(I%,J%) / TEMP
1790   NEXT J%
1800   '
1810   ' ... Eliminate using the pivot
1820   '
1830   FOR J%  = 1 TO NEQ%
1840     IF J% = I% THEN GOTO 1920
1850     TEMP  = A(J%,I%)
1860     '
1870     ' ... Loop for all elements in the row
1880     '
1890     FOR K%  = 1 TO NEQ%+1
1900       A(J%,K%) = A(J%,K%) - TEMP*A(I%,K%)
1910     NEXT K%
1920   NEXT J%
1930 NEXT I%
1940 RETURN
1950 '
1960 '========================    GAUSSJ.BAS    ============================
```

Sample Run

The following sample run shows how example 4.1 would be solved using the program GAUSSJ.BAS. Input from the keyboard is shown underlined.

```
=================================================================

PROGRAM GAUSSJ.BAS                        Copyright (c) J.Balfour 1991

       Program for the solution of simultaneous equations
            by Gauss-Jordan with no row interchange

                  For further information contact
 James A.D.Balfour, Heriot-Watt University, Riccarton, Edinburgh
              Tel 031-449-5111, Fax 031-451-3170

=================================================================

Number of equations = 3

Input the augmented matrix

Element ( 1 , 1 ) = 2
Element ( 1 , 2 ) = 1
Element ( 1 , 3 ) = 0
Element ( 1 , 4 ) = 4
Element ( 2 , 1 ) = 1
Element ( 2 , 2 ) = 3
Element ( 2 , 3 ) = -2
Element ( 2 , 4 ) = 1
Element ( 3 , 1 ) = 0
Element ( 3 , 2 ) = -2
Element ( 3 , 3 ) = 5
Element ( 3 , 4 ) = 11

+ + + + + + + + + + + + +
+   AUGMENTED MATRIX   +
+ + + + + + + + + + + + +

 2              1              0              4

 1              3             -2              1

 0             -2              5              11

* * * * * * * * * * * * * *
*    SOLUTION VECTOR      *
* * * * * * * * * * * * * *

 1

 2

 3
```

4.3 Multiple Right-Hand Sides

It is a feature of direct methods that very little extra labour is required to solve for additional right-hand sides.

When analysing structures using the stiffness method the right-hand side vector is the vector of applied loads. Consequently to analyse a structure for a number of loads cases costs very little more than the analysis for a single load case. If Gauss-Jordan elimination is used then multiple right-hand sides are dealt with by operating on all right-hand sides simultaneously. This is best illustrated by an example.

Example 4.2

Solve the following two systems of equations using Gauss-Jordan elimination.

$$\begin{bmatrix} 2 & 1 & 0 \\ 1 & 3 & -2 \\ 0 & -2 & 5 \end{bmatrix} \begin{bmatrix} x_1 \\ x_2 \\ x_3 \end{bmatrix} = \begin{bmatrix} 25 \\ -20 \\ 60 \end{bmatrix}$$

and

$$\begin{bmatrix} 2 & 1 & 0 \\ 1 & 3 & -2 \\ 0 & -2 & 5 \end{bmatrix} \begin{bmatrix} x_1 \\ x_2 \\ x_3 \end{bmatrix} = \begin{bmatrix} -1.0 \\ -8.5 \\ 11.5 \end{bmatrix}$$

The augmented matrix is

$$\begin{bmatrix} 2 & 1 & 0 & 25 & -1.0 \\ 1 & 3 & -2 & -20 & -8.5 \\ 0 & -2 & 5 & 60 & 11.5 \end{bmatrix}$$

First pivot - Divide row 1 by the pivot

$$\begin{bmatrix} 1 & 0.5 & 0 & 12.5 & -0.5 \\ 1 & 3 & -2 & -20 & -8.5 \\ 0 & -2 & 5 & 60 & 11.5 \end{bmatrix}$$

Subtract - 1 x row 1 from row 2
0 x row 1 from row 3

$$\begin{bmatrix} 1 & 0.5 & 0 & 12.5 & -0.5 \\ 0 & 2.5 & -2 & -32.5 & -8.0 \\ 0 & -2 & 5 & 60 & 11.5 \end{bmatrix}$$

Second pivot - Divide row 2 by the pivot

$$\begin{bmatrix} 1 & 0.5 & 0 & 12.5 & -0.5 \\ 0 & 1 & -0.8 & -13.0 & -3.2 \\ 0 & -2 & 5 & 60 & 11.5 \end{bmatrix}$$

Subtract - 0.5 x row 2 from row 1
-2 x row 2 from row 3

$$\begin{bmatrix} 1 & 0 & 0.4 & 19.0 & 1.1 \\ 0 & 1 & -0.8 & -13.0 & -3.2 \\ 0 & 0 & 3.4 & 34.0 & 5.1 \end{bmatrix}$$

Third pivot - Divide row 3 by the pivot

$$\begin{bmatrix} 1 & 0 & 0.4 & 19.0 & 1.1 \\ 0 & 1 & -0.8 & -13.0 & -3.2 \\ 0 & 0 & 1 & 10.0 & 1.5 \end{bmatrix}$$

Subtract - 0.4 x row 3 from row 1
-0.8 x row 3 from row 2

$$\begin{bmatrix} 1 & 0 & 0 & 15.0 & 0.5 \\ 0 & 1 & 0 & -5.0 & -2.0 \\ 0 & 0 & 1 & 10.0 & 1.5 \end{bmatrix}$$

Hence the two solution vectors are

$$\begin{bmatrix} x_1 \\ x_2 \\ x_3 \end{bmatrix} = \begin{bmatrix} 15 \\ -5 \\ 10 \end{bmatrix} \quad \text{and} \quad \begin{bmatrix} x_1 \\ x_2 \\ x_3 \end{bmatrix} = \begin{bmatrix} 0.5 \\ -2.0 \\ 1.5 \end{bmatrix}$$

4.4 Gauss Elimination - Program GAUSS.BAS

Many commercial structural analysis programs use Gauss elimination to solve the system of equations produced by the stiffness method. The method has two distinct phases.

1. Reduction of the coefficient matrix to triangular form.
2. Back substitution to find the solution vector.

Normalisation is used to avoid the pivots becoming uncomfortably large or small (i.e. the pivotal row is divided by the pivot prior to elimination). These procedures are best illustrated by an example.

Example 4.3

Solve the following system of equations by Gauss elimination

$$\begin{bmatrix} 2 & 1 & 0 & 0 \\ 1 & 3 & -2 & 0 \\ 0 & -2 & 5 & -1 \\ 0 & 0 & -1 & 3 \end{bmatrix} \begin{bmatrix} x_1 \\ x_2 \\ x_3 \\ x_4 \end{bmatrix} = \begin{bmatrix} 4 \\ 1 \\ 7 \\ 9 \end{bmatrix}$$

The augmented matrix is

$$\begin{bmatrix} 2 & 1 & 0 & 0 & 4 \\ 1 & 3 & -2 & 0 & 1 \\ 0 & -2 & 5 & -1 & 7 \\ 0 & 0 & -1 & 3 & 9 \end{bmatrix}$$

First pivot - Normalise row 1 of the augmented matrix, and then subtract
1 x row 1 from row 2
0 x row 1 from row 3
0 x row 1 from row 4

$$\begin{bmatrix} 1 & 0.5 & 0 & 0 & 2 \\ 0 & 2.5 & -2 & 0 & -1 \\ 0 & -2 & 5 & -1 & 7 \\ 0 & 0 & -1 & 3 & 9 \end{bmatrix}$$

Second pivot - Normalise row 2 and then subtract

-2 x row 2 from row 3
0 x row 2 from row 4

$$\begin{bmatrix} 1 & 0.5 & 0 & 0 & 2.0 \\ 0 & 1 & -0.8 & 0 & -0.4 \\ 0 & 0 & 3.4 & -1 & 6.2 \\ 0 & 0 & -1 & 3 & 9.0 \end{bmatrix}$$

Third pivot - Normalise row 3 and then subtract
-1 x row 3 from row 4

$$\begin{bmatrix} 1 & 0.5 & 0 & 0 & 2.0 \\ 0 & 1 & -0.8 & 0 & -0.4 \\ 0 & 0 & 1 & -0.294 & 1.824 \\ 0 & 0 & 0 & 2.706 & 10.824 \end{bmatrix}$$

Fourth pivot - Normalise row 4

$$\begin{bmatrix} 1 & 0.5 & 0 & 0 & 2.0 \\ 0 & 1 & -0.8 & 0 & -0.4 \\ 0 & 0 & 1 & -0.294 & 1.824 \\ 0 & 0 & 0 & 1 & 4.0 \end{bmatrix}$$

The coefficient matrix has now been reduced to triangular form (i.e. all sub-diagonal elements have been eliminated). The solution can now be found by back substitution.

$$\begin{aligned} x_4 &= & &= 4.000 \\ x_3 &= 1.824 - (-0.294)\,x_4 & &= 3.000 \\ x_2 &= -0.400 - (-0.8)\,x_3 - (0.0)\,x_4 & &= 2.000 \\ x_1 &= 2.000 - (0.5)\,x_2 - (0.0)\,x_3 - (0.0)\,x_4 & &= 1.000 \end{aligned}$$

This example illustrates two important characteristics of Gauss elimination.

1. Banding is maintained in the upper triangle.
2. The submatrix below the pivot (shown by the broken lines) remains symmetric.

Advantage can be taken of these facts by storing only the upper semi-band of the coefficient matrix as illustrated in fig 3.15.

Program GAUSS.BAS

This program solves a system of linear simultaneous equations with a symmetric, banded coefficient matrix by Gauss elimination.

Input

Only the upper semi-band of the coefficient matrix is input and the matrix is stored as shown in fig 3.15. The right-hand side vector(s) is stored in a separate array.

Comments on the Algorithm

To save space the coefficient matrix is overwritten with the decomposed matrix and the right-hand side vector(s) is overwritten by the solution vector(s) during computation.

Output

The upper semi-band of the coefficient matrix and the right-hand side vector(s) are output. After solution the solution vector(s) is output.

Listing

```
1000 '========================    GAUSS.BAS    ============================
1010 '
1020 OPTION BASE 1
1030 CLS
1040 PRINT
1050 PRINT "================================================================="
1060 PRINT
1070 PRINT "PROGRAM GAUSS.BAS                        Copyright (c) J.Balfour 1991"
1080 PRINT
1090 PRINT "    Program for the solution of simultaneous linear equations"
1100 PRINT "            by Gauss elimination with no row interchange"
1110 PRINT
1120 PRINT "                  For further information contact"
1130 PRINT " James A.D.Balfour, Heriot-Watt University, Riccarton, Edinburgh"
1140 PRINT "              Tel 031-449-5111, Fax 031-451-3170"
1150 PRINT
1160 PRINT "================================================================="
1170 '
1180 ' ... Note that variables are defined in Appendix A
1190 '
1200 MAXNEQ% = 40 : MAXBAND% = 20 : MAXNRHS% = 10
1210 DIM A(MAXNEQ%,MAXBAND%), B(MAXNEQ%,MAXNRHS%)
1220 '
1230 ' ... Maximum number of equations is set by the above statement
1240 '
1250 PRINT
1260 INPUT "Number of equations         = ", NEQ%
1270 IF (NEQ%>0) AND (NEQ%<=MAXNEQ%) THEN GOTO 1300
1280 PRINT "Maximum number equations is currently set at"; MAXNEQ% : GOTO 1700
1290 '
1300 INPUT "Semi-bandwidth              = ", BAND%
1310 IF (BAND%>0) AND (BAND%<=MAXBAND%) THEN GOTO 1340
1320 PRINT "Maximum bandwidth is currently set at"; MAXBAND% : GOTO 1700
1330 '
1340 INPUT "Number of right-hand sides = ", NRHS%
1350 IF (NRHS%>0) AND (NRHS%<=MAXNRHS%) THEN GOTO 1380
1360 PRINT "Maximum number of right-hand sides is currently set at"; MAXNRHS%
1370 GOTO 1700
1380 PRINT
1390 PRINT "Input the upper semi-band of the coefficient matrix"
1400 PRINT
1410 FOR I% = 1 TO NEQ%
1420   FOR J% = 1 TO BAND%
1430     K% = I% + J% - 1
1440     IF K% > NEQ% THEN GOTO 1470
1450     PRINT "Element("; I%; ","; K%; ")  =  "; : INPUT "", A(I%,J%)
1460   NEXT J%
1470 NEXT I%
1480 '
1490 ' ... Input the right-hand side vector(s)
1500 '
1510 FOR J% = 1 TO NRHS%
1520   PRINT
1530   PRINT "Enter the right-hand side vector number"; J%
1540   PRINT
1550   FOR I% = 1 TO NEQ%
1560     PRINT "Element("; I%; ")  =  "; : INPUT "", B(I%,J%)
1570   NEXT I%
1580 NEXT J%
1590 '
1600 ' ... Echo the coefficient matrix
1610 '
1620 PRINT
1630 PRINT "+ + + + + + + + + + + + + + + + + + + + + + + +"
1640 PRINT "+   UPPER SEMI-BAND OF THE COEFFICIENT MATRIX   +"
```

```
1650 PRINT "+ + + + + + + + + + + + + + + + + + + + + + + + + + + + +"
1660 PRINT
1670 FOR I% = 1 TO NEQ%
1680   FOR J% = 1 TO BAND%
1690     PRINT A(I%,J%),
1700   NEXT J%
1710   PRINT : PRINT
1720 NEXT I%
1730 '
1740 ' ... Echo the right-hand side vector(s)
1750 '
1760 PRINT "+ + + + + + + + + + + + + + + + + + + + +"
1770 PRINT "+   RIGHT-HAND SIDE VECTOR(S)   +"
1780 PRINT "+ + + + + + + + + + + + + + + + + + + + +"
1790 PRINT
1800 FOR I% = 1 TO NEQ%
1810   FOR J% = 1 TO NRHS%
1820     PRINT B(I%,J%),
1830   NEXT J%
1840   PRINT : PRINT
1850 NEXT I%
1860 '
1870 ' ... Call the subroutine to solve the equations
1880 '
1890 GOSUB 2090
1900 '
1910 ' ... Print the solution vector(s)
1920 '
1930 PRINT "* * * * * * * * * * * * * *"
1940 PRINT "*   SOLUTION VECTOR(S)   *"
1950 PRINT "* * * * * * * * * * * * * *"
1960 PRINT
1970 FOR I% = 1 TO NEQ%
1980   FOR J% = 1 TO NRHS%
1990     PRINT B(I%,J%),
2000   NEXT J%
2010   PRINT : PRINT
2020 NEXT I%
2030 END
2040 '
2050 '****************    SOLVE SIMULTANEOUS EQUATIONS    ********************
2060 '
2070 ' ... Eliminate using each pivot in turn
2080 '
2090 BB% = BAND%
2100 FOR I% = 1 TO NEQ%
2110   '
2120   ' ... Check if in the unused triangle
2130   '
2140   IF I% > NEQ%-BAND%+1 THEN BB% = NEQ%-I%+1
2150   PIVOT  = A(I%,1)
2160   '
2170   ' ... Normalise
2180   '
2190   FOR J% = 1 TO BB%
2200     A(I%,J%) = A(I%,J%) / PIVOT
2210   NEXT J%
2220   FOR J%  = 1 TO NRHS%
2230     B(I%,J%) = B(I%,J%) / PIVOT
2240   NEXT J%
2250   '
2260   ' ... Check if in the last row
2270   '
2280   IF BB% = 1 THEN GOTO 2490
2290   '
2300   ' ... Eliminate in all row above the pivot and within the band
2310   '
2320   FOR K% = 2 TO BB%
2330     '
2340     ' ... Calculate the row number and then evaluate the multiplier
2350     '
2360     L% = I% + K% - 1
2370     MULT = A(I%,K%) * PIVOT
2380     '
2390     ' ... Loop for all elements in the elimination row
2400     '
2410     FOR J% = K% TO BB%
```

```
2420        M% = J% - K% + 1
2430        A(L%,M%) = A(L%,M%) - MULT*A(I%,J%)
2440      NEXT J%
2450      FOR J% = 1 TO NRHS%
2460        B(L%,J%) = B(L%,J%) - MULT*B(I%,J%)
2470      NEXT J%
2480    NEXT K%
2490 NEXT I%
2500 '
2510 ' ... Back substitute
2520 '
2530 FOR I% = 1 TO NEQ%-1
2540    BB% = I%
2550    IF I% > BAND%-1 THEN BB% = BAND%-1
2560    FOR K% = 1 TO NRHS%
2570      FOR J% = 1 TO BB%
2580        B(NEQ%-I%,K%) = B(NEQ%-I%,K%) - A(NEQ%-I%,J%+1)*B(NEQ%-I%+J%,K%)
2590      NEXT J%
2600    NEXT K%
2610 NEXT I%
2620 RETURN
2630 '
2640 '========================    GAUSS.BAS    ==============================
```

Sample Run

The sample run that follows shows the program GAUSS.BAS being used to solve example 4.3. Input from the keyboard is shown underlined.

```
==================================================================

PROGRAM GAUSS.BAS                    Copyright (c) J.Balfour 1991

    Program for the solution of simultaneous linear equations
          by Gauss elimination with no row interchange

               For further information contact
  James A.D.Balfour, Heriot-Watt University, Riccarton, Edinburgh
            Tel 031-449-5111, Fax 031-451-3170

==================================================================

Number of equations          = 4
Semi-bandwidth               = 2
Number of right-hand sides = 1

Input the upper semi-band of the coefficient matrix

Element( 1 , 1 )  =  2
Element( 1 , 2 )  =  1
Element( 2 , 2 )  =  3
Element( 2 , 3 )  =  -2
Element( 3 , 3 )  =  5
Element( 3 , 4 )  =  -1
Element( 4 , 4 )  =  3

Enter the right-hand side vector number 1

Element( 1 )  =  4
Element( 2 )  =  1
Element( 3 )  =  7
Element( 4 )  =  9

+ + + + + + + + + + + + + + + + + + + + + + + + +
+   UPPER SEMI-BAND OF THE COEFFICIENT MATRIX   +
+ + + + + + + + + + + + + + + + + + + + + + + + +

 2            1

 3           -2

 5           -1

 3            0
```

```
+ + + + + + + + + + + + + + + + + +
+   RIGHT-HAND SIDE VECTOR(S)     +
+ + + + + + + + + + + + + + + + + +

 4

 1

 7

 9

* * * * * * * * * * * * * *
*   SOLUTION VECTOR(S)    *
* * * * * * * * * * * * * *

 1

 2

 3

 4
```

4.5 Triangular Decomposition - Program UDU.BAS

Crout reduction decomposes a non-singular coefficient matrix to the product of two triangular matrices, as shown in fig 4.1. In matrix notation

$$A = L\,U$$

where L is a unit lower triangular matrix and U is an upper triangular matrix.

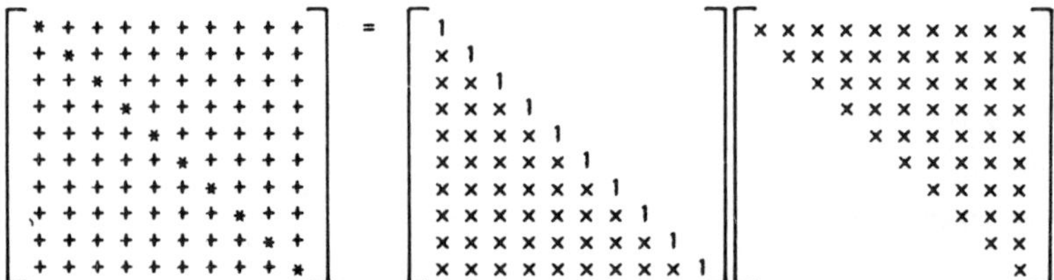

Figure 4.1 Crout reduction

The solution algorithm is as follows

1. Use triangular decomposition to express the coefficient matrix as a product of two triangular matrices. Hence the system of equations $A\,X = B$ becomes

$$L\,U\,X = B \tag{4.1}$$

2. Let $U\,X = Y$ (4.2)

   Hence equation (4.1) can be rewritten as follows

$$L\,Y = B \tag{4.3}$$

3. Following triangular decomposition, Y is found by forward substitution (equation (4.3)). X is then found by backward substitution (equation (4.2)).

If the coefficient matrix A is symmetric and positive definite then it can be decomposed into a product of an upper triangular matrix and its transpose, i.e.

$$A = \bar{U}^T \bar{U} \tag{4.4}$$

where $\bar{U}$ is an upper triangular matrix. This technique is known as Cholesky decomposition. An alternative decomposition is

$$A = U^T D U \tag{4.5}$$

where U is a unit upper triangular matrix, and D is a diagonal matrix.

For linear elastic structures the stiffness matrix is positive definite and symmetric. Hence either of the decompositions shown in equations (4.4) and (4.5) can be employed. $U^T D U$ decomposition is slightly more efficient than $\bar{U}^T \bar{U}$ decomposition. This chapter will consider only $U^T D U$ decomposition.

$$A X = U^T D U X = B$$

Let

$$U X = Y \tag{4.6}$$

then

$$U^T D Y = B \tag{4.7}$$

Once A has been decomposed to $U^T D U$, Y is found by forward substitution (equation (4.7)). Finally the solution vector X is found by backward substitution (equation (4.6)).

The $U^T D U$ decomposition algorithm is as follows

Decomposition

$$
\begin{aligned}
d_{ij} &= 0 && \text{if } i \neq j \\
&= a_{ii} && \text{if } i = 1 \text{ and } j = 1 \\
&= a_{ii} - \sum_{k=1}^{(i-1)} u_{ki}^2 d_k && \text{if } i > 1 \text{ and } j = i \\
u_{ij} &= 0 && \text{if } i > j \\
&= 1 && \text{if } i = j \\
&= \frac{a_{ij}}{a_{ii}} && \text{if } i = 1 \text{ and } j > i
\end{aligned}
$$

$$= \frac{a_{ij} - \sum_{k=1}^{(i-1)} u_{ki}\, u_{kj}\, d_k}{d_i} \qquad \text{if} \quad i > 1 \quad \text{and} \quad j > i$$

Forward Substitution

$$y_i = \frac{b_i}{d_i} \qquad \text{if} \quad i = 1$$

$$= \frac{b_i - \sum_{k=1}^{(i-1)} u_{ki}\, d_k\, y_k}{d_i} \qquad \text{if} \quad i > 1$$

Backward Substitution

$$x_i = y_i \qquad \text{if} \quad i = n$$

$$= y_i - \sum_{k=(i+1)}^{n} u_{ik}\, x_k \qquad \text{if} \quad i < n$$

Example 4.4

Decompose the matrix shown below using $U^T D\, U$ decomposition.

$$\begin{bmatrix} 2 & 1 & 1 & 0 \\ 1 & 6 & 2 & 1 \\ 1 & 2 & 7 & 2 \\ 0 & 1 & 2 & 4 \end{bmatrix}$$

$$d_1 = 2.0$$

$$u_{12} = 1 / 2 = 0.5$$

$$u_{13} = 1 / 2 = 0.5$$

$$u_{14} = 0 / 2 = 0.0$$

$$d_2 = 6 - 0.5^2 \times 2 = 5.50$$

$$u_{23} = (2 - 0.5 \times 0.5 \times 2) / 5.5 = 0.273$$

$$u_{24} = (1 - 0.5 \times 0.0 \times 2) / 5.5 = 0.181$$

$$d_3 = 7 - 0.5^2 \times 2 - 0.273^2 \times 5.5 = 6.090$$

$$u_{34} = (2 - 0.5 \times 0.0 \times 2 - 0.273 \times 0.181 \times 5.5) / 6.090$$

$$= 0.284$$

$$d_4 = 4 - 0.0^2 \times 2 - 0.181^2 \times 5.5 - 0.284^2 \times 6.090$$

$$= 3.329$$

Note that the decomposition of a $n \times n$ coefficient matrix involves "n" major steps. Step "i" entails finding d_{ii} and row "i" of U.

Check that $U^T D\, U = A$

$$\begin{bmatrix} 1 & 0 & 0 & 0 \\ 0.5 & 1 & 0 & 0 \\ 0.5 & 0.273 & 1 & 0 \\ 0 & 0.181 & 0.284 & 1 \end{bmatrix} \begin{bmatrix} 2 & 0 & 0 & 0 \\ 0 & 5.5 & 0 & 0 \\ 0 & 0 & 6.090 & 0 \\ 0 & 0 & 0 & 3.329 \end{bmatrix} \begin{bmatrix} 1 & 0.5 & 0.5 & 0 \\ 0 & 1 & 0.273 & 0.181 \\ 0 & 0 & 1 & 0.284 \\ 0 & 0 & 0 & 1 \end{bmatrix} =$$

$$\begin{bmatrix} 2.000 & 1.000 & 1.000 & 0.000 \\ 1.000 & 6.000 & 2.002 & 0.996 \\ 1.000 & 2.002 & 7.000 & 2.001 \\ 0.000 & 0.996 & 2.002 & 4.001 \end{bmatrix}$$

Example 4.5

Using the decomposed matrix from example 4.4 solve the following system of equations.

$$\begin{bmatrix} 2 & 1 & 1 & 0 \\ 1 & 6 & 2 & 1 \\ 1 & 2 & 7 & 2 \\ 0 & 1 & 2 & 4 \end{bmatrix} \begin{bmatrix} x_1 \\ x_2 \\ x_3 \\ x_4 \end{bmatrix} = \begin{bmatrix} 1.5 \\ -5.5 \\ 16.0 \\ 7.0 \end{bmatrix}$$

i.e. $A\,X = B$, which becomes $U^T D\, U\, X = B$ on decomposition

$U\,X = Y$, hence $U^T D\, Y = B$. Using U^T and D from example 4.4 allows Y to be found by forward substitution as follows.

$$\begin{bmatrix} 2 & 0 & 0 & 0 \\ 1 & 5.5 & 0 & 0 \\ 1 & 1.502 & 6.090 & 0 \\ 0 & 0.996 & 1.730 & 3.329 \end{bmatrix} \begin{bmatrix} y_1 \\ y_2 \\ y_3 \\ y_4 \end{bmatrix} = \begin{bmatrix} 1.5 \\ -5.5 \\ 16.0 \\ 7.0 \end{bmatrix}$$

$$y_1 = 1.5 / 2 = 0.75$$

$$y_2 = (-5.5 - y_1) / 5.5 = -1.136$$

$$y_3 = (16.0 - y_1 - 1.502\, y_2) / 6.090 = 2.784$$

$$y_4 = (7.0 - 0.996\, y_2 - 1.730\, y_3) / 3.329 = 0.996$$

Finally the solution is found by backward substitution of $\boldsymbol{U}\,\boldsymbol{X} = \boldsymbol{Y}$

$$\begin{bmatrix} 1 & 0.5 & 0.5 & 0 \\ 0 & 1 & 0.273 & 0.181 \\ 0 & 0 & 1 & 0.284 \\ 0 & 0 & 0 & 1 \end{bmatrix} \begin{bmatrix} x_1 \\ x_2 \\ x_3 \\ x_4 \end{bmatrix} = \begin{bmatrix} 0.750 \\ -1.136 \\ 2.784 \\ 0.996 \end{bmatrix}$$

which yields

$$\begin{aligned} x_4 & & &= 0.996 \\ x_3 &= 2.784 - 0.284\,x_4 & &= 2.501 \\ x_2 &= -1.136 - 0.181\,x_4 - 0.273\,x_3 & &= -1.992 \\ x_1 &= 0.750 - 0.000\,x_4 - 0.500\,x_3 - 0.500\,x_2 & &= 0.496 \end{aligned}$$

Program UDU.BAS

This program solves a system of linear simultaneous equations with a symmetric, positive definite, banded coefficient matrix by $\boldsymbol{U}^T\boldsymbol{D}\,\boldsymbol{U}$ decomposition.

Input

Only the upper semi-band of the coefficient matrix is input and the matrix is stored as shown in fig 3.15. The right-hand side vector(s) is stored in a separate array.

Comments on the Algorithm

To save space the coefficient matrix is overwritten with the decomposed matrix and the right-hand side vector(s) is overwritten by the solution vector(s) during computation.

Output

The upper semi-band of the coefficient matrix and the right-hand side vector(s) are output. Finally the solution vector(s) is output.

Listing

```
1000 '=========================       UDU.BAS      ==============================
1010 '
1020 OPTION BASE 1
1030 CLS
1040 PRINT
```

```
1050 PRINT "=================================================================="
1060 PRINT
1070 PRINT "PROGRAM UDU.BAS                         Copyright (c) J.Balfour 1991"
1080 PRINT
1090 PRINT "         Program for the solution of simultaneous linear"
1100 PRINT "               equations by triangular decomposition"
1110 PRINT
1120 PRINT "                 For further information contact"
1130 PRINT " James A.D.Balfour, Heriot-Watt University, Riccarton, Edinburgh"
1140 PRINT "               Tel 031-449-5111, Fax 031-451-3170"
1150 PRINT
1160 PRINT "=================================================================="
1170 '
1180 ' ... Note that variables are defined in Appendix A
1190 '
1200 MAXNEQ% = 40 : MAXBAND% = 20 : MAXNRHS% = 10
1210 DIM A(MAXNEQ%,MAXBAND%), B(MAXNEQ%,MAXNRHS%)
1220 '
1230 ' ... Maximum number of equations is set by the above statement
1240 '
1250 PRINT
1260 INPUT "Number of equations        = ", NEQ%
1270 IF (NEQ%>0) AND (NEQ%<=MAXNEQ%) THEN GOTO 1300
1280 PRINT "Maximum number equations is currently set at"; MAXNEQ% : GOTO 1700
1290 '
1300 INPUT "Semi-bandwidth             = ", BAND%
1310 IF (BAND%>0) AND (BAND%<=MAXBAND%) THEN GOTO 1340
1320 PRINT "Maximum bandwidth is currently set at"; MAXBAND% : GOTO 1700
1330 '
1340 INPUT "Number of right-hand sides = ", NRHS%
1350 IF (NRHS%>0) AND (NRHS%<=MAXNRHS%) THEN GOTO 1380
1360 PRINT "Maximum number of right-hand sides is currently set at"; MAXNRHS%
1370 GOTO 1700
1380 PRINT
1390 PRINT "Input the upper semi-band of the coefficient matrix"
1400 PRINT
1410 FOR I% = 1 TO NEQ%
1420   FOR J% = 1 TO BAND%
1430     K% = I% + J% - 1
1440     IF K% > NEQ% THEN GOTO 1470
1450     PRINT "Element("; I%; ","; K%; ")  =  "; : INPUT "", A(I%,J%)
1460   NEXT J%
1470 NEXT I%
1480 '
1490 ' ... Input the right-hand side vector(s)
1500 '
1510 FOR J% = 1 TO NRHS%
1520   PRINT
1530   PRINT "Enter the right-hand side vector number"; J%
1540   PRINT
1550   FOR I% = 1 TO NEQ%
1560     PRINT "Element("; I%; ")  =  "; : INPUT "", B(I%,J%)
1570   NEXT I%
1580 NEXT J%
1590 '
1600 ' ... Echo the coefficient matrix
1610 '
1620 PRINT
1630 PRINT "+ + + + + + + + + + + + + + + + + + + + + + + + + + + +"
1640 PRINT "+   UPPER SEMI-BAND OF THE COEFFICIENT MATRIX   +"
1650 PRINT "+ + + + + + + + + + + + + + + + + + + + + + + + + + + +"
1660 PRINT
1670 FOR I% = 1 TO NEQ%
1680   FOR J% = 1 TO BAND%
1690     PRINT A(I%,J%),
1700   NEXT J%
1710   PRINT : PRINT
1720 NEXT I%
1730 '
1740 ' ... Echo the right-hand side vector(s)
1750 '
1760 PRINT "+ + + + + + + + + + + + + + + + + + + +"
1770 PRINT "+   RIGHT-HAND SIDE VECTOR(S)   +"
1780 PRINT "+ + + + + + + + + + + + + + + + + + + +"
1790 PRINT
1800 FOR I% = 1 TO NEQ%
1810   FOR J% = 1 TO NRHS%
```

```
1820       PRINT B(I%,J%),
1830     NEXT J%
1840     PRINT : PRINT
1850 NEXT I%
1860 '
1870 ' ... Factor the coefficient matrix
1880 '
1890 GOSUB 2100
1900 '
1910 ' ... Solve the equations
1920 '
1930 GOSUB 2420
1940 '
1950 ' ... Print the solution vector(s)
1960 '
1970 PRINT
1980 PRINT "* * * * * * * * * * * * * * *"
1990 PRINT "*    SOLUTION VECTOR(S)     *"
2000 PRINT "* * * * * * * * * * * * * * *"
2010 PRINT
2020 FOR I% = 1 TO NEQ%
2030   FOR J% = 1 TO NRHS%
2040     PRINT B(I%,J%),
2050   NEXT J%
2060   PRINT : PRINT
2070 NEXT I%
2080 END
2090 '
2100 '*********************      TRIANGULAR DECOMPOSITION      ******************
2110 '
2120 ' This subroutine will decompose a symmetric, banded, positive definite
2130 ' matrix [A] using triangular decomposition. On return [A] is reduced
2140 ' to the form [A] = [Ut][D][U] where the first column of [A] contains the
2150 ' diagonal matrix [D] and the remaining columns contain [U].
2160 '
2170 FOR I% = 1 TO NEQ%
2180   IF I% = 1 THEN GOTO 2300
2190   BB%   = BAND% - 1
2200   IF I% < BAND% THEN BB% = I%-1
2210   '
2220   ' ... Evaluate the diagonal element
2230   '
2240   FOR L% = 1 TO BB%
2250     A(I%,1) = A(I%,1) - A(I%-L%,L%+1)*A(I%-L%,L%+1)*A(I%-L%,1)
2260   NEXT L%
2270   '
2280   ' ... Evaluate the off-diagonal elements
2290   '
2300   FOR J% = 2 TO BAND%
2310     IF I% = 1 THEN GOTO 2370
2320     BB%   = BAND% - J%
2330     IF I% < BAND% THEN BB% = I%-1
2340     FOR L% = 1 TO BB%
2350       A(I%,J%) = A(I%,J%) - A(I%-L%,L%+1)*A(I%-L%,J%+L%)*A(I%-L%,1)
2360     NEXT L%
2370     A(I%,J%) = A(I%,J%) / A(I%,1)
2380   NEXT J%
2390 NEXT I%
2400 RETURN
2410 '
2420 '*****************     SOLVE SIMULTANEOUS EQUATIONS     ********************
2430 '
2440 ' This subroutine solves a system of linear simultaneous equations
2450 ' where the coefficient matrix [A] has been composed into [Ut][D][U] form
2460 '
2470 ' ... Forward substitution
2480 '
2490 FOR I% = 1 TO NEQ%
2500   BB% = I%-1
2510   IF I% > BAND%-1 THEN BB% = BAND%-1
2520   FOR J% = 1 TO NRHS%
2530    SUM = 0
2540    IF I% = 1 THEN GOTO 2580
2550    FOR L% = 1 TO BB%
2560      SUM = SUM + A(I%-L%,L%+1)*A(I%-L%,1)*B(I%-L%,J%)
2570    NEXT L%
2580    B(I%,J%) = (B(I%,J%) - SUM) / A(I%,1)
```

```
2590    NEXT J%
2600 NEXT I%
2610 '
2620 ' ... Backward substitution
2630 '
2640 FOR I% = 1 TO NEQ%-1
2650    BB% = I%
2660    IF I% > BAND%-1 THEN BB% = BAND%-1
2670    FOR J% = 1 TO NRHS%
2680       FOR L% = 1 TO BB%
2690          B(NEQ%-I%,J%) = B(NEQ%-I%,J%) - A(NEQ%-I%,L%+1)*B(NEQ%-I%+L%,J%)
2700       NEXT L%
2710    NEXT J%
2720 NEXT I%
2730 RETURN
2740 '
2750 '==========================    UDU.BAS    ==============================
```

Sample Run

There follows a sample run that shows the program UDU.BAS being used to solve example 4.5. Input from the keyboard is shown underlined.

```
=================================================================

PROGRAM UDU.BAS                    Copyright (c) J.Balfour 1991

        Program for the solution of simultaneous linear
             equations by triangular decomposition

               For further information contact
 James A.D.Balfour, Heriot-Watt University, Riccarton, Edinburgh
             Tel 031-449-5111, Fax 031-451-3170

=================================================================

Number of equations        = 4
Semi-bandwidth             = 3
Number of right-hand sides = 1

Input the upper semi-band of the coefficient matrix

Element( 1 , 1 )  =  2
Element( 1 , 2 )  =  1
Element( 1 , 3 )  =  1
Element( 2 , 2 )  =  6
Element( 2 , 3 )  =  2
Element( 2 , 4 )  =  1
Element( 3 , 3 )  =  7
Element( 3 , 4 )  =  2
Element( 4 , 4 )  =  4

Enter the right-hand side vector number 1

Element( 1 )  =  1.5
Element( 2 )  =  -5.5
Element( 3 )  =  16
Element( 4 )  =  7

+ + + + + + + + + + + + + + + + + + + + + + + + +
+   UPPER SEMI-BAND OF THE COEFFICIENT MATRIX   +
+ + + + + + + + + + + + + + + + + + + + + + + + +

 2             1             1

 6             2             1

 7             2             0

 4             0             0

+ + + + + + + + + + + + + + + + +
+   RIGHT-HAND SIDE VECTOR(S)   +
+ + + + + + + + + + + + + + + + +
```

```
 1.5
-5.5
 16
 7

* * * * * * * * * * * * * * *
*   SOLUTION VECTOR(S)   *
* * * * * * * * * * * * * * *

 .4999999
-2
 2.5
 .9999999
```

4.6 Matrix Inversion by Gauss-Jordan Elimination

The inverse of the square matrix A is defined as A^{-1} where

$$A\,A^{-1} = I \tag{4.8}$$

I is the identity matrix. The above multiplication is commutative, i.e.

$$A^{-1}A = I$$

Multiplication of both sides of a system of simultaneous equations by the inverse of the coefficient matrix yields

$$A^{-1}A\,X = A^{-1}B$$

or

$$X = A^{-1}B$$

Hence a solution to a system of simultaneous linear equations can be obtained by multiplication of the right-hand side vector by the inverse of the coefficient matrix.

The inverse of the coefficient matrix is a convenient concept in matrix algebra. However, matrix inversion has a number of practical disadvantages. The evaluation of the inverse of a matrix is a numerically intensive procedure and this makes the solution of simultaneous equations by matrix inversion uncompetitive. If the coefficient matrix is banded then the banding is destroyed during inversion. This is an obvious disadvantage in the stiffness method which tends to produce narrowly banded matrices. Also, because matrix inversion is more numerically intensive than rival methods of equation solving, the effects of ill-conditioning are more likely to be problematic (see section 4.7). The large number of arithmetic operations associated with matrix inversion is, however, a useful property when investigating the effects of ill-conditioning. A matrix inversion program has, therefore, been included to facilitate the investigation of ill-conditioned coefficient matrices. A matrix inversion program can, of course, be used to solve simultaneous equations although the practice is to be discouraged for the reasons outlined above.

By definition, multiplication of the first column of the inverse of the coefficient matrix by the coefficient matrix itself must produce the first column of the identity matrix (see equation (4.8)).

$$\boldsymbol{A}\,\boldsymbol{X}_1 \quad = \quad \boldsymbol{B}_1$$

where

$\boldsymbol{A}$ is the coefficient matrix
$\boldsymbol{X}_1$ is the first column of the inverse of $\boldsymbol{A}$
$\boldsymbol{B}_1$ is the first column of the identity matrix

Similarly

$$\boldsymbol{A}\,\boldsymbol{X}_2 \quad = \quad \boldsymbol{B}_2$$

where

$\boldsymbol{A}$ is the coefficient matrix
$\boldsymbol{X}_2$ is the second column of the inverse of $\boldsymbol{A}$
$\boldsymbol{B}_2$ is the second column of the identity matrix

This means that solving for a set of right-hand side vectors which, collectively, form the identity matrix, produces a set of solution vectors that collectively comprise the inverse of the coefficient matrix. Any solution method can be used to find an inverse in this way, but Gauss-Jordan elimination is, perhaps, the most straightforward.

Example 4.6

Use Gauss-Jordan elimination to find the solution to the equations in example 4.1 by matrix inversion. The augmented matrix for that problem is

$$\begin{bmatrix} 2 & 1 & 0 & 1 & 0 & 0 \\ 1 & 3 & -2 & 0 & 1 & 0 \\ 0 & -2 & 5 & 0 & 0 & 1 \end{bmatrix}$$

First pivot - Normalise row 1

$$\begin{bmatrix} 1 & 0.5 & 0 & 0.5 & 0 & 0 \\ 1 & 3 & -2 & 0 & 1 & 0 \\ 0 & -2 & 5 & 0 & 0 & 1 \end{bmatrix}$$

Subtract 1 x row 1 from row 2
0 x row 1 from row 3

$$\begin{bmatrix} 1 & 0.5 & 0 & 0.5 & 0 & 0 \\ 0 & 2.5 & -2 & -0.5 & 1 & 0 \\ 0 & -2 & 5 & 0 & 0 & 1 \end{bmatrix}$$

Second pivot - Normalise row 2

$$\begin{bmatrix} 1 & 0.5 & 0 & 0.5 & 0 & 0 \\ 0 & 1 & -0.8 & -0.2 & 0.4 & 0 \\ 0 & -2 & 5 & 0 & 0 & 1 \end{bmatrix}$$

Subtract 0.5 x row 2 from row 1
-2 x row 2 from row 3

$$\begin{bmatrix} 1 & 0 & 0.4 & 0.6 & -0.2 & 0 \\ 0 & 1 & -0.8 & -0.2 & 0.4 & 0 \\ 0 & 0 & 3.4 & -0.4 & 0.8 & 1 \end{bmatrix}$$

Third pivot - Normalise row 3

$$\begin{bmatrix} 1 & 0 & 0.4 & 0.6 & -0.2 & 0 \\ 0 & 1 & -0.8 & -0.2 & 0.4 & 0 \\ 0 & 0 & 1 & -0.118 & 0.235 & 0.294 \end{bmatrix}$$

Subtract 0.4 x row 3 from row 1
-0.8 x row 3 from row 2

$$\begin{bmatrix} 1 & 0 & 0 & 0.647 & -0.294 & -0.118 \\ 0 & 1 & 0 & -0.294 & 0.588 & 0.235 \\ 0 & 0 & 1 & -0.118 & 0.235 & 0.294 \end{bmatrix}$$

Using the inverse to solve the equations (i.e. $X = A^{-1}B$, where B is taken from example 4.1) yields

$$\begin{bmatrix} x_1 \\ x_2 \\ x_3 \end{bmatrix} = \begin{bmatrix} 0.647 & -0.294 & -0.118 \\ -0.294 & 0.588 & 0.235 \\ -0.118 & 0.235 & 0.294 \end{bmatrix} \begin{bmatrix} 4 \\ 1 \\ 11 \end{bmatrix} = \begin{bmatrix} 0.996 \\ 1.997 \\ 2.997 \end{bmatrix}$$

Note that the solution is not exact as only three significant figures have been used in the calculations.

4.7 Ill-Conditioned Equations - Program INVERT.BAS

If arithmetical operations could be conducted using an infinite number of significant figures, then any system of simultaneous equations could be classified as being either

1. Singular (i.e. having no unique solution), or
2. Non-singular (i.e. having a unique solution).

A singular system of equations is characterised by having a coefficient matrix with a determinant of zero, whereas a non-singular system of equations has a coefficient matrix with a non-zero determinant.

In practice, however, computer arithmetic is conducted using a finite number of significant figures. Typically, single precision arithmetic uses about seven significant figures and double precision arithmetic uses about 14 significant figures. This means that arithmetical results obtained using a computer have only a limited degree of accuracy. Hence, if the determinant of a coefficient matrix is evaluated using a computer, and the result turns out to be zero, then it is not possible to say that the equations are singular. Due to the limited accuracy of the computation a number that is actually very close to zero may be held as zero. Consequently it is only possible to conclude that if a determinant is calculated to be zero, or very close to zero, then the equations are singular or almost singular.

Example 4.7

Solve the system of equations shown below, (i) using all of the available significant figures and (ii) using only five significant figures.

$$\begin{bmatrix} 1.000000 & 1.356589 \\ 1.356589 & 1.840000 \end{bmatrix} \begin{bmatrix} x_1 \\ x_2 \end{bmatrix} = \begin{bmatrix} 0.006516518 \\ 0.003737453 \end{bmatrix}$$

(i) *Solution using all of the significant figures.*

$$\begin{bmatrix} 1.000000 & 1.356589 \\ 0.0 & -0.0003337150 \end{bmatrix} \begin{bmatrix} x_1 \\ x_2 \end{bmatrix} = \begin{bmatrix} 0.006516518 \\ -0.005102784 \end{bmatrix}$$

hence

$$\begin{bmatrix} x_1 \\ x_2 \end{bmatrix} = \begin{bmatrix} -20.73687 \\ 15.29084 \end{bmatrix}$$

(ii) *Solution using only five significant figures*

$$\begin{bmatrix} 1.0000 & 1.3565 \\ 0.0 & -0.000092250 \end{bmatrix} \begin{bmatrix} x_1 \\ x_2 \end{bmatrix} = \begin{bmatrix} 0.0065165 \\ -0.0051021 \end{bmatrix}$$

hence

$$\begin{bmatrix} x_1 \\ x_2 \end{bmatrix} = \begin{bmatrix} -75.019 \\ 55.308 \end{bmatrix}$$

Example 4.7 shows quite clearly that the number of significant figures used during computation can have a dramatic effect upon the results obtained.

If small changes to the coefficient matrix produce large changes in the solution, then the equations are said to be ill-conditioned. The reader should note that example 4.7 demonstrates the effects of ill-conditioning; it does not investigate ill-conditioning itself, and a clear distinction should be made between the presence and the effects of ill-conditioning.

In general, the stiffness method produces well-conditioned systems of equations to which a sufficiently accurate solution can be found using single precision computer arithmetic. The governing equations only become badly behaved under "special" circumstances (e.g. when a structure is close to elastic instability). The reader should note that ill-conditioned equations have physical significance, and if ill-conditioned equations appear unexpectedly, then the cause of the ill-conditioning needs to be found and understood.

The Presence of Ill-conditioning

There is no absolute definition of an ill-conditioned matrix. It can be argued that a matrix is only ill-conditioned if a satisfactorily accurate solution cannot be obtained using the number of significant figures offered by the computer being used (i.e. a matrix is only ill-conditioned if the effects of ill-conditioning are unacceptable). Such a definition is not very useful. It is better, although equally vague, to say that ill-conditioning is present if the solution vector is likely to be very sensitive to the number of significant figures used during computation. There are a number of tests that can be applied to a coefficient matrix to investigate its condition. Of the available tests, inspection of the inverse of the coefficient matrix is perhaps the most useful. In general it can be said that the bigger the elements of the inverse are in comparison with the elements of the original matrix, the poorer is the condition of the matrix. If the elements of the inverse are many orders of magnitude greater than the elements of the coefficient matrix itself, then ill-conditioning may be a problem and should be investigated.

Example 4.8

Use inspection of the inverse to investigate the condition of the coefficient matrix used in example 4.7. The coefficient matrix is as follows

$$A = \begin{bmatrix} 1.000000 & 1.356589 \\ 1.356589 & 1.840000 \end{bmatrix}$$

and its inverse is

$$A^{-1} = -\frac{1}{0.000333715}\begin{bmatrix} 1.840000 & -1.356589 \\ -1.356589 & 1.000000 \end{bmatrix}$$

$$= \begin{bmatrix} -5513.686 & 4065.112 \\ 4065.112 & -2996.568 \end{bmatrix}$$

The elements of the inverse are approximately 3000 times bigger than the elements of the coefficient matrix, indicating that ill-conditioning is a potential problem.

Example 4.8 shows that the elements of the inverse of an ill-conditioned matrix are large because its determinant is small. As a system of equations approaches singularity the determinant of the coefficient matrix approaches, zero causing the elements of the inverse to be magnified.

Example 4.9

The symmetric matrix shown below is known as the Hilbert matrix.

$$\begin{bmatrix} 1 & \frac{1}{2} & \frac{1}{3} & \cdots & \frac{1}{n} \\ \frac{1}{2} & \frac{1}{3} & \frac{1}{4} & \cdots & \frac{1}{n+1} \\ \frac{1}{3} & \frac{1}{4} & \frac{1}{5} & \cdots & \frac{1}{n+2} \\ \cdot & & & & \cdot \\ \cdot & & & & \cdot \\ \frac{1}{n} & \frac{1}{n+1} & \frac{1}{n+2} & \cdots & \frac{1}{2n-1} \end{bmatrix}$$

Investigate the condition of 3 x 3 and 5 x 5 Hilbert matrices using the magnitude of the elements of the inverse as a measure of condition.

$$H_3 = \begin{bmatrix} 1.000000 & 0.500000 & 0.333333 \\ 0.500000 & 0.333333 & 0.2500000 \\ 0.333333 & 0.250000 & 0.200000 \end{bmatrix}$$

and

$$H_3^{-1} = \begin{bmatrix} 9.00065 & -36.0034 & 30.0032 \\ -36.0034 & 192.0036 & -180.016 \\ 30.0032 & -180.016 & 180.015 \end{bmatrix}$$

$$H_5 = \begin{bmatrix} 1.000000 & 0.500000 & 0.333333 & 0.250000 & 0.200000 \\ 0.500000 & 0.333333 & 0.250000 & 0.200000 & 0.166667 \\ 0.333333 & 0.250000 & 0.200000 & 0.166667 & 0.142857 \\ 0.250000 & 0.200000 & 0.166667 & 0.142857 & 0.125000 \\ 0.200000 & 0.166667 & 0.142857 & 0.125000 & 0.111111 \end{bmatrix}$$

and

$$H_5^{-1} = \begin{bmatrix} 26.9278 & -335.949 & 1204.81 & -1633.57 & 744.183 \\ -335.949 & 5469.83 & -21783.1 & 31228.4 & -14725.2 \\ 1204.80 & -21783.2 & 91785.8 & -136306 & 65840.8 \\ -1633.57 & 31228.6 & -136307 & 207403 & -101980 \\ 744.186 & -14725.4 & 65841.3 & -101980 & 50832.2 \end{bmatrix}$$

Both inverses exhibit elements that are much bigger than the elements of the original matrices, indicating that the original matrices are poorly conditioned. The Hilbert matrix is, in fact, well known to be extremely poorly conditioned.

The Effects of Ill-conditioning

The inverse of the coefficient matrix can be used to investigate the severity of the effects of ill-conditioning. There are two commonly employed tests

1. Compare $A A^{-1}$ with the identity matrix I.
2. Compare $[A^{-1}]^{-1}$ with the original coefficient matrix A.

If the coefficient matrix is well-conditioned then the above comparisons will be almost exact. If, on the other hand, the coefficient matrix is poorly conditioned then the effect of ill-conditioning will be observed as a significant difference between the two matrices. Of the two tests, inversion of the inverse is the more severe as it entails a greater number of arithmetical operations.

For instance, the inverse of the inverse of a 5 x 5 Hilbert matrix (shown below) is noticeably different from the original matrix (given in example 4.9).

$$\begin{bmatrix} 0.999968 & 0.499755 & 0.333139 & 0.249839 & 0.199863 \\ 0.499760 & 0.333158 & 0.249863 & 0.199887 & 0.166571 \\ 0.333142 & 0.249863 & 0.199893 & 0.166579 & 0.142783 \\ 0.249841 & 0.199886 & 0.166579 & 0.142785 & 0.124939 \\ 0.199864 & 0.166570 & 0.142782 & 0.124939 & 0.111059 \end{bmatrix}$$

It should also be noted that although any technique for solving simultaneous equations can be used to invert the coefficient matrix, it is better to invert using the same computational procedure as will be used to solve the equations (e.g. if Gauss elimination is to be used to solve the equations then the inverse should also be obtained using Gauss elimination).

As the effects of ill-conditioning arise purely from the truncation of real numbers during computation, it follows that the best defence against the effects of ill-conditioning is to use many significant figures. Most computer languages have the facility to conduct arithmetic to a number of levels of precision. For most calculations single precision arithmetic, which uses about seven significant figures, is sufficiently accurate.

Calculations requiring a higher degree of accuracy are usually conducted using double precision arithmetic, which offers, as the name suggests, approximately twice as many significant figures as single precision arithmetic. Some machines and languages have a tertiary level of accuracy known as extended precision, allowing calculations to be performed using about twice as many significant figures as double precision arithmetic (perhaps 28 significant figures). While high precision arithmetic offers better accuracy than single precision arithmetic, it takes longer and uses more computer memory. In general, high precision arithmetic should only be used if satisfactory results cannot be obtained using single precision arithmetic.

If ill-conditioning is suspected, then comparison of the solutions obtained using two different levels of precision gives some insight into the arithmetical stability of the problem in hand.

Program INVERT.BAS

This program inverts a coefficient matrix by Gauss-Jordan elimination without row interchange. The matrix passed to the inversion routine is overwritten by the inverse. The program offers the following three options for the study of ill-conditioning

1. Matrix inversion.
2. Inversion of the inverse of the coefficient matrix.
3. Multiplication of the inverse of the coefficient matrix by the coefficient matrix itself.

Input

The coefficient matrix is input row by row.

Comments on the Algorithm

The inversion routine overwrites the coefficient matrix with its inverse. To allow comparison of the product of the coefficient matrix and its inverse with the identity matrix, a copy of the coefficient matrix is taken before inversion. As the inversion routine does not employ row interchange a zero pivot will cause the algorithm to break down.

Output

The coefficient matrix is echoed and the user stipulates whether the inverse, the inverse of the inverse, or the product of the inverse and the coefficient matrix, is to be output.

Listing

```
1000 '========================    INVERT.BAS    ============================
1010 '
1020 OPTION BASE 1
1030 CLS
1040 PRINT
1050 PRINT "================================================================="
1060 PRINT
1070 PRINT "PROGRAM INVERT.BAS                        Copyright (c) J.Balfour 1991"
1080 PRINT
1090 PRINT "        Program to invert a matrix by Gauss-Jordan elimination"
1100 PRINT "                    with no row interchange"
1110 PRINT
1120 PRINT "                    For further information contact"
1130 PRINT " James A.D.Balfour, Heriot-Watt University, Riccarton, Edinburgh"
1140 PRINT "               Tel 031-449-5111, Fax 031-451-3170"
1150 PRINT
1160 PRINT "================================================================="
1170 '
1180 ' ... Note that variables are defined in Appendix A
1190 '
1200 MAXNEQ% = 25
1210 DIM A(MAXNEQ%,MAXNEQ%), C(MAXNEQ%,MAXNEQ%), D(MAXNEQ%,MAXNEQ%)
1220 '
1230 ' ... Maximum number of equations is set by the above statement
1240 '
1250 PRINT
1260 INPUT "Size of the matrix = ", NEQ%
```

```
1270 IF (NEQ%>0) AND (NEQ%<=MAXNEQ%) THEN GOTO 1290
1280 PRINT "Maximum number equations is currently set at"; MAXNEQ% : GOTO 1630
1290 PRINT
1300 PRINT "Input the augmented matrix" : PRINT
1310 FOR I% = 1 TO NEQ%
1320   FOR J% = 1 TO NEQ%
1330     PRINT "Element ("; I%; ","; J%; ") = "; : INPUT "", A(I%,J%)
1340   NEXT J%
1350 NEXT I%
1360 '
1370 ' ... Echo the and copy the coefficient matrix
1380 '
1390 PRINT
1400 PRINT "+ + + + + + + + + + + + + +"
1410 PRINT "+   COEFFICIENT MATRIX  +"
1420 PRINT "+ + + + + + + + + + + + + +"
1430 PRINT
1440 FOR I% = 1 TO NEQ%
1450   FOR J% = 1 TO NEQ%
1460     PRINT USING "##.######^^^^  "; A(I%,J%),
1470     C(I%,J%) = A(I%,J%)
1480   NEXT J%
1490   PRINT : PRINT
1500 NEXT I%
1510 WHILE ANS%<>4
1520   PRINT "Do you wish to
1530   PRINT "(1) Invert the matrix once
1540   PRINT "(2) Invert the matrix twice
1550   PRINT "(3) Invert the matrix once and multiply by the original matrix"
1560   PRINT "(4) Stop"
1570   INPUT ANS%
1580   '
1590   ' ... Invert the coefficient matrix
1600   '
1610   GOSUB 2030
1620   PRINT
1630   ON ANS% GOTO 1640, 1680, 1730, 2010
1640   PRINT "* * * * * * * * * * * * * * * * * * * * *"
1650   PRINT "*   INVERSE OF THE COEFFICIENT MATRIX   *"
1660   PRINT "* * * * * * * * * * * * * * * * * * * * *"
1670   GOTO  1920
1680   GOSUB 2030
1690   PRINT "* * * * * * * * * * * * * * * * * * * * * * * * * * * * * *"
1700   PRINT "*   INVERSE OF THE INVERSE OF THE COEFFICIENT MATRIX    *"
1710   PRINT "* * * * * * * * * * * * * * * * * * * * * * * * * * * * * *"
1720   GOTO  1920
1730   PRINT "* * * * * * * * * * * * * * * * * * * * * * * * * * * * * *"
1740   PRINT "*   PRODUCT OF THE INVERSE AND THE COEFFICIENT MATRIX   *"
1750   PRINT "* * * * * * * * * * * * * * * * * * * * * * * * * * * * * *"
1760   FOR I%  = 1 TO NEQ%
1770     FOR J%  = 1 TO NEQ%
1780       D(I%,J%) = A(I%,J%)
1790     NEXT J%
1800   NEXT I%
1810   FOR I%  = 1 TO NEQ%
1820     FOR J%  = 1 TO NEQ%
1830       A(I%,J%) = 0
1840       FOR K%  = 1 TO NEQ%
1850         A(I%,J%) = A(I%,J%) + C(I%,K%)*D(K%,J%)
1860       NEXT K%
1870     NEXT J%
1880   NEXT I%
1890   '
1900   ' ... Print results and reset the coefficient matrix
1910   '
1920   PRINT
1930   FOR I%  = 1 TO NEQ%
1940     FOR J%  = 1 TO NEQ%
1950       PRINT USING "##.######^^^^  "; A(I%,J%),
1960       A(I%,J%) = C(I%,J%)
1970     NEXT J%
1980     PRINT : PRINT
1990   NEXT I%
2000 WEND
2010 END
2020 '
2030 '************     INVERT MATRIX BY GAUSS-JORDAN ELIMINATION     **********
```

```
2040 '
2050 ' ... Loop for all pivots
2060 '
2070 FOR I% = 1 TO NEQ%
2080   TEMP = A(I%,I%)
2090   '
2100   ' ... Divide the pivotal row by the pivot
2110   '
2120   FOR J% = 1 TO NEQ%
2130     A(I%,J%) = A(I%,J%) / TEMP
2140   NEXT J%
2150   A(I%,I%) = 1 / TEMP
2160   '
2170   ' ... Loop for all rows
2180   '
2190   FOR J% = 1 TO NEQ%
2200     IF J% = I% THEN GOTO 2300
2210     FACT  = A(J%,I%)
2220     '
2230     ' ... Loop for all elements in the row
2240     '
2250     FOR K% = 1 TO NEQ%
2260       TEMP = A(J%,K%)
2270       IF K% = I% THEN TEMP = 0
2280       A(J%,K%) = TEMP - FACT*A(I%,K%)
2290     NEXT K%
2300   NEXT J%
2310 NEXT I%
2320 RETURN
2330 '
2340 '========================    INVERT.BAS    ==============================
```

Sample Run

The following sample run shows the program INVERT.BAS being used to investigate the presence of ill-conditioning in a Hilbert matrix of order 5. Input from the keyboard is shown underlined.

```
=================================================================

PROGRAM INVERT.BAS                       Copyright (c) J.Balfour 1991

       Program to invert a matrix by Gauss-Jordan elimination
                   with no row interchange

               For further information contact
 James A.D.Balfour, Heriot-Watt University, Riccarton, Edinburgh
            Tel 031-449-5111, Fax 031-451-3170

=================================================================

Size of the matrix = 5

Input the augmented matrix

Element ( 1 , 1 ) = 1
Element ( 1 , 2 ) = .5
Element ( 1 , 3 ) = .333333
Element ( 1 , 4 ) = .25
Element ( 1 , 5 ) = .2
Element ( 2 , 1 ) = .5
Element ( 2 , 2 ) = .333333
Element ( 2 , 3 ) = .25
Element ( 2 , 4 ) = .2
Element ( 2 , 5 ) = .166667
Element ( 3 , 1 ) = .333333
Element ( 3 , 2 ) = .25
Element ( 3 , 3 ) = .2
Element ( 3 , 4 ) = .166667
Element ( 3 , 5 ) = .142857
Element ( 4 , 1 ) = .25
```

```
Element ( 4 , 2 ) = .2
Element ( 4 , 3 ) = .166667
Element ( 4 , 4 ) = .142857
Element ( 4 , 5 ) = .125
Element ( 5 , 1 ) = .2
Element ( 5 , 2 ) = .166667
Element ( 5 , 3 ) = .142857
Element ( 5 , 4 ) = .125
Element ( 5 , 5 ) = .111111

+ + + + + + + + + + + + + 
+   COEFFICIENT MATRIX  +
+ + + + + + + + + + + + + 

 1.000000E+00   5.000000E-01   3.333330E-01   2.500000E-01   2.000000E-01

 5.000000E-01   3.333330E-01   2.500000E-01   2.000000E-01   1.666670E-01

 3.333330E-01   2.500000E-01   2.000000E-01   1.666670E-01   1.428570E-01

 2.500000E-01   2.000000E-01   1.666670E-01   1.428570E-01   1.250000E-01

 2.000000E-01   1.666670E-01   1.428570E-01   1.250000E-01   1.111110E-01

Do you wish to
(1) Invert the matrix once
(2) Invert the matrix twice
(3) Invert the matrix once and multiply by the original matrix
(4) Stop
? 1

* * * * * * * * * * * * * * * * * * * * * 
*   INVERSE OF THE COEFFICIENT MATRIX   *
* * * * * * * * * * * * * * * * * * * * * 

 2.692783E+01  -3.359481E+02   1.204805E+03  -1.633565E+03   7.441802E+02

-3.359481E+02   5.469836E+03  -2.178312E+04   3.122841E+04  -1.472519E+04

 1.204809E+03  -2.178318E+04   9.178576E+04  -1.363064E+05   6.584081E+04

-1.633577E+03   3.122863E+04  -1.363070E+05   2.074033E+05  -1.019798E+05

 7.441893E+02  -1.472536E+04   6.584134E+04  -1.019802E+05   5.083223E+04

Do you wish to
(1) Invert the matrix once
(2) Invert the matrix twice
(3) Invert the matrix once and multiply by the original matrix
(4) Stop
? 2

* * * * * * * * * * * * * * * * * * * * * * * * * * * * 
*   INVERSE OF THE INVERSE OF THE COEFFICIENT MATRIX   *
* * * * * * * * * * * * * * * * * * * * * * * * * * * * 

 9.986662E-01   4.080200E-01   3.324368E-01   2.492289E-01   1.993232E-01

 4.989859E-01   3.325229E-01   2.493239E-01   1.994194E-01   1.661581E-01

 3.325143E-01   2.493479E-01   1.994566E-01   1.662009E-01   1.424487E-01

 2.493131E-01   1.994540E-01   1.662125E-01   1.424674E-01   1.246589E-01

 1.994082E-01   1.661972E-01   1.424662E-01   1.246652E-01   1.108180E-01

Do you wish to
(1) Invert the matrix once
(2) Invert the matrix twice
(3) Invert the matrix once and multiply by the original matrix
(4) Stop
? 3
```

```
* * * * * * * * * * * * * * * * * * * * * * * * * * * * * *
*   PRODUCT OF THE INVERSE AND THE COEFFICIENT MATRIX   *
* * * * * * * * * * * * * * * * * * * * * * * * * * * * * *

 9.999389E-01   1.464844E-03  -3.906250E-03   1.562500E-02  -2.929688E-03

-3.814697E-05   1.000488E+00  -2.929688E-03   9.765625E-03  -1.953125E-03

-3.814697E-05   4.882813E-04   9.980469E-01   8.789062E-03  -9.765625E-04

-7.629395E-06  -1.220703E-04  -1.953125E-03   1.003906E+00   0.000000E+00

-7.629395E-06   2.441406E-04  -4.882813E-04   2.929688E-03   1.000000E+00

Do you wish to
(1) Invert the matrix once
(2) Invert the matrix twice
(3) Invert the matrix once and multiply by the original matrix
(4) Stop
? 4
```

Chapter 5

Plane Trusses

5.1 Introduction

In practice most trusses are connected together by welding or by two or more bolts, so that the joints are effectively rigid. However, even when rigidly jointed, triangulated frameworks transmit load primarily by axial force in the elements. Hence bending stresses are usually of minor importance compared with axial stresses, thus allowing the assumption of pinned joints to be safely made. Before the use of computers became widespread in design offices, trusses were normally assumed to be pin jointed to simplify the structural analysis. Today computers allow advantage to be taken of the additional stiffness that comes from rigid connections, and trusses are usually analysed as rigidly jointed frameworks. Pin jointed trusses are therefore no longer of service in the design office. They do, however, still have a role to play in engineering education.

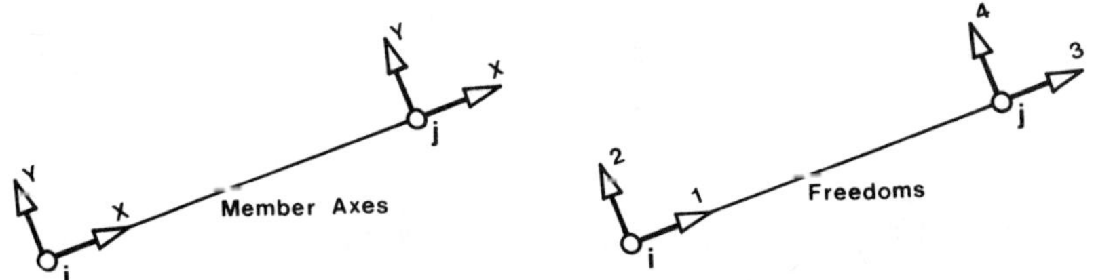

Figure 5.1 Element axes and freedoms

The techniques employed when using the stiffness method to analyse pin jointed trusses are identical to those used in the analysis of rigidly jointed frameworks. For educational purposes pin jointed trusses have the advantage of having a structural action that is inherently simpler than that of rigidly jointed frames, and this makes the understanding of the fundamentals of the stiffness method easier. The reader should be aware that in this text pin jointed trusses are used purely as an introduction to the stiffness method. In practice trusses that are rigidly jointed are almost always better analysed as rigidly jointed frameworks.

5.2 The Element Stiffness Matrix in Element Axes

The forces at the ends of a truss member are related to the displacements at the ends by the element stiffness matrix. Figure 5.1 shows a typical plane truss element "i,j". By treating the element as a simple structure with four freedoms, the element stiffness matrix can be found. To obtain the element stiffness matrix the force system required to maintain unit displacement at each freedom in turn must be evaluated.

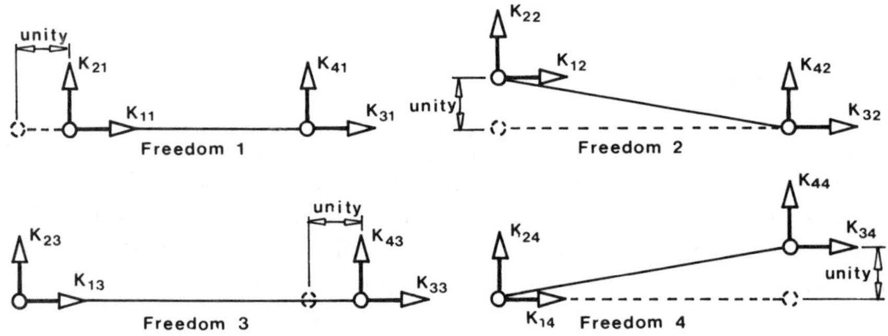

Figure 5.2 Evaluation of stiffness terms

For a pin jointed member only axial forces are involved and the stiffness terms can be found by inspection of fig 5.2. Note also that the axial stiffness of a line element is considered in section 2.12. All stiffness coefficients are zero with the exception of

$$k_{11} = k_{33} = \frac{EA}{L} \quad \text{and} \quad k_{13} = k_{31} = -\frac{EA}{L}$$

hence

$$\begin{bmatrix} f_{ijx} \\ f_{ijy} \\ \hdashline f_{jix} \\ f_{jiy} \end{bmatrix} = \left[\begin{array}{cc:cc} \frac{EA}{L} & 0 & \frac{-EA}{L} & 0 \\ 0 & 0 & 0 & 0 \\ \hdashline \frac{-EA}{L} & 0 & \frac{EA}{L} & 0 \\ 0 & 0 & 0 & 0 \end{array}\right] \begin{bmatrix} \delta_{ijx} \\ \delta_{ijy} \\ \hdashline \delta_{jix} \\ \delta_{jiy} \end{bmatrix} \tag{5.1}$$

The subscripts for force and displacement can be decoded as follows. The first two characters identify the element by the nodes at its ends, with the first character indicating the end under consideration. The third subscript defines the direction of the force or displacement (in the element axes system). If equation (5.1) is split into submatrices as indicated by the broken lines then

$$\begin{bmatrix} f_{ij} \\ \hline f_{ji} \end{bmatrix} = \left[\begin{array}{c|c} k_{ii}^{j} & k_{ij} \\ \hline k_{ji} & k_{jj}^{i} \end{array}\right] \begin{bmatrix} \delta_{ij} \\ \hline \delta_{ji} \end{bmatrix}$$

hence

$$f_{ij} = k_{ii}^{j}\,\delta_{ij} + k_{ij}\,\delta_{ji} \tag{5.2}$$

and

$$f_{ji} = k_{ji}\,\delta_{ij} + k_{jj}^{i}\,\delta_{ji} \tag{5.3}$$

The product $k_{ii}^{j}\,\delta_{ij}$ gives the forces at end *"i"* of element *"i,j"* due to displacements at end *"i"*, and $k_{ij}\,\delta_{ji}$ gives the forces at end *"i"* due to the displacements at end *"j"*. Of course the resultant force vector f_{ij} is the matrix sum of these products.

5.3 Transformation of Force and Displacement

Equations (5.2) and (5.3) show how the element end forces are related to the element end displacements by the element stiffness matrix. In general the element axes associated with these equations will not coincide with the axes used to describe nodal forces and displacements (i.e. the nodal axes).

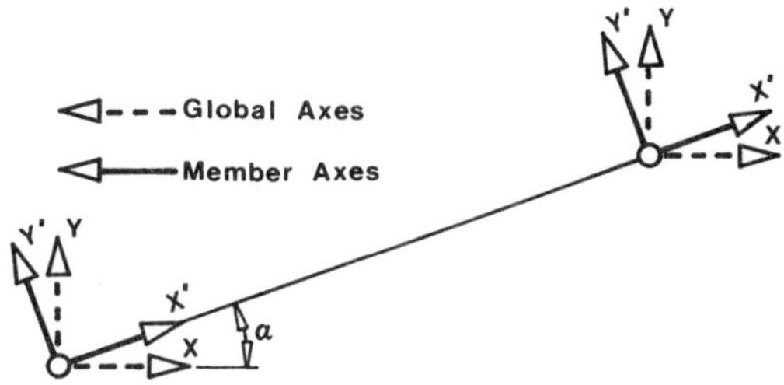

Figure 5.3 Truss element lying at an angle to the global *"x"* axis

Solution of the stiffness equation yields the nodal displacements in terms of the global axes, and these displacements must be transformed into the element axes system shown in fig 5.1 before equations (5.2) and (5.3) can be used to find the element end forces. Figure 5.3 shows a typical plane truss element. Note that α *is the anti-clockwise rotation of the global axes that will make them coincide with the element axes.*

Section 3.8 showed how the the transformation matrix T is used to transform force and displacement between two sets of orthogonal, right-handed axes systems sharing a common origin. In this text plane trusses are assumed to lie in "*x,y*" plane which means that the element and global "*z*" axes coincide and the required transformation from global to element axes is

$$\begin{bmatrix} x \\ y \\ z \end{bmatrix} = \begin{bmatrix} l_x & m_x & 0 \\ l_y & m_y & 0 \\ 0 & 0 & 1 \end{bmatrix} \begin{bmatrix} x' \\ y' \\ z' \end{bmatrix}$$

As there are no forces or displacements in the "*z*" direction the required transformation in two dimensions is

$$\begin{bmatrix} x \\ y \end{bmatrix} = \begin{bmatrix} l_x & m_x \\ l_y & m_y \end{bmatrix} \begin{bmatrix} x' \\ y' \end{bmatrix}$$

$$l_x = \cos\alpha$$

$$m_x = \cos(90 - \alpha) = \sin\alpha$$

$$l_y = \cos(90 + \alpha) = -\sin\alpha$$

$$m_y = \cos\alpha$$

Hence the required transformation is

$$\boldsymbol{T} = \begin{bmatrix} \cos\alpha & \sin\alpha \\ -\sin\alpha & \cos\alpha \end{bmatrix}$$

The stiffness method yields displacements in the global axes axes system, and the above transformation matrix can be used to transform global displacements into the element axes systems as follows.

Node "i"

$$\begin{bmatrix} \delta_{ijx} \\ \delta_{ijy} \end{bmatrix} = \begin{bmatrix} \cos\alpha & \sin\alpha \\ -\sin\alpha & \cos\alpha \end{bmatrix} \begin{bmatrix} \Delta_{ix} \\ \Delta_{iy} \end{bmatrix}$$

i.e.

$$\boldsymbol{\delta}_{ij} = \boldsymbol{T}_{ij}\ \boldsymbol{\Delta}_i \qquad (5.4)$$

Node "j"

$$\begin{bmatrix} \delta_{jix} \\ \delta_{jiy} \end{bmatrix} = \begin{bmatrix} \cos\alpha & \sin\alpha \\ -\sin\alpha & \cos\alpha \end{bmatrix} \begin{bmatrix} \Delta_{jx} \\ \Delta_{jy} \end{bmatrix}$$

i.e.

$$\boldsymbol{\delta}_{ji} = \boldsymbol{T}_{ij}\ \boldsymbol{\Delta}_j \qquad (5.5)$$

where

$\boldsymbol{T}_{ij}$ is the transformation matrix for element "*i,j*".

δ_{ij} is the displacement vector at node "*i*" in the element axes system for element "*i,j*".

δ_{ji} is the displacement vector at node "*j*" in the element axes system for element "*i,j*".

Δ_i is the displacement vector at node "*i*" in the global axes system.

Δ_j is the displacement vector at node "*j*" in the global axes system.

The transformation matrix T_{ij} can also be used to transform forces, e.g.

$$\begin{bmatrix} f_{ijx} \\ f_{ijy} \end{bmatrix} = \begin{bmatrix} cos\alpha & sin\alpha \\ -sin\alpha & cos\alpha \end{bmatrix} \begin{bmatrix} F_{ijx} \\ F_{ijy} \end{bmatrix}$$

i.e.

$$f_{ij} = T_{ij} F_{ij} \tag{5.6}$$

and at end "*j*"

$$f_{ji} = T_{ij} F_{ji} \tag{5.7}$$

where

T_{ij} is the transformation matrix for element "*i,j*".

f_{ij} is the force vector at end "*i*" of element "*i,j*" in the element axes system.

f_{ji} is the force vector at end "*j*" of element "*i,j*" in the element axes system.

F_{ij} is the force vector at end "*i*" of element "*i,j*" in the global axes system.

F_{ji} is the force vector at end "*j*" of element "*i,j*" in the global axes system.

Note that in global axes the nodal displacement components have only two subscripts, yet the force components have three subscripts. This is because it is necessary to be able refer to the force components from different elements.

The reverse transformation (i.e. from element axes to global axes) is effected by the inverse of T_{ij}. Say that the displacements at node "*i*" are required in the global axes system, but are known in the element axes system for element "*i,j*". Then multiplying both sides of equation (5.4) yields

$$T_{ij}^{-1} \delta_{ij} = T_{ij}^{-1} T_{ij} \Delta_i$$

hence

$$\Delta_i = T_{ij}^{-1} \delta_{ij}$$

5.4 The Element Stiffness Matrix in Global Axes

In section 5.2 the relationship between the forces at end "*i*" of element "*i,j*" and the displacements at its ends was shown to be

$$f_{ij} = k^j_{ii}\,\delta_{ij} + k_{ij}\,\delta_{ji}$$

Section 5.3 showed that the inverse of the transformation matrix T^{-1}_{ij} could transform element end forces from the element axes system used in the above equation to the global axes system. Hence the element end forces can be expressed in terms of the global axes system as follows

$$\begin{aligned} F_{ij} &= T^{-1}_{ij}\,f_{ij} \\ &= T^{-1}_{ij}\,k^j_{ii}\,\delta_{ij} + T^{-1}_{ij}\,k_{ij}\,\delta_{ji} \end{aligned} \qquad (5.8)$$

When generating the structure stiffness equation the forces and displacements at each node must be expressed in a common axes system. Consequently the element end displacements, which have hitherto been expressed in the element axes system, must be expressed in the nodal axes system using the transformations given by equations (5.4) and (5.5)

$$\delta_{ij} = T_{ij}\,\Delta_i \qquad \text{and} \qquad \delta_{ji} = T_{ij}\,\Delta_j$$

Substituting the preceding equation into equation (5.8)

$$F_{ij} = T^{-1}_{ij}\,k^j_{ii}\,T_{ij}\,\Delta_i + T^{-1}_{ij}\,k_{ij}\,T_{ij}\,\Delta_j$$

or

$$F_{ij} = K^j_{ii}\,\Delta_i + K_{ij}\,\Delta_j \qquad (5.9)$$

where

$$K^j_{ii} = T^{-1}_{ij}\,k^j_{ii}\,T_{ij}$$

and

$$K_{ij} = T^{-1}_{ij}\,k_{ij}\,T_{ij}$$

The matrix product $K^j_{ii}\,\Delta_i$ gives the forces (in the global axes system) at end "*i*" of element "*i,j*" associated with displacements at end "*i*" (also in the global axes system). Similarly the matrix product $K_{ij}\,\Delta_j$ gives the element end forces at end "*i*" of element "*i,j*" associated with displacements at end "*j*". The matrices $K^j_{ii}\,\Delta_i$ and K_{ij} are known as global element stiffness submatrices. It is computationally advantageous to expand these matrices as follows

$$\begin{aligned} K^j_{ii} &= T^{-1}_{ij}\,k^j_{ii}\,T_{ij} \\ &= \begin{bmatrix} \cos\alpha & -\sin\alpha \\ \sin\alpha & \cos\alpha \end{bmatrix} \begin{bmatrix} EA/L & 0 \\ 0 & 0 \end{bmatrix} \begin{bmatrix} \cos\alpha & \sin\alpha \\ -\sin\alpha & \cos\alpha \end{bmatrix} \end{aligned}$$

$$= \frac{EA}{L}\begin{bmatrix} cos^2\alpha & cos\alpha\ sin\alpha \\ cos\alpha\ sin\alpha & sin^2\alpha \end{bmatrix} \tag{5.10}$$

similarly

$$K_{ij} = -\frac{EA}{L}\begin{bmatrix} cos^2\alpha & cos\alpha\ sin\alpha \\ cos\alpha\ sin\alpha & sin^2\alpha \end{bmatrix} \tag{5.11}$$

Similar expressions can be developed for end "j"

$$F_{ji} = T^{-1}_{ij}\, f_{ji}$$

$$F_{ji} = T^{-1}_{ij}\, k_{ji}\, T_{ij}\, \Delta_i + T^{-1}_{ij}\, k^{i}_{jj}\, T_{ij}\, \Delta_j$$

or

$$F_{ji} = K_{ji}\, \Delta_i + K^{i}_{jj}\, \Delta_j$$

where

$$K^{i}_{jj} = T^{-1}_{ij}\, k^{i}_{jj}\, T_{ij}$$

$$= \frac{EA}{L}\begin{bmatrix} cos^2\alpha & cos\alpha\ sin\alpha \\ cos\alpha\ sin\alpha & sin^2\alpha \end{bmatrix} \tag{5.12}$$

similarly

$$K_{ji} = -\frac{EA}{L}\begin{bmatrix} cos^2\alpha & cos\alpha\ sin\alpha \\ cos\alpha\ sin\alpha & sin^2\alpha \end{bmatrix} \tag{5.13}$$

Equations (5.10) to (5.13) show that only the K^{j}_{ii} submatrix needs to be evaluated as the other submatrices can be obtained by multiplying the corresponding elements of K^{j}_{ii} as follows

Find K_{ij} and K_{ji} by multiplying corresponding elements of K^{j}_{ii} by

$$\begin{bmatrix} -1 & -1 \\ -1 & -1 \end{bmatrix} \tag{5.14}$$

Find K^{i}_{jj} by multiplying corresponding elements of K^{j}_{ii} by

$$\begin{bmatrix} 1 & 1 \\ 1 & 1 \end{bmatrix} \tag{5.15}$$

Note that this is not a matrix multiplication, but simply the multiplication of corresponding elements.

5.5 Nodal Equilibrium

Figure 5.4(a) shows a simple triangulated framework, and fig 5.4(b) shows how it might be idealised.

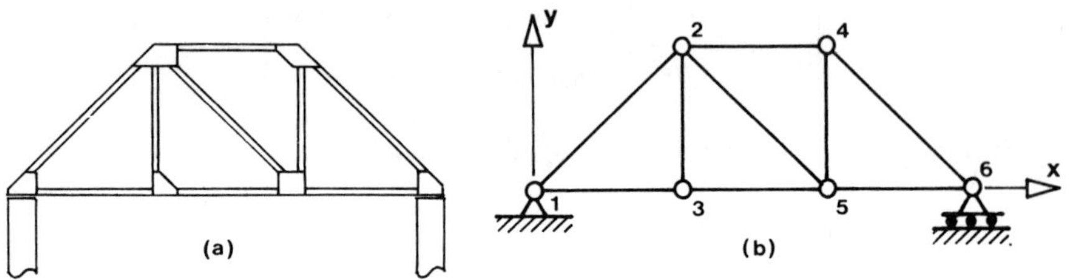

Figure 5.4 A simple triangulated framework

If node 2 is removed from the truss its freebody diagram is as shown in fig 5.5(a), and if the node is in equilibrium then there can be no net force acting upon it. Resolving each element force into components as shown in fig 5.5(b), and summing horizontal and vertical forces acting on the node produces the following equations

$$P_{2x} + F'_{21x} + F'_{23x} + F'_{24x} + F'_{25x} = 0$$

and

$$P_{2y} + F'_{21y} + F'_{23y} + F'_{24y} + F'_{25y} = 0$$

or

$$\mathbf{P}_2 + \mathbf{F}'_{21} + \mathbf{F}'_{23} + \mathbf{F}'_{24} + \mathbf{F}'_{25} = \mathbf{0}$$

where

$$\mathbf{P}_2 = \begin{bmatrix} P_{2x} \\ P_{2y} \end{bmatrix} \quad \mathbf{F}'_{21} = \begin{bmatrix} F'_{21x} \\ F'_{21y} \end{bmatrix} \quad \cdots\cdots \quad \mathbf{F}'_{25} = \begin{bmatrix} F'_{25x} \\ F'_{25y} \end{bmatrix}$$

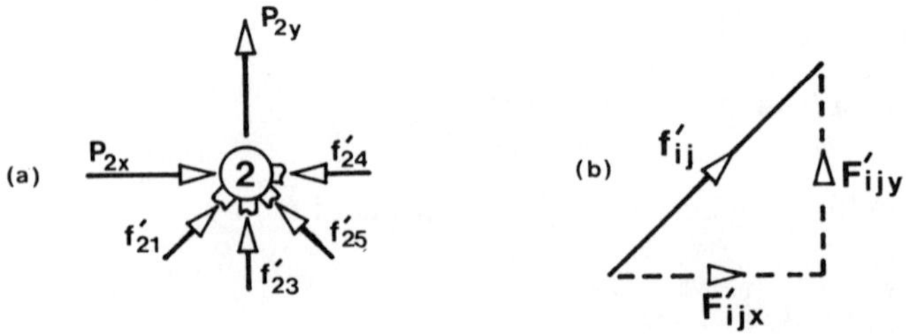

Figure 5.5 Freebody diagram for node 2

Hence, the equilibrium equation for node "*i*", which has elements "*i,a*", "*i,b*", "*i,c*", ... "*i,n*" framing into it, is

$$P_i + F'_{ia} + F'_{ib} + F'_{ic} + \ldots F'_{in} = 0$$

One of the fundamental principles of static equilibrium is that for every action there must be an equal and opposite reaction. Consequently the force exerted by an element on a node is equal and opposite to the force the node exerts upon that element. Figure 5.6 illustrates this effect.

$$f_{21} = -f'_{21} \qquad f_{23} = -f'_{23}$$
$$f_{24} = -f'_{24} \qquad f_{25} = -f'_{25}$$

Expressing the forces as components in the global axes system

$$F_{21x} = -F'_{21x} \quad \ldots \quad F_{25x} = -F'_{25x}$$
$$F_{21y} = -F'_{21y} \quad \ldots \quad F_{25y} = -F'_{25y}$$

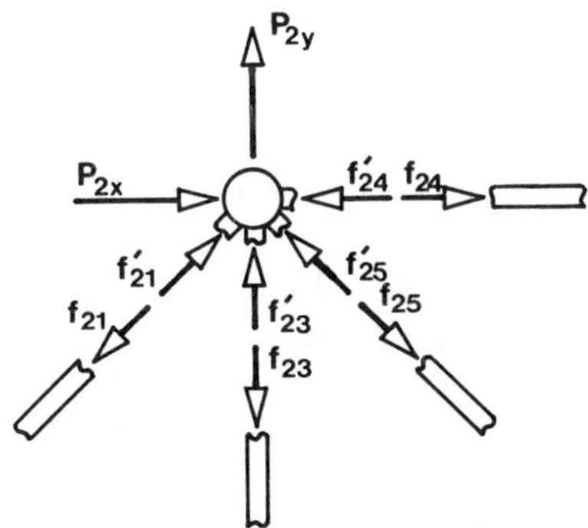

Figure 5.6 Nodal forces and element end forces

and the equation of equilibrium for node 2 becomes

$$P_2 = F_{21} + F_{23} + F_{24} + F_{25}$$

For node "i" with elements "i,a", "i,b", "i,c", ... "i,n" framing into it, the above equation becomes

$$P_i = F_{ia} + F_{ib} + F_{ic} + \ldots + F_{in} \tag{5.16}$$

Equation (5.16) simply states that the external applied load vector P_i at node "i" must be balanced by the vectors of internal element end forces F_{ia}, F_{ib}, F_{ic}, ... F_{in}. Obviously for the summation to be meaningful the force components must be described in a consistent axes system (see section 3.6). Equation (5.9) gives the relationship between element end forces and the nodal displacements in the gobal axes system

$$F_{ij} = K^j_{ii}\,\Delta_i + K_{ij}\,\Delta_j$$

where

$$K^j_{ii} = T^{-1}_{ij}\,k^j_{ii}\,T_{ij}$$

and

$$K_{ij} = T^{-1}_{ij} \, k_{ij} \, T_{ij}$$

Substituting for the element end forces in equation (5.16) gives

$$P_i = K^a_{ii} \, \Delta_i + K_{ia} \, \Delta_a + K^b_{ii} \, \Delta_i + K_{ib} \, \Delta_b + \, ... \, + K^n_{ii} \, \Delta_i + K_{in} \, \Delta_n$$

or

$$P_i = K_{ii} \, \Delta_i + K_{ia} \, \Delta_a + K_{ib} \, \Delta_b + \, ... \, + K_{in} \, \Delta_n \qquad (5.17)$$

where

$$K_{ii} = \sum_{j=a}^{n} K^j_{ii}$$

5.6 The Initial Structure Stiffness Matrix

In section 5.5 the equation of nodal equilibrium was shown to be

$$P_i = K_{ii} \, \Delta_i + K_{ia} \, \Delta_a + K_{ib} \, \Delta_b + \, ... \, + K_{in} \, \Delta_n$$

The nodal displacement vectors Δ_1, Δ_2, ... Δ_n, consist of both known and unknown displacements (i.e. the boundary conditions and the unknown nodal displacements). Similarly the nodal force vectors P_1, P_2, ... P_n will contain known and unknown forces (i.e. the known applied loads and the unknown reactive forces). Hence it is evident that the above equation takes no account of the boundary conditions.

Presentation of the remainder of the theory in this section is simplified if the boundary conditions are not applied until after the structure stiffness matrix has been assembled. Prior to the application of the boundary conditions the equations of nodal equilibrium for all nodes comprise a system of simultaneous equations which is collectively known as *the initial structure stiffness equation.*

$$\begin{bmatrix} P_1 \\ P_2 \\ \cdot \\ \cdot \\ \cdot \\ P_n \end{bmatrix} = \begin{bmatrix} K_{11} & K_{12} & K_{13} & .... & K_{1n} \\ K_{21} & K_{22} & K_{23} & .... & K_{2n} \\ \cdot & & & & \cdot \\ \cdot & & & & \cdot \\ \cdot & & & & \cdot \\ K_{n1} & K_{n2} & K_{n3} & .... & K_{nn} \end{bmatrix} \begin{bmatrix} \Delta_1 \\ \Delta_2 \\ \cdot \\ \cdot \\ \cdot \\ \Delta_n \end{bmatrix}$$

The above system of simultaneous equations simply expresses the equation of nodal equilibrium for all nodes in matrix form. Note that the equations are written in terms of submatrices where the vectors of nodal force and displacement are of length 2 and each stiffness submatrix is 2 x 2 square matrix. Hence the dimension of the initial structure stiffness matrix is two times the number of nodes. The procedure to assemble the initial structure stiffness matrix is as follows

1. Set all elements of the structure stiffness matrix equal to zero.

2. For element "*i,j*" calculate the K_{ii}^{j} using equation (5.10) and then use equations (5.14) and (5.15) to determine the K_{jj}^{i}, K_{ij} and K_{ji} stiffness submatrices.

3. Add these stiffness submatrices to the following locations in the initial structure stiffness matrix

$$K_{ii}^{j} \text{ add to locations } \begin{bmatrix} (2i\text{-}1)(2i\text{-}1) & (2i\text{-}1)(2i) \\ (2i)(2i\text{-}1) & (2i,2i) \end{bmatrix}$$

$$K_{ij} \text{ add to locations } \begin{bmatrix} (2i\text{-}1)(2j\text{-}1) & (2i\text{-}1)(2j) \\ (2i)(2j\text{-}1) & (2i,2j) \end{bmatrix}$$

$$K_{ji} \text{ add to locations } \begin{bmatrix} (2j\text{-}1)(2i\text{-}1) & (2j\text{-}1)(2i) \\ (2j)(2i\text{-}1) & (2j,2i) \end{bmatrix}$$

$$K_{jj}^{i} \text{ add to locations } \begin{bmatrix} (2j\text{-}1)(2j\text{-}1) & (2j\text{-}1)(2j) \\ (2j)(2j\text{-}1) & (2j,2j) \end{bmatrix}$$

4. Repeat steps 2 and 3 for all other elements.

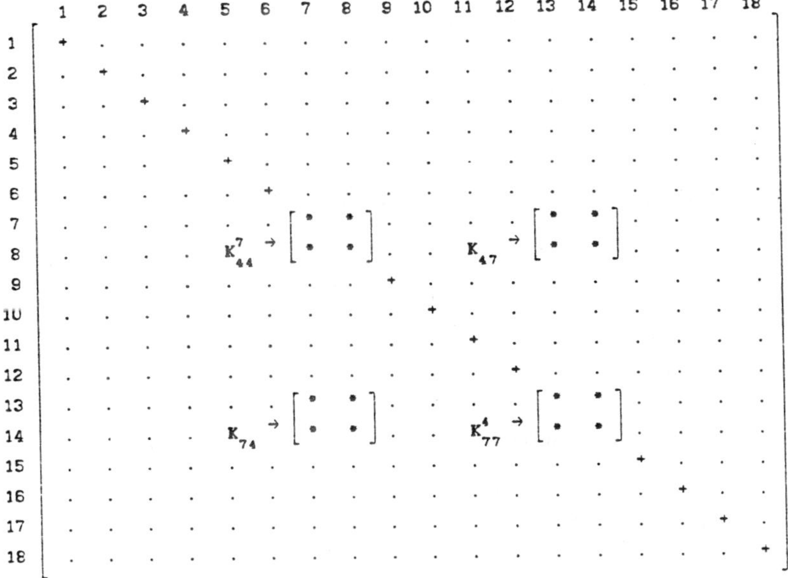

Figure 5.7 Adding element 4,7 to the initial structure stiffness matrix

Figure 5.7 shows the locations where the global element stiffness submatrices for element 4,7 would be added to the initial structure stiffness matrix. Note that the initial structure stiffness matrix is 18 x 18 indicating that the structure has 9 nodes.

Example 5.1

Assemble the initial structure stiffness equation, in terms of submatrices, for the truss shown below.

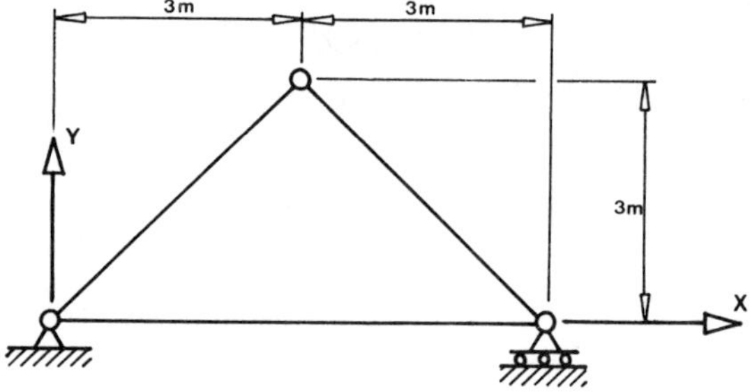

First remove the boundary conditions, leaving the structure completely unrestrained.

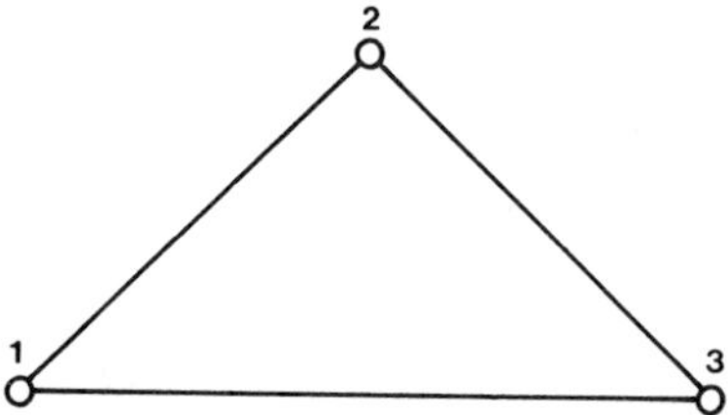

Zero the structure stiffness matrix (note that each zero represents a 2 x 2 submatrix of zeros).

$$\begin{bmatrix} 0 & 0 & 0 \\ 0 & 0 & 0 \\ 0 & 0 & 0 \end{bmatrix}$$

Add the element stiffness submatrices for element 1,2

$$\begin{bmatrix} K_{11} & K_{12} & 0 \\ K_{21} & K_{22} & 0 \\ 0 & 0 & 0 \end{bmatrix} \qquad \text{where} \quad \begin{aligned} K_{11} &= K_{11}^{2} \\ K_{22} &= K_{22}^{1} \end{aligned}$$

(note that the superscripts, taken in conjunction with subscripts, identify the element under consideration).

Add the element stiffness submatrices for element 1,3

$$\begin{bmatrix} K_{11} & K_{12} & K_{13} \\ K_{21} & K_{22} & 0 \\ K_{31} & 0 & K_{33} \end{bmatrix} \quad \text{where} \quad \begin{aligned} K_{11} &= K_{11}^{2} + K_{11}^{3} \\ K_{22} &= K_{22}^{1} \\ K_{33} &= K_{33}^{1} \end{aligned}$$

Add the element stiffness submatrices for member 2,3

$$\begin{bmatrix} K_{11} & K_{12} & K_{13} \\ K_{21} & K_{22} & K_{23} \\ K_{31} & K_{32} & K_{33} \end{bmatrix} \quad \text{where} \quad \begin{aligned} K_{11} &= K_{11}^{2} + K_{11}^{3} \\ K_{22} &= K_{22}^{1} + K_{22}^{3} \\ K_{33} &= K_{33}^{1} + K_{33}^{2} \end{aligned}$$

Hence the initial structure stiffness equation, in terms of submatrices, is

$$\begin{bmatrix} K_{11} & K_{12} & K_{13} \\ K_{21} & K_{22} & K_{23} \\ K_{31} & K_{32} & K_{33} \end{bmatrix} \begin{bmatrix} \Delta_1 \\ \Delta_2 \\ \Delta_3 \end{bmatrix} = \begin{bmatrix} P_1 \\ P_2 \\ P_3 \end{bmatrix}$$

Note that the initial structure stiffness matrix is singular (i.e. there is no unique solution to the initial stiffness relationship). This is because until the structure is adequately restrained there are an infinite number of possible equilibrium configurations.

Take, for example, the truss analysed in the previous example (example 5.1). Imagine the truss to be floating, completely unrestrained and unloaded, at some location in space. Now apply the loads and the reactions necessary for equilibrium. Under the action of these forces the truss would assume its true displaced shape. However, because the restraints have not been specified the truss and its loads could be moved to any other point in space without disturbing its equilibrium. Hence, until a structure is adequately restrained it can undergo rigid body translation without disturbing its equilibrium. This condition is characterised by a singular structure stiffness matrix. Note also that the element stiffness matrix is always singular because it is, in effect, the initial structure stiffness matrix for a single element structure.

Example 5.2

If the frame considered in example 5.1 is constructed from steel angle sections having a cross-sectional area of 3000 mm^2, numerically evaluate the initial structure stiffness matrix. Use 200 kN/mm^2 as the elastic constant for steel. As the structure has three nodes the initial structure stiffness matrix will be 6 x 6. The initial structure stiffness matrix is first set to zero, and then the element stiffness submatrices are added for each element in turn.

Element 1,2 $A = 3000$ mm^2, $L = 4243$ mm, $E = 200$ kN/mm^2, $\alpha = 45°$

$$K_{11}^2 = T_{12}^{-1} k_{11}^2 T_{12}$$

$$= \frac{EA}{L}\begin{bmatrix} cos^2\alpha & cos\alpha\, sin\alpha \\ cos\alpha\, sin\alpha & sin^2\alpha \end{bmatrix}$$

$$= \begin{bmatrix} 70.7 & 70.7 \\ 70.7 & 70.7 \end{bmatrix}$$

Adding K_{11}^2, K_{12} and K_{22}^1 to the structure stiffness matrix gives

$$\left[\begin{array}{cc|cc|cc} 70.7 & 70.7 & -70.7 & -70.7 & 0 & 0 \\ & 70.7 & -70.7 & -70.7 & 0 & 0 \\ \hline & & 70.7 & 70.7 & 0 & 0 \\ & \text{symmetric} & & 70.7 & 0 & 0 \\ \hline & & & & 0 & 0 \\ & & & & & 0 \end{array}\right]$$

Note that K_{21} does not have to be added because the structure stiffness matrix is known to be symmetric, and that K_{12} and K_{22}^1 were found from K_{11}^2 using the relationships developed in section 5.4.

Element 1,3 $A = 3000$ mm^2, $L = 6000$ mm, $E = 200$ kN/mm^2, $\alpha = 0°$

$$K_{11}^2 = \begin{bmatrix} 100.0 & 0 \\ 0 & 0 \end{bmatrix}$$

Adding K_{11}^3, K_{13} and K_{33}^1 to the structure stiffness matrix gives

$$\left[\begin{array}{cc|cc|cc} 170.7 & 70.7 & -70.7 & -70.7 & -100.0 & 0 \\ & 70.7 & -70.7 & -70.7 & 0 & 0 \\ \hline & & 70.7 & 70.7 & 0 & 0 \\ & \text{symmetric} & & 70.7 & 0 & 0 \\ \hline & & & & 100.0 & 0 \\ & & & & & 0 \end{array}\right]$$

Element 2,3 $A = 3000 \text{ mm}^2$, $L = 4243$ mm, $E = 200 \text{ kN/mm}^2$, $\alpha = 315°$

$$K_{22}^{3} = \begin{vmatrix} 70.7 & -70.7 \\ -70.7 & 70.7 \end{vmatrix}$$

Adding K_{22}^{3}, K_{23} and K_{33}^{2} to the structure stiffness matrix gives

$$\begin{vmatrix} 170.7 & 70.7 & -70.7 & -70.7 & -100.0 & 0 \\ & 70.7 & -70.7 & -70.7 & 0 & 0 \\ & & 141.4 & 0 & -70.7 & 70.7 \\ & \text{symmetric} & & 141.4 & 70.7 & -70.7 \\ & & & & 170.0 & -70.7 \\ & & & & & 70.7 \end{vmatrix}$$

The calculation of element stiffness submatrices is a tedious and error prone procedure that is best done with the aid of a computer.

Program ESTIFFPT.BAS

This program calculates the element stiffness matrix for a plane truss element in terms of the element and the global axes systems. The program loops for each element in turn and requests the following data.

- A the cross-sectional area of the element
- L the length of the element
- E the elastic constant for the element
- α the angle that the element makes with the global "x" axis defined as shown

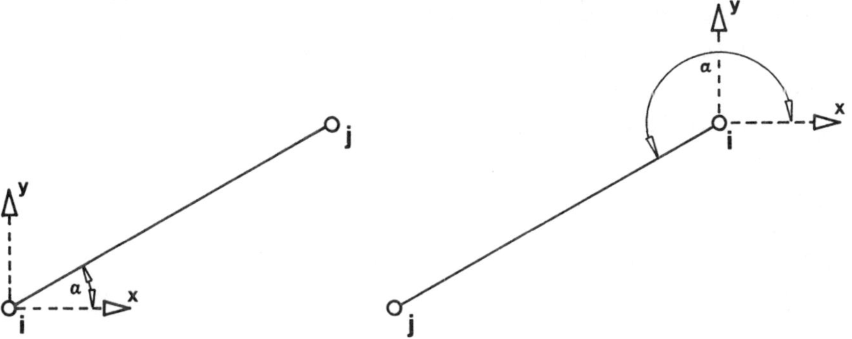

"A", "L", and "E" must be in *consistent* units, and the unit of force used used for "E" must correspond to the unit of force used for the applied loading. "α"

is in degrees. The program loops until it receives a zero for the lower node number.

Output

The input is echoed and the element stiffness matrix is then output, first in terms of the element axes system, and then in terms of the global axes system, as illustrated in the sample run that follows the program listing.

Listing

```
1000 '========================      ESTIFFPT.BAS      ===========================
1010 '
1020 OPTION BASE 1
1030 DIM ESTIFF(4,4)
1040 CLS
1050 PRINT
1060 PRINT "=================================================================="
1070 PRINT
1080 PRINT "PROGRAM ESTIFFPT.BAS                       Copyright (c) J.Balfour 1991"
1090 PRINT
1100 PRINT "    Program for the calculation of element stiffness matrix"
1110 PRINT "      in element and global axes for a plane truss elements"
1120 PRINT
1130 PRINT "                 For further information contact"
1140 PRINT " James A.D.Balfour, Heriot-Watt University, Riccarton, Edinburgh"
1150 PRINT "               Tel 031-449-5111, Fax 031-451-3170"
1160 PRINT
1170 PRINT "=================================================================="
1180 '
1190 IN% = 1 : K% = 1
1200 WHILE IN% <> 0
1210   PRINT
1220   PRINT "ELEMENT "; K%
1230   PRINT
1240   INPUT "Lower node number (0 to stop) = ", IN%
1250   IF IN%=0 THEN GOTO 1400
1260   INPUT "Higher node number            = ", JN%
1262   PRINT
1264   PRINT        "+ + + + + + + + + + + +"
1266   PRINT USING "+   ELEMENT ## ##    +"; IN%, JN%
1268   PRINT        "+ + + + + + + + + + + +"
1269   PRINT
1270   INPUT "Area (A)                     = ", A
1280   INPUT "Length (L)                   = ", L
1290   INPUT "Elastic constant (E)         = ", E
1300   INPUT "Angle with the global x-axis  = ", ALPHA
1310   '
1320   ' ... Calculate the element stiffness in the element axes system
1330   '
1340   GOSUB 1430
1350   '
1360   ' ... Calculate the element stiffness in the global axes system
1370   '
1380   GOSUB 1740
1390   K% = K% + 1
1400 WEND
1410 END
1420 '
1430 '********     ELEMENT STIFFNESS IN THE ELEMENT AXES SYSTEM     **********
1440 '
1450 FOR I% = 1 TO 4
1460   FOR J% = 1 TO 4
1470     ESTIFF(I%,J%) = 0
1480   NEXT J%
1490 NEXT I%
1500 '
1510 ' ... Calculate the element stiffness terms (upper triangle only)
1520 '
```

```
1530 ESTIFF(1,1) =   A * E / L
1540 ESTIFF(1,3) = - ESTIFF(1,1)
1550 ESTIFF(3,3) =   ESTIFF(1,1)
1560 '
1570 ' ... Output the element stiffness matrix
1580 '
1590 PRINT
1600 PRINT "* * * * * * * * * * * * * * * * * * * **"
1610 PRINT "*   ELEMENT STIFFNESS MATRIX IN    *"
1620 PRINT "*   THE MEMBER AXES SYSTEM         *"
1630 PRINT "* * * * * * * * * * * * * * * * * * * **"
1640 PRINT
1650 FOR I% = 1 TO 4
1660   FOR J% = 1 TO 4
1670     IF I% <= J% THEN PRINT USING "#.####^^^^   "; ESTIFF(I%,J%);
1680     IF I% >  J% THEN PRINT USING "#.####^^^^   "; ESTIFF(J%,I%);
1690   NEXT J%
1700   PRINT
1710 NEXT I%
1720 RETURN
1730 '
1740 '**********    ELEMENT STIFFNESS IN THE GLOBAL AXES SYSTEM    **********
1750 '
1760 ' ... Convert the angle to radians
1770 '
1780 AR = 4 * ATN(1) * ALPHA / 180
1790 C  = COS(AR)
1800 S  = SIN(AR)
1810 FOR I% = 1 TO 4
1820   FOR J% = 1 TO 4
1830     ESTIFF(I%,J%) = 0
1840   NEXT J%
1850 NEXT I%
1860 '
1870 ' ... Calculate the element stiffness terms (upper triangle only)
1880 '
1890 ESTIFF(1,1) =   E * A * C * C / L
1900 ESTIFF(1,2) =   E * A * C * S / L
1910 ESTIFF(1,3) = - ESTIFF(1,1)
1920 ESTIFF(1,4) = - ESTIFF(1,2)
1930 ESTIFF(2,2) =   E * A * S * S / L
1940 ESTIFF(2,3) = - ESTIFF(1,2)
1950 ESTIFF(2,4) = - ESTIFF(2,2)
1960 ESTIFF(3,3) =   ESTIFF(1,1)
1970 ESTIFF(3,4) =   ESTIFF(1,2)
1980 ESTIFF(4,4) =   ESTIFF(2,2)
1990 '
2000 ' ... Output the element stiffness matrix
2010 '
2020 PRINT
2030 PRINT "* * * * * * * * * * * * * * * * * * * **"
2040 PRINT "*   ELEMENT STIFFNESS MATRIX IN    *"
2050 PRINT "*   THE GLOBAL AXES SYSTEM         *"
2060 PRINT "* * * * * * * * * * * * * * * * * * * **"
2070 PRINT
2080 FOR I% = 1 TO 4
2090   FOR J% = 1 TO 4
2100     IF I% <= J% THEN PRINT USING "#.####^^^^   "; ESTIFF(I%,J%);
2110     IF I% >  J% THEN PRINT USING "#.####^^^^   "; ESTIFF(J%,I%);
2120   NEXT J%
2130   PRINT
2140 NEXT I%
2150 RETURN
2160 '
2170 '=======================    ESTIFFPT.BAS    ============================
```

Sample Run

The following run demonstrates program ESTIFFPT.BAS being used to find the element stiffness matrices for the elements of the structure analysed in example 5.1. Input from the keyboard is shown underlined.

```
==================================================================

PROGRAM ESTIFFPT.BAS                    Copyright (c) J.Balfour 1991

    Program for the calculation of element stiffness matrix
      in element and global axes for a plane truss elements

                 For further information contact
 James A.D.Balfour, Heriot-Watt University, Riccarton, Edinburgh
             Tel 031-449-5111, Fax 031-451-3170

==================================================================

ELEMENT  1

Lower node number (0 to stop) = 1
Higher node number            = 2

+ + + + + + + + + + +
+   ELEMENT  1  2   +
+ + + + + + + + + + +

Area (A)                      = 3000
Length (L)                    = 4243
Elastic constant (E)          = 200
Angle with the global x-axis  = 45

* * * * * * * * * * * * * * * * * *
*   ELEMENT STIFFNESS MATRIX IN   *
*   THE MEMBER AXES SYSTEM        *
* * * * * * * * * * * * * * * * * *

0.1414E+03   0.0000E+00   -.1414E+03   0.0000E+00
0.0000E+00   0.0000E+00   0.0000E+00   0.0000E+00
-.1414E+03   0.0000E+00   0.1414E+03   0.0000E+00
0.0000E+00   0.0000E+00   0.0000E+00   0.0000E+00

* * * * * * * * * * * * * * * * * *
*   ELEMENT STIFFNESS MATRIX IN   *
*   THE GLOBAL AXES SYSTEM        *
* * * * * * * * * * * * * * * * * *

0.7070E+02   0.7070E+02   -.7070E+02   -.7070E+02
0.7070E+02   0.7070E+02   -.7070E+02   -.7070E+02
-.7070E+02   -.7070E+02   0.7070E+02   0.7070E+02
-.7070E+02   -.7070E+02   0.7070E+02   0.7070E+02

ELEMENT  2

Lower node number (0 to stop) = 1
Higher node number            = 3

+ + + + + + + + + + +
+   ELEMENT  1  3   +
+ + + + + + + + + + +

Area (A)                      = 3000
Length (L)                    = 6000
Elastic constant (E)          = 200
Angle with the global x-axis  = 0

* * * * * * * * * * * * * * * * * *
*   ELEMENT STIFFNESS MATRIX IN   *
*   THE MEMBER AXES SYSTEM        *
* * * * * * * * * * * * * * * * * *

0.1000E+03   0.0000E+00   -.1000E+03   0.0000E+00
0.0000E+00   0.0000E+00   0.0000E+00   0.0000E+00
-.1000E+03   0.0000E+00   0.1000E+03   0.0000E+00
0.0000E+00   0.0000E+00   0.0000E+00   0.0000E+00

* * * * * * * * * * * * * * * * * *
*   ELEMENT STIFFNESS MATRIX IN   *
*   THE GLOBAL AXES SYSTEM        *
* * * * * * * * * * * * * * * * * *

0.1000E+03   0.0000E+00   -.1000E+03   0.0000E+00
```

```
0.0000E+00   0.0000E+00   0.0000E+00   0.0000E+00
-.1000E+03   0.0000E+00   0.1000E+03   0.0000E+00
0.0000E+00   0.0000E+00   0.0000E+00   0.0000E+00

ELEMENT  3

Lower node number (0 to stop) = 2
Higher node number            = 3

+ + + + + + + + + + +
+   ELEMENT  2  3   +
+ + + + + + + + + + +

Area (A)                      = 3000
Length (L)                    = 4243
Elastic constant (E)          = 200
Angle with the global x-axis  = 315

* * * * * * * * * * * * * * * * * *
*   ELEMENT STIFFNESS MATRIX IN   *
*   THE MEMBER AXES SYSTEM        *
* * * * * * * * * * * * * * * * * *

0.1414E+03   0.0000E+00   -.1414E+03   0.0000E+00
0.0000E+00   0.0000E+00   0.0000E+00   0.0000E+00
-.1414E+03   0.0000E+00   0.1414E+03   0.0000E+00
0.0000E+00   0.0000E+00   0.0000E+00   0.0000E+00

* * * * * * * * * * * * * * * * * *
*   ELEMENT STIFFNESS MATRIX IN   *
*   THE GLOBAL AXES SYSTEM        *
* * * * * * * * * * * * * * * * * *

0.7070E+02   -.7070E+02   -.7070E+02   0.7070E+02
-.7070E+02   0.7070E+02   0.7070E+02   -.7070E+02
-.7070E+02   0.7070E+02   0.7070E+02   -.7070E+02
0.7070E+02   -.7070E+02   -.7070E+02   0.7070E+02

ELEMENT  4

Lower node number (0 to stop) = 0
```

Program ISTIFFPT.BAS

This program illustrates how the program ESTIFFPT.BAS can be developed to produce a program which will automatically assemble the initial structure stiffness matrix.

Input

The program loops requesting data for each element in turn. Once the data for an element has been input the global element stiffness matrix is calculated and added to the initial structure stiffness matrix and this procedure is repeated until all elements have been processed.

Comments on the Algorithm

Only the upper triangle of the element stiffness matrices is calculated and only the upper triangle of the initial structure stiffness matrix is assembled.

Output

The symmetry of the initial structure stiffness matrix is exploited to allow the complete matrix to be output.

Listing

```
1000 '=======================    ISTIFFPT.BAS    ============================
1010 '
1020 ' ... Set maximum number of nodes
1030 '
1040 NNODE% = 20 : TRUE% = 1 : FALSE% = 0
1050 OPTION BASE 1
1060 DIM K(2*NNODE%,2*NNODE%), ESTIFF(4,4)
1070 CLS
1080 PRINT
1090 PRINT "=================================================================="
1100 PRINT
1110 PRINT "PROGRAM ISTIFFPT.BAS                    Copyright (c) J.Balfour 1991"
1120 PRINT
1130 PRINT "     Program for the calculation of element stiffness matrices"
1140 PRINT "        in element and global axes for a plane truss element"
1150 PRINT
1160 PRINT "                  For further information contact"
1170 PRINT " James A.D.Balfour, Heriot-Watt University, Riccarton, Edinburgh"
1180 PRINT "                Tel 031-449-5111, Fax 031-451-3170"
1190 PRINT
1200 PRINT "=================================================================="
1210 PRINT
1220 PRINT "Maximum number of nodes = "NNODE%
1230 '
1240 ' ... Zero the initial stiffness matrix
1250 '
1260 NDOF%  = 2 * NNODE%
1270 FOR I% = 1 TO NDOF%
1280   FOR J% = 1 TO NDOF%
1290     K(I%,J%) = 0
1300   NEXT J%
1310 NEXT I%
1320 '
1330 ' ... Input elements and add stiffness to the initial stiffness matrix
1340 '
1350 GOSUB 1420
1360 '
1370 ' ... Output the initial structure stiffness matrix
1380 '
1390 GOSUB 1920
1400 END
1410 '
1420 '********************    INPUT THE ELEMENT DATA    *********************
1430 '
1440 K% = 1 : NDOF% = 0 : DONE% = FALSE%
1450 WHILE DONE% = FALSE%
1460   PRINT
1470   PRINT "ELEMENT "; K%
1480   PRINT
1490   INPUT "Lower node number (0 to stop) = ", IN%
1500   IF IN% = 0 THEN DONE% = TRUE% : GOTO 1890
1510   INPUT "Higher node number            = ", JN%
1520   PRINT
1530   PRINT "+ + + + + + + + + + + + +"
1540   PRINT USING "+   ELEMENT ## ##   +"; IN%; JN%
1550   PRINT "+ + + + + + + + + + + + +"
1560   PRINT
1570   INPUT "Area (A)                      = ", A
1580   INPUT "Length (L)                    = ", L
1590   INPUT "Elastic constant (E)          = ", E
1600   INPUT "Angle with the global x-axis  = ", ALPHA
1610   IF 2*JN% > NDOF% THEN NDOF% = 2*JN%
1620   '
1630   ' ... Convert the angle to radians
1640   '
1650   AR = 4 * ATN(1) * ALPHA / 180
1660   C  = COS(AR)
1670   S  = SIN(AR)
1680   '
1690   ' ... Calculate the upper triangle of the element stiffness matrix
1700   '
1710   ESTIFF(1,1) = E * A * C * C / L
1720   ESTIFF(1,2) = E * A * C * S / L
1730   ESTIFF(2,2) = E * A * S * S / L
```

```
1740   '
1750   ' ... Add the element stiffness terms to the upper triangle of
1760   '     the structure stiffness matrix
1770   '
1780   K(2*IN%-1,2*IN%-1) =   K(2*IN%-1,2*IN%-1) + ESTIFF(1,1)
1790   K(2*IN%-1,2*IN%  ) =   K(2*IN%-1,2*IN%  ) + ESTIFF(1,2)
1800   K(2*IN%  ,2*IN%  ) =   K(2*IN%  ,2*IN%  ) + ESTIFF(2,2)
1810   K(2*IN%-1,2*JN%-1) = - ESTIFF(1,1)
1820   K(2*IN%-1,2*JN%  ) = - ESTIFF(1,2)
1830   K(2*IN%  ,2*JN%-1) = - ESTIFF(1,2)
1840   K(2*IN%  ,2*JN%  ) = - ESTIFF(2,2)
1850   K(2*JN%-1,2*JN%-1) =   K(2*JN%-1,2*JN%-1) + ESTIFF(1,1)
1860   K(2*JN%-1,2*JN%  ) =   K(2*JN%-1,2*JN%  ) + ESTIFF(1,2)
1870   K(2*JN%  ,2*JN%  ) =   K(2*JN%  ,2*JN%  ) + ESTIFF(2,2)
1880   K% = K% + 1
1890 WEND
1900 RETURN
1910 '
1920 '**********    OUTPUT THE INITIAL STRUCTURE STIFFNESS MATRIX    ********
1930 '
1940 PRINT
1950 PRINT "* * * * * * * * * * * * * * *"
1960 PRINT "*   INITIAL STRUCTURE     *"
1970 PRINT "*   STIFFNESS MATRIX      *"
1980 PRINT "* * * * * * * * * * * * * * *"
1990 PRINT
2000 FOR I% = 1 TO NDOF%
2010   FOR J% = 1 TO NDOF%
2020     IF I% <= J% THEN PRINT USING "#.####^^^^   "; K(I%,J%);
2030     IF I% >  J% THEN PRINT USING "#.####^^^^   "; K(J%,I%);
2040     IF J% MOD 6 = 0 THEN PRINT
2050   NEXT J%
2060   IF J% MOD 6 <> 0 THEN PRINT
2070   PRINT
2080 NEXT I%
2090 RETURN
2100 '
2110 '=======================    ISTIFFPT.BAS    ===========================
```

Sample Run

The following run demonstrates program ISTIFFPT.BAS being used to find the initial structure stiffness for the structure analysed in example 5.1. Input from the keyboard is shown underlined.

```
=================================================================

PROGRAM ISTIFFPT.BAS                     Copyright (c) J.Balfour 1991

     Program for the calculation of element stiffness matrices
        in element and global axes for a plane truss element

               For further information contact
 James A.D.Balfour, Heriot-Watt University, Riccarton, Edinburgh
           Tel 031-449-5111, Fax 031 451 3170

=================================================================

Maximum number of nodes =  20

ELEMENT  1

Lower node number (0 to stop) = 1
Higher node number            = 2

+ + + + + + + + + + + +
+   ELEMENT  1  2    +
+ + + + + + + + + + + +

Area (A)                      = 3000
Length (L)                    = 4243
Elastic constant (E)          = 200
Angle with the global x-axis  = 45
```

```
ELEMENT  2

Lower node number (0 to stop) = 1
Higher node number            = 3

+ + + + + + + + + + +
+   ELEMENT  1  3   +
+ + + + + + + + + + +

Area (A)                      = 3000
Length (L)                    = 6000
Elastic constant (E)          = 200
Angle with the global x-axis  = 0

ELEMENT  3

Lower node number (0 to stop) = 2
Higher node number            = 3

+ + + + + + + + + + +
+   ELEMENT  2  3   +
+ + + + + + + + + + +

Area (A)                      = 3000
Length (L)                    = 4243
Elastic constant (E)          = 200
Angle with the global x-axis  = 315

ELEMENT  4

Lower node number (0 to stop) = 0

* * * * * * * * * * * * *
*   INITIAL STRUCTURE   *
*   STIFFNESS MATRIX    *
* * * * * * * * * * * * *

0.1707E+03   0.7070E+02   -.7070E+02   -.7070E+02   -.1000E+03   0.0000E+00

0.7070E+02   0.7070E+02   -.7070E+02   -.7070E+02   0.0000E+00   0.0000E+00

-.7070E+02   -.7070E+02   0.1414E+03   -.7629E-05   -.7070E+02   0.7070E+02

-.7070E+02   -.7070E+02   -.7629E-05   0.1414E+03   0.7070E+02   -.7070E+02

-.1000E+03   0.0000E+00   -.7070E+02   0.7070E+02   0.1707E+03   -.7070E+02

0.0000E+00   0.0000E+00   0.7070E+02   -.7070E+02   -.7070E+02   0.7070E+02
```

5.7 Application of the Boundary Conditions

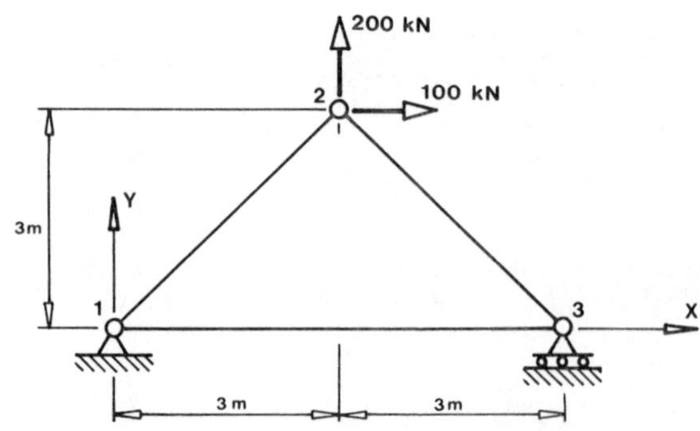

Figure 5.8 A three element truss

Prior to the application of the boundary conditions the initial structure stiffness equation is a mixed system of simultaneous equations (i.e. there are known and unknown quantities on each side of the equality). Consider, for example, the truss shown in fig 5.8.

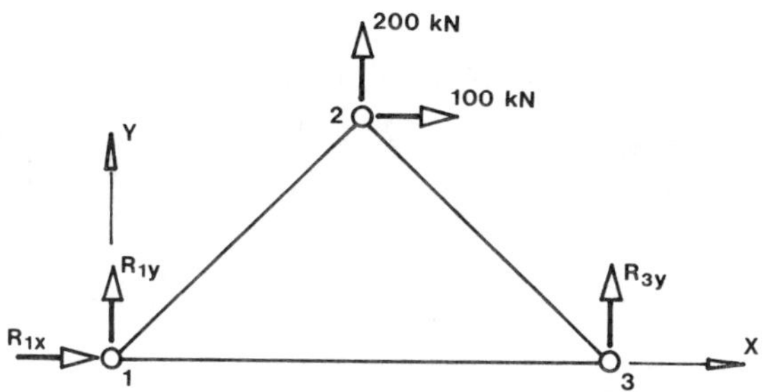

Figure 5.9 Freebody diagram for the complete structure

The freebody diagram for the complete structure is shown in fig 5.9. Inspection of fig. 5.9 shows that the force vector $\boldsymbol{P}$ contains the known applied forces and the unknown reactive forces. Further, the boundary conditions demand that $\Delta_{1x} = \Delta_{1y} = \Delta_{3y} = 0$. Hence the vectors of force and displacement both contain known and unknown quantities, the initial stiffness relationship is

$$\begin{bmatrix} K_{11} & K_{12} & K_{13} & K_{14} & K_{15} & K_{16} \\ K_{21} & K_{22} & K_{23} & K_{24} & K_{25} & K_{26} \\ K_{31} & K_{32} & K_{33} & K_{34} & K_{35} & K_{36} \\ K_{41} & K_{42} & K_{43} & K_{44} & K_{45} & K_{46} \\ K_{51} & K_{52} & K_{53} & K_{54} & K_{55} & K_{56} \\ K_{61} & K_{62} & K_{63} & K_{64} & K_{65} & K_{66} \end{bmatrix} \begin{bmatrix} \Delta_{1x} \\ \Delta_{1y} \\ \Delta_{2x} \\ \Delta_{2y} \\ \Delta_{3x} \\ \Delta_{3y} \end{bmatrix} = \begin{bmatrix} P_{1x} \\ P_{1y} \\ P_{2x} \\ P_{2y} \\ P_{3x} \\ P_{3y} \end{bmatrix} \tag{5.18}$$

Inserting the known displacements and applied loads gives

$$\begin{bmatrix} K_{11} & K_{12} & K_{13} & K_{14} & K_{15} & K_{16} \\ K_{21} & K_{22} & K_{23} & K_{24} & K_{25} & K_{26} \\ K_{31} & K_{32} & K_{33} & K_{34} & K_{35} & K_{36} \\ K_{41} & K_{42} & K_{43} & K_{44} & K_{45} & K_{46} \\ K_{51} & K_{52} & K_{53} & K_{54} & K_{55} & K_{56} \\ K_{61} & K_{62} & K_{63} & K_{64} & K_{65} & K_{66} \end{bmatrix} \begin{bmatrix} 0 \\ 0 \\ \Delta_{2x} \\ \Delta_{2y} \\ \Delta_{3x} \\ 0 \end{bmatrix} = \begin{bmatrix} R_{1x} \\ R_{2y} \\ 100 \\ 200 \\ 0 \\ R_{3y} \end{bmatrix}$$

As the order in which these equations are written is of no importance they can be rewritten as follows

$$\begin{bmatrix} K_{33} & K_{34} & K_{35} & K_{31} & K_{32} & K_{36} \\ K_{43} & K_{44} & K_{45} & K_{41} & K_{42} & K_{46} \\ K_{53} & K_{54} & K_{55} & K_{51} & K_{52} & K_{56} \\ K_{13} & K_{14} & K_{15} & K_{11} & K_{12} & K_{16} \\ K_{23} & K_{24} & K_{25} & K_{21} & K_{22} & K_{26} \\ K_{63} & K_{64} & K_{65} & K_{61} & K_{62} & K_{66} \end{bmatrix} \begin{bmatrix} \Delta_{2x} \\ \Delta_{2y} \\ \Delta_{3x} \\ 0 \\ 0 \\ 0 \end{bmatrix} = \begin{bmatrix} 100 \\ 200 \\ 0 \\ R_{1x} \\ R_{2y} \\ R_{3y} \end{bmatrix}$$

Or, in terms of the submatrices indicated by the broken lines

$$\begin{bmatrix} K_{I,I} & K_{I,II} \\ K_{II,I} & K_{II,II} \end{bmatrix} \begin{bmatrix} \boldsymbol{\Delta} \\ \boldsymbol{0} \end{bmatrix} = \begin{bmatrix} \boldsymbol{P} \\ \boldsymbol{R} \end{bmatrix} \tag{5.19}$$

hence

$$K_{I,I}\ \boldsymbol{\Delta} = \boldsymbol{P} \tag{5.20}$$

Equation (5.20) is known as the final structure stiffness equation. It relates the unknown nodal displacements and the known applied loads and it is, in fact, the equation that was solved when the stiffness method was applied directly in section 3.5. Equation (5.20) can be obtained from the initial structure stiffness equation by simply deleting the row and column centred on the diagonal for each boundary condition in turn, as shown below.

$$\begin{bmatrix} \cancel{K_{11}} & \cancel{K_{12}} & \cancel{K_{13}} & \cancel{K_{14}} & \cancel{K_{15}} & \cancel{K_{16}} \\ \cancel{K_{21}} & \cancel{K_{22}} & \cancel{K_{23}} & \cancel{K_{24}} & \cancel{K_{25}} & \cancel{K_{26}} \\ \cancel{K_{31}} & \cancel{K_{32}} & K_{33} & K_{34} & K_{35} & \cancel{K_{36}} \\ \cancel{K_{41}} & \cancel{K_{42}} & K_{43} & K_{44} & K_{45} & \cancel{K_{46}} \\ \cancel{K_{51}} & \cancel{K_{52}} & K_{53} & K_{54} & K_{55} & \cancel{K_{56}} \\ \cancel{K_{61}} & \cancel{K_{62}} & \cancel{K_{63}} & \cancel{K_{64}} & \cancel{K_{65}} & \cancel{K_{66}} \end{bmatrix} \begin{bmatrix} \cancel{0} \\ \cancel{0} \\ \Delta_{2x} \\ \Delta_{2y} \\ \Delta_{3x} \\ \cancel{0} \end{bmatrix} = \begin{bmatrix} \cancel{R_{1x}} \\ \cancel{R_{2y}} \\ 100 \\ 200 \\ 0 \\ \cancel{R_{3y}} \end{bmatrix}$$

Equation (5.19) also yields the following equation

$$K_{II,I}\ \boldsymbol{\Delta} = \boldsymbol{R} \tag{5.21}$$

The solution of equation (5.20) yields the unknown structural displacements which can then be used in equation (5.21) to find the reactions.

Example 5.3

Find the displacements for the truss shown. As the initial stiffness matrix for this structure was found in example 5.2, the initial structure stiffness equation can be written down without calculation, as follows

$$\begin{bmatrix} 170.7 & 70.7 & -70.7 & 70.7 & -100.0 & 0 \\ & 70.7 & -70.7 & -70.7 & 0 & 0 \\ & & 141.4 & 0 & -70.7 & 70.7 \\ & \textit{symmetric} & & 141.4 & 70.7 & -70.7 \\ & & & & 170.7 & -70.7 \\ & & & & & 70.7 \end{bmatrix} \begin{bmatrix} \Delta_{1x} \\ \Delta_{1y} \\ \Delta_{2x} \\ \Delta_{2y} \\ \Delta_{3x} \\ \Delta_{3y} \end{bmatrix} = \begin{bmatrix} P_{1x} \\ P_{1y} \\ P_{2x} \\ P_{2y} \\ P_{3x} \\ P_{3y} \end{bmatrix}$$

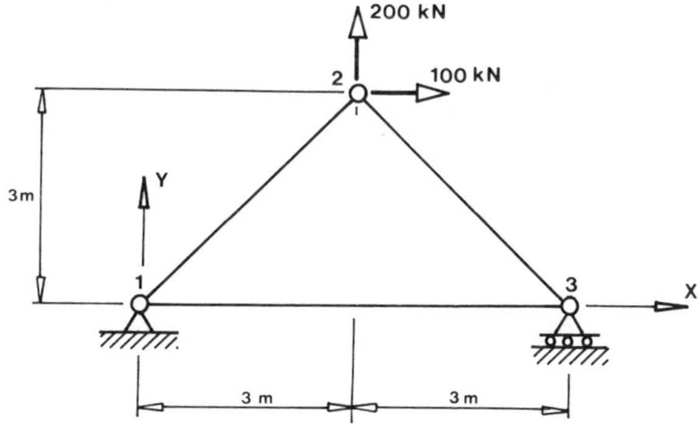

Applying the boundary conditions

$$\Delta_{1y} = \Delta_{2x} = \Delta_{3y} = 0$$

as shown, and inserting the values of the applied loads produces the final structure stiffness equation.

$$\begin{bmatrix} 141.4 & 0 & -70.7 \\ 0 & 141.4 & 70.7 \\ -70.7 & 70.7 & 170.7 \end{bmatrix} \begin{bmatrix} \Delta_{2x} \\ \Delta_{2y} \\ \Delta_{3x} \end{bmatrix} = \begin{bmatrix} 100 \\ 200 \\ 0 \end{bmatrix}$$

Solving these equations by hand, or preferably by using any one of the equation solving programs given in Chapter 4 yields

$$\begin{bmatrix} \Delta_{2x} \\ \Delta_{2y} \\ \Delta_{3x} \end{bmatrix} = \begin{bmatrix} 0.457 \text{ mm} \\ 1.664 \text{ mm} \\ -0.500 \text{ mm} \end{bmatrix}$$

Example 5.4

Find the reactions to the frame analysed in example 5.3.

If the boundary conditions are imposed by crossing out rows and columns in the initial structure stiffness matrix then the $K_{II,I}$ matrix comprises the elements eliminated by a horizontal line only. These elements are shown boxed in the following equation

$$\begin{bmatrix} 170.7 & 70.7 & -70.7 & -70.7 & -100.0 & 0 \\ & 70.7 & -70.7 & -70.7 & 0 & 0 \\ & & 141.4 & 0 & -70.7 & 70.7 \\ & \textit{symmetric} & & 141.4 & 70.7 & -70.7 \\ & & & & 170.7 & -70.7 \\ & & & & & 70.7 \end{bmatrix} \begin{bmatrix} 0 \\ 0 \\ \Delta_{2x} \\ \Delta_{2y} \\ \Delta_{3x} \\ 0 \end{bmatrix} = \begin{bmatrix} R_{1x} \\ R_{1y} \\ P_{2x} \\ P_{2y} \\ P_{3x} \\ R_{3y} \end{bmatrix}$$

The reactions can be found by substituting the displacements found in example 5.3 into equation (5.21)

$$\begin{bmatrix} -70.7 & -70.7 & -100.0 \\ -70.7 & -70.7 & 0 \\ 70.7 & -70.7 & -70.7 \end{bmatrix} \begin{bmatrix} \Delta_{2x} \\ \Delta_{2y} \\ \Delta_{3x} \end{bmatrix} = \begin{bmatrix} R_{1x} \\ R_{1y} \\ R_{3y} \end{bmatrix}$$

hence

$$\begin{bmatrix} R_{1x} \\ R_{1y} \\ R_{3y} \end{bmatrix} = \begin{bmatrix} -70.7 & -70.7 & -100.0 \\ -70.7 & -70.7 & 0 \\ 70.7 & -70.7 & -70.7 \end{bmatrix} \begin{bmatrix} 0.457 \\ 1.664 \\ -0.500 \end{bmatrix} = \begin{bmatrix} -100 \text{ kN} \\ -150 \text{ kN} \\ -50 \text{ kN} \end{bmatrix}$$

Note that the $K_{II,I}$ matrix is not normally square or symmetric.

5.8 The Final Structure Stiffness Equation

In practice it is uneconomical to assemble the initial structure stiffness matrix and apply the boundary conditions afterwards. It is far better to take account of the boundary conditions during the assembly of the stiffness matrix so that only the final structure stiffness matrix is assembled. The $K_{II,I}$ submatrix in equation (5.19) need not be assembled as the reactions can be found by summing the element end forces (in the global axes system) at each restrained node (see section 5.9).

To assemble the final structure stiffness matrix requires a knowledge of the boundary conditions. Restraints are conveniently described by the number of the node which is restrained, plus the direction of the restraint. Programming is simplified if the directions of restraint are given numbers. For example, in the case of a space frame structure where the nodal degree of freedom is six, the restraint directions might be numbered thus

1 translation restrained in the global "x" direction
2 translation restrained in the global "y" direction
3 translation restrained in the global "z" direction
4 rotation restrained about the global "x" axis
5 rotation restrained about the global "y" axis
6 rotation restrained about the global "z" axis.

In the case of a plane truss, restraint is only meaningful in directions 1 and 2. The boundary conditions for the truss shown in fig 5.10 would be input as follows

NODE	DIRECTION
1	1
1	2
3	2

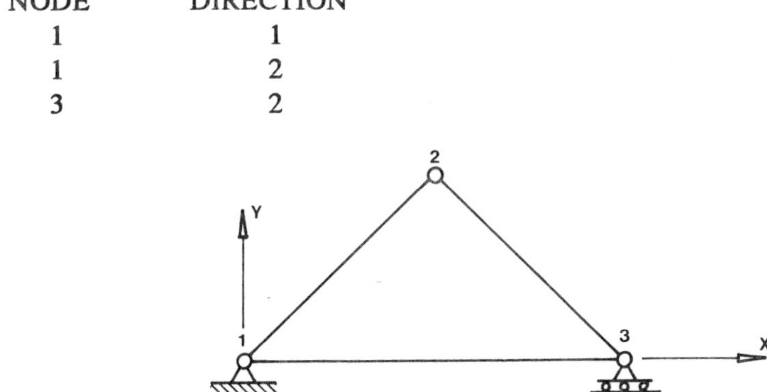

Figure 5.10 Three element truss

Input can be reduced by entering the restraint directions at each node as a composite number, and then decoding the composite number within the program. This technique is particularly useful when each node has many possible directions of restraint (e.g. in space frame structures). Using composite numbers the restraints for the truss shown in fig 5.10 are

NODE	DIRECTION
1	12
3	2

A one-dimensional array called the *freedom vector* will be used to facilitate the direct generation of the final structure stiffness matrix. The length of the freedom vector is equal to initial degree of freedom (i.e. the number of nodes times the nodal degree of freedom). It is developed from the boundary conditions as follows

1. Zero all the elements of the freedom vector.
2. Replace the element corresponding to each restrained freedom by a one.
3. Starting from the beginning of the freedom vector replace ones with zeros, and zeros with consecutive integers.

For example, the freedom vector for the truss shown in fig 5.10 would be found as follows

$$\text{Zero the freedom vector} \begin{bmatrix} 0 \\ 0 \\ 0 \\ 0 \\ 0 \\ 0 \end{bmatrix} \quad \text{Apply the boundary conditions} \begin{bmatrix} 1 \\ 1 \\ 0 \\ 0 \\ 0 \\ 1 \end{bmatrix} \quad \text{Number the freedoms} \begin{bmatrix} 0 \\ 0 \\ 1 \\ 2 \\ 3 \\ 0 \end{bmatrix}$$

Before the stiffness of an element can be added to the final structure stiffness matrix its *code number* must be developed from the freedom vector. The element code number comprises the freedom numbers associated with the displacements at the ends of the element. Hence the element code number for a plane truss element "i,j" comprises the elements ($2i$-1), ($2i$), ($2j$-1), and ($2j$) of the freedom vector.

The truss shown in fig 5.10 will be used to illustrate how element code numbers are used to develop the final structure stiffness matrix. The freedom vector has already been shown to be

$$\{\boxed{0 \quad 0} \; 1 \quad 2 \; \boxed{3 \quad 0}\}$$

The freedom numbers associated with the displacements at the ends of element 1,3 are shown boxed. Hence the code number for element 1,3 is [0 0 3 0].

$$\begin{array}{c} \\ 0 \\ 0 \\ 3 \\ 0 \end{array} \begin{array}{c} \begin{array}{cccc} 0 & 0 & 3 & 0 \end{array} \\ \begin{bmatrix} K_{11} & K_{12} & K_{13} & K_{14} \\ K_{21} & K_{22} & K_{23} & K_{24} \\ K_{31} & K_{32} & K_{33} & K_{34} \\ K_{41} & K_{42} & K_{43} & K_{44} \end{bmatrix} \end{array}$$

When the code number is applied to the global element stiffness matrix as shown it is evident that the only stiffness term to contribute to the final stiffness matrix is K_{33} which should be added to location (3,3). The reader should note that the global element stiffness matrix comprises the four global element stiffness submatrices as defined in equations (5.10) to (5.13). Applying the code number for element 1,2 to the global element stiffness matrix for element 1,2 gives

$$
\begin{array}{c|cccc}
 & 0 & 0 & 1 & 2 \\
\hline
0 & K_{11} & K_{12} & K_{13} & K_{14} \\
0 & K_{21} & K_{22} & K_{23} & K_{24} \\
1 & K_{31} & K_{32} & K_{33} & K_{34} \\
2 & K_{41} & K_{42} & K_{43} & K_{44}
\end{array}
$$

Hence element stiffness K_{33} should be added to location (1,1) in the final structure stiffness matrix, element stiffness K_{34} should be added to location (1,2), etc. Note that a zero in either the "horizontal" or the "vertical" code number means that that stiffness coefficient does not contribute to the final structure stiffness matrix.

Example 5.5

Assemble the final structure stiffness matrix for the truss shown below using the structure freedom vector and element code numbers. Note the frame is identical to that analysed in example 5.2.

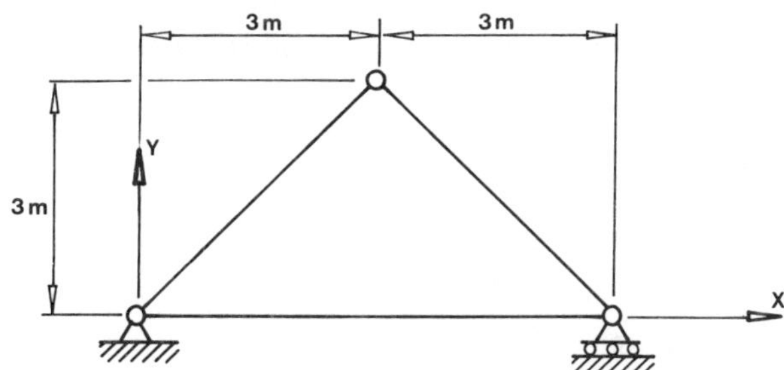

The freedom vector has already been developed for this structure (see fig 5.10) and is known to be {0 0 1 2 3 0}.

Element 1,2 - Code number [0 0 1 2]

Apply the code number to the global element stiffness matrix (from example 5.2)

$$
\begin{array}{c|cccc}
 & 0 & 0 & 1 & 2 \\
\hline
0 & 70.7 & 70.7 & -70.7 & -70.7 \\
0 & 70.7 & 70.7 & -70.7 & -70.7 \\
1 & -70.7 & -70.7 & 70.7 & 70.7 \\
2 & -70.7 & -70.7 & 70.7 & 70.7
\end{array}
$$

Adding the element stiffness to the final structure stiffness matrix (currently set to zero) gives

$$\begin{bmatrix} 70.7 & 70.7 & 0 \\ 70.7 & 70.7 & 0 \\ 0 & 0 & 0 \end{bmatrix}$$

Element 1,3 - Code number [0 0 3 0]

Apply the code number to the global element stiffness matrix (from example 5.2)

$$\begin{array}{c} \\ 0 \\ 0 \\ 3 \\ 0 \end{array} \begin{array}{c} \begin{array}{cccc} 0 & 0 & 3 & 0 \end{array} \\ \begin{bmatrix} 100.0 & 0 & -100.0 & 0 \\ 0 & 0 & 0 & 0 \\ -100.0 & 0 & 100.0 & 0 \\ 0 & 0 & 0 & 0 \end{bmatrix} \end{array}$$

Adding the element stiffness to the final structure stiffness matrix gives

$$\begin{bmatrix} 70.7 & 70.7 & 0 \\ 70.7 & 70.7 & 0 \\ 0 & 0 & 100 \end{bmatrix}$$

Element 2,3 - Code number [1 2 3 0]

Apply the code number to the global element stiffness matrix (from example 5.2).

$$\begin{array}{c} \\ 1 \\ 2 \\ 3 \\ 0 \end{array} \begin{array}{c} \begin{array}{cccc} 1 & 2 & 3 & 0 \end{array} \\ \begin{bmatrix} 70.7 & -70.7 & -70.7 & 70.7 \\ -70.7 & 70.7 & 70.7 & -70.7 \\ -70.7 & 70.7 & 70.7 & -70.7 \\ 70.7 & -70.7 & -70.7 & 70.7 \end{bmatrix} \end{array}$$

Adding the element stiffness to the final structure stiffness matrix gives

$$\begin{bmatrix} 141.4 & 0 & -70.7 \\ 0 & 141.4 & 70.7 \\ -70.7 & 70.7 & 70.7 \end{bmatrix}$$

Program FSTIFFPT.BAS

This program shows how the freedom vector is set up and then used to assemble the final structure stiffness matrix from the global element stiffness matrices using element code numbers. Note that no attempt has been made to save space by taking advantage of either the symmetry or the bandwidth of the structure stiffness matrix.

Input

The program first requests details of the boundary conditions, followed by the data for each element in turn.

Comments on the Algorithm

Once the boundary conditions have been established the freedom vector is developed. The program then loops for each element in turn. The global element stiffness matrix is evaluated and the element code number is used to add the element stiffness to the upper triangle of the structure stiffness matrix.

Output

The final structure stiffness matrix is output, and the reader should observe that a special piece of code is required to output the complete array as only the upper triangle of the structure stiffness matrix is available.

Listing

```
1000 '=======================    FSTIFFPT.BAS     ==========================
1010 '
1020 ' ... Set maximum number of nodes
1030 '
1040 MAXNNODE% = 20 : TRUE% = 1 : FALSE% = 0
1050 OPTION BASE 1
1060 DIM K(2*MAXNNODE%,2*MAXNNODE%), FREE%(2*MAXNNODE%), ESTIFF(4,4)
1070 CLS
1080 PRINT
1090 PRINT "============================================================="
1100 PRINT
1110 PRINT "PROGRAM FSTIFFPT.BAS                    Copyright (c) J.Balfour 1991"
1120 PRINT
1130 PRINT "     Program for the calculation of element stiffness matrices"
1140 PRINT "        in element and global axes for a plane truss element"
1150 PRINT
1160 PRINT "                 For further information contact"
1170 PRINT " James A.D.Balfour, Heriot-Watt University, Riccarton, Edinburgh"
1180 PRINT "               Tel 031-449-5111, Fax 031-451-3170"
1190 PRINT
1200 PRINT "============================================================="
1210 PRINT
1220 PRINT "Maximum number of nodes ="; MAXNNODE%
1230 INPUT "Actual number of nodes  = ", NNODE%
1240 '
1250 ' ... Zero the initial stiffness matrix and the freedom vector
1260 '
1270 FOR I% = 1 TO 2*NNODE%
1280   FOR J% = 1 TO 2*NNODE%
1290     K(I%,J%) = 0
1300   NEXT J%
1310   FREE%(I%) = 0
```

```
1320 NEXT I%
1330 '
1340 ' ... Input the boundary conditions and generate the freedom vector
1350 '
1360 GOSUB 1460
1370 '
1380 ' ... Input the element data and assemble the stiffness matrix
1390 '
1400 GOSUB 1840
1410 '
1420 ' ... Output the final structure stiffness matrix
1430 '
1440 GOSUB 2420
1450 END
1460 '
1470 '*****************    INPUT THE BOUNDARY CONDITIONS    ******************
1480 '
1490 CLS : PRINT
1500 PRINT "+ + + + + + + + + + + + + + +"
1510 PRINT "+  BOUNDARY  CONDITIONS   +"
1520 PRINT "+ + + + + + + + + + + + + + +"
1530 PRINT
1540 PRINT "Restraint in the global x-direction =  1"
1550 PRINT "Restraint in the global y-direction =  2"
1560 PRINT "Enter restraints as a composite number"
1570 PRINT "(e.g. if node is restrained in both directions enter 12)"
1580 DONE% = FALSE%
1590 WHILE DONE% =FALSE
1600   PRINT
1610   INPUT "Node number (0 to stop) = ", IN%
1620   IF IN% = 0 THEN DONE% = TRUE% : GOTO 1710
1630   INPUT "Direction(s)            = ", DIRN%
1640   '
1650   ' ... Evaluate the freedom to be restrained
1660   '
1670   J% = 2*IN% - 2 + (DIRN% MOD 10)
1680   FREE%(J%) = 1
1690   DIRN%     = INT(DIRN%/10)
1700   IF DIRN% > 0 THEN GOTO 1670
1710 WEND
1720 '
1730 ' ... NUMBER THE FREEDOMS (RESTRAINTS SET TO ZERO)
1740 '
1750 FOR I%    = 1 TO 2*NNODE%
1760   IF FREE%(I%) = 1 THEN GOTO 1800
1770   NDOF%     = NDOF% + 1
1780   FREE%(I%) = NDOF%
1790   GOTO 1810
1800   FREE%(I%) = 0
1810 NEXT I%
1820 RETURN
1830 '
1840 '********   ASSEMBLE THE INITIAL STRUCTURE STIFFNESS MATRIX    *********
1850 '
1860 CLS : PRINT
1870 PRINT "+ + + + + +  + +"
1880 PRINT "+   ELEMENTS   +"
1890 PRINT "+ + + + + +  + +"
1900 PRINT
1910 '
1920 FOR K% = 1 TO NNODE%
1930   PRINT "Element "; K% : PRINT
1940   INPUT "Lower node number (0 to stop)   = ", IN%
1950   IF IN% = 0 THEN DONE% = TRUE% : GOTO 2400
1960   INPUT "Higher node number              = ", JN%
1970   INPUT "Area (A)                        = ", A
1980   INPUT "Length (L)                      = ", L
1990   INPUT "Elastic constant (E)            = ", E
2000   INPUT "Angle with the global x-axis    = ", ALPHA
2010   '
2020   ' ... CONVERT THE ANGLE TO RADIANS
2030   '
2040   AR = 4 * ATN(1) * ALPHA / 180
2050   C  = COS(AR)
2060   S  = SIN(AR)
2070   '
2080   ' ... Generate the upper triangle of the element stiffness matrix
```

```
2090     '
2100     ESTIFF(1,1) =   E * A * C * C / L
2110     ESTIFF(1,2) =   E * A * C * S / L
2120     ESTIFF(1,3) = - ESTIFF(1,1)
2130     ESTIFF(1,4) = - ESTIFF(1,2)
2140     ESTIFF(2,2) =   E * A * S * S / L
2150     ESTIFF(2,3) = - ESTIFF(1,2)
2160     ESTIFF(2,4) = - ESTIFF(2,2)
2170     ESTIFF(3,3) =   ESTIFF(1,1)
2180     ESTIFF(3,4) =   ESTIFF(1,2)
2190     ESTIFF(4,4) =   ESTIFF(2,2)
2200     '
2210     ' ... Set up the element code number
2220     '
2230     EFREE%(1) = FREE%(2*IN%-1)
2240     EFREE%(2) = FREE%(2*IN%)
2250     EFREE%(3) = FREE%(2*JN%-1)
2260     EFREE%(4) = FREE%(2*JN%)
2270     '
2280     ' ... Add the element stiffness terms
2290     '
2300     FOR I%      = 1 TO 4
2310       IF EFREE%(I%) = 0 THEN GOTO 2380
2320       FOR J%      = I% TO 4
2330         IF EFREE%(J%) = 0 THEN GOTO 2370
2340         L% = EFREE%(I%)
2350         M% = EFREE%(J%)
2360         K(L%,M%) = K(L%,M%) + ESTIFF(I%,J%)
2370       NEXT J%
2380     NEXT I%
2390     PRINT
2400 NEXT K%
2410 RETURN
2420 '
2430 '**********    OUTPUT THE FINAL STRUCTURE STIFFNESS MATRIX    **********
2440 '
2450 PRINT : PRINT
2460 PRINT "* * * * * * * * * * * * * *"
2470 PRINT "*   FINAL STRUCTURE      *"
2480 PRINT "*   STIFFNESS MATRIX     *"
2490 PRINT "* * * * * * * * * * * * * *"
2500 PRINT
2510 FOR I% = 1 TO NDOF%
2520   FOR J% = 1 TO NDOF%
2530     IF I% <= J% THEN PRINT USING "#.####^^^^   "; K(I%,J%);
2540     IF I% >  J% THEN PRINT USING "#.####^^^^   "; K(J%,I%);
2550     IF J% MOD 6 = 0 THEN PRINT
2560   NEXT J%
2570   IF J% MOD 6 <> 0 THEN PRINT
2580   PRINT
2590 NEXT I%
2600 RETURN
2610 '
2620 '=======================    FSTIFFPT.BAS    ===========================
```

Sample Run

The following run demonstrates program FSTIFFPT.BAS being used to find the final structure stiffness for the structure analysed in example 5.5. Input from the keyboard is shown underlined.

```
=================================================================

PROGRAM FSTIFFPT.BAS                 Copyright (c) J.Balfour 1991

     Program for the calculation of element stiffness matrices
        in element and global axes for a plane truss element

                  For further information contact
 James A.D.Balfour, Heriot-Watt University, Riccarton, Edinburgh
            Tel 031-449-5111, Fax 031-451-3170

=================================================================
```

```
Maximum number of nodes = 20
Actual number of nodes  = 3


+ + + + + + + + + + + + + + 
+  BOUNDARY  CONDITIONS   +
+ + + + + + + + + + + + + + 

Restraint in the global x-direction =  1
Restraint in the global y-direction =  2
Enter restraints as a composite number
(e.g. if node is restrained in both directions enter 12)

Node number (0 to stop) = 1
Direction(s)            = 12

Node number (0 to stop) = 3
Direction(s)            = 2

Node number (0 to stop) = 0


+ + + + + +  + +
+   ELEMENTS   +
+ + + + + +  + +

Element  1

Lower node number (0 to stop)  = 1
Higher node number             = 2
Area (A)                       = 3000
Length (L)                     = 4243
Elastic constant (E)           = 200
Angle with the global x-axis   = 45

Element  2

Lower node number (0 to stop)  = 1
Higher node number             = 3
Area (A)                       = 3000
Length (L)                     = 6000
Elastic constant (E)           = 200
Angle with the global x-axis   = 0

Element  3

Lower node number (0 to stop)  = 2
Higher node number             = 3
Area (A)                       = 3000
Length (L)                     = 4243
Elastic constant (E)           = 200
Angle with the global x-axis   = 315


* * * * * * * * * * * * * 
*   FINAL STRUCTURE      *
*   STIFFNESS MATRIX     *
* * * * * * * * * * * * * 

0.1414E+03   -.7629E-05   -.7070E+02

-.7629E-05   0.1414E+03   0.7070E+02

-.7070E+02   0.7070E+02   0.1707E+03
```

5.9 Element End Forces and Reactions

The element stiffness submatrices in the element axes system are used to calculate the element end forces from the element end displacements using the following equation (see section 5.2)

$$f_{ij} = k_{ii}^{j}\,\delta_{ij} + k_{ij}\,\delta_{ji}$$

Solution of the final structure stiffness equation yields the nodal displacements in terms of the nodal axes systems. To find the element end forces using the preceding equation the nodal displacements must first be transformed into the element axes system.

$$f_{ij} = k_{ii}^{j}\,T_{ij}\,\Delta_i + k_{ij}\,T_{ij}\,\Delta_j$$

hence

$$\begin{bmatrix} f_{ijx} \\ f_{ijy} \end{bmatrix} = \begin{bmatrix} \frac{EA}{L} & 0 \\ 0 & 0 \end{bmatrix} \begin{bmatrix} cos\alpha & sin\alpha \\ -sin\alpha & cos\alpha \end{bmatrix} \begin{bmatrix} \Delta_{ix} \\ \Delta_{iy} \end{bmatrix} + \begin{bmatrix} -\frac{EA}{L} & 0 \\ 0 & 0 \end{bmatrix} \begin{bmatrix} cos\alpha & sin\alpha \\ -sin\alpha & cos\alpha \end{bmatrix} \begin{bmatrix} \Delta_{jx} \\ \Delta_{jy} \end{bmatrix}$$

hence

$$f_{ijx} = \frac{EA}{L}\left[cos\alpha\,(\Delta_{ix} - \Delta_{jx}) + sin\alpha\,(\Delta_{iy} - \Delta_{jy})\right] \tag{5.22}$$

$$f_{ijy} = 0$$

Similarly it can be shown that

$$f_{jix} = -f_{ijx} \qquad \text{and} \qquad f_{jiy} = 0$$

Note also that if f_{ijx} is positive then the element is in *compression.*

Reactions

When loads are applied to a structure the reactive forces assume values that will keep the structure in equilibrium. A restrained node experiences forces from the following three sources.

1. Forces from the elements.
2. Reactive forces.
3. Loads applied to the restrained node (if any).

Figure 5.11(a) shows a restrained node *"i"* which has elements *"i,a"* and *"i,b"* framing into it. The freebody diagram for node *"i"* is shown in fig 5.11(b).

Taking components of the axial forces in elements *"i,a"* and *"i,b"* in the directions of the global axes system allows the equations of equilibrium for node *"i"* to be written as follows

$$P_{ix} + R_{ix} + F'_{iax} + F'_{ibx} = 0$$

and

$$P_{iy} + R_{iy} + F'_{iay} + F'_{iby} = 0$$

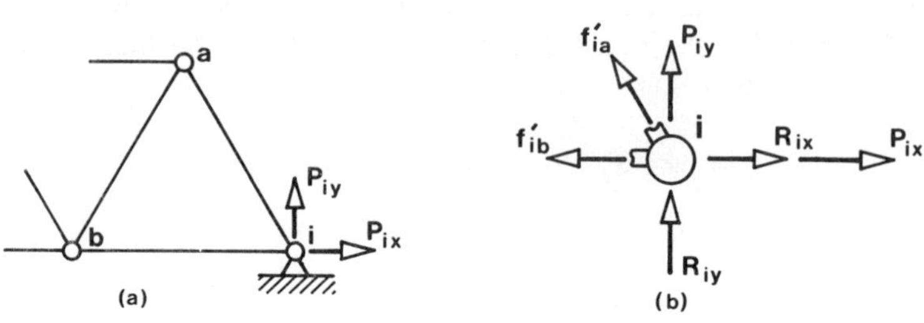

Figure 5.11 An example of a restrained node

Note that F'_{iax} is the component of f'_{ia} in the direction of the global "x" axis.

In section 5.5 the nodes at the ends of an element were shown to experience forces that were equal and opposite to the element end forces, i.e.

$$F'_{ijx} = -F_{ijx}$$

and

$$F'_{ijy} = -F_{ijy}$$

The equations of nodal equilibrium can, therefore, be written as follows

$$R_{ix} = F_{iax} + F_{ibx} - P_{ix}$$

and

$$R_{iy} = F_{iay} + F_{iby} - P_{iy}$$

or, in matrix form

$$R_i = F_{ia} + F_{ib} - P_i$$

where

$$R_i = \begin{bmatrix} R_{ix} \\ R_{iy} \end{bmatrix} \quad F_{ia} = \begin{bmatrix} F_{iax} \\ F_{iay} \end{bmatrix} \quad F_{ib} = \begin{bmatrix} F_{ibx} \\ F_{iby} \end{bmatrix} \quad P_i = \begin{bmatrix} P_{ix} \\ P_{iy} \end{bmatrix}$$

In general

$$R_i = \sum_{j=a}^{n} F_{ij} - P_i$$

where the summation is for all elements connected to the restrained node "i".

Note that the element end forces will have been found in terms of the element axes system and must be transformed into the global axes system prior to summation. Hence the above equation should be written as follows

$$R_i = \sum_{j=a}^{n} T_{ij}^{-1} f_{ij} - P_i \tag{5.23}$$

Program EFORCEPT.BAS

This program calculates the force in a pin ended element from the element end displacements.

Input

The program loops for all elements requesting the element properties plus end displacements are input for each element in turn. The axial force in the element is calculated.

Comments on the Algorithm

Equation (5.22) is used to calculate the element forces.

Output

The force in each element is output.

Listing

```
1000 '=======================     EFORCEPT.BAS     ===========================
1010 '
1020 TRUE% = 1 : FALSE% = 0
1030 CLS
1040 PRINT
1050 PRINT "=================================================================="
1060 PRINT
1070 PRINT "PROGRAM EFORCEPT.BAS                      Copyright (c) J.Balfour 1991"
1080 PRINT
1090 PRINT "          Program for the calculation of element end forces"
1100 PRINT "                      for a plane truss element"
1110 PRINT
1120 PRINT "                    For further information contact"
1130 PRINT " James A.D.Balfour, Heriot-Watt University, Riccarton, Edinburgh"
1140 PRINT "              Tel 031-449-5111, Fax 031-451-3170"
1150 PRINT
1160 PRINT "=================================================================="
1170 PRINT
1180 '
1190 ' ... Loop for all members
1200 '
1210 DONE% = FALSE% : K% = 1
1220 WHILE DONE% = FALSE%
1230   PRINT "ELEMENT "; K%
1240   PRINT
1250   INPUT "Lower node number (0 to stop) = ", IN%
1260   IF IN%=0 THEN DONE% = TRUE% : GOTO 1580
1270   INPUT "Higher node number            = ", JN%
1280   INPUT "Area (A)                      = ", A
1290   INPUT "Length (L)                    = ", L
1300   INPUT "Elastic constant (E)          = ", E
1310   INPUT "Angle with the global x-axis  = ", ALPHA
1320   PRINT
1330   PRINT "ENTER THE NODAL DISPLACEMENTS (GLOBAL AXES)"
1340   PRINT
1350   PRINT USING "X-displacement at node ##     = "; IN%;
1360   INPUT "", XI
1370   PRINT USING "Y-displacement at node ##     = "; IN%;
1380   INPUT "", YI
1390   PRINT USING "X-displacement at node ##     = "; JN%;
1400   INPUT "", XJ
```

```
1410    PRINT USING "Y-displacement at node ##      = "; JN%;
1420    INPUT "", YJ
1430    '
1440    ' ... CONVERT THE ANGLE TO RADIANS
1450    '
1460    AR     = 4 * ATN(1) * ALPHA / 180
1470    '
1480    ' ... CALCULATE AND OUTPUT THE MEMBER FORCE
1490    '
1500    FORCE = (-E*A/L) * (COS(AR)*XI + SIN(AR)*YI - COS(AR)*XJ - SIN(AR)*YJ)
1510    PRINT
1520    PRINT USING "FORCE IN ELEMENT    ## ##      = "; IN%, JN%;
1530    PRINT ABS(FORCE);
1540    IF FORCE > 0 THEN PRINT "    (TENSION)"
1550    IF FORCE < 0 THEN PRINT "    (COMPRESSION)"
1560    K% = K% + 1
1570    PRINT
1580 WEND
1590 END
1600 '
1610 '========================    EFORCEPT.BAS    ============================
```

Sample Run

The following run demonstrates program EFORCEPT.BAS being used to find the forces in the elements of the structure analysed in example 5.3. Input from the keyboard is shown underlined.

```
=================================================================

PROGRAM EFORCEPT.BAS                  Copyright (c) J.Balfour 1991

         Program for the calculation of element end forces
                    for a plane truss element

               For further information contact
 James A.D.Balfour, Heriot-Watt University, Riccarton, Edinburgh
           Tel 031-449-5111, Fax 031-451-3170

=================================================================

ELEMENT  1

Lower node number (0 to stop) = 1
Higher node number            = 2
Area (A)                      = 3000
Length (L)                    = 4243
Elastic constant (E)          = 200
Angle with the global x-axis  = 45

ENTER THE NODAL DISPLACEMENTS (GLOBAL AXES)

X-displacement at node  1     = 0
Y-displacement at node  1     = 0
X-displacement at node  2     = 0.457
Y-displacement at node  2     = 1.66

FORCE IN ELEMENT    1  2      =  211.6821    (TENSION)

ELEMENT  2

Lower node number (0 to stop) = 1
Higher node number            = 3
Area (A)                      = 3000
Length (L)                    = 6000
Elastic constant (E)          = 200
Angle with the global x-axis  = 0

ENTER THE NODAL DISPLACEMENTS (GLOBAL AXES)

X-displacement at node  1     = 0
Y-displacement at node  1     = 0
```

```
X-displacement at node  3     = -0.5
Y-displacement at node  3     = 0

FORCE IN ELEMENT    1  3      =  50     (COMPRESSION)


ELEMENT  3

Lower node number (0 to stop) = 2
Higher node number            = 3
Area (A)                      = 3000
Length (L)                    = 4243
Elastic constant (E)          = 200
Angle with the global x-axis  = 315

ENTER THE NODAL DISPLACEMENTS (GLOBAL AXES)

X-displacement at node  2     = 0.457
Y-displacement at node  2     = 1.66
X-displacement at node  3     = -0.5
Y-displacement at node  3     = 0

FORCE IN ELEMENT    2  3      =  70.29383     (TENSION)


ELEMENT  4

Lower node number (0 to stop) = 0
```

5.10 Putting it all Together - Program PTRUSS.BAS

This chapter has now covered all of the procedures necessary for the analysis of plane trusses using the stiffness method. In this section the short computer programs to assemble the final structure stiffness matrix (FSTIFFPT.BAS) and to calculate element end forces (EFORCEPT.BAS) are combined with the Gauss-Jordan equation solver (GAUSSJ.BAS) to produce a simple automatic truss analysis program PTRUSS.BAS.

The volume of input required by all previous programs has been small enough for it to be entered from the keyboard without undue difficulty. This program requires the following information to be input.

1. Job title and units.
2. Nodal coordinates.
3. Element topology.
4. Element properties.
5. Restraints.
6. Loads.

The volume of input is likely to be so large that error-free data entry becomes difficult. Further, the program user may well wish to re-run the program using data that is only slightly different from the previous run. For these reasons it is desirable that the data should be read from a data file. This, of course, introduces another level of difficulty to the programming. The advantage of using data files will soon, however, become apparent. Finally it should be noted that commercial structural analysis software invariably makes use of data files.

Program PTRUSS.BAS

This program, which will automatically analyse plane trusses, is essentially an amalgamation of the programs FSTIFFPT.BAS, EFORCEPT.BAS and GAUSSJ.BAS.

Input

Unless PTRUSS.BAS has been chained from PRE.BAS or DCHECK.BAS the user is prompted for the name of the data file to be read by the program. PTRUSS.BAS then offers the user the following menu which drives the program.

Do you wish to:
(1) Analyse with output to the screen
(2) Analyse with output to the printer
(3) Analyse with output to a file
(4) Edit the data
(5) Check the data
(6) Stop

This data file should be constructed using the data preprocessor program PRE.BAS which is described in section 5.11. The program DCHECK.BAS can be used to check the data for obvious errors. DCHECK.BAS is presented in section 5.12.

Comments on the Algorithm

The program first uses the boundary conditions to establish the freedom vector. It then adds the stiffness of each element to the final structure stiffness matrix in turn. The upper triangle only of the final structure stiffness matrix is assembled. Symmetry is later used to complete the matrix. No attempt is made to minimise storage by taking advantage of either the symmetry or the bandwidth of the structure stiffness matrix. The loading vector is developed and then the equations are solved using Gauss-Jordan elimination. Once the unknown nodal displacements have been found they are used to calculate the element end forces and the reactions.

Output

The data from the data file is echoed and messages are output to indicate the beginning of each solution phase. After solution of the stiffness equation nodal displacements are output, followed by the axial forces in the elements, and then the reactions.

Listing

```
1000 '========================    PTRUSS.BAS    ==============================
1010 '
1020 COMMON DATAFILE$
1030 OPTION BASE 1 : KEY OFF
1040 TRUE% = -1 : FALSE% = 0 : DEVICE$ = "SCRN:" : PAGELEN% = 23
1050 CLS
1060 PRINT
1070 PRINT "=================================================================="
1080 PRINT
1090 PRINT " PROGRAM PTRUSS                         Copyright (c) J.Balfour 1991"
1100 PRINT
1110 PRINT "      Program for the automatic analysis of plane trusses"
1120 PRINT "         Data generated with the data preprocessor PRE"
1130 PRINT
1140 PRINT "               For further information contact"
1150 PRINT " James A.D.Balfour, Heriot-Watt University, Riccarton, Edinburgh"
1160 PRINT "                 Tel 031-449-5111, Fax 031-451-3170"
1170 PRINT
1180 PRINT "=================================================================="
1190 '
1200 ' ... Note that variables are defined in Appendix A and the
1210 '     following statements set the maximum problem size
1220 '
1230 MAXNODE%  = 45                        '... Max no of nodes
1240 MAXELEM%  = 70                        '... Max no of elements
1250 MAXPROP%  = 20                        '... Max no of element properties
1260 MAXREST%  = 20                        '... Max no of restrained nodes
1270 MAXNLOAD% = 50                        '... Max no of nodal loads
1280 '
1290 DIM NODE(MAXNODE%,3), ELEM(MAXELEM%,6),   PROP(MAXPROP%,3)
1300 DIM REST(MAXREST%,4), NLOAD(MAXNLOAD%,3), FREE%(2*MAXNODE%), WKSP$(8)
1310 DIM K(2*MAXNODE%,2*MAXNODE%+1),            ESTIFF(4,4),        EFREE%(4)
1320 '
1330 ' ... Check if this program has been chained from PRE
1340 '
1350 IF DATAFILE$ <> "" THEN GOSUB 7810 : GOTO 1380
1360 PRINT
1370 INPUT "Name of the data file  =  ", DATAFILE$
1380 GOSUB 2060
1390 NOPTIONS% = 7
1400 WHILE (REPLY% <> 7)
1410   CLS : PRINT
1420   PRINT "Do you wish to:"
1430   PRINT "   (1) Analyse with output to the screen"
1440   PRINT "   (2) Analyse with output to the printer"
1450   PRINT "   (3) Analyse with output to a file"
1460   PRINT "   (4) Edit the data"
1470   PRINT "   (5) Check the data"
1480   PRINT "   (6) Read a new file"
1490   PRINT "   (7) Stop"
1500   NOPTIONS% = 7 : GOSUB 7100
1510   ON REPLY% GOTO 1530, 1560, 1640, 1710, 1740, 1770, 1790
1520     '
1530     OPEN "SCRN:" FOR OUTPUT AS #2                        ' Reply = 1
1540     GOSUB 1840 : GOTO 1790
1550     '
1560     ON ERROR GOTO 1610                                   ' Reply = 2
1570     OPEN "LPT1:" FOR OUTPUT AS #2
1580     PRINT : PRINT "Wait - looking for printer" : PRINT #2,
1590     ON ERROR GOTO 0 : DEVICE$ = "LPT1:" : GOSUB 1840
1600     PRINT #2, CHR$(12); : GOTO 1790
1610     CLS : PRINT : PRINT "** ERROR ** Failed to find the printer"
1620     GOSUB 7810 : RESUME 1790
1630     '
1640     ON ERROR GOTO 1680                                   ' Reply = 3
1650     INPUT "Name of file for output = ", DEVICE$
1660     OPEN DEVICE$ FOR OUTPUT AS #2
1670     ON ERROR GOTO 0 : GOSUB 1840 : GOTO 1790
1680     CLS : PRINT : PRINT "** ERROR ** Failed to open output file"
1690     GOSUB 7810 : RESUME 1790
1700     '
1710     PRINT "Wait - chaining PRE"                          ' Reply = 4
1720     CHAIN "PRE"
1730     '
```

```
1740      PRINT "Wait - chaining DCHECK"                    ' Reply = 5
1750      CHAIN "DCHECK"
1760      '
1770      INPUT "Name of the data file  =  ", DATAFILE$     ' Reply = 6
1780      GOSUB 2060
1790      CLOSE #2 : DEVICE$ = "SCRN:" : ON ERROR GOTO 0
1800 WEND
1810 KEY ON
1820 END
1830 '
1840 ' ************************     ANALYSE     ******************************
1850 '
1860 IF DATAFILE%=OK% THEN GOTO 1910
1870 CLS : PRINT
1880 PRINT "Analysis cannot proceed due to errors in data file";
1890 GOSUB 7810          'Pause
1900 GOTO  2040
1910 GOSUB 3390          'Print data
1920 PRINT : PRINT "Generating the freedom vector                  "
1930 GOSUB 4590
1940 PRINT : PRINT "Assembling the stiffness matrix, adding element :- ";
1950 ROW% = CSRLIN: COL% = POS(0)
1960 GOSUB 4920
1970 PRINT : PRINT "Generating the loading vector                  "
1980 GOSUB 5650
1990 PRINT : PRINT "Solving equations - no. of equations = "; NDOF%
2000 GOSUB 5760
2010 GOSUB 7810          'Pause
2020 GOSUB 6080          'Output the nodal displacements
2030 GOSUB 6310          'Output the element forces and reactions
2040 RETURN
2050 '
2060 '************************   READ DATA FILE    ***************************
2070 '
2080 ' ...  This subroutine reads the data from a data file
2090 '
2100 PRINT
2110 PRINT "Wait - Reading file "; DATAFILE$; " line";
2120 ROW% = CSRLIN : COL% = POS(0) : LCOUNT% = 0
2130 DATAFILEOK% = FALSE% : DATAEND% = FALSE%
2140 NNODE% = 0 : NELEM% = 0 : NPROP% = 0 : NREST% = 0 : NNLOAD% = 0
2150 ON ERROR GOTO 2170 : OPEN DATAFILE$ FOR INPUT AS #1
2160 ON ERROR GOTO 0 : GOTO 2230
2170 CLS : PRINT
2180 PRINT "** ERROR ** Failed to open file "; DATAFILE$; " for input"
2190 GOSUB 7810 : RESUME 3360
2200 '
2210 ' ... Read check for problem type
2220 '
2230 TARGET$ = "PTRUSS" : GOSUB 7570
2240 IF STRFOUND% THEN GOTO 2310
2250 CLS : PRINT
2260 PRINT "** ERROR ** Not a file for PTRUSS"
2270 GOSUB 7810 : GOTO 3360
2280 '
2290 ' ... Read title and units strings
2300 '
2310 NLINES% = 2 : GOSUB 7720 : IF DATAEND% THEN GOTO 2380
2320 WHILE (MID$(WRKSTR$,I%,1) = " ")
2330   I% = I% + 1
2340 WEND
2350 TITLE$  = MID$(WRKSTR$,I%)
2360 TARGET$ = "Units" : GOSUB 7570
2370 IF STRFOUND% THEN GOTO 2410
2380 CLS : PRINT
2390 PRINT "** ERROR ** Error reading Title/Units strings"
2400 GOSUB 7810 : GOTO 2430
2410 UNITS$ = MID$(WRKSTR$,INSTR(WRKSTR$,":- ")+3,24)
2420 DATAFILEOK% = TRUE%
2430 WHILE NOT DATAEND%
2440   '
2450   ' ... Get module string
2460   '
2470   IF LEFT$(WRKSTR$,4)="+    " THEN GOTO 2490
2480   TARGET$ = "+    " : GOSUB 7570 : IF NOT STRFOUND% THEN GOTO 3350
2490   MODSTR$ = WRKSTR$
2500   '
```

```
2510    ' ...  Identify module
2520    '
2530    IF INSTR(MODSTR$,"NODAL COORDINATES")=0 THEN GOTO 2680
2540    '
2550    ' ... Read nodal coordinates
2560    '
2570    TARGET$ = "NODE" : GOSUB 7570 : IF NOT STRFOUND% THEN GOTO 3280
2580    GOSUB 7240
2590    WHILE NNOS% > 0
2600      NNODE% = NNODE% + 1
2610      FOR J% = 1 TO NNOS%
2620        NODE(NNODE%,J%) = VAL(WKSP$(J%))
2630      NEXT J%
2640      GOSUB 7240
2650    WEND
2660    GOTO 3350
2670    '
2680    IF INSTR(MODSTR$,"ELEMENTS")=0 THEN GOTO 2830
2690    '
2700    ' ... Read elements
2710    '
2720    TARGET$ = "ELEMENT" : GOSUB 7570 : IF NOT STRFOUND% THEN GOTO 3280
2730    GOSUB 7240
2740    WHILE NNOS% > 0
2750      NELEM% = NELEM% + 1
2760      FOR J% = 1 TO NNOS%
2770        ELEM(NELEM%,J%) = VAL(WKSP$(J%))
2780      NEXT J%
2790      GOSUB 7240
2800    WEND
2810    GOTO 3350
2820    '
2830    IF INSTR(MODSTR$,"PROPERTIES")=0 THEN GOTO 2980
2840    '
2850    ' ... Read properties
2860    '
2870    TARGET$ = "PROPERTY" : GOSUB 7570 : IF NOT STRFOUND% THEN GOTO 3280
2880    GOSUB 7240
2890    WHILE NNOS% > 0
2900      NPROP% = NPROP% + 1
2910      FOR J% = 1 TO NNOS%
2920        PROP(NPROP%,J%) = VAL(WKSP$(J%))
2930      NEXT J%
2940      GOSUB 7240
2950    WEND
2960    GOTO 3350
2970    '
2980    IF INSTR(MODSTR$,"RESTRAINTS")=0 THEN GOTO 3130
2990    '
3000    ' ... Read restraints
3010    '
3020    TARGET$ = "NODE" : GOSUB 7570 : IF NOT STRFOUND% THEN GOTO 3280
3030    GOSUB 7240
3040    WHILE NNOS% > 0
3050      NREST% = NREST% + 1
3060      FOR J% = 1 TO NNOS%
3070        REST(NREST%,J%) = VAL(WKSP$(J%))
3080      NEXT J%
3090      GOSUB 7240
3100    WEND
3110    GOTO 3350
3120    '
3130    IF INSTR(MODSTR$,"NODAL LOADS")=0 THEN GOTO 3280
3140    '
3150    ' ... Read nodal loads
3160    '
3170    TARGET$ = "NODE" : GOSUB 7570 : IF NOT STRFOUND% THEN GOTO 3280
3180    GOSUB 7240
3190    WHILE NNOS% > 0
3200      NNLOAD% = NNLOAD% + 1
3210      FOR J% = 1 TO NNOS%
3220        NLOAD(NNLOAD%,J%) = VAL(WKSP$(J%))
3230      NEXT J%
3240      GOSUB 7240
3250    WEND
3260    GOTO 3350
3270    '
```

```
3280   CLS : PRINT
3290   PRINT "** ERROR ** in module "; MODSTR$ "run DCHECK for more info";
3300   GOSUB 7810
3310   PRINT
3320   PRINT "Wait - Reading file "; DATAFILE$; " line";
3330   ROW% = CSRLIN : COL% = POS(0)
3340   DATAFILEOK% = FALSE% : WRKSTR$ = ""
3350 WEND
3360 CLOSE #1 : ON ERROR GOTO 0
3370 RETURN
3380 '
3390 ' ***********************    PRINT DATA    *****************************
3400 '
3410 GOSUB 3490                          '... List header
3420 GOSUB 3810                          '... List nodal coordinates
3430 GOSUB 3960                          '... List element data
3440 GOSUB 4120                          '... List list element properties
3450 GOSUB 4280                          '... List restraint data
3460 GOSUB 4430                          '... List nodal loads
3470 RETURN
3480 '
3490 ' +++++++++++++++++++++++   LIST HEADER    ++++++++++++++++++++++++++++
3500 '
3510 CLS : PRINT #2,
3520 PRINT #2, "======================================";
3530 PRINT #2, "======================================"
3540 PRINT #2,
3550 PRINT #2, "     PROGRAM PTRUSS";STRING$(24,32);
3560 PRINT #2, "Copyright(c) J.A.D.Balfour 1991"
3570 PRINT #2,
3580 FOR I% = 1 TO (76-LEN(TITLE$))/2
3590   PRINT #2, " ";
3600 NEXT I%
3610 PRINT #2, TITLE$
3620 PRINT #2,
3630 PRINT #2, "     File Name :- "; DATAFILE$;
3640 FOR I% = 1 TO (40-LEN(DATAFILE$))
3650   PRINT #2, " ";
3660 NEXT I%
3670 PRINT #2, "Date :- "; MID$(DATE$,4,3); MID$(DATE$,1,3);
3680 PRINT #2, MID$(DATE$, 9, 2)
3690 PRINT #2, "     Units     :- "; UNITS$;
3700 FOR I% = 1 TO (40-LEN(UNITS$))
3710   PRINT #2, " ";
3720 NEXT I%
3730 PRINT #2, "Time :- "; TIME$
3740 PRINT #2,
3750 PRINT #2, "======================================";
3760 PRINT #2, "======================================"
3770 PRINT #2,
3780 GOSUB 7810
3790 RETURN
3800 '
3810 ' +++++++++++++++++    LIST NODAL COORDINATES    +++++++++++++++++++++
3820 '
3830 PRINT #2,
3840 PRINT #2, "+ + + + + + + + + + + + + +"
3850 PRINT #2, "+    NODAL COORDINATES    +"
3860 PRINT #2, "+ + + + + + + + + + + + + +"
3870 PRINT #2,
3880 PRINT #2, "NODE            X              Y"
3890 FOR I% = 1 TO NNODE%
3900   GOSUB 7900
3910   PRINT #2, " "; NODE(I%,1), NODE(I%,2), NODE(I%,3)
3920 NEXT I%
3930 GOSUB 7810
3940 RETURN
3950 '
3960 ' ++++++++++++++++++++++   LIST ELEMENTS    +++++++++++++++++++++++++
3970 '
3980 GOSUB 7900
3990 PRINT #2,
4000 PRINT #2, "+ + + + + + + + + +"
4010 PRINT #2, "+    ELEMENTS    +"
4020 PRINT #2, "+ + + + + + + + + +"
4030 PRINT #2,
4040 PRINT #2, "ELEMENT    PROPERTY"
```

```
4050 FOR I% = 1 TO NELEM%
4060   GOSUB 7900
4070   PRINT #2, USING "### ###        ##"; ELEM(I%,1); ELEM(I%,2); ELEM(I%,3)
4080 NEXT I%
4090 GOSUB 7810
4100 RETURN
4110 '
4120 ' +++++++++++++++++++++++    LIST PROPERTIES    +++++++++++++++++++++++++
4130 '
4140 PRINT #2,
4150 PRINT #2, "+ + + + + + +  + + +"
4160 PRINT #2, "+   PROPERTIES   +"
4170 PRINT #2, "+ + + + + + +  + + +"
4180 PRINT #2,
4190 PRINT #2, "PROPERTY         A              E"
4200 FOR I% = 1 TO NPROP%
4210   GOSUB 7900
4220   PRINT #2, USING "  ###      #.####^^^^"; PROP(I%,1); PROP(I%,2);
4230   PRINT #2, USING "      #.####^^^^  "; PROP(I%,3)
4240 NEXT I%
4250 GOSUB 7810
4260 RETURN
4270 '
4280 ' ++++++++++++++++++++++++    LIST RESTRAINTS    ++++++++++++++++++++++++
4290 '
4300 PRINT #2,
4310 PRINT #2, "+ + + + + + + + + + + "
4320 PRINT #2, "+   RESTRAINTS    +"
4330 PRINT #2, "+ + + + + + + + + + + "
4340 PRINT #2,
4350 PRINT #2, "NODE    DIRECTION(S)"
4360 FOR I% = 1 TO NREST%
4370   GOSUB 7900
4380   PRINT #2, USING "###      ######"; REST(I%,1); REST(I%,2)
4390 NEXT I%
4400 GOSUB 7810
4410 RETURN
4420 '
4430 ' ++++++++++++++++++++    LIST NODAL LOADS    +++++++++++++++++++++++++
4440 '
4450 PRINT #2,
4460 PRINT #2, "+ + + + + + + + + + +"
4470 PRINT #2, "+   NODAL LOADS   +"
4480 PRINT #2, "+ + + + + + + + + + +"
4490 PRINT #2,
4500 PRINT #2, "NODE    DIRECTION         VALUE"
4510 FOR I% = 1 TO NNLOAD%
4520   GOSUB 7900
4530   PRINT #2, USING "###        ###";  NLOAD(I%,1); NLOAD(I%,2);
4540   PRINT #2, USING "          #.####^^^^"; NLOAD(I%,3)
4550 NEXT I%
4560 GOSUB 7810
4570 RETURN
4580 '
4590 ' *****************    GENERATE THE FREEDOM VECTOR    *******************
4600 '
4610 ' ... Zero the freedom vector
4620 '
4630 FOR I% = 1 TO 2*NNODE%
4640   FREE%(I%) = 0
4650 NEXT I%
4660 '
4670 ' ... Loop for all restrained nodes
4680 '
4690 FOR I% = 1 TO NREST%
4700   K% = REST(I%,2)
4710   '
4720   ' ... Evaluate the freedom to be restrained from K%
4730   '
4740   J% = 2 * REST(I%,1) - 2 + (K% MOD 10)
4750   FREE%(J%) = 1
4760   K% = INT(K%/10)
4770   IF K% > 0 THEN GOTO 4740
4780 NEXT I%
4790 '
4800 ' ... Number the freedoms (restraints set to zero)
4810 '
```

```
4820 NDOF% = 0
4830 FOR I% = 1 TO 2*NNODE%
4840   IF FREE%(I%) = 1 THEN GOTO 4880
4850   NDOF% = NDOF% + 1
4860   FREE%(I%) = NDOF%
4870   GOTO 4890
4880   FREE%(I%) = 0
4890 NEXT I%
4900 RETURN
4910 '
4920 ' ***************    ASSEMBLE THE STIFFNESS MATRIX    ******************
4930 '
4940 ' ... Zero the augmented matrix
4950 '
4960 FOR I% = 1 TO NDOF%
4970   FOR J% = 1 TO NDOF%+1
4980     K(I%,J%) = 0
4990   NEXT J%
5000 NEXT I%
5010 '
5020 ' ... Loop for each element
5030 '
5040 FOR K% = 1 TO NELEM%
5050   LOCATE ROW%, COL%: PRINT K%;
5060   IN% = ELEM(K%,1)
5070   JN% = ELEM(K%,2)
5080   PN% = ELEM(K%,3)
5090   '
5100   ' ... Calculate the element length (store in ELEM(K%,4))
5110   '
5120   TEMP(1) = (NODE(JN%,2) - NODE(IN%,2))^2 + (NODE(JN%,3) - NODE(IN%,3))^2
5130   ELEM(K%,4) = SQR(TEMP(1))
5140   '
5150   ' ... Calculate the cosine and sine of the angle that the element makes
5160   ' ... with the coordinate x-axis (store in ELEM(K%,5) AND ELEM(K%,6)
5170   '
5180   ELEM(K%,5) = (NODE(JN%,2) - NODE(IN%,2)) / ELEM(K%,4)
5190   ELEM(K%,6) = (NODE(JN%,3) - NODE(IN%,3)) / ELEM(K%,4)
5200   '
5210   ' ... Assemble the upper triangle of the element stiffness matrix
5220   '
5230   ESTIFF(1,1) =  PROP(PN%,3)*PROP(PN%,2)*ELEM(K%,5)*ELEM(K%,5)/ELEM(K%,4)
5240   ESTIFF(1,2) =  PROP(PN%,3)*PROP(PN%,2)*ELEM(K%,5)*ELEM(K%,6)/ELEM(K%,4)
5250   ESTIFF(1,3) = -ESTIFF(1,1)
5260   ESTIFF(1,4) = -ESTIFF(1,2)
5270   ESTIFF(2,2) =  PROP(PN%,3)*PROP(PN%,2)*ELEM(K%,6)*ELEM(K%,6)/ELEM(K%,4)
5280   ESTIFF(2,3) = -ESTIFF(1,2)
5290   ESTIFF(2,4) = -ESTIFF(2,2)
5300   ESTIFF(3,3) =  ESTIFF(1,1)
5310   ESTIFF(3,4) =  ESTIFF(1,2)
5320   ESTIFF(4,4) =  ESTIFF(2,2)
5330   '
5340   ' ... Set up the element code number
5350   '
5360   EFREE%(1) = FREE%(2*IN% - 1)
5370   EFREE%(2) = FREE%(2*IN%)
5380   EFREE%(3) = FREE%(2*JN% - 1)
5390   EFREE%(4) = FREE%(2*JN%)
5400   '
5410   ' ... Add the element stiffness to the structure stiffness matrix
5420   '
5430   FOR I% = 1 TO 4
5440     IF EFREE%(I%) = 0 THEN GOTO 5510
5450     FOR J% = I% TO 4
5460       IF EFREE%(J%) = 0 THEN GOTO 5500
5470       L% = EFREE%(I%)
5480       M% = EFREE%(J%)
5490       K(L%,M%) = K(L%,M%) + ESTIFF(I%,J%)
5500     NEXT J%
5510   NEXT I%
5520 NEXT K%
5530 PRINT
5540 '
5550 ' ... Use symmetry to fill out the stiffness matrix
5560 '
5570 FOR I% = 1 TO NDOF% - 1
5580   FOR J% = I% + 1 TO NDOF%
```

```
5590      K(J%,I%) = K(I%,J%)
5600    NEXT J%
5610 NEXT I%
5620 '
5630 RETURN
5640 '
5650 '******************     SET UP THE LOADING VECTOR     ********************
5660 '
5670 ' ... Loop for all loads
5680 '
5690 FOR I% = 1 TO NNLOAD%
5700   J% = 2 * NLOAD(I%,1) - 2 + NLOAD(I%,2)
5710   J% = FREE%(J%)
5720   IF J% > 0 THEN K(J%,NDOF%+1) = NLOAD(I%,3)
5730 NEXT I%
5740 RETURN
5750 '
5760 ' **********************   SOLVE EQUATIONS     **************************
5770 '
5780 ' ... Loop for all pivots
5790 '
5800 PRINT "                        - current pivot    = ";
5810 ROW% = CSRLIN: COL% = POS(0):
5820 FOR I% = 1 TO NDOF%
5830   LOCATE ROW%, COL% : PRINT I%;
5840   TEMP(1) = K(I%,I%)
5850   '
5860   ' ... Normalise
5870   '
5880   FOR J% = 1 TO NDOF% + 1
5890     K(I%,J%) = K(I%,J%) / TEMP(1)
5900   NEXT J%
5910   '
5920   ' ... Eliminate for all rows using the current pivot
5930   '
5940   FOR J% = 1 TO NDOF%
5950     IF J% = I% THEN GOTO 6030
5960     TEMP(1) = K(J%,I%)
5970     '
5980     ' ... Loop for all elements in the current row
5990     '
6000     FOR K% = 1 TO NDOF% + 1
6010       K(J%,K%) = K(J%,K%) - TEMP(1)*K(I%,K%)
6020     NEXT K%
6030   NEXT J%
6040 NEXT I%
6050 PRINT
6060 RETURN
6070 '
6080 '*********************    OUTPUT THE DISPLACEMENTS  *********************
6090 '
6100 PRINT #2,
6110 PRINT #2, "* * * * * * * * * * * * *"
6120 PRINT #2, "*   DISPLACEMENTS   *           * :-  RESTRAINT"
6130 PRINT #2, "* * * * * * * * * * * * *"
6140 PRINT #2,
6150 PRINT #2, "NODE     X-DISPLACEMENT  Y-DISPLACEMENT"
6160 FOR I% = 1 TO NNODE%
6170   GOSUB 7900
6180   PRINT #2, USING "###     "; I%;
6190   FOR J% = 1 TO 2
6200     K% = FREE%(2*I% + J% - 2)
6210     IF K% = 0 THEN GOTO 6240
6220     PRINT #2, USING "      #.####^^^^"; K(K%,NDOF%+1);
6230     GOTO 6250
6240     PRINT #2, "         *      ";
6250   NEXT J%
6260   PRINT #2,
6270 NEXT I%
6280 GOSUB 7810                                '... Pause
6290 RETURN
6300 '
6310 ' ****** CALCULATE AND OUTPUT THE ELEMENT FORCES AND REACTIONS    ******
6320 '
6330 ' ... Zero the reaction components
6340 '
6350 FOR I% = 1 TO NREST%
```

```
6360   REST(I%,3) = 0
6370   REST(I%,4) = 0
6380 NEXT I%
6390 PRINT #2,
6400 PRINT #2, "* * * * * * * * * * * * **"
6410 PRINT #2, "*   ELEMENT FORCES    *          TENSION POSITIVE"
6420 PRINT #2, "* * * * * * * * * * * * **"
6430 PRINT #2,
6440 PRINT #2, "ELEMENT           FORCE"
6450 FOR I% = 1 TO NELEM%
6460   II% = 2 * ELEM(I%,1) - 1
6470   JJ% = 2 * ELEM(I%,2) - 1
6480   PN% = ELEM(I%,3)
6490   K% = FREE%(II%)
6500   IF K% = 0 THEN XI = 0 ELSE XI = K(K%,NDOF%+1)
6510   K% = FREE%(II%+1)
6520   IF K% = 0 THEN YI = 0 ELSE YI = K(K%,NDOF%+1)
6530   K% = FREE%(JJ%)
6540   IF K% = 0 THEN XJ = 0 ELSE XJ = K(K%,NDOF%+1)
6550   K% = FREE%(JJ%+1)
6560   IF K% = 0 THEN YJ = 0 ELSE YJ = K(K%, NDOF% + 1)
6570   FORCE = ELEM(I%,5) * (XJ - XI) + ELEM(I%,6) * (YJ - YI)
6580   FORCE = FORCE * PROP(PN%,3) * PROP(PN%,2) / ELEM(I%,4)
6590   GOSUB 7900
6600   PRINT #2, USING "## ##        "; ELEM(I%,1); ELEM(I%,2);
6610   PRINT #2, USING "#.####^^^^"; FORCE
6620   '
6630   ' ... Add any contribution from the element forces to the reactions
6640   '
6650   FOR J% = 1 TO NREST%
6660     K% = 0
6670     IF ELEM(I%,1) = REST(J%,1) THEN K% = -1
6680     IF ELEM(I%,2) = REST(J%,1) THEN K% =  1
6690     IF K% = 0 THEN GOTO 6720
6700     REST(J%,3) = REST(J%,3) + K%*ELEM(I%,5) * FORCE
6710     REST(J%,4) = REST(J%,4) + K%*ELEM(I%,6) * FORCE
6720   NEXT J%
6730 NEXT I%
6740 GOSUB 7810                                  '... Pause
6750 '
6760 ' ... Add any applied loads
6770 '
6780 FOR I% = 1 TO NREST%
6790   FOR J% = 1 TO NNLOAD%
6800     IF REST(I%,1) <> NLOAD(J%,1) THEN GOTO 6830
6810     K% = NLOAD(J%,2)
6820     REST(I%, 2 + K%) = REST(I%,2+K%) - NLOAD(J%,3)
6830   NEXT J%
6840 NEXT I%
6850 '
6860 ' ... Output the reactions
6870 '
6880 PRINT #2,
6890 PRINT #2, "* * * * * * * * * **"
6900 PRINT #2, "*   REACTIONS    *"
6910 PRINT #2, "* * * * * * * * * **"
6920 PRINT #2,
6930 PRINT #2, "                X-FORCE         Y-FORCE"
6940 TEMP(1) = 0 : TEMP(2) = 0
6950 FOR I% = 1 TO NREST%
6960   GOSUB 7900
6970   PRINT #2, USING "NODE ##"; REST(I%,1);
6980   PRINT #2, USING "      #.####^^^^"; REST(I%,3); REST(I%,4)
6990   TEMP(1) = TEMP(1) + REST(I%,3)
7000   TEMP(2) = TEMP(2) + REST(I%,4)
7010 NEXT I%
7020 GOSUB 7900
7030 PRINT #2, "                ---------       ---------"
7040 GOSUB 7900
7050 PRINT #2, "TOTAL  ";
7060 PRINT #2, USING "      #.####^^^^"; TEMP(1); TEMP(2)
7070 GOSUB 7810                                  '... Pause
7080 RETURN
7090 '
7100 ' *******************     GET KEYBOARD RESPONSE     *********************
7110 '
7120 ROW% = CSRLIN: COL% = POS(0)
```

```
7130 REPLY% = 0
7140 WHILE REPLY% = 0
7150   LOCATE ROW%, COL%: INPUT "", WRKSTR$: REPLY% = VAL(WRKSTR$)
7160   IF (REPLY% < 1) OR (REPLY% > NOPTIONS%) THEN GOTO 7190
7170   IF (REPLY% < 10) AND (LEN(WRKSTR$) = 1) THEN GOTO 7200
7180   IF (REPLY% > 9)  AND (LEN(WRKSTR$) = 2) THEN GOTO 7200
7190   SOUND 700, 4: REPLY% = 0: LOCATE ROW%, COL%
7200 WEND
7210 PRINT
7220 RETURN
7230 '
7240 '***********************   RECORD NUMBERS   *****************************
7250 '
7260 NNOS% = 0 : DONE% = TRUE%
7270 IF EOF(1)<>0 THEN DATAEND% = TRUE% : GOTO 7550
7280 '
7290 ' ... Read next line and check if it is blank
7300 '
7310 LINE INPUT #1, WRKSTR$
7320 LCOUNT% = LCOUNT% + 1 : LOCATE ROW%, COL% : PRINT LCOUNT%;
7330 FOR I% = 1 TO LEN(WRKSTR$)
7340   IF MID$(WRKSTR$,I%,1)<> " " THEN DONE% = FALSE% : I% = LEN(WRKSTR$)
7350 NEXT I%
7360 IF DONE% = TRUE% THEN GOTO 7550
7370 FOR I% = 1 TO 8
7380   WKSP$(I%) = ""
7390 NEXT I%
7400 '
7410 ' ... Search for numbers
7420 '
7430 FOR I% = 1 TO LEN(WRKSTR$)
7440   WHILE (MID$(WRKSTR$, I%, 1) = " ") AND I% <= LEN(WRKSTR$)
7450     I% = I% + 1
7460   WEND
7470   IF (I% > LEN(WRKSTR$)) THEN GOTO 7530
7480   IF(TARGET$="PTRUSS") AND LCOUNT% > 4 THEN DONE% = TRUE%
7490   NNOS% = NNOS% + 1: WKSP$(NNOS%) = ""
7500   WHILE (MID$(WRKSTR$, I%, 1) <> " ") AND I% <= LEN(WRKSTR$)
7510     WKSP$(NNOS%) = WKSP$(NNOS%) + MID$(WRKSTR$,I%,1): I% = I% + 1
7520   WEND
7530 NEXT I%
7540 WRKSTR$ = ""
7550 RETURN
7560 '
7570 '***************************   FIND STRING   ***************************
7580 '
7590 STRFOUND% = FALSE% : DONE% = FALSE%
7600 WHILE DONE% = FALSE%
7610   IF EOF(1)<>0 THEN DATAEND% = TRUE% : DONE% = TRUE% : GOTO 7690
7620   LINE INPUT #1, WRKSTR$
7630   LCOUNT% = LCOUNT% + 1 : LOCATE ROW%, COL% : PRINT LCOUNT%;
7640   IF INSTR(WRKSTR$,TARGET$)>0 THEN STRFOUND% = TRUE% : DONE% = TRUE%
7650   '
7660   ' ... Check for new module
7670   '
7680   IF LEFT$(WRKSTR$,4) ="+   " THEN DONE% = TRUE%
7690 WEND
7700 RETURN
7710 '
7720 '***************************   READ LINES   ****************************
7730 '
7740 FOR I% = 1 TO NLINES%
7750   IF EOF(1)<>0 THEN DATEND%=TRUE% : I% = NLINES% : GOTO 7780
7760   LINE INPUT #1, WRKSTR$
7770   LCOUNT% = LCOUNT% + 1 : LOCATE ROW%, COL% : PRINT LCOUNT%;
7780 NEXT I%
7790 RETURN
7800 '
7810 '********************   ANY KEY TO CONTINUE   **********************
7820 '
7830 IF DEVICE$<>"SCRN:" THEN GOTO 7880
7840 LOCATE PAGELEN%+1, 1
7850 PRINT SPACE$(40); "Press any key to continue";
7860 WHILE INKEY$ = "" : WEND
7870 CLS
7880 RETURN
7890 '
```

```
7900 '**********************    CHECK FOR PAGE END    *********************
7910 '
7920 IF CSRLIN>=PAGELEN% THEN GOSUB 7810
7930 RETURN
7940 '
7950 ' ======================    END - PTRUSS.BAS    =======================
```

Sample Run 1

The following run demonstrates program PTRUSS.BAS being used to analyse the truss in example 5.3.

```
==============================================================================

    PROGRAM PTRUSS                            Copyright(c) J.A.D.Balfour 1991

                        CASF 2nd Edition EXAMPLE 5.3

    File Name :- PT1.DAT                                  Date :- 26-02-92
    Units     :- jN and mm                                Time :- 20:06:08

==============================================================================


+ + + + + + + + + + + + + +
+   NODAL COORDINATES    +
+ + + + + + + + + + + + + +

NODE           X             Y
  1            0             0
  2            3000          3000
  3            6000          0

+ + + + + + + + + +
+   ELEMENTS     +
+ + + + + + + + + +

ELEMENT     PROPERTY
  1   2        1
  1   3        1
  2   3        1

+ + + + + + + + + +
+   PROPERTIES    +
+ + + + + + + + + +

PROPERTY          A             E
    1       0.3000E+04     0.2000E+03

+ + + + + + + + + +
+   RESTRAINTS    +
+ + + + + + + + + +

NODE    DIRECTION(S)
  1         12
  3          2

+ + + + + + + + + +
+   NODAL LOADS   +
+ + + + + + + + + +

NODE    DIRECTION          VALUE
  2          1          0.1000E+03
  2          2          0.2000E+03

* * * * * * * * * * *
*   DISPLACEMENTS    *              * :-   RESTRAINT
* * * * * * * * * * *

NODE     X-DISPLACEMENT  Y-DISPLACEMENT
  1             *               *
  2         0.4571E+00      0.1664E+01
  3         -.5000E+00          *
```

```
* * * * * * * * * * * * *
*   ELEMENT FORCES    *           TENSION POSITIVE
* * * * * * * * * * * * *

ELEMENT             FORCE
 1  2          0.2121E+03
 1  3          -.5000E+02
 2  3          0.7071E+02

* * * * * * * * *
*   REACTIONS   *
* * * * * * * * *

                  X-FORCE         Y-FORCE
NODE  1        -.1000E+03      -.1500E+03
NODE  3        0.3815E-05      -.5000E+02
               ----------      ----------
TOTAL          -.1000E+03      -.2000E+03
```

Sample Run 2

This run shows output from PTRUSS.PT while solving example 3.2.

```
=============================================================================

    PROGRAM PTRUSS                        Copyright(c) J.A.D.Balfour 1991

                      CASF 2nd Edition - Example 3.2

    File Name :- PT2.DAT                                 Date :- 26-02-92
    Units     :- jN mm & jg                              Time :- 20:06:37

=============================================================================

+ + + + + + + + + + + + + +
+   NODAL COORDINATES    +
+ + + + + + + + + + + + + +

NODE             X            Y
  1              0            0
  2              0            2000
  3              2000         2000
  4              2000         0

+ + + + + + + + +
+   ELEMENTS    +
+ + + + + + + + +

ELEMENT     PROPERTY
  1   2         1
  1   3         1
  1   4         1
  2   3         1
  2   4         1
  3   4         1

+ + + + + + + + +
+   PROPERTIES   +
+ + + + + + + + +

PROPERTY           A               E
     1      0.4000E+03      0.1000E+02

+ + + + + + + + + +
+   RESTRAINTS    +
+ + + + + + + + + +

NODE     DIRECTION(S)
  1          12
  4           2

+ + + + + + + + + +
+   NODAL LOADS   +
+ + + + + + + + + +
```

```
NODE     DIRECTION          VALUE
  3          1           0.1000E+02
  3          2           0.5000E+01


* * * * * * * * * * * *
*    DISPLACEMENTS    *               * :-   RESTRAINT
* * * * * * * * * * * *

NODE       X-DISPLACEMENT  Y-DISPLACEMENT
  1              *               *
  2          0.8321E+01      0.1723E+01
  3          0.1004E+02      -.7767E+00
  4          0.1723E+01          *


* * * * * * * * * * * * *
*    ELEMENT FORCES     *              TENSION POSITIVE
* * * * * * * * * * * * *

ELEMENT              FORCE
 1  2          0.3447E+01
 1  3          0.9268E+01
 1  4          0.3447E+01
 2  3          0.3447E+01
 2  4          -.4874E+01
 3  4          -.1553E+01


* * * * * * * * *
*    REACTIONS  *
* * * * * * * * *

                   X-FORCE           Y-FORCE
NODE  1          -.1000E+02        -.1000E+02
NODE  4          0.2384E-06        0.5000E+01
                 ---------         ---------
TOTAL            -.1000E+02        -.5000E+01
```

5.11 A Data Preprocessor - Program PRE.BAS

This program preprocesses the data for the plane truss analysis program PTRUSS.BAS. It will also preprocess data for the following programs, which are described in subsequent chapters

STRUSS.BAS	a space truss static analysis program.
PFRAME.BAS	a plane frame static analysis program.
PFDYNAM.BAS	a plane frame dynamic analysis program.
GRID.BAS	a grillage static analysis program.
SFRAME.BAS	a space frame static analysis program.

The data preprocessor program PRE.BAS allows either the creation of a new data file or the modification of an existing data file.

It is interesting to note that the preprocessor is far longer than the analysis program PTRUSS.BAS itself. This is perhaps testimony to the compactness and efficiency of structural analysis programs based upon the stiffness method. The program operates by a question and answer sequence that is intended to be self-explanatory.

Program Organisation

PRE.BAS first establishes whether an existing file is to be modified or a new file is to be created. If a file is to be modified then the existing file is read. If a new file is to be created then the program requests information about the job title, the units used, the nodal coordinates, the element locations, element properties, the restraints, and the loads. After each data set has been input PRE.BAS allows the user to edit the data.

Once the data is complete PRE.BAS treats the new data file exactly as a file to modified and allows the user to add, delete, or modify the data. Prior to modifying a block of data (e.g. loads) the block is displayed. Once the modifications have been made, that block of data is tidied and the revised data displayed. When the data is complete the user can either exit the program, check the data using the program DCHECK.BAS (described in section 5.12), or run PTRUSS.BAS with the current data file. In all cases the current data is copied to disk.

Data File Structure

PRE.BAS data files are plain ASCII file which can be listed on the screen or modified with a text editor (this is not recommended for beginners). In general, it is better to build/modify the file from within PRE.BAS as this avoids the possibility of corrupting the data file. The following information is intend to help the user who wishes to build or modify a data file using a text editior.

The data file starts with a header which must have contain the following information.

PROGRAM <PROGRAM NAME>
<ignored>
<TITLE>
<ignored>
<ignored on input - contains the file name and date on output>
<........>:-◇ <units string>

Notes

PROGRAM <PROGRAM NAME> must be in one the first four lines of the data file and <PROGRAM NAME> it must be one of the following strings "PTRUSS", "STRUSS", "PFRAME", "PFDYNAM", "GRID", "SFRAME". Lines preceding this line are ignored.

<TITLE> can contain any string

.... represents any character string and ◇ represents a space

The remainder of the data follows the header and is assumed to consist of a number of modules. A line containing a module name has this format

+ ◇ ◇ <MODULE NAME>

Where ◇ represents a space. The following module names are recognised.

NODAL COORDINATES
ELEMENTS
PROPERTIES
RESTRAINTS
LOCAL AXES
NODAL LOADS
ELEMENT LOADS
ADDED MASSES

Note that not all programs recognise all modules. For instance, only the dynamics program PFDYNAM.BAS recognises the module "ADDED MASS". The order of the modules is immaterial and on recognising a module name the file is read until a line containing numeric data is read. The module is then read line by line until a line containing no numeric data is encountered. At that point PRE.BAS then looks for the next module name. The numeric data must be in the order expected by PRE.BAS. For example when reading nodal coordinates for a space truss PRE.BAS expects to find three numbers per line with the first number being the *"x"* coordinate, the second the *"y"* coordinate and the third the *"z"* coordinate. The best way to find out what data PRE.BAS expects is to build a dummy file using PRE.BAS and then use that file as a template.

Listing

```
1000 '=========================      PRE.BAS      ================================
1010 '
1020 COMMON DATAFILE$
1030 OPTION BASE 1 : KEY OFF
1040 TRUE% = -1 : FALSE% = 0 : OK% = 1 : FAIL% = 0
1050 EDITING% = FALSE% : DEVICE$ = "SCRN:" : PAGELEN% = 23
1060 CLS
1070 PRINT
1080 PRINT "=================================================================="
1090 PRINT
1100 PRINT " PROGRAM PRE                              Copyright (c) J.Balfour 1991"
1110 PRINT
1120 PRINT "    This program preprocesses data for the analysis programs"
1130 PRINT "    PTRUSS, STRUSS, PFRAME, PFDYNAM, GRID, and SFRAME"
1140 PRINT
1150 PRINT "                 For further information contact"
1160 PRINT " James A.D.Balfour, Heriot-Watt University, Riccarton, Edinburgh"
1170 PRINT "                Tel 031-449-5111, Fax 031-451-3170"
1180 PRINT
1190 PRINT "=================================================================="
1200 '
1210 ' ... Note that variables are defined in Appendix A
1220 '
1230 DIM NODE(50,4)      '... Node array
1240 DIM ELEM(100,4)     '... Element array
```

```
1250 DIM PROP(20,7)      '... Property array
1260 DIM REST(20,2)      '... Restraint array
1270 DIM LAXES(20,2)     '... Local axes array
1280 DIM NLOAD(50,3)     '... Nodal load array
1290 DIM ELOAD(30,5)     '... Element load array
1300 DIM AMASS(30,2)     '... Added mass array
1310 DIM WKSP$(7)        '... Workspace array
1320 '
1330 PRINT
1340 IF DATAFILE$ <> "" THEN GOSUB 18300 : GOSUB 2730 : GOTO 1360
1350 GOSUB 2050
1360 WHILE (REPLY% <> 7)
1370   CLS : PRINT
1380   PRINT "Do you wish to"
1390   PRINT "  (1) Edit data"
1400   PRINT "  (2) Check the data (data will be written to file first)"
1410   PRINT "  (3) List all of the current data to the screen"
1420   PRINT "  (4) List all of the current data to the printer"
1430   PRINT "  (5) Analyse (write data to file first)"
1440   PRINT "  (6) Create/edit a new file (data will be written to file first)"
1450   PRINT "  (7) Stop (data will be written to file first)"
1460   NOPTIONS% = 7 : GOSUB 17440
1470   IF REPLY%>5 OR DATAFILEOK% THEN GOTO 1520
1480   CLS : PRINT
1490   PRINT "** ERROR ** Invalid choice - no valid data file loaded";
1500   GOSUB 18300
1510   GOTO  1530
1520   ON REPLY% GOSUB 4520, 4760, 4760, 4760, 4760, 4760, 4760
1530 WEND
1540 KEY ON
1550 END
1560 '
1570 ' ******************    SET MAXIMUM PROBLEM SIZE    ********************
1580 '
1590 IF INSTR(WRKSTR$, "PTRUSS") = 0 THEN GOTO 1660
1600 PTYPE$    = "PTRUSS"
1610 MAXNODE%  = 45 : MAXELEM%  = 70 : MAXPROP%  = 20 : MAXREST%  = 20
1620 MAXLAXES% =  1 : MAXNLOAD% = 50 : MAXELOAD% =  1 : MAXAMASS% =  1
1630 MAXBAND%  = 90
1640 GOTO 2030
1650 '
1660 IF INSTR(WRKSTR$, "STRUSS") = 0 THEN GOTO 1730
1670 PTYPE$    = "STRUSS"
1680 MAXNODE%  = 30 : MAXELEM%  = 50 : MAXPROP%  = 20 : MAXREST%  = 20
1690 MAXLAXES% =  1 : MAXNLOAD% = 50 : MAXELOAD% =  1 : MAXAMASS% =  1
1700 MAXBAND%  = 90
1710 GOTO 2030
1720 '
1730 IF INSTR(WRKSTR$, "PFRAME") = 0 THEN GOTO 1800
1740 PTYPE$    = "PFRAME"
1750 MAXNODE%  = 45 : MAXELEM%  =100 : MAXPROP%  = 20 : MAXREST%  = 20
1760 MAXLAXES% = 20 : MAXNLOAD% = 45 : MAXELOAD% = 20 : MAXAMASS% =  1
1770 MAXBAND%  = 40
1780 GOTO 2030
1790 '
1800 IF INSTR(WRKSTR$, "PFDYNAM") = 0 THEN GOTO 1870
1810 PTYPE$    = "PFDYNAM"
1820 MAXNODE%  = 17 : MAXELEM%  = 25 : MAXPROP%  = 10 : MAXREST%  = 10
1830 MAXLAXES% = 10 : MAXNLOAD% =  1 : MAXELOAD% =  1 : MAXAMASS% = 00
1840 MAXBAND%  = 51
1850 GOTO 2030
1860 '
1870 IF INSTR(WRKSTR$, "GRID") = 0 THEN GOTO 1940
1880 PTYPE$    = "GRID"
1890 MAXNODE%  = 50 : MAXELEM%  =100 : MAXPROP%  = 20 : MAXREST%  = 20
1900 MAXLAXES% =  1 : MAXNLOAD% = 60 : MAXELOAD% = 30 : MAXAMASS% =  1
1910 MAXBAND%  = 40
1920 GOTO 2030
1930 '
1940 IF INSTR(WRKSTR$, "SFRAME") = 0 THEN GOTO 2000
1950 PTYPE$    = "SFRAME"
1960 MAXNODE%  = 25 : MAXELEM%  = 50 : MAXPROP%  = 20 : MAXREST%  = 20
1970 MAXLAXES% =  1 : MAXNLOAD% = 40 : MAXELOAD% =  1 : MAXAMASS% =  1
1980 MAXBAND%  = 45
1990 GOTO 2030
2000 CLS : PRINT
2010 PRINT "** ERROR ** Not a PRE data file" : DATAFILEOK% = FALSE%
```

```
2020 GOSUB 18300
2030 RETURN
2040 '
2050 '**********************    CREATE/EDIT NEW FILE    **********************
2060 '
2070 PRINT "Do you wish to"
2080 PRINT "  (1)  Create a new data file"
2090 PRINT "  (2)  Modify an existing data file"
2100 NOPTIONS% = 2 : GOSUB 17440
2110 INPUT "Name of the data file  =  ", DATAFILE$
2120 FOR I% = 1 TO 7
2130   WKSP$(I%) = ""
2140 NEXT I%
2150 ON REPLY% GOSUB 2180, 2730           '...  Input or read data
2160 RETURN
2170 '
2180 '************************   INPUT DATA    *****************************
2190 '
2200 ' ...  This subroutine calls the routines that allow the problem data
2210 '      to be input
2220 '
2230 ' ... Check that data file can be opened
2240 '
2250 DATAFILEOK% = FALSE%
2260 ON ERROR GOTO 2330
2270 OPEN DATAFILE$ FOR INPUT AS #2
2280 ON ERROR GOTO 0 : CLOSE #2 : PRINT
2290 PRINT "File "; DATAFILE$; " aleady exists, overwrite (Y)es or (N)o ";
2300 INPUT WKSP$(1)
2310 IF (WKSP$(1)="y") OR (WKSP$(1)="Y") THEN GOTO 2340
2320 GOTO 2700
2330 RESUME 2340
2340 ON ERROR GOTO 2370
2350 OPEN DATAFILE$ FOR OUTPUT AS #2 : ON ERROR GOTO 0
2360 CLOSE #2 : GOTO 2410
2370 CLS : PRINT
2380 PRINT "** ERROR ** Failed to open file: "; DATAFILE$;
2390 GOSUB 18300
2400 CLOSE #2 : ON ERROR GOTO 0 : RESUME 2690
2410 OPEN "SCRN:" FOR OUTPUT AS #2
2420 CLS : LOCATE 4, 6
2430 PRINT "Note that pressing 'return' sets the current value equal to the"
2440 LOCATE 5, 6
2450 PRINT "corresponding value for the previously entered data item. For"
2460 LOCATE 6, 6
2470 PRINT "example pressing the return key when prompted for, say, an"
2480 LOCATE 7, 6
2490 PRINT "x-coordinate will set that coordinate equal to the x-coordinate"
2500 LOCATE 8, 6
2510 PRINT "of the previous node."
2520 GOSUB 18300
2530 REPLY%  = 0 : NNODE%  = 0 : NELEM%   = 0 : NPROP%  = 0 : NREST% = 0
2540 NLAXES% = 0 : NNLOAD% = 0 : NELOAD%  = 0 : NAMASS% = 0
2550 GOSUB 5310                                '... Problem type
2560 GOSUB 5520                                '... Job title, and units
2570 GOSUB 5730 : EDITING%=TRUE% : GOSUB 5730    '... Nodal coordinates
2580 GOSUB 6390 : EDITING%=TRUE% : GOSUB 6390    '... Input elements
2590 GOSUB 7040 : EDITING%=TRUE% : GOSUB 7040    '... Properties
2600 GOSUB 8090 : EDITING%=TRUE% : GOSUB 8090    '... Restraints
2610 IF (PTYPE$<>"PFRAME") AND (PTYPE$<>"PFDYNAM") THEN GOTO 2630
2620 GOSUB 9010 : EDITING%=TRUE% : GOSUB 9010    '... Local axes
2630 IF (PTYPE$="PFDYNAM") THEN GOTO 2650
2640 GOSUB 9580 : EDITING%=TRUE% : GOSUB 9580    '... Nodal loads
2650 IF (PTYPE$<>"PFRAME") AND (PTYPE$<>"GRID") THEN GOTO 2670
2660 GOSUB 10580 : EDITING%=TRUE%: GOSUB 10580   '... Element loads
2670 IF (PTYPE$<>"PFDYNAM") THEN GOTO 2690
2680 GOSUB 11330 : EDITING%=TRUE%:GOSUB 11330    '... Added masses
2690 DATAFILEOK% = TRUE%
2700 CLOSE #2
2710 RETURN
2720 '
2730 '**********************   READ DATA FILE    **************************
2740 '
2750 ' ...  This subroutine reads the data from a data file
2760 '
2770 PRINT
2780 PRINT "Wait - Reading file "; DATAFILE$; " line";
```

```
2790 ROW% = CSRLIN : COL% = POS(0) : LCOUNT% = 0
2800 DATAFILEOK% = FALSE% : DATAEND% = FALSE%
2810 NNODE%  = 0 : NELEM%  = 0 : NPROP%  = 0 : NREST% = 0 : NLAXES% = 0
2820 NNLOAD% = 0 : NELOAD% = 0 : NAMASS% = 0
2830 ON ERROR GOTO 2850 : OPEN DATAFILE$ FOR INPUT AS #1
2840 ON ERROR GOTO 0 : GOTO 2920
2850 CLS : PRINT
2860 PRINT "** ERROR ** Failed to open file "; DATAFILE$; " for input"
2870 GOSUB 18300
2880 RESUME 4490
2890 '
2900 ' ... Read problem type, set up array sizes and no of nos in each row
2910 '
2920 TARGET$ = "PROGRAM" : GOSUB 18050
2930 IF STRFOUND% THEN GOTO 2970
2940 CLS : PRINT
2950 PRINT "** ERROR ** Not a valid data file" : GOSUB 18300
2960 GOTO 4490
2970 GOSUB 1570                                   '...  Set up array sizes
2980 '
2990 ' ... Read title and units strings
3000 '
3010 NLINES% = 2 : GOSUB 18210 : IF STRFOUND% THEN GOTO 3050
3020 PRINT : PRINT
3030 PRINT "*** WARNING *** Unexpected end to data" : GOSUB 18300
3040 GOTO 4480
3050 WHILE (MID$(WRKSTR$,I%,1) = " ")
3060   I% = I% + 1
3070 WEND
3080 TITLE$  = MID$(WRKSTR$,I%)
3090 NLINES% = 3 : GOSUB 18210 : IF NOT STRFOUND% THEN GOTO 4470
3100 UNITS$ = MID$(WRKSTR$,INSTR(WRKSTR$,":- ")+3,24)
3110 GOTO 3120
3120 DATAFILEOK% = TRUE%
3130 WHILE NOT DATAEND%
3140   '
3150   ' ... Get module string
3160   '
3170   IF LEFT$(WRKSTR$,4)="+   " THEN GOTO 3190
3180   TARGET$ = "+   " : GOSUB 18050 : IF NOT STRFOUND% THEN GOTO 4470
3190   MODSTR$ = WRKSTR$
3200   '
3210   ' ...  Identify module
3220   '
3230   IF INSTR(MODSTR$,"NODAL COORDINATES")=0 THEN GOTO 3380
3240   '
3250   ' ... Read nodal coordinates
3260   '
3270   TARGET$ = "NODE" : GOSUB 18050 : IF NOT STRFOUND% THEN GOTO 4420
3280   GOSUB 17690
3290   WHILE NNOS%<>0
3300     NNODE% = NNODE% + 1
3310     FOR J% = 1 TO NNOS%
3320       NODE(NNODE%,J%) = VAL(WKSP$(J%))
3330     NEXT J%
3340     GOSUB 17690
3350   WEND
3360   GOTO 4470
3370   '
3380   IF INSTR(MODSTR$,"ELEMENTS")=0 THEN GOTO 3530
3390   '
3400   ' ... Read elements
3410   '
3420   TARGET$ = "ELEMENT" : GOSUB 18050 : IF NOT STRFOUND% THEN GOTO 4420
3430   GOSUB 17690
3440   WHILE NNOS%<>0
3450     NELEM% = NELEM% + 1
3460     FOR J% = 1 TO NNOS%
3470       ELEM(NELEM%,J%) = VAL(WKSP$(J%))
3480     NEXT J%
3490     GOSUB 17690
3500   WEND
3510   GOTO 4470
3520   '
3530   IF INSTR(MODSTR$,"PROPERTIES")=0 THEN GOTO 3680
3540   '
3550   ' ... Read properties
```

```
3560    '
3570    TARGET$ = "PROPERTY" : GOSUB 18050 : IF NOT STRFOUND% THEN GOTO 4420
3580    GOSUB 17690
3590    WHILE NNOS%<>0
3600      NPROP% = NPROP% + 1
3610      FOR J% = 1 TO NNOS%
3620        PROP(NPROP%,J%) = VAL(WKSP$(J%))
3630      NEXT J%
3640      GOSUB 17690
3650    WEND
3660    GOTO 4470
3670    '
3680    IF INSTR(MODSTR$,"RESTRAINTS")=0 THEN GOTO 3830
3690    '
3700    ' ... Read restraints
3710    '
3720    TARGET$ = "NODE" : GOSUB 18050 : IF NOT STRFOUND% THEN GOTO 4420
3730    GOSUB 17690
3740    WHILE NNOS%<>0
3750      NREST% = NREST% + 1
3760      FOR J% = 1 TO NNOS%
3770        REST(NREST%,J%) = VAL(WKSP$(J%))
3780      NEXT J%
3790      GOSUB 17690
3800    WEND
3810    GOTO 4470
3820    '
3830    IF INSTR(MODSTR$,"LOCAL AXES")=0 THEN GOTO 3980
3840    '
3850    ' ... Read local axes
3860    '
3870    TARGET$ = "NODE" : GOSUB 18050 : IF NOT STRFOUND% THEN GOTO 4420
3880    GOSUB 17690
3890    WHILE NNOS%<>0
3900      NLAXES% = NLAXES% + 1
3910      FOR J% = 1 TO NNOS%
3920        LAXES(NLAXES%,J%) = VAL(WKSP$(J%))
3930      NEXT J%
3940      GOSUB 17690
3950    WEND
3960    GOTO 4470
3970    '
3980    IF INSTR(MODSTR$,"NODAL LOADS")=0 THEN GOTO 4130
3990    '
4000    ' ... Read nodal loads
4010    '
4020    TARGET$ = "NODE" : GOSUB 18050 : IF NOT STRFOUND% THEN GOTO 4420
4030    GOSUB 17690
4040    WHILE NNOS%<>0
4050      NNLOAD% = NNLOAD% + 1
4060      FOR J% = 1 TO NNOS%
4070        NLOAD(NNLOAD%,J%) = VAL(WKSP$(J%))
4080      NEXT J%
4090      GOSUB 17690
4100    WEND
4110    GOTO 4470
4120    '
4130    IF INSTR(MODSTR$,"ELEMENT LOADS")=0 THEN GOTO 4280
4140    '
4150    ' ... Read element loads
4160    '
4170    TARGET$ = "TYPE" : GOSUB 18050 : IF NOT STRFOUND% THEN GOTO 4420
4180    GOSUB 17690
4190    WHILE NNOS%<>0
4200      NELOAD% = NELOAD% + 1
4210      FOR J% = 1 TO NNOS%
4220        ELOAD(NELOAD%,J%) = VAL(WKSP$(J%))
4230      NEXT J%
4240      GOSUB 17690
4250    WEND
4260    GOTO 4470
4270    '
4280    IF INSTR(MODSTR$,"ADDED MASS")=0 THEN GOTO 4420
4290    '
4300    ' ... Read added mass
4310    '
4320    TARGET$ = "NODE" : GOSUB 18050 : IF NOT STRFOUND% THEN GOTO 4420
```

```
4330   GOSUB 17690
4340   WHILE NNOS%<>0
4350     NAMASS% = NAMASS% + 1
4360     FOR J% = 1 TO NNOS%
4370       AMASS(NAMASS%,J%) = VAL(WKSP$(J%))
4380     NEXT J%
4390     GOSUB 17690
4400   WEND
4410   GOTO 4470
4420   CLS : PRINT
4430   PRINT "** ERROR ** in module " MODSTR$; " run DCHECK for more info.";
4440   GOSUB 18300
4450   PRINT "Wait - Reading file "; DATAFILE$; " line";
4460   ROW% = CSRLIN : COL% = POS(0)
4470 WEND
4480 GOSUB 17140                        '... List problem size
4490 ON ERROR GOTO 0 : CLOSE #1
4500 RETURN
4510 '
4520 '*************************    EDIT DATA    ******************************
4530 '
4540 ' ...  This subroutine calls the data editing subroutines
4550 '
4560 OPEN "SCRN:" FOR OUTPUT AS #2
4570 WHILE REPLY%<>10
4580   CLS : PRINT
4590   PRINT "Do you wish to edit"
4600   PRINT "  (1) Job title and units"
4610   PRINT "  (2) Nodal coordinates"
4620   PRINT "  (3) Elements"
4630   PRINT "  (4) Properties"
4640   PRINT "  (5) Restraints"
4650   PRINT "  (6) Local axes"
4660   PRINT "  (7) Nodal loads"
4670   PRINT "  (8) Element loads"
4680   PRINT "  (9) Added masses"
4690   PRINT " (10) Stop editing data"
4700   NOPTIONS% = 10 : GOSUB 17440 : IF REPLY%<10 THEN EDITING%=TRUE%
4710   ON REPLY% GOSUB 11910,5730,6390,7040,8090,9010,9580,10580,11330
4720 WEND
4730 CLOSE #2
4740 RETURN
4750 '
4760 '**********************    WRITE THE DATA    ***************************
4770 '
4780 ' ... This subroutine calls the subroutines that write the data to
4790 '     the screen, disk file or the printer
4800 '
4810 IF DATAFILEOK% THEN GOTO 4870
4820 IF REPLY%=6 THEN GOTO 5140
4830 CLS : PRINT
4840 PRINT "** ERROR ** No valid data to write to the data file"
4850 GOSUB 18300
4860 GOTO  5150
4870 ON REPLY% GOTO 4890, 4980, 4890, 4910, 4980, 4980, 4980
4880 '
4890 OPEN "SCRN:" FOR OUTPUT AS #2   : GOTO 5010
4900 '
4910 CLS : PRINT : PRINT "Wait - looking for printer"
4920 ON ERROR GOTO 4950
4930 OPEN "LPT1:" FOR OUTPUT AS #2 : PRINT #2,
4940 PRINT "Wait - printing data" : DEVICE$ = "LPT1:" : GOTO 5010
4950 PRINT : PRINT "** ERROR ** Failed to find the printer" :
4960 GOSUB 18300 : RESUME 5110
4970 '
4980 CLS : PRINT
4990 PRINT "Wait - writing data to file "; DATAFILE$
5000 OPEN DATAFILE$ FOR OUTPUT AS #2 : DEVICE$ = DATAFILE$
5010 GOSUB 14590          '... List header
5020 GOSUB 14910          '... List nodal coordinates
5030 GOSUB 15140          '... List element data
5040 GOSUB 15410          '... List list element properties
5050 GOSUB 16140          '... List restraint data
5060 GOSUB 16330          '... List local axes
5070 GOSUB 16520          '... List nodal loads
5080 GOSUB 16720          '... List element loads
5090 GOSUB 16950          '... List added masses
```

```
5100 IF REPLY%=4 THEN PRINT #2, CHR$(12);
5110 ON ERROR GOTO 0 : CLOSE #2 : DEVICE$ = "SCRN:"
5120 IF REPLY%=2 THEN GOSUB 5240
5130 IF REPLY%=5 THEN GOSUB 5170
5140 IF REPLY%=6 THEN CLS : GOSUB 2050
5150 RETURN
5160 '
5170 '**********************     RUN ANALYSIS PROGRAM     **********************
5180 '
5190 PRINT : PRINT
5200 PRINT "Wait - chaining "; PTYPE$;
5210 CHAIN PTYPE$
5220 END
5230 '
5240 '**********************     RUN DATA CHECK     **************************
5250 '
5260 PRINT : PRINT
5270 PRINT "Wait - chaining DCHECK"
5280 CHAIN "DCHECK"
5290 END
5300 '
5310 '**********************     INPUT STRUCTURE TYPE     **********************
5320 '
5330 CLS : PRINT : PRINT
5340 PRINT "Select problem type"
5350 PRINT "  (1) Plane truss"
5360 PRINT "  (2) Space truss"
5370 PRINT "  (3) Plane frame - static  analysis"
5380 PRINT "  (4) Plane frame - dynamic analysis"
5390 PRINT "  (5) Grillage"
5400 PRINT "  (6) Space frame"
5410 NOPTIONS%  = 6 : GOSUB 17440
5420 IF (REPLY% = 1) THEN WRKSTR$ = "PTRUSS"
5430 IF (REPLY% = 2) THEN WRKSTR$ = "STRUSS"
5440 IF (REPLY% = 3) THEN WRKSTR$ = "PFRAME"
5450 IF (REPLY% = 4) THEN WRKSTR$ = "PFDYNAM"
5460 IF (REPLY% = 5) THEN WRKSTR$ = "GRID"
5470 IF (REPLY% = 6) THEN WRKSTR$ = "SFRAME"
5480 GOSUB 1570                          '... Set maximum problem size
5490 GOSUB 17140                         '... List maximum problem size
5500 RETURN
5510 '
5520 '**********************     INPUT JOB TITLE AND UNITS     ******************
5530 '
5540 CLS : PRINT
5550 PRINT "Input job title and units - job title is simply a string to"
5560 PRINT "                              to identify the problem"
5570 PRINT
5580 PRINT "                            - note that the units string is for"
5590 PRINT "                              information only and is ignored"
5600 PRINT "                              during calculation. Consistent units
5610 PRINT "                              must be used throughout."
5620 PRINT
5630 PRINT "                            - use units of Newtons, metres and"
5640 PRINT "                              kilograms for dynamic analysis"
5650 PRINT
5660 LINE INPUT "Job title                        = ", TITLE$
5670 PRINT
5680 IF PTYPE$="PFDYNAM" THEN GOTO 5700
5690 LINE INPUT "Units (force and length)         = ", UNITS$ : GOTO 5710
5700 LINE INPUT "Units (force, length and mass) = ", UNITS$
5710 RETURN
5720 '
5730 '********************     INPUT/EDIT NODAL COORDINATES     *****************
5740 '
5750 DONE% = FALSE% : NODENO% = 0
5760 IF EDITING% THEN GOTO 5790
5770 CLS : PRINT
5780 PRINT "Input nodal coordinates" : GOSUB 6270 : GOTO 5810
5790 GOSUB 14910 : PRINT
5800 PRINT "Edit nodal coordinates - set node number to H for help"
5810 DONE% = FALSE% : NODENO% = 0
5820 WHILE NOT DONE%
5830   PRINT
5840   PRINT "Node number      =  ";
5850   IF EDITING% THEN J% = 1 : GOSUB 17580     : GOTO 5910
5860   NODENO% = NODENO%+1
```

```
5870    IF NODENO%<0 THEN LOCATE CSRLIN,21 ELSE LOCATE CSRLIN,20
5880    PRINT; NODENO%;
5890    LOCATE CSRLIN,21
5900    INPUT "", WKSP$(1)
5910    IF (WKSP$(1)="H") OR   (WKSP$(1)="h")  THEN GOSUB 6270: GOTO 5950
5920    IF (WKSP$(1)<>"L") AND (WKSP$(1)<>"l") THEN GOTO 5960
5930    IF EDITING% THEN GOSUB 14910
5940    IF NOT EDITING% THEN EDITING%=TRUE% : GOSUB 14910 : EDITING%=FALSE%
5950    NODENO% = NODENO% - 1 : GOTO 6230
5960    IF WKSP$(1)="" THEN WKSP$(1) = STR$(NODENO%)
5970    NODENO% = FIX(VAL(WKSP$(1)))
5980    IF NODENO% = 0 THEN DONE% = TRUE%       : GOTO  6230
5990    IF NODENO% < 0 THEN GOTO 6070
6000    PRINT "X-coordinate      =  "; : J% = 2  : GOSUB 17580
6010    PRINT "Y-coordinate      =  "; : J% = 3  : GOSUB 17580
6020    IF (PTYPE$ <> "STRUSS") AND (PTYPE$ <> "SFRAME") THEN GOTO  6070
6030    PRINT "Z-coordinate      =  "; : J% = 4  : GOSUB 17580
6040    '
6050    ' ... Check for replacement/deleted nodes
6060    '
6070    K% = 0
6080    FOR I% = 1 TO NNODE%
6090      IF NODE(I%,1) = ABS(NODENO%) THEN K% = I% : I% = NNODE%
6100    NEXT I%
6110    '
6120    ' ... Check if maximum number of nodes exceeded
6130    '
6140    IF K%=0 THEN K% = NNODE% + 1 : NNODE% = K%
6150    IF (NNODE%<=MAXNODE%) THEN GOTO 6190
6160    PRINT : PRINT "** ERROR ** Maximum number of nodes exceeed"
6170    GOSUB 18300
6180    NNODE% = NNODE% - 1 : DONE%=TRUE% : GOTO 6230
6190    NODE(K%,1) = FIX(VAL(WKSP$(1)))
6200    NODE(K%,2) = VAL(WKSP$(2))
6210    NODE(K%,3) = VAL(WKSP$(3))
6220    NODE(K%,4) = VAL(WKSP$(4))
6230 WEND
6240 EDITING% = FALSE%
6250 RETURN
6260 '
6270 '++++++++++++++   NODAL COORDINATE HELP SCREEN   ++++++++++++++++++++
6280 '
6290 PRINT
6300 PRINT "Nodal coordinates - set node number to
6310 PRINT "                         0 to stop"
6320 PRINT "                         H for help"
6330 PRINT "                         L to list nodal coordinates data"
6340 PRINT "                       - node numbers must be consecutive integers"
6350 PRINT "                         starting at 1 (can be entered in any order)"
6360 PRINT "                       - set node number negative to cancel node"
6370 RETURN
6380 '
6390 '********************   INPUT/EDIT ELEMENT DATA   ********************
6400 '
6410 IF EDITING% THEN GOTO 6440
6420 CLS : PRINT
6430 PRINT "Input element data" : GOSUB 6890 : GOTO 6460
6440 GOSUB 15140 : PRINT
6450 PRINT "Edit element data - set lower node number to H for help"
6460 DONE% = FALSE%
6470 WHILE NOT DONE%
6480   PRINT
6490   PRINT "Lower  node number (i)  =  "; : J% = 1 : GOSUB 17580
6500   IF (WKSP$(1)="H")  THEN GOSUB 6890            : GOTO 6860
6510   IF (WKSP$(1)<>"L") THEN GOTO 6550
6520   IF EDITING%        THEN GOSUB 15140
6530   IF NOT EDITING%    THEN EDITING%=TRUE% : GOSUB 15140 : EDITING%=FALSE%
6540   GOTO 6860
6550   IF (FIX(VAL(WKSP$(1))) = 0) THEN DONE%=TRUE%  : GOTO 6860
6560   PRINT "Higher node number (j)  =  "; : J% = 2 : GOSUB 17580
6570   IF FIX(VAL(WKSP$(1)))<FIX(VAL(WKSP$(2))) THEN GOTO 6600
6580   PRINT : PRINT "Please input lower node number first"
6590   GOTO 6480
6600   PRINT "Property number          =  "; : J% = 3 : GOSUB 17580
6610   IF (PTYPE$ <> "PFRAME") AND (PTYPE$ <> "PFDYNAM") THEN GOTO 6630
6620   PRINT "Element type             =  "; : J% = 4 : GOSUB 17580 : GOTO 6680
6630   IF (PTYPE$ <> "SFRAME") THEN GOTO 6680
```

```
6640    PRINT "Beta (degrees)          =  "; : J% = 4 : GOSUB 17580 : GOTO 6680
6650    '
6660    ' ...  Check if existing element is to be overwriten
6670    '
6680    K% = 0
6690    FOR I% = 1 TO NELEM%
6700      IF ELEM(I%,1) <> FIX(VAL(WKSP$(1))) THEN GOTO 6720
6710      IF ELEM(I%,2) =  FIX(VAL(WKSP$(2))) THEN K% = I% : GOTO 6760
6720    NEXT I%
6730    '
6740    ' ... Check if maximum number of elements exceeded
6750    '
6760    IF K%=0 THEN K% = NELEM% + 1 : NELEM% = K%
6770    IF (NELEM%<=MAXELEM%) THEN GOTO 6830
6780    PRINT
6790    PRINT "** ERROR ** Maximum number of nodes exceeed"
6800    PRINT "            Try listing data to remove any deleted elements"
6810    GOSUB 18300
6820    NELEM% = NELEM% - 1 : DONE%=TRUE% : GOTO 6860
6830    FOR J% = 1 TO 4
6840      ELEM(K%,J%) = FIX(VAL(WKSP$(J%)))
6850    NEXT J%
6860 WEND
6870 EDITING% = FALSE%
6880 RETURN
6890 '++++++++++++++++++++   ELEMENT HELP SCREEN   +++++++++++++++++++++++++
6900 '
6910 PRINT
6920 PRINT "Element data - set lower node number to"
6930 PRINT "                        0 to stop"
6940 PRINT "                        H for help"
6950 PRINT "                        L to list element data"
6960 PRINT "                - set property no. to zero to cancel element"
6970 IF (PTYPE$ <> "PFRAME") AND (PTYPE$ <> "PFDYNAM") THEN GOTO 7020
6980 PRINT "                - element types :  no pins          = 1,    ";
6990 PRINT "pin at end 'j'     = 2"
7000 PRINT "                                   pin at end 'i' = 3,    ";
7010 PRINT "pin at 'i' and 'j' = 4"
7020 RETURN
7030 '
7040 '**********************   INPUT/EDIT PROPERTIES   *********************
7050 '
7060 IF EDITING% THEN GOTO 7090
7070 CLS : PRINT
7080 PRINT "Input property data" : GOSUB 7950 : GOTO 7110
7090 GOSUB 15410 : PRINT
7100 PRINT "Edit property data - set property number to H for help"
7110 DONE% = FALSE%
7120 WHILE NOT DONE%
7130    PRINT
7140    PRINT "Property number            ";
7150    IF (PTYPE$="SFRAME") THEN PRINT "        =  "; ELSE PRINT "=  ";
7160    J% = 1 : GOSUB 17580
7170    IF (WKSP$(1)="H")   THEN GOSUB 7950 : GOTO 7580
7180    IF (WKSP$(1)<>"L")  THEN GOTO 7220
7190    IF EDITING%         THEN GOSUB 15410
7200    IF NOT EDITING%     THEN EDITING%=TRUE% : GOSUB 15410 : EDITING%=FALSE%
7210    GOTO 7580
7220    IF (FIX(VAL(WKSP$(1)))=0) THEN DONE%=TRUE% : GOTO 7580
7230    '
7240    ' ... If editing an existing set of properties values copy into WKSP$
7250    '
7260    FOR I% = 1 TO NPROP%
7270      IF PROP(I%,1) <> FIX(VAL(WKSP$(1))) THEN GOTO 7310
7280      FOR J% = 1 TO 7
7290        WKSP$(J%) = MID$(STR$(PROP(I%,J%)),2,255)
7300      NEXT J%
7310    NEXT I%
7320    IF PTYPE$ = "PTRUSS"  THEN GOSUB 7620
7330    IF PTYPE$ = "STRUSS"  THEN GOSUB 7620
7340    IF PTYPE$ = "PFRAME"  THEN GOSUB 7680
7350    IF PTYPE$ = "PFDYNAM" THEN GOSUB 7680
7360    IF PTYPE$ = "GRID"    THEN GOSUB 7770
7370    IF PTYPE$ = "SFRAME"  THEN GOSUB 7850
7380    '
7390    ' ...  Check if existing property to be overwritten
7400    '
```

```
7410   K% = 0
7420   FOR I% = 1 TO NPROP%
7430     IF PROP(I%,1) = VAL(WKSP$(1)) THEN K% = I% : GOTO 7480
7440   NEXT I%
7450   '
7460   ' ... Check if maximum number of properties exceeded
7470   '
7480   IF K%=0 THEN K% = NPROP% + 1 : NPROP% = K%
7490   IF (NPROP%<=MAXPROP%) THEN GOTO 7530
7500   PRINT : PRINT "** ERROR ** Maximum number of properties exceeed"
7510   GOSUB 18300
7520   NPROP% = NPROP% - 1 : DONE%=TRUE% : GOTO 7580
7530   FOR J% = 1 TO 7
7540     PROP(K%,J%) = VAL(WKSP$(J%))
7550   NEXT J%
7560   PROP(K%,1) = FIX(PROP(K%,1))
7570   PROP(K%,2) = PROP(K%,2)
7580 WEND
7590 EDITING% = FALSE%
7600 RETURN
7610 '
7620 '+++++++++++++++++++++++   TRUSS MEMBER   ++++++++++++++++++++++++++++
7630 '
7640 PRINT "Cross-sectional area (A)   =  "; : J% = 2 : GOSUB 17580
7650 PRINT "Elastic constant     (E)   =  "; : J% = 3 : GOSUB 17580
7660 RETURN
7670 '
7680 '+++++++++++++++++++++   PLANE FRAME MEMBER   ++++++++++++++++++++++++
7690 '
7700 PRINT "Cross-sectional area  (A)  =  "; : J% = 2 : GOSUB 17580
7710 PRINT "Second moment of area (I)  =  "; : J% = 3 : GOSUB 17580
7720 PRINT "Elastic constant      (E)  =  "; : J% = 4 : GOSUB 17580
7730 IF (PTYPE$ = "PFRAME") THEN GOTO 7750
7740 PRINT "Mass per unit length  (m)  =  "; : J% = 5 : GOSUB 17580
7750 RETURN
7760 '
7770 '+++++++++++++++++++++++   GRILLAGE MEMBER   +++++++++++++++++++++++++
7780 '
7790 PRINT "Second moment of area (I)  =  "; : J% = 2 : GOSUB 17580
7800 PRINT "Torsional constant    (J)  =  "; : J% = 3 : GOSUB 17580
7810 PRINT "Modulus of rigidity   (G)  =  "; : J% = 4 : GOSUB 17580
7820 PRINT "Elastic constant      (E)  =  "; : J% = 5 : GOSUB 17580
7830 RETURN
7840 '
7850 '+++++++++++++++++++++++   SPACE FRAME MEMBER   ++++++++++++++++++++++
7860 '
7870 PRINT "Cross-sectional area          (A)    =  "; : J% = 2 : GOSUB 17580
7880 PRINT "Second moment of area Y-Y     (Iyy)  =  "; : J% = 3 : GOSUB 17580
7890 PRINT "Second moment of area Z-Z     (Izz)  =  "; : J% = 4 : GOSUB 17580
7900 PRINT "Torsional second mt of area   (Jxx)  =  "; : J% = 5 : GOSUB 17580
7910 PRINT "Modulus of rigidity           (G)    =  "; : J% = 6 : GOSUB 17580
7920 PRINT "Elastic constant              (E)    =  "; : J% = 7 : GOSUB 17580
7930 RETURN
7940 '
7950 '+++++++++++++++   ELEMENT PROPERTY HELP SCREEN   +++++++++++++++++++++
7960 '
7970 PRINT
7980 PRINT "Element properties - set property number to"
7990 PRINT "                          0 to stop"
8000 PRINT "                          H for help"
8010 PRINT "                          L to list property data"
8020 IF PTYPE$ = "GRID" THEN GOTO 8050
8030 PRINT "                        - set cross sectional area to 0 to cancel
property"
8040 GOTO 8060
8050 PRINT "                        - set second moment of area to 0 to cancel
property"
8060 PRINT "                        - hit return to retain any value"
8070 RETURN
8080 '
8090 '*********************   INPUT/EDIT RESTRAINTS   *********************
8100 '
8110 IF EDITING% THEN GOTO 8140
8120 CLS : PRINT
8130 PRINT "Input restraints" : GOSUB 8480 : GOTO 8160
8140 GOSUB 16140 : PRINT
8150 PRINT "Edit restraints - set node number to H for help"
```

```
8160 DONE% = FALSE%
8170 WHILE NOT DONE%
8180 PRINT
8190   PRINT "Node to be restrained  =  "; : J% = 1 : GOSUB 17580
8200   IF (WKSP$(1)="H")  THEN GOSUB 8480           : GOTO 8440
8210   IF (WKSP$(1)<>"L") THEN GOTO 8250
8220   IF EDITING%        THEN GOSUB 16140
8230   IF NOT EDITING%    THEN EDITING%=TRUE% : GOSUB 16140 : EDITING%=FALSE%
8240   GOTO 8440
8250   IF (FIX(VAL(WKSP$(1)))=0) THEN DONE%=TRUE%   : GOTO 8440
8260   PRINT "Direction(s)            =  "; : J% = 2 : GOSUB 17580
8270   '
8280   '  ...  Check if existing restraint is to be overwritten
8290   '
8300   K% = 0
8310   FOR I% = 1 TO NREST%
8320     IF REST(I%,1) = FIX(VAL(WKSP$(1))) THEN K% = I% : GOTO 8370
8330   NEXT I%
8340   '
8350   ' ... Check if maximum number of restraints exceeded
8360   '
8370   IF K%=0 THEN K% = NREST% + 1 : NREST% = K%
8380   IF (NREST%<=MAXREST%) THEN GOTO 8420
8390   PRINT : PRINT "** ERROR ** Maximum number of restraints exceeed"
8400   GOSUB 18300
8410   NREST% = NREST% - 1 : DONE%=TRUE% : GOTO 8440
8420   REST(K%,1) = FIX(VAL(WKSP$(1)))
8430   REST(K%,2) = FIX(VAL(WKSP$(2)))
8440 WEND
8450 EDITING% = FALSE%
8460 RETURN
8470 '
8480 '++++++++++++++++   RESTRAINTS HELP SCREEN   +++++++++++++++++++++++++
8490 '
8500 PRINT
8510 PRINT "Restraints - set node number to
8520 PRINT "                 0 to stop
8530 PRINT "                 H for help"
8540 PRINT "                 L to list restraint data"
8550 PRINT "           - set restraint direction(s) to 0 to cancel"
8560 PRINT "           - restraint directions are"
8570 IF PTYPE$ = "PTRUSS"  THEN GOSUB 8670
8580 IF PTYPE$ = "STRUSS"  THEN GOSUB 8730
8590 IF PTYPE$ = "PFRAME"  THEN GOSUB 8800
8600 IF PTYPE$ = "PFDYNAM" THEN GOSUB 8800
8610 IF PTYPE$ = "GRID"    THEN GOSUB 8870
8620 IF PTYPE$ = "SFRAME"  THEN GOSUB 8940
8630 PRINT "           - enter restraints as a composite number e.g. if a";
8640 PRINT "             node is restrained in directions 1 and 2 enter 12"
8650 RETURN
8660 '
8670 '+++++++++++++++++++++++++   PLANE TRUSS   ++++++++++++++++++++++++++
8680 '
8690 PRINT "                 x-translation = 1"
8700 PRINT "                 y-translation = 2"
8710 RETURN
8720 '
8730 '+++++++++++++++++++++++++   SPACE TRUSS   ++++++++++++++++++++++++++
8740 '
8750 PRINT "                 x-translation = 1"
8760 PRINT "                 y-translation = 2"
8770 PRINT "                 z-translation = 3"
8780 RETURN
8790 '
8800 '+++++++++++++++++++++++++   PLANE FRAME   ++++++++++++++++++++++++++
8810 '
8820 PRINT "                 x-translation = 1"
8830 PRINT "                 y-translation = 2"
8840 PRINT "                 rotation      = 3"
8850 RETURN
8860 '
8870 '++++++++++++++++++++++++++   GRILLAGE   +++++++++++++++++++++++++++
8880 '
8890 PRINT "                 z-translation = 1"
8900 PRINT "                 x-rotation    = 2"
8910 PRINT "                 y-rotation    = 3"
8920 RETURN
```

```
8930 '
8940 '+++++++++++++++++++++++++++   SPACE FRAME   +++++++++++++++++++++++++++
8950 '
8960 PRINT "                    x-translation = 1,   y-translation = 2"
8970 PRINT "                    z-translation = 3,   x-rotation    = 4"
8980 PRINT "                    y-rotation    = 5,   z-rotation    = 6"
8990 RETURN
9000 '
9010 '**********************   INPUT/EDIT LOCAL AXES   **********************
9020 '
9030 '
9040 IF (PTYPE$="PFRAME") OR (PTYPE$="PFDYNAM") THEN GOTO 9090
9050 IF INPUTING% THEN GOTO 10050
9060 PRINT : PRINT "Local axes are not allowed in program "; PTYPE$;
9070 GOSUB 18300
9080 GOTO 9440
9090 IF EDITING% THEN GOTO 9120
9100 CLS : PRINT
9110 PRINT "Input local axes" : GOSUB 9460 : GOTO 9140
9120 GOSUB 16330 : PRINT
9130 PRINT "Edit local axes - set node to H for help"
9140 DONE% = FALSE%
9150 WHILE NOT DONE%
9160   PRINT
9170   PRINT "Node            =  "; : J% = 1        : GOSUB 17580
9180   IF (WKSP$(1)="H")  THEN GOSUB 9460  : GOTO 9420
9190   IF (WKSP$(1)<>"L") THEN GOTO 9230
9200   IF EDITING%        THEN GOSUB 15410
9210   IF NOT EDITING%    THEN EDITING%=TRUE%:GOSUB 15410:EDITING%=FALSE%
9220   GOTO 9420
9230   IF (FIX(VAL(WKSP$(1)))=0) THEN DONE%=TRUE% : GOTO 9420
9240   PRINT "Rotation (deg) =  "; : J% = 2        : GOSUB 17580
9250   '
9260   ' ... Check if existing axes to be overwritten
9270   '
9280   K% = 0
9290   FOR I% = 1 TO NLAXES%
9300     IF LAXES(I%,1) = FIX(VAL(WKSP$(1))) THEN K% = I% : GOTO 9350
9310   NEXT I%
9320   '
9330   ' ... Check if maximum number of local axes exceeded
9340   '
9350   IF K%=0 THEN K% = NLAXES% + 1 : NLAXES% = K%
9360   IF (NLAXES%<=MAXLAXES%) THEN GOTO 9400
9370   PRINT : PRINT "** ERROR ** Maximum number of local axes exceeed"
9380   GOSUB 18300
9390   NLAXES% = NLAXES% - 1 : DONE%=TRUE% : GOTO 9420
9400   LAXES(K%,1) = FIX(VAL(WKSP$(1)))
9410   LAXES(K%,2) = VAL(WKSP$(2))
9420 WEND
9430 EDITING% = FALSE%
9440 RETURN
9450 '
9460 '+++++++++++++++++++++   LOCAL AXES HELP SCREEN   +++++++++++++++++++++
9470 '
9480 PRINT
9490 PRINT "Local axes - set node number to"
9500 PRINT "                    0 to stop
9510 PRINT "                    H for help"
9520 PRINT "                    L to list local axes data"
9530 PRINT "             - set rotation to zero to cancel local axes"
9540 PRINT "             - note that anticlockwise rotation from the global";
9550 PRINT "               axes is +ve"
9560 RETURN
9570 '
9580 '*********************   INPUT/EDIT NODAL LOADS   **********************
9590 '
9600 IF (PTYPE$<>"PFDYNAM") THEN GOTO 9650
9610 IF INPUTING% THEN GOTO 10050
9620 PRINT : PRINT "Nodal loads are not allowed in program "; PTYPE$;
9630 GOSUB 18300
9640 GOTO 10050
9650 IF EDITING% THEN GOTO 9680
9660 CLS : PRINT
9670 PRINT "Input nodal loads" : GOSUB 10070 : GOTO 9700
9680 GOSUB 16520 : PRINT
9690 PRINT "Edit nodal loads - set node to H for help"
```

```
9700 DONE% = FALSE%
9710 WHILE NOT DONE%
9720   PRINT
9730   PRINT "Node         =  "; : J% = 1          : GOSUB 17580
9740   IF (WKSP$(1)="H")  THEN GOSUB 10070         : GOTO  10030
9750   IF (WKSP$(1)<>"L") THEN GOTO 9790
9760   IF EDITING%        THEN GOSUB 16520
9770   IF NOT EDITING%    THEN EDITING%=TRUE% : GOSUB 16520 : EDITING%=FALSE%
9780   GOTO 10030
9790   IF (FIX(VAL(WKSP$(1)))=0) THEN DONE%=TRUE% : GOTO  10030
9800   PRINT "Direction    =  "; : J% = 2          : GOSUB 17580
9810   PRINT "Magnitude    =  "; : J% = 3          : GOSUB 17580
9820   '
9830   ' ... Overwrite if this load already input
9840   '
9850   K% = 0
9860   FOR I% = 1 TO NNLOAD%
9870     IF NLOAD(I%,1) <> FIX(VAL(WKSP$(1))) THEN GOTO 9890
9880     IF NLOAD(I%,2) =  ABS(FIX(VAL(WKSP$(2)))) THEN K% = I% : GOTO 9930
9890   NEXT I%
9900   '
9910   ' ... Check if maximum number of nodal loads exceeded
9920   '
9930   IF K%=0 THEN K% = NNLOAD% + 1 : NNLOAD% = K%
9940   IF (NNLOAD%<=MAXNLOAD%) THEN GOTO 9980
9950   PRINT : PRINT "** ERROR ** Maximum number of nodal loads exceeed"
9960   GOSUB 18300
9970   NNLOAD% = NNLOAD% - 1 : DONE%=TRUE% : GOTO 10030
9980   NLOAD(K%,1) = FIX(VAL(WKSP$(1)))
9990   NLOAD(K%,2) = FIX(VAL(WKSP$(2)))
10000   NLOAD(K%,3) = VAL(WKSP$(3))
10010   IF NLOAD(K%,2)>0 THEN GOTO 10030
10020   NLOAD(K%,2)=ABS(NLOAD(K%,2)) : NLOAD(K%,3)=-NLOAD(K%,3)
10030 WEND
10040 EDITING% = FALSE%
10050 RETURN
10060 '
10070 '+++++++++++++++++++   NODAL LOADS HELP SCREEN   ++++++++++++++++++++
10080 '
10090 PRINT
10100 PRINT "Nodal loads - set node number to zero to stop"
10110 PRINT "                0 to stop
10120 PRINT "                H for help"
10130 PRINT "                L to list nodal load data"
10140 PRINT "            - set load magnitude to zero to cancel load"
10150 PRINT "            - nodal load directions (global axes) are"
10160 IF PTYPE$ = "PTRUSS"  THEN GOSUB 10240
10170 IF PTYPE$ = "STRUSS"  THEN GOSUB 10300
10180 IF PTYPE$ = "PFRAME"  THEN GOSUB 10370
10190 IF PTYPE$ = "PFDYNAM" THEN GOTO  10190
10200 IF PTYPE$ = "GRID"    THEN GOSUB 10440
10210 IF PTYPE$ = "SFRAME"  THEN GOSUB 10510
10220 RETURN
10230 '
10240 '++++++++++++++++++++++++   PLANE TRUSS   ++++++++++++++++++++++++++
10250 '
10260 PRINT "                x-direction = 1"
10270 PRINT "                y-direction = 2"
10280 RETURN
10290 '
10300 '++++++++++++++++++++++++   SPACE TRUSS   ++++++++++++++++++++++++++
10310 '
10320 PRINT "                x-direction = 1"
10330 PRINT "                y-direction = 2"
10340 PRINT "                z-direction = 3"
10350 RETURN
10360 '
10370 '++++++++++++++++++++++++   PLANE FRAME   ++++++++++++++++++++++++++++
10380 '
10390 PRINT "                x-direction = 1"
10400 PRINT "                y-direction = 2"
10410 PRINT "                moment      = 3"
10420 RETURN
10430 '
10440 '++++++++++++++++++++++++   GRILLAGE   ++++++++++++++++++++++++++++++
10450 '
10460 PRINT "                z-direction = 1"
```

```
10470 PRINT "                      x-moment     = 2"
10480 PRINT "                      y-moment     = 3"
10490 RETURN
10500 '
10510 '+++++++++++++++++++++++++   SPACE FRAME   ++++++++++++++++++++++++++++
10520 '
10530 PRINT "                      x-direction = 1,    y-direction = 2"
10540 PRINT "                      z-direction = 3,    x-moment    = 4"
10550 PRINT "                      y-moment    = 5,    z-moment    = 6"
10560 RETURN
10570 '
10580 '********************   INPUT/EDIT ELEMENT LOADS   ********************
10590 '
10600 IF (PTYPE$="PFRAME") OR (PTYPE$="GRID") THEN GOTO 10650
10610 IF  INPUTING% THEN GOTO 11150
10620 PRINT : PRINT "Element loads not allowed in program "; PTYPE$
10630 GOSUB 18300
10640 GOTO 11150
10650 IF EDITING% THEN GOTO 10680
10660 CLS : PRINT
10670 PRINT "Input element loads" : GOSUB 11170 : GOTO 10700
10680 GOSUB 16720 : PRINT
10690 PRINT "Edit element loads - set lower node to H for help"
10700 DONE% = FALSE%
10710 WHILE (DONE% = FALSE%)
10720   PRINT
10730   PRINT "Lower  node                  =  "; : J% = 1 : GOSUB 17580
10740   IF (WKSP$(1)="H")  THEN GOSUB 11170              : GOTO  11130
10750   IF (WKSP$(1)<>"L") THEN GOTO 10790
10760   IF EDITING%        THEN GOSUB 16720
10770   IF NOT EDITING% THEN EDITING%=TRUE%:GOSUB 16720 : EDITING%=FALSE%
10780   GOTO 11130
10790   IF (FIX(VAL(WKSP$(1)))=0) THEN DONE%=TRUE%       : GOTO  11130
10800   PRINT "Higher node                  =  "; : J% = 2 : GOSUB 17580
10810   PRINT "Load type                    =  "; : J% = 3 : GOSUB 17580
10820   IF((FIX(VAL(WKSP$(3))=1)) OR FIX((VAL(WKSP$(3))=2))) THEN GOTO 10850
10830   PRINT "** ERROR ** Invalid load type - must be 1 or 2 - redo"
10840   GOTO 10720
10850   PRINT "Magnitude                    =  "; : J% = 4 : GOSUB 17580
10860   IF (FIX(VAL(WKSP$(3))) = 1) THEN GOTO 10910
10870   PRINT "Distance from lower node  =  "; : J% = 5 : GOSUB 17580
10880   '
10890   ' ... Check this is an existing load to be overwritten
10900   '
10910   K% = 0
10920   FOR I% = 1 TO NELOAD%
10930     IF ELOAD(I%,1) <>  FIX(VAL(WKSP$(1))) THEN GOTO 10980
10940     IF ELOAD(I%,2) <>  FIX(VAL(WKSP$(2))) THEN GOTO 10980
10950     IF ELOAD(I%,3) <>  FIX(VAL(WKSP$(3))) THEN GOTO 10980
10960     IF ELOAD(I%,3) <>2 THEN K% = I% :     GOTO 11020
10970     IF ABS(ELOAD(I%,5)-VAL(WKSP$(5)))<.001 THEN K% = I% : GOTO 11020
10980   NEXT I%
10990   '
11000   ' ... Check if maximum number of element loads exceeded
11010   '
11020   IF K%=0 THEN K% = NELOAD% + 1 : NELOAD% = K%
11030   IF (NELOAD%<=MAXELOAD%) THEN GOTO 11070
11040   PRINT : PRINT "** ERROR ** Maximum number of element loads exceeed"
11050   GOSUB 18300
11060   NELOAD% = NELOAD% - 1 : DONE%=TRUE% : GOTO 11130
11070   FOR J% = 1 TO 5
11080     ELOAD(K%,J%) = VAL(WKSP$(J%))
11090   NEXT J%
11100   ELOAD(K%,1) = FIX(ELOAD(K%,1))
11110   ELOAD(K%,2) = FIX(ELOAD(K%,2))
11120   ELOAD(K%,3) = FIX(ELOAD(K%,3))
11130 WEND
11140 EDITING% = FALSE%
11150 RETURN
11160 '
11170 '+++++++++++++++++++   ELEMENT LOAD HELP SCREEN   +++++++++++++++++++
11180 '
11190 PRINT
11200 PRINT "Element loads - set lower node number to"
11210 PRINT "                        0 to stop
11220 PRINT "                        H for help"
11230 PRINT "                        L to list element load data"
```

```
11240 PRINT "                  - set magnitude to zero to cancel load"
11250 PRINT "                  - load types are"
11260 PRINT "                       UDL        =  1"
11270 PRINT "                       point load =  2"
11280 PRINT "                  - a +ve element load acts in the +ve ";
11290 IF PTYPE$="PFRAME" THEN PRINT "element 'y' direction"
11300 IF PTYPE$="GRID"   THEN PRINT "global 'z' direction"
11310 RETURN
11320 '
11330 '*******************   INPUT/EDIT ADDED MASS   **********************
11340 '
11350 IF PTYPE$="PFDYNAM" THEN GOTO 11400
11360 IF INPUTING% THEN GOTO 11770
11370 PRINT : PRINT "Added masses are not allowed in program "; PTYPE$
11380 GOSUB 18300
11390 GOTO 11770
11400 IF EDITING% THEN GOTO 11430
11410 CLS : PRINT
11420 PRINT "Input added masses" : GOSUB 11790 : GOTO 11450
11430 GOSUB 16950 : PRINT
11440 PRINT "Edit added masses - set node to H for help"
11450 DONE% = FALSE%
11460 WHILE NOT DONE%
11470   PRINT
11480   PRINT "Node =  "; : J% = 1                 : GOSUB 17580
11490   IF (WKSP$(1)="H")  THEN GOSUB 11790        : GOTO  11130
11500   IF (WKSP$(1) = "L")    THEN GOSUB 16950    : GOTO  11130
11510   IF (WKSP$(1)<>"L") THEN GOTO 11550
11520   IF EDITING%        THEN GOSUB 16950
11530   IF NOT EDITING% THEN EDITING%=TRUE% : GOSUB 16950 : EDITING%=FALSE%
11540   GOTO 11130
11550   IF (FIX(VAL(WKSP$(1)))=0) THEN DONE%=TRUE% : GOTO  11750
11560   NAMASS% = NAMASS% + 1
11570   PRINT "Mass =  "; : J% = 2 : GOSUB 17580
11580   '
11590   ' ... Check if an existing added mass to be overwritten
11600   '
11610   K% = 0
11620   FOR I% = 1 TO NAMASS%
11630     IF AMASS(I%,1) = FIX(VAL(WKSP$(1))) THEN K% = I% : GOTO 11680
11640   NEXT I%
11650   '
11660   ' ... Check if maximum number of added masses exceeded
11670   '
11680   IF K%=0 THEN K% = NAMASS% + 1 : NAMASS% = K%
11690   IF (NAMASS%<=MAXAMASS%) THEN GOTO 11730
11700   PRINT : PRINT "** ERROR ** Maximum number of added masses exceeed"
11710   GOSUB 18300
11720   NAMASS% = NAMASS% - 1 : DONE%=TRUE% : GOTO 11750
11730   AMASS(K%,1) = FIX(VAL(WKSP$(1)))
11740   AMASS(K%,2) = VAL(WKSP$(2))
11750 WEND
11760 EDITING% = FALSE%
11770 RETURN
11780 '
11790 '++++++++++++++++++++   ADDED MASS HELP SCREEN   +++++++++++++++++++
11800 '
11810 PRINT
11820 PRINT "Added masses - set node number to"
11830 PRINT "                    0 to stop
11840 PRINT "                    H for help"
11850 PRINT "                    L to list added mass data"
11860 PRINT "              - set mass to 0 to cancel"
11870 PRINT "              - lumped masses are assumed to have no rotational";
11880 PRINT "inertia"
11890 RETURN
11900 '
11910 '******************   EDIT JOB TITLE AND UNITS   ********************
11920 '
11930 DONE% = FALSE%
11940 WHILE DONE% = FALSE%
11950   DONE% = TRUE% : GOSUB 14590
11960   PRINT
11970   PRINT "Hit return to retain existing title/units"
11980   PRINT
11990   INPUT "New job title  =  ", WRKSTR$
12000   IF (LEN(WRKSTR$) <> 0) THEN TITLE$ = WRKSTR$ : DONE% = FALSE%
```

```
12010   PRINT
12020   INPUT "New units        = ", WRKSTR$
12030   IF (LEN(WRKSTR$) <> 0) THEN UNITS$ = WRKSTR$ : DONE% = FALSE%
12040 WEND
12050 RETURN
12060 '
12070 '**********************    SORT NODAL COORDINATES    ********************
12080 '
12090 ' ... Remove any deleted nodes
12100 '
12110 COUNT% = 0
12120 FOR I% = 1 TO NNODE%
12130   IF NODE(I%,1) <= 0 THEN GOTO 12180
12140   COUNT% = COUNT% + 1
12150   FOR J% = 1 TO 4
12160     NODE(COUNT%,J%) = NODE(I%,J%)
12170   NEXT J%
12180 NEXT I%
12190 NNODE% = COUNT%
12200 '
12210 ' ...  Sort nodes into order
12220 '
12230 SORTED% = FALSE%
12240 WHILE NOT SORTED%
12250  SORTED% = TRUE%
12260   FOR I% = 1 TO NNODE% - 1
12270     IF NODE(I%,1) <  NODE(I%+1,1) THEN GOTO 12340
12280     SORTED% = FALSE%
12290     FOR J% = 1 TO 4
12300       TEMP           = NODE(I%,  J%)
12310       NODE(I%,J%)    = NODE(I%+1,J%)
12320       NODE(I%+1,J%) = TEMP
12330     NEXT J%
12340   NEXT I%
12350 WEND
12360 RETURN
12370 '
12380 '*********************     SORT ELEMENTS      ****************************
12390 '
12400 ' ... Remove any deleted elements
12410 '
12420 COUNT% = 0
12430 FOR I% = 1 TO NELEM%
12440   IF ELEM(I%,3) = 0 THEN GOTO 12490
12450   COUNT% = COUNT% + 1
12460   FOR J% = 1 TO 4
12470     ELEM(COUNT%,J%) = ELEM(I%,J%)
12480   NEXT J%
12490 NEXT I%
12500 NELEM% = COUNT%
12510 '
12520 ' ...  Sort elements into order
12530 '
12540 SORTED% = FALSE%
12550 WHILE NOT SORTED%
12560  SORTED% = TRUE%
12570   FOR I% = 1 TO NELEM% - 1
12580     IF ELEM(I%,1) >  ELEM(I%+1,1) THEN GOTO 12610
12590     IF ELEM(I%,1) <  ELEM(I%+1,1) THEN GOTO 12670
12600     IF ELEM(I%,2) <= ELEM(I%+1,2) THEN GOTO 12670
12610     SORTED% = FALSE%
12620     FOR J% = 1 TO 4
12630       TEMP           = ELEM(I%,  J%)
12640       ELEM(I%,J%)    = ELEM(I%+1,J%)
12650       ELEM(I%+1,J%) = TEMP
12660     NEXT J%
12670   NEXT I%
12680 WEND
12690 RETURN
12700 '
12710 '************************    SORT PROPERTIES    ************************
12720 '
12730 ' ... Remove any deleted properties
12740 '
12750 COUNT% = 0
12760 FOR I% = 1 TO NPROP%
12770   IF PROP(I%,2) = 0 THEN GOTO 12820
```

```
12780   COUNT% = COUNT% + 1
12790   FOR J% = 1 TO 7
12800     PROP(COUNT%,J%) = PROP(I%,J%)
12810   NEXT J%
12820 NEXT I%
12830 NPROP% = COUNT%
12840 '
12850 ' ...  Sort properties into order
12860 '
12870 SORTED% = FALSE%
12880 WHILE NOT SORTED%
12890  SORTED% = TRUE%
12900   FOR I% = 1 TO NPROP% - 1
12910     IF (PROP(I%,1) <= PROP(I%+1,1)) THEN GOTO 12980
12920     SORTED% = FALSE%
12930     FOR J% = 1 TO 7
12940       TEMP          = PROP(I%,  J%)
12950       PROP(I%,J%)   = PROP(I%+1,J%)
12960       PROP(I%+1,J%) = TEMP
12970     NEXT J%
12980   NEXT I%
12990 WEND
13000 RETURN
13010 '
13020 '***********************   SORT RESTRAINTS   *************************
13030 '
13040 '  ---  Remove any deleted restraints (ie directions set to zero)
13050 '
13060 COUNT% = 0
13070 FOR I% = 1 TO NREST%
13080   IF REST(I%, 2) = 0 THEN GOTO 13110
13090   COUNT% = COUNT% + 1
13100   REST(COUNT%,1) = REST(I%,1) : REST(COUNT%,2) = REST(I%,2)
13110 NEXT I%
13120 NREST% = COUNT%
13130 '
13140 '  ---  Put restraints into order
13150 '
13160 SORTED% = FALSE%
13170 WHILE NOT SORTED%
13180   SORTED% = TRUE%
13190   FOR I% = 1 TO NREST% - 1
13200     IF (REST(I%,1) <= REST(I%+1,1)) THEN GOTO 13270
13210     SORTED% = FALSE%
13220     FOR J% = 1 TO 2
13230       TEMP          = REST(I%,  J%)
13240       REST(I%,J%)   = REST(I%+1,J%)
13250       REST(I%+1,J%) = TEMP
13260     NEXT J%
13270   NEXT I%
13280 WEND
13290 RETURN
13300 '
13310 '************************   SORT LOCAL AXES   ************************
13320 '
13330 '  ---  Remove any deleted local axes
13340 '
13350 COUNT% = 0
13360 FOR I% = 1 TO NLAXES%
13370   IF (LAXES(I%, 2) = 0) THEN GOTO 13400
13380   COUNT% = COUNT% + 1
13390   LAXES(COUNT%,1) = LAXES(I%,1) : LAXES(COUNT%,2) = LAXES(I%,2)
13400 NEXT I%
13410 NLAXES% = COUNT%
13420 '
13430 ' ... Put local axes into order
13440 '
13450 SORTED% = FALSE%
13460 WHILE NOT SORTED%
13470   SORTED% = TRUE%
13480   FOR I% = 1 TO NLAXES% - 1
13490     IF (LAXES(I%,1) <= LAXES(I%+1,1)) THEN GOTO 13560
13500     SORTED% = FALSE%
13510     FOR J% = 1 TO 2
13520       TEMP           = LAXES(I%,  J%)
13530       LAXES(I%,J%)   = LAXES(I%+1,J%)
13540       LAXES(I%+1,J%) = TEMP
```

```
13550      NEXT J%
13560    NEXT I%
13570 WEND
13580 RETURN
13590 '
13600 '************************    SORT NODAL LOADS    ************************
13610 '
13620 ' ... Remove any zero loads
13630 '
13640 COUNT% = 0
13650 FOR I% = 1 TO NNLOAD%
13660   IF NLOAD(I%, 3) = 0 THEN GOTO 13710
13670   COUNT% = COUNT% + 1
13680   FOR J% = 1 TO 3
13690     NLOAD(COUNT%,J%) = NLOAD(I%,J%)
13700   NEXT J%
13710 NEXT I%
13720 NNLOAD% = COUNT%
13730 '
13740 ' ... Put nodal loads into order
13750 '
13760 SORTED% = FALSE%
13770 WHILE NOT SORTED%
13780   SORTED% = TRUE%
13790   FOR I% = 1 TO NNLOAD% - 1
13800     IF (NLOAD(I%,1) >  NLOAD(I%+1,1)) THEN GOTO 13830
13810     IF (NLOAD(I%,1) <  NLOAD(I%+1,1)) THEN GOTO 13890
13820     IF (NLOAD(I%,2) <= NLOAD(I%+1,2)) THEN GOTO 13890
13830     SORTED% = FALSE%
13840     FOR J% = 1 TO 3
13850       TEMP           = NLOAD(I%,  J%)
13860       NLOAD(I%,J%)   = NLOAD(I%+1,J%)
13870       NLOAD(I%+1,J%) = TEMP
13880     NEXT J%
13890   NEXT I%
13900 WEND
13910 RETURN
13920 '
13930 '***********************    SORT ELEMENT LOADS    ************************
13940 '
13950 ' ... Remove any zero member loads
13960 '
13970 COUNT% = 0
13980 FOR I% = 1 TO NELOAD%
13990   IF ELOAD(I%,4)=0 THEN GOTO 14040
14000   COUNT% = COUNT% + 1
14010   FOR J% = 1 TO 5
14020     ELOAD(COUNT%,J%) = ELOAD(I%,J%)
14030   NEXT J%
14040 NEXT I%
14050 NELOAD% = COUNT%
14060 '
14070 ' ... Put element loads into order
14080 '
14090 SORTED% = FALSE%
14100 WHILE NOT SORTED%
14110   SORTED% = TRUE%
14120   FOR I% = 1 TO NELOAD% - 1
14130     IF ELOAD(I%,1) >  ELOAD(I%+1,1) THEN GOTO 14200
14140     IF ELOAD(I%,1) <  ELOAD(I%+1,1) THEN GOTO 14260
14150     IF ELOAD(I%,2) >  ELOAD(I%+1,2) THEN GOTO 14200
14160     IF ELOAD(I%,2) <  ELOAD(I%+1,2) THEN GOTO 14260
14170     IF ELOAD(I%,3) >  ELOAD(I%+1,3) THEN GOTO 14200
14180     IF ELOAD(I%,3) <  ELOAD(I%+1,3) THEN GOTO 14260
14190     IF ELOAD(I%,5) <= ELOAD(I%+1,5) THEN GOTO 14260
14200     SORTED% = FALSE%
14210     FOR J% = 1 TO 5
14220       TEMP           = ELOAD(I%,  J%)
14230       ELOAD(I%,J%)   = ELOAD(I%+1,J%)
14240       ELOAD(I%+1,J%) = TEMP
14250     NEXT J%
14260   NEXT I%
14270 WEND
14280 RETURN
14290 '
14300 '***********************    SORT ADDED MASSES    ************************
14310 '
```

```
14320 ' ... Remove any deleted added masses (ie mass set to zero)
14330 '
14340 COUNT% = 0
14350 FOR I% = 1 TO NAMASS%
14360   IF AMASS(I%, 2) = 0 THEN GOTO 14390
14370   COUNT% = COUNT% + 1
14380   AMASS(COUNT%,1) = AMASS(I%,1) : AMASS(COUNT%,2) = AMASS(I%,2)
14390 NEXT I%
14400 NAMASS% = COUNT%
14410 '
14420 ' ... Put added masses into order
14430 '
14440 SORTED% = FALSE%
14450 WHILE NOT SORTED%
14460   SORTED% = TRUE%
14470   FOR I% = 1 TO NAMASS% - 1
14480     IF (AMASS(I%,1) <= AMASS(I%+1,1)) THEN GOTO 13270
14490     SORTED% = FALSE%
14500     FOR J% = 1 TO 2
14510       TEMP            = AMASS(I%,  J%)
14520       AMASS(I%,J%)    = AMASS(I%+1,J%)
14530       AMASS(I%+1,J%) = TEMP
14540     NEXT J%
14550   NEXT I%
14560 WEND
14570 RETURN
14580 '
14590 '**************************   LIST HEADER   *****************************+
14600 '
14610 CLS : PRINT #2,
14620 PRINT #2, "========================================";
14630 PRINT #2, "========================================"
14640 PRINT #2,
14650 PRINT #2, "    PROGRAM ";PTYPE$;STRING$(35-LEN(PTYPE$),32);
14660 PRINT #2, "Copyright(c) J.A.D.Balfour"
14670 PRINT #2,
14680 FOR I% = 1 TO (76-LEN(TITLE$))/2
14690   PRINT #2, " ";
14700 NEXT I%
14710 PRINT #2, TITLE$
14720 PRINT #2,
14730 PRINT #2, "    File Name :- "; DATAFILE$;
14740 FOR I% = 1 TO (40-LEN(DATAFILE$))
14750   PRINT #2, " ";
14760 NEXT I%
14770 PRINT #2, "Date :- "; MID$(DATE$,4,3); MID$(DATE$,1,3);
14780 PRINT #2, MID$(DATE$, 9, 2)
14790 PRINT #2, "    Units     :- "; UNITS$;
14800 FOR I% = 1 TO (40-LEN(UNITS$))
14810   PRINT #2, " ";
14820 NEXT I%
14830 PRINT #2, "Time :- "; TIME$
14840 PRINT #2,
14850 PRINT #2, "========================================";
14860 PRINT #2, "========================================"
14870 PRINT #2,
14880 GOSUB 18300
14890 RETURN
14900 '
14910 '*********************   LIST NODAL COORDINATES   ***********************+
14920 '
14930 IF (NNODE%<>0) THEN GOSUB 12070            '... Sort nodes
14940 IF NOT EDITING% THEN CLS
14950 PRINT #2,
14960 PRINT #2, "+ + + + + + + + + + + + + +"
14970 PRINT #2, "+   NODAL COORDINATES    +"
14980 PRINT #2, "+ + + + + + + + + + + + + +"
14990 PRINT #2,
15000 PRINT #2, "NODE             X              Y              ";
15010 IF (NNODE% = 0) THEN GOTO 15120
15020 LCOUNT% = 6
15030 IF (PTYPE$ = "STRUSS") OR (PTYPE$ = "SFRAME") THEN PRINT #2, "Z";
15040 PRINT #2,
15050 FOR I% = 1 TO NNODE%
15060   GOSUB 18430
15070   PRINT #2, " "; NODE(I%,1), NODE(I%,2), NODE(I%,3),
15080   IF (PTYPE$="STRUSS") OR (PTYPE$="SFRAME") THEN PRINT #2, NODE(I%,4),
```

```
15090   PRINT #2,
15100 NEXT I%
15110 GOSUB 18300
15120 RETURN
15130 '
15140 '************************   LIST ELEMENTS   ***************************
15150 '
15160 IF (NELEM% = 0) THEN GOTO 15390
15170 GOSUB 12380                                 '... Sort elements
15180 CLS : PRINT #2,
15190 PRINT #2, "+ + + + + + + + + +"
15200 PRINT #2, "+   ELEMENTS    +"
15210 PRINT #2, "+ + + + + + + + + +"
15220 PRINT #2,
15230 PRINT #2, "ELEMENT    PROPERTY";
15240 LCOUNT% = 6
15250 IF (PTYPE$ = "PFRAME")  THEN PRINT #2, "      TYPE";
15260 IF (PTYPE$ = "PFDYNAM") THEN PRINT #2, "      TYPE";
15270 IF (PTYPE$ = "SFRAME")  THEN PRINT #2, "     BETA";
15280 PRINT #2,
15290 IF (NELEM% = 0) THEN GOTO 15390
15300 FOR I% = 1 TO NELEM%
15310   GOSUB 18430
15320   PRINT #2, USING "### ###      ###"; ELEM(I%,1); ELEM(I%,2); ELEM(I%,3);
15330   IF PTYPE$ = "PFRAME"  THEN PRINT #2, USING "           ##";  ELEM(I%,4);
15340   IF PTYPE$ = "PFDYNAM" THEN PRINT #2, USING "           ##";  ELEM(I%,4);
15350   IF PTYPE$ = "SFRAME"  THEN PRINT #2, USING "        ###.#"; ELEM(I%,4);
15360   PRINT #2,
15370 NEXT I%
15380 GOSUB 18300
15390 RETURN
15400 '
15410 '************************   LIST PROPERTIES   *************************
15420 '
15430 IF (NPROP% = 0) THEN GOTO 15580
15440 GOSUB 12710                                 '... Sort properties
15450 CLS : PRINT #2,
15460 PRINT #2, "+ + + + + + + + + + +"
15470 PRINT #2, "+   PROPERTIES    +"
15480 PRINT #2, "+ + + + + + + + + + +"
15490 PRINT #2,
15500 LCOUNT% = 6
15510 IF (PTYPE$ = "PTRUSS")  THEN GOSUB 15600
15520 IF (PTYPE$ = "STRUSS")  THEN GOSUB 15600
15530 IF (PTYPE$ = "PFRAME")  THEN GOSUB 15710
15540 IF (PTYPE$ = "PFDYNAM") THEN GOSUB 15710
15550 IF (PTYPE$ = "GRID")    THEN GOSUB 15850
15560 IF (PTYPE$ = "SFRAME")  THEN GOSUB 15990
15570 GOSUB 18300
15580 RETURN
15590 '
15600 '++++++++++++++++++++++++++++   TRUSS   ++++++++++++++++++++++++++++++
15610 '
15620 PRINT #2, "PROPERTY        A            E"
15630 IF (NPROP% = 0) THEN GOTO 15690
15640 FOR I% = 1 TO NPROP%
15650   GOSUB 18430
15660   PRINT #2, USING " ###     #.####^^^^"; PROP(I%,1); PROP(I%,2);
15670   PRINT #2, USING "   #.####^^^^"; PROP(I%,3)
15680 NEXT I%
15690 RETURN
15700 '
15710 '+++++++++++++++++++++++++   PLANE FRAME   ++++++++++++++++++++++++++
15720 '
15730 PRINT #2, "PROPERTY        A            I            E";
15740 IF PTYPE$="PFDYNAM" THEN PRINT #2, "           m" ELSE PRINT #2,
15750 IF (NPROP% = 0) THEN GOTO 15830
15760 FOR I% = 1 TO NPROP%
15770   GOSUB 18430
15780   PRINT #2, USING " ###     #.####^^^^";     PROP(I%,1); PROP(I%,2);
15790   PRINT #2, USING "   #.####^^^^   #.####^^^^"; PROP(I%,3); PROP(I%,4);
15800   IF PTYPE$="PFRAME"  THEN PRINT #2,
15810   IF PTYPE$="PFDYNAM" THEN PRINT #2, USING "   #.####^^^^"; PROP(I%,5)
15820 NEXT I%
15830 RETURN
15840 '
15850 '+++++++++++++++++++++++++++   GRILLAGE   +++++++++++++++++++++++++++
```

```
15860 '
15870 PRINT #2, "PROPERTY         I              J              G              E"
15880 IF (NPROP% = 0) THEN GOTO 15970
15890 FOR I% = 1 TO NPROP%
15900   GOSUB 18430
15910   PRINT #2, USING "  ###  "; PROP(I%,1);
15920   FOR J% = 2 TO 5
15930     PRINT #2, USING "    #.####^^^^"; PROP(I%,J%);
15940   NEXT J%
15950   PRINT #2,
15960 NEXT I%
15970 RETURN
15980 '
15990 '+++++++++++++++++++++++++++   SPACE FRAME   +++++++++++++++++++++++++
16000 '
16010 PRINT #2, "PROPERTY       A           Iyy          Izz           J";
16020 PRINT #2, "               G            E"
16030 IF (NPROP% = 0) THEN GOTO 15970
16040 FOR I% = 1 TO NPROP%
16050   GOSUB 18430
16060   PRINT #2, USING "  ##   "; PROP(I%,1);
16070   FOR J% = 2 TO 7
16080     PRINT #2, USING "  #.####^^^^"; PROP(I%,J%);
16090   NEXT J%
16100   PRINT #2,
16110 NEXT I%
16120 RETURN
16130 '
16140 '*************************   LIST RESTRAINTS   ************************
16150 '
16160 IF (NREST% = 0) THEN GOTO 16310
16170 GOSUB 13020                                  '... Sort restraints
16180 CLS : PRINT #2,
16190 PRINT #2, "+ + + + + + + + + + +"
16200 PRINT #2, "+   RESTRAINTS    +"
16210 PRINT #2, "+ + + + + + + + + + +"
16220 PRINT #2,
16230 PRINT #2, "NODE    DIRECTION(S)"
16240 IF (NREST% = 0) THEN GOTO 16310
16250 LCOUNT% = 6
16260 FOR I% = 1 TO NREST%
16270   GOSUB 18430
16280   PRINT #2, USING "###     ######"; REST(I%,1); REST(I%,2)
16290 NEXT I%
16300 GOSUB 18300
16310 RETURN
16320 '
16330 '************************   LIST LOCAL AXES   *************************
16340 '
16350 IF (NLAXES% = 0) THEN GOTO 16500
16360 CLS : GOSUB 13310                                  '... Sort local axes
16370 PRINT #2,
16380 PRINT #2, "+ + + + + + + + + + + + "
16390 PRINT #2, "+   LOCAL AXES    +"
16400 PRINT #2, "+ + + + + + + + + + + + "
16410 PRINT #2,
16420 PRINT #2, "NODE       ANGLE"
16430 IF (NLAXES% = 0) THEN GOTO 16500
16440 COUNT% = 6
16450 FOR I% = 1 TO NLAXES%
16460   GOSUB 18430
16470   PRINT #2, USING "  ##      ####.#"; LAXES(I%,1); LAXES(I%,2)
16480 NEXT I%
16490 GOSUB 18300
16500 RETURN
16510 '
16520 '**********************   LIST NODAL LOADS   **************************
16530 '
16540 IF (NNLOAD% = 0) THEN GOTO 16700
16550 GOSUB 13600                                  '... Sort nodal loads
16560 CLS : PRINT #2,
16570 PRINT #2, "+ + + + + + + + + + +"
16580 PRINT #2, "+   NODAL LOADS   +"
16590 PRINT #2, "+ + + + + + + + + + +"
16600 PRINT #2,
16610 PRINT #2, "NODE    DIRECTION        VALUE"
16620 IF (NNLOAD% = 0) THEN GOTO 16700
```

```
16630 LCOUNT% = 6
16640 FOR I% = 1 TO NNLOAD%
16650   GOSUB 18430
16660   PRINT #2, USING "###        ###"; NLOAD(I%,1); NLOAD(I%,2);
16670   PRINT #2, USING "        #.####^^^^"; NLOAD(I%,3)
16680 NEXT I%
16690 GOSUB 18300
16700 RETURN
16710 '
16720 '**********************   LIST ELEMENT LOADS   ************************
16730 '
16740 IF (NELOAD% = 0) THEN GOTO 16930
16750 GOSUB 13930                                   '... Sort element loads
16760 CLS : PRINT #2,
16770 PRINT #2, "+ + + + + + + + + + + + +"
16780 PRINT #2, "+   ELEMENT LOADS    +    -    UDL = 1,    POINT LOAD = 2"
16790 PRINT #2, "+ + + + + + + + + + + + +"
16800 PRINT #2,
16810 PRINT #2, "ELEMENT   LOAD     MAGNITUDE    DISTANCE FROM"
16820 PRINT #2, "          TYPE                  LOWER NODE"
16830 IF (NELOAD% = 0) THEN GOTO 16930
16840 LCOUNT% = 7
16850 FOR I% = 1 TO NELOAD%
16860   GOSUB 18430
16870   PRINT #2, USING "### ###    ###"; ELOAD(I%,1); ELOAD(I%,2); ELOAD(I%,3);
16880   PRINT #2, USING "     #.####^^^^"; ELOAD(I%,4);
16890   IF ELOAD(I%,3) = 2  THEN PRINT #2, USING "    #.####^^^^"; ELOAD(I%,5)
16900   IF ELOAD(I%,3) <> 2 THEN PRINT #2, "         N/A"
16910 NEXT I%
16920 GOSUB 18300
16930 RETURN
16940 '
16950 '************************   LIST ADDED MASSES   **********************
16960 '
16970 IF NAMASS%=0 THEN GOTO 17120
16980 GOSUB 14300                                   '... Sort added masses
16990 CLS : PRINT #2,
17000 PRINT #2, "+ + + + + + + + + + + + +"
17010 PRINT #2, "+   ADDED MASSES     +"
17020 PRINT #2, "+ + + + + + + + + + + + +"
17030 PRINT #2,
17040 PRINT #2, "NODE        MASS"
17050 IF (NAMASS% = 0) THEN GOTO 17120
17060 LCOUNT% = 6
17070 FOR I% = 1 TO NAMASS%
17080   GOSUB 18430
17090   PRINT #2, USING "###      #.###^^^^"; AMASS(I%,1); AMASS(I%,2)
17100 NEXT I%
17110 GOSUB 18300
17120 RETURN
17130 '
17140 ' +++++++++++++++++++++   LIST ARRAY SIZES   +++++++++++++++++++++++
17150 '
17160 CLS : PRINT
17170 PRINT "Maximum problem size for program "; PTYPE$; " is as follows:"
17180 PRINT
17190 PRINT "    maximum number of nodes          = "; MAXNODE%
17200 PRINT "    maximum number of elements       = "; MAXELEM%
17210 PRINT "    maximum number of properties     = "; MAXPROP%
17220 PRINT "    maximum number of restraints     = "; MAXREST%
17230 IF MAXLAXES% = 1 THEN GOTO 17250
17240 PRINT "    maximum number of local axes     = "; MAXLAXES%
17250 PRINT "    maximum number of nodal loads    = "; MAXNLOAD%
17260 IF MAXELOAD% = 1 THEN GOTO 17280
17270 PRINT "    maximum number of element loads = "; MAXELOAD%
17280 IF MAXAMASS% = 1 THEN GOTO 17300
17290 PRINT "    maximum number of added masses  = "; MAXAMASS%
17300 PRINT "    maximum semi-bandwidth           = "; MAXBAND%
17310 PRINT : PRINT
17320 PRINT "The semi-bandwidth is determined by the element with the biggest"
17330 PRINT "difference between the node numbers at its ends (MaxDiff) and"
17340 PRINT "the number and location of the restraints.
17350 PRINT
17360 PRINT "An overestimate of the semi-bandwidth can be made from the"
17370 PRINT "following expression"
17380 PRINT "                            (MaxDiff+1) * ";
17390 IF (PTYPE$="PTRUSS") THEN PRINT "2" : GOTO 17410
```

```
17400 IF (PTYPE$="SFRAME") THEN PRINT "6" ELSE PRINT "3"
17410 GOSUB 18300                        ' Any key to continue
17420 RETURN
17430 '
17440 '**********************     GET MENU RESPONSE     **********************
17450 '
17460 ROW% = CSRLIN : COL% = POS(0)
17470 REPLY% = 0
17480 WHILE REPLY% = 0
17490   LOCATE ROW%, COL% : INPUT "", WRKSTR$ : REPLY% = VAL(WRKSTR$)
17500   IF (REPLY% < 1) OR (REPLY% > NOPTIONS%) THEN GOTO 17530
17510   IF (REPLY% < 10) AND (LEN(WRKSTR$) = 1) THEN GOTO 17540
17520   IF (REPLY% >  9) AND (LEN(WRKSTR$) = 2) THEN GOTO 17540
17530   SOUND 700, 4 : REPLY% = 0 : LOCATE ROW%, COL%
17540 WEND
17550 PRINT
17560 RETURN
17570 '
17580 '*****************     PROCESS KEYBOARD INPUT     *********************
17590 '
17600 ROW% = CSRLIN : COL% = POS(0)
17610 INPUT ; "", T$
17620 IF LEN(T$) = 0 THEN LOCATE ROW%, COL% : PRINT WKSP$(J%) : GOTO 17670
17630 IF T$ = "h" THEN T$ = "H"
17640 IF T$ = "l" THEN T$ = "L"
17650 WKSP$(J%) = T$
17660 PRINT
17670 RETURN
17680 '
17690 '**********************     RECORD NUMBERS     ***************************
17700 '
17710 NNOS% = 0 : DONE% = TRUE%
17720 IF EOF(1)<>0 THEN DATAEND%=TRUE% : GOTO 18030
17730 '
17740 ' ... Read next line and check if it is blank
17750 '
17760 LINE INPUT #1, WRKSTR$
17770 LCOUNT% = LCOUNT% + 1 : LOCATE ROW%, COL% : PRINT LCOUNT%;
17780 FOR I% = 1 TO LEN(WRKSTR$)
17790   IF MID$(WRKSTR$,I%,1)<> " " THEN DONE% = FALSE% : GOTO 17810
17800 NEXT I%
17810 IF DONE% = TRUE% THEN WRKSTR$ = "" : GOTO 18030
17820 FOR I% = 1 TO 7
17830   WKSP$(I%) = ""
17840 NEXT I%
17850 '
17860 ' ... Search for numbers
17870 '
17880 FOR I% = 1 TO LEN(WRKSTR$)
17890   WHILE (MID$(WRKSTR$, I%, 1) = " ") AND I% <= LEN(WRKSTR$)
17900     I% = I%+1
17910   WEND
17920   IF NNOS% < 8 THEN GOTO 17960
17930   PRINT
17940   PRINT "*** ERROR *** Too many nos. on line - run DCHECK"
17950   GOTO 18030
17960   IF (I% > LEN(WRKSTR$)) THEN GOTO 18010
17970   NNOS% = NNOS%+1: WKSP$(NNOS%) = ""
17980   WHILE (MID$(WRKSTR$, I%, 1) <> " ") AND I% <= LEN(WRKSTR$)
17990     WKSP$(NNOS%) = WKSP$(NNOS%) + MID$(WRKSTR$,I%,1): I% = I% + 1
18000   WEND
18010 NEXT I%
18020 WRKSTR$ = ""
18030 RETURN
18040 '
18050 '*************************     FIND STRING     *************************
18060 '
18070 STRFOUND%=FALSE% : DONE% = FALSE%
18080 WHILE DONE% = FALSE%
18090   IF EOF(1)<>0 THEN DATAEND% = TRUE% : DONE% = TRUE% : GOTO 18180
18100   LINE INPUT #1, WRKSTR$
18110   LCOUNT% = LCOUNT% + 1 : LOCATE ROW%, COL% : PRINT LCOUNT%;
18120   IF(TARGET$="PROGRAM") AND LCOUNT% > 4 THEN DONE% = TRUE%
18130   IF INSTR(WRKSTR$,TARGET$)>0 THEN STRFOUND% = TRUE% : DONE% = TRUE%
18140   '
18150   ' ... Check for new module
18160   '
```

```
18170   IF LEFT$(WRKSTR$,4) ="+    " THEN DONE% = TRUE%
18180 WEND
18190 RETURN
18200 '
18210 '***************************   READ LINES    *****************************
18220 '
18230 FOR I% = 1 TO NLINES%
18240   IF EOF(1)<>0 THEN DATAEND% = TRUE% : I% = NLINES% : GOTO 18270
18250   LINE INPUT #1, WRKSTR$
18260   LCOUNT% = LCOUNT% + 1 : LOCATE ROW%, COL% : PRINT LCOUNT%;
18270 NEXT I%
18280 RETURN
18290 '
18300 '*********************   ANY KEY TO CONTINUE    ***********************
18310 '
18320 IF DEVICE$<>"SCRN:" THEN GOTO 18410
18330 IF EDITING% THEN PRINT ELSE LOCATE PAGELEN%+1, 1
18340 PRINT SPACE$(40); "Press any key to continue";
18350 WHILE INKEY$ = "" : WEND
18360 LOCATE ,1
18370 PRINT SPACE$(40); "                         ";
18380 LCOUNT% = 0
18390 IF EDITING% THEN ROW%=CSRLIN : IF(ROW%>0) THEN LOCATE ROW%-1,1
18400 IF NOT EDITING% THEN CLS
18410 RETURN
18420 '
18430 '**********************   CHECK FOR PAGE END    *********************
18440 '
18450 IF LCOUNT%>=PAGELEN% THEN GOSUB 18300
18460 LCOUNT% = LCOUNT% + 1
18470 RETURN
18480 '
18490 '==========================   PRE.BAS    ============================
```

Sample Run

The following run demonstrates PRE.BAS being used to assemble the data necessary to solve example 5.3 using the program PTRUSS.BAS. Input from the keyboard is shown underlined.

```
=================================================================

 PROGRAM PRE                        Copyright (c) J.Balfour 1991

    This program preprocesses data for the analysis programs
    PTRUSS, STRUSS, PFRAME, PFDYNAM, GRID, and SFRAME

               For further information contact
 James A.D.Balfour, Heriot-Watt University, Riccarton, Edinburgh
              Tel 031-449-5111, Fax 031-451-3170

=================================================================

Do you wish to
  (1)  Create a new data file
  (2)  Modify an existing data file
1

Name of the data file  =  PT1.DAT


Note that pressing 'return' sets the current value equal to the
corresponding value for the previously entered data item. For
example pressing the return key when prompted for, say, an
x-coordinate will set that coordinate equal to the x-coordinate
of the previous node.


Select problem type
  (1) Plane truss
  (2) Space truss
  (3) Plane frame - static  analysis
```

```
   (4) Plane frame - dynamic analysis
   (5) Grillage
   (6) Space frame
1


Maximum problem size for program PTRUSS is as follows:

     maximum number of nodes          =  45
     maximum number of elements       =  70
     maximum number of properties     =  20
     maximum number of restraints     =  20
     maximum number of nodal loads    =  50
     maximum semi-bandwidth           =  90


The semi-bandwidth is determined by the element with the biggest
difference between the node numbers at its ends (MaxDiff) and
the number and location of the restraints.

An overestimate of the semi-bandwidth can be made from the
following expression
                         (MaxDiff+1) * 2


Input job title and units - note that the units string is for
                            information only and is ignored
                            during calculation. Consistent units
                            must be used throughout.

Job title                     = Example 5.3

Units (force and length)      = kN and mm


Input nodal coordinates

Nodal coordinates - set node number to
                       0 to stop
                       H for help
                       L to list nodal coordinates data
                  - node numbers must be consecutive integers
                    starting at 1 (can be entered in any order)
                  - set node number negative to cancel node

Node number       =  1
X-coordinate      =  0
Y-coordinate      =  0

Node number       =  2
X-coordinate      =  3000
Y-coordinate      =  3000

Node number       =  3
X-coordinate      =  6000
Y-coordinate      =  0

Node number       =  0

+ + + + + + + + + + + + +
+   NODAL COORDINATES   +
+ + + + + + + + + + + + +

NODE             X              Y
  1              0              0
  2              3000           3000
  3              6000           0

Edit nodal coordinates - set node number to H for help

Node number       =  0


Input element data

Element data - set lower node number to
                   0 to stop
```

```
                   H for help
                   L to list element data
              - set property no. to zero to cancel element

Lower  node number (i)  =  1
Higher node number (j)  =  2
Property number         =  1

Lower  node number (i)  =  1
Higher node number (j)  =  3
Property number         =  1

Lower  node number (i)  =  2
Higher node number (j)  =  3
Property number         =  1

Lower  node number (i)  =  0

+ + + + + + + + +
+   ELEMENTS    +
+ + + + + + + + +

ELEMENT     PROPERTY
  1   2        1
  1   3        1
  2   3        1


Edit element data - set lower node number to H for help

Lower  node number (i)  =  0


Input property data

Element properties - set property number to
                        0 to stop
                        H for help
                        L to list property data
                   - set cross sectional area to 0 to cancel property
                   - hit return to retain any value

Property number            =  1
Cross-sectional area (A)   =  200
Elastic constant     (E)   =  200

Property number            =  0

+ + + + + + + + + +
+   PROPERTIES    +
+ + + + + + + + + +

PROPERTY          A            E
    1     0.2000E+03   0.2000E+03

Edit property data - set property number to H for help

Property number            =  1
Cross-sectional area (A)   =  3000
Elastic constant     (E)   =  200

Property number            =  L

+ + + + + + + + + +
+   PROPERTIES    +
+ + + + + + + + + +

PROPERTY          A            E
    1     0.3000E+04   0.2000E+03

Property number            =  0


Input restraints

Restraints - set node number to
               0 to stop
```

```
                H for help
                L to list restraint data
           - set restraint direction(s) to 0 to cancel
           - restraint directions are
                x-translation = 1
                y-translation = 2
           - enter restraints as a composite number e.g. if a
             node is restrained in directions 1 and 2 enter 12

Node to be restrained  =  3
Direction(s)           =  2

Node to be restrained  =  1
Direction(s)           =  12

Node to be restrained  =  0

+ + + + + + + + + + +
+   RESTRAINTS      +
+ + + + + + + + + + +

NODE    DIRECTION(S)
  1         12
  3          2

Edit restraints - set node number to H for help

Node to be restrained  =  0


Input nodal loads

Nodal loads - set node number to zero to stop
                0 to stop
                H for help
                L to list nodal load data
            - set load magnitude to zero to cancel load
            - nodal load directions (global axes) are
                x-direction = 1
                y-direction = 2

Node         =  3
Direction    =  1
Magnitude    =  100

Node         =  2
Direction    =  2
Magnitude    =  200

Node         =  0

+ + + + + + + + + + +
+   NODAL LOADS     +
+ + + + + + + + + + +

NODE    DIRECTION       VALUE
  2         2        0.2000E+03
  3         1        0.1000E+03

Edit nodal loads - set node to H for help

Node         =  0

Do you wish to
  (1) Edit data
  (2) Check the data (data will be written to file first)
  (3) List all of the current data to the screen
  (4) List all of the current data to the printer
  (5) Analyse (write data to file first)
  (6) Create/edit a new file (data will be written to file first)
  (7) Stop (data will be written to file first)
1


Do you wish to edit
  (1) Job title and units
  (2) Nodal coordinates
```

```
   (3) Elements
   (4) Properties
   (5) Restraints
   (6) Local axes
   (7) Nodal loads
   (8) Element loads
   (9) Added masses
  (10) Stop editing data
7

+ + + + + + + + + +
+   NODAL LOADS   +
+ + + + + + + + + +

NODE    DIRECTION       VALUE
  2         2        0.2000E+03
  3         1        0.1000E+03

Edit nodal loads - set node to H for help

Node         =  3
Direction    =  1
Magnitude    =  0

Node         =  2
Direction    =  1
Magnitude    =  100

Node         =  L

+ + + + + + + + + +
+   NODAL LOADS   +
+ + + + + + + + + +

NODE    DIRECTION       VALUE
  2         1        0.1000E+03
  2         2        0.2000E+03

Node         =  0

Do you wish to edit
   (1) Job title and units
   (2) Nodal coordinates
   (3) Elements
   (4) Properties
   (5) Restraints
   (6) Local axes
   (7) Nodal loads
   (8) Element loads
   (9) Added masses
  (10) Stop editing data
10

Do you wish to
   (1) Edit data
   (2) Check the data (data will be written to file first)
   (3) List all of the current data to the screen
   (4) List all of the current data to the printer
   (5) Analyse (write data to file first)
   (6) Create/edit a new file (data will be written to file first)
   (7) Stop (data will be written to file first)
7

Wait - writing data to file PT1.DAT
```

5.12 A Data Checking Program - DCHECK.BAS

This program checks data prepared for the program PTRUSS.BAS. It does not contain every conceivable check; the reader might like to consider additional checks that could be built in. DCHECK.BAS will also check data for the following programs

STRUSS.BAS	a space truss static analysis program.
PFRAME.BAS	a plane frame static analysis program.
PFDYNAM.BAS	a plane frame dynamic analysis program.
GRID.BAS	a grillage static analysis program.
SFRAME.BAS	a space frame static analysis program.

These programs are described in subsequent chapters.

Listing

```
1000 '========================    DCHECK.BAS    ============================
1010 '
1020 COMMON DATAFILE$
1030 OPTION BASE 1 : KEY OFF
1040 TRUE% = -1 : FALSE% = 0 : OK% = 1 : FAIL% = 0
1050 DEVICE$ = "SCRN:" : PAGELEN% = 23
1060 CLS : PRINT
1070 PRINT "================================================================"
1080 PRINT
1090 PRINT " PROGRAM DCHECK                          Copyright (c) J.Balfour 1991"
1100 PRINT
1110 PRINT "        This program checks data for the analysis programs"
1120 PRINT "        PTRUSS, STRUSS, PFRAME, PFDYNAM, GRID, AND SFRAME"
1130 PRINT
1140 PRINT "                For further information contact"
1150 PRINT " James A.D.Balfour, Heriot-Watt University, Riccarton, Edinburgh"
1160 PRINT "               Tel 031-449-5111, Fax 031-451-3170"
1170 PRINT
1180 PRINT "================================================================"
1190 PRINT
1200 '
1210 ' ... Note that variables are defined in Appendix A
1220 '
1230 DIM NODE(50,4)                    '... Node array
1240 DIM ELEM(100,4)                   '... Element array
1250 DIM PROP(20,7)                    '... Property array
1260 DIM REST(20,2)                    '... Restraint array
1270 DIM LAXES(20,2)                   '... Local axes array
1280 DIM NLOAD(50,3)                   '... Nodal load array
1290 DIM ELOAD(30,5)                   '... Element load array
1300 DIM AMASS(30,2)                   '... Added mass array
1310 DIM WKSP$(10)                     '... Workspace array
1320 PRINT
1330 PRINT "Note: lines terminated with '~~' have been ignored"
1340 IF DATAFILE$ <> "" THEN GOSUB 10640 : GOTO 1400
1350 '
1360 ' ... Prompt for data file name
1370 '
1380 PRINT
1390 INPUT "Name of the data file  =  ", DATAFILE$
1400 GOSUB 2090
1410 WHILE REPLY% <> 7
1420   ERRORS% = 0: WARNINGS% = 0
1430   CLS : PRINT
1440   PRINT "Do you wish to"
1450   PRINT "  (1) Check data - output to the screen"
1460   PRINT "  (2) Check data - output to the printer"
1470   PRINT "  (3) Check data - output to a file"
1480   PRINT "  (4) Edit data"
1490   PRINT "  (5) Analyse"
```

```
1500    PRINT "  (6) Read a new file"
1510    PRINT "  (7) Stop"
1520    NOPTIONS% = 7 : GOSUB 9780
1530    ON REPLY% GOTO 1570, 1610, 1720, 1950, 1980, 2040, 2050
1540    '
1550    ' ...  Open screen for output
1560    '
1570    OPEN "SCRN:" FOR OUTPUT AS #2 : CLS : GOTO 1830     '... Reply = 1
1580    '
1590    ' ...  Open the printer for output
1600    '
1610    ON ERROR GOTO 1660                                  '... Reply = 2
1620    OPEN "LPT1:" FOR OUTPUT AS #2 : PRINT #2,
1630    ON ERROR GOTO 0
1640    PRINT
1650    PRINT "Wait - printing data check" : DEVICE$ = "LPT1:" : GOTO 1830
1660    CLS : PRINT
1670    PRINT "Failed to find the printer"
1680    GOSUB 10640 : RESUME 2050
1690    '
1700    ' ... Open a file for output
1710    '                                                   '... Reply = 3
1720    INPUT "Name of file for output = ", DEVICE$
1730    ON ERROR GOTO 1770 : OPEN DEVICE$ FOR OUTPUT AS #2
1740    ON ERROR GOTO 0
1750    PRINT
1760    PRINT "Wait - writing data check to file "; DEVICE$ : GOTO 1830
1770    CLS : PRINT
1780    PRINT "Failed to open output file"
1790    GOSUB 10640 : RESUME 2050
1800    '
1810    ' ... Read and check the data file
1820    '
1830    GOSUB 2200
1840    IF DATAFILEOK% THEN GOTO 1860
1850    LCOUNT% = LCOUNT% + 2 : GOSUB 10730
1860    PRINT #2,
1870    PRINT #2, "DATA CHECKED:-  ";
1880    PRINT #2, ERRORS%; " Errors    "; WARNINGS%; " Warnings"
1890    IF DEVICE$ = "SCRN:" THEN GOTO 1930 ELSE DEVICE$ = "SCRN:"
1900    PRINT
1910    PRINT "DATA CHECKED:-  ";
1920    PRINT ERRORS%; " Errors    "; WARNINGS%; " Warnings"
1930    PRINT : GOSUB 10640 : GOTO 2010
1940    '
1950    CLS : PRINT : PRINT                                 '... Reply = 4
1960    PRINT "Wait - chaining PRE" : CHAIN "PRE"
1970    '
1980    CLS : PRINT : PRINT                                 '... Reply = 5
1990    PRINT "Wait - chaining "; PTYPE$; : CHAIN PTYPE$
2000    '
2010    IF REPLY% = 2 THEN PRINT #2, CHR$(12);
2020    GOTO 2050
2030    '
2040    INPUT "Name of the data file  =  ", DATAFILE$       '... Reply = 6
2050    GOSUB 2090 : CLOSE #2 : ON ERROR GOTO 0
2060 WEND
2070 END
2080 '
2090 '********************    CHECK THAT DATA FILE CAN BE OPENED    **************
2100 '
2110 ON ERROR GOTO 2140
2120 OPEN DATAFILE$ FOR INPUT AS #1
2130 ON ERROR GOTO 0 : DATAFILEOK% = TRUE% : CLOSE #1 : GOTO 2180
2140 CLS : PRINT
2150 PRINT "** ERROR ** Failed to open file: "; DATAFILE$; " for input"
2160 GOSUB 10640 : DATAFILEOK% = FALSE% : RESUME 2170
2170 ON ERROR GOTO 0
2180 RETURN
2190 '
2200 '******************    READS AND CHECKS THE DATA FILE    *****************
2210 '
2220 ' ...  This subroutine reads and checks the data from a data file
2230 '
2240 ERRORS% = 0 : WARNINGS% = 0 : DATAEND% = FALSE%
2250 NNODE%  = 0 : NELEM%   = 0 : NPROP%  =    0 : NREST%  = 0
2260 NLAXES% = 0 : NNLOAD%  = 0 : NELOAD% =    0 : NAMASS% = 0
```

```
2270 '
2280 ' ... Open file for input
2290 '
2300 OPEN DATAFILE$ FOR INPUT AS #1
2310 '
2320 ' ... Read problem type
2330 '
2340 TARGET$ = "PROGRAM" : GOSUB 10380
2350 IF NOT STRFOUND% THEN GOTO 2850
2360 '
2370 ' ... Set up array sizes and no of nos in each row based on problem type
2380 '
2390 IF INSTR(WRKSTR$, "PTRUSS") = 0 THEN GOTO 2460
2400 PTYPE$     = "PTRUSS"
2410 MAXNODE%  = 45 : MAXELEM%  = 70 : MAXPROP%  = 20 : MAXREST%  = 20
2420 MAXLAXES% =  0 : MAXNLOAD% = 50 : MAXELOAD% =  0 : MAXAMASS% =  0
2430 NNODENO%  =  3 : NELEMNO%  =  3 : NPROPNO%  =  3 : NRESTNO%  =  2
2440 NLAXESNO% =  0 : NNLOADNO% =  3 : NELOADNO% =  0 : NAMASSNO% =  0
2450 GOTO 2900
2460 IF INSTR(WRKSTR$, "STRUSS") = 0 THEN GOTO 2530
2470 PTYPE$     = "STRUSS"
2480 MAXNODE%  = 30 : MAXELEM%  = 50 : MAXPROP%  = 20 : MAXREST%  = 20
2490 MAXLAXES% =  0 : MAXNLOAD% = 50 : MAXELOAD% =  0 : MAXAMASS% =  0
2500 NNODENO%  =  4 : NELEMNO%  =  3 : NPROPNO%  =  3 : NRESTNO%  =  2
2510 NLAXESNO% =  0 : NNLOADNO% =  3 : NELOADNO% =  0 : NAMASSNO% =  0
2520 GOTO 2900
2530 IF INSTR(WRKSTR$, "PFRAME") = 0 THEN GOTO 2610
2540 PTYPE$     = "PFRAME"
2550 MAXNODE%  = 45 : MAXELEM%  = 70 : MAXPROP%  = 20 : MAXREST%  = 20
2560 MAXLAXES% = 20 : MAXNLOAD% = 45 : MAXELOAD% = 20 : MAXAMASS% =  1
2570 NNODENO%  =  3 : NELEMNO%  =  4 : NPROPNO%  =  4 : NRESTNO%  =  2
2580 NLAXESNO% =  2 : NNLOADNO% =  3 : NELOADNO% =  5 : NAMASSNO% =  3
2590 MAXBAND%  = 40 : NFREE%    =  3
2600 GOTO 2900
2610 IF INSTR(WRKSTR$, "PFDYNAM") = 0 THEN GOTO 2690
2620 PTYPE$     = "PFDYNAM"
2630 MAXNODE%  = 17 : MAXELEM%  = 25 : MAXPROP%  = 10 : MAXREST%  = 10
2640 MAXLAXES% = 10 : MAXNLOAD% =  1 : MAXELOAD% =  1 : MAXAMASS% = 30
2650 NNODENO%  =  3 : NELEMNO%  =  4 : NPROPNO%  =  5 : NRESTNO%  =  2
2660 NLAXESNO% =  2 : NNLOADNO% =  3 : NELOADNO% =  5 : NAMASSNO% =  3
2670 MAXBAND%  = 51 : NFREE%    =  3
2680 GOTO 2900
2690 IF INSTR(WRKSTR$, "GRID") = 0 THEN GOTO 2770
2700 PTYPE$     = "GRID"
2710 MAXNODE%  = 50 : MAXELEM%  = 70 : MAXPROP%  = 20 : MAXREST%  = 20
2720 MAXLAXES% =  0 : MAXNLOAD% = 50 : MAXELOAD% = 30 : MAXAMASS% =  0
2730 NNODENO%  =  3 : NELEMNO%  =  3 : NPROPNO%  =  5 : NRESTNO%  =  2
2740 NLAXESNO% =  0 : NNLOADNO% =  3 : NELOADNO% =  5 : NAMASSNO% =  0
2750 MAXBAND%  = 40 : NFREE%    =  3
2760 GOTO 2900
2770 IF INSTR(WRKSTR$, "SFRAME") = 0 THEN GOTO 2850
2780 PTYPE$     = "SFRAME"
2790 MAXNODE%  = 30: MAXELEM%  = 50: MAXPROP%  = 20: MAXREST%  = 20
2800 MAXLAXES% =  0: MAXNLOAD% = 50: MAXELOAD% =  0: MAXAMASS% =  0
2810 NNODENO%  =  4: NELEMNO%  =  4: NPROPNO%  =  7: NRESTNO%  =  2
2820 NLAXESNO% =  0: NNLOADNO% =  3: NELOADNO% =  0: NAMASSNO% =  0
2830 MAXBAND%  = 40 : NFREE%    =  6
2840 GOTO 2900
2850 CLS : PRINT
2860 PRINT "** ERROR ** Not a valid DCHECK data file - checking aborted"
2870 GOSUB 10640 : GOTO 3960
2880 LCOUNT% = LCOUNT% + 1 : GOSUB 10730
2890 '
2900 GOSUB 3990                          '... Read title and units strings
2910 WHILE NOT DATAEND%
2920   '
2930   ' ... Get module string
2940   '
2950   IF LEFT$(WRKSTR$,4)<>"+   " THEN GOTO 2980
2960   PRINT #2, "** WARNING ** Previous module was empty"
2970   WARNINGS% = WARNINGS%+1 : GOTO 2990
2980   TARGET$ = "+   " : GOSUB 10380 : IF NOT STRFOUND% THEN GOTO 3700
2990   MODSTR$ = WRKSTR$
3000   '
3010   ' ...  Identify module
3020   '
3030   IF INSTR(MODSTR$,"NODAL COORDINATES")=0 THEN GOTO 3110
```

```
3040    TARGET$ = "NODE" : GOSUB 10380 : IF NOT STRFOUND% THEN GOTO 3700
3050    IF NNODE%=0 THEN GOTO 3090
3060    '
3070    LCOUNT% = LCOUNT% + 1 : GOSUB 10730
3080    PRINT #2, "** ERROR ** Duplicate module" : ERRORS% = ERRORS% + 1
3090    GOSUB 4160 : GOTO 3700                  ' Read and check nodal data
3100    '
3110    IF INSTR(MODSTR$,"ELEMENTS")=0 THEN GOTO 3190
3120    TARGET$ = "ELEMENT" : GOSUB 10380 : IF NOT STRFOUND% THEN GOTO 3700
3130    IF NELEM%=0 THEN GOTO 3170
3140    '
3150    LCOUNT% = LCOUNT% + 1 : GOSUB 10730
3160    PRINT #2, "** ERROR ** Duplicate module" : ERRORS% = ERRORS% + 1
3170    GOSUB 4690 : GOTO 3700                  ' Read and check element data
3180    '
3190    IF INSTR(MODSTR$,"PROPERTIES")=0 THEN GOTO 3270
3200    TARGET$ = "PROPERTY" : GOSUB 10380 : IF NOT STRFOUND% THEN GOTO 3700
3210    IF NPROP%=0 THEN GOTO 3250
3220    '
3230    LCOUNT% = LCOUNT% + 1 : GOSUB 10730
3240    PRINT #2, "** ERROR ** Duplicate module" : ERRORS% = ERRORS% + 1
3250    GOSUB 4920 : GOTO 3700                  ' Read and check property data
3260    '
3270    IF INSTR(MODSTR$,"RESTRAINTS")=0 THEN GOTO 3350
3280    TARGET$ = "NODE" : GOSUB 10380 : IF NOT STRFOUND% THEN GOTO 3700
3290    IF NREST%=0 THEN GOTO 3330
3300    '
3310    LCOUNT% = LCOUNT% + 1 : GOSUB 10730
3320    PRINT #2, "** ERROR ** Duplicate module" : ERRORS% = ERRORS% + 1
3330    GOSUB 5470 : GOTO 3700                  ' Read and check restraint data
3340    '
3350    IF INSTR(MODSTR$,"LOCAL AXES")=0 THEN GOTO 3430
3360    TARGET$ = "NODE" : GOSUB 10380 : IF NOT STRFOUND% THEN GOTO 3700
3370    IF NLAXES%=0 THEN GOTO 3410
3380    '
3390    LCOUNT% = LCOUNT% + 1 : GOSUB 10730
3400    PRINT #2, "** ERROR ** Duplicate module" : ERRORS% = ERRORS% + 1
3410    GOSUB 5960 : GOTO 3700                  ' Read and check local axes
3420    '
3430    IF INSTR(MODSTR$,"NODAL LOADS")=0 THEN GOTO 3510
3440    TARGET$ = "NODE" : GOSUB 10380 : IF NOT STRFOUND% THEN GOTO 3700
3450    IF NNLOAD%=0 THEN GOTO 3490
3460    '
3470    LCOUNT% = LCOUNT% + 1 : GOSUB 10730
3480    PRINT #2, "** ERROR ** Duplicate module" : ERRORS% = ERRORS% + 1
3490    GOSUB 6270 : GOTO 3700                  ' Read and check nodal loads
3500    '
3510    IF INSTR(MODSTR$,"ELEMENT LOADS")=0 THEN GOTO 3590
3520    TARGET$ = "TYPE" : GOSUB 10380 : IF NOT STRFOUND% THEN GOTO 3700
3530    IF NELOAD%=0 THEN GOTO 3570
3540    '
3550    LCOUNT% = LCOUNT% + 1 : GOSUB 10730
3560    PRINT #2, "** ERROR ** Duplicate module" : ERRORS% = ERRORS% + 1
3570    GOSUB 6670 : GOTO 3700                  ' Read and check element loads
3580    '
3590    IF INSTR(MODSTR$,"ADDED MASS")=0 THEN GOTO 3680
3600    TARGET$ = "NODE" : GOSUB 10380 : IF NOT STRFOUND% THEN GOTO 3700
3610    IF NAMASS%=0 THEN GOTO 3650
3620    '
3630    LCOUNT% = LCOUNT% + 1 : GOSUB 10730
3640    PRINT #2, "** ERROR ** Duplicate module" : ERRORS% = ERRORS% + 1
3650    GOSUB 7010 : GOTO 3700                  ' Read and check added masses
3660    '
3670    LCOUNT% = LCOUNT% + 1 : GOSUB 10730
3680    PRINT #2, "** ERROR ** Unrecognised module" : GOTO 3700
3690    PRINT #2, "** ERROR ** Unexpected end to data" : ERRORS% = ERRORS%+1
3700 WEND
3710 IF NNODE% > 0 THEN GOTO 3740
3720 LCOUNT% = LCOUNT% + 1 : GOSUB 10730
3730 PRINT #2, "** ERROR ** NODES module missing"      : ERRORS%=ERRORS%+1
3740 IF NELEM% > 0 THEN GOTO 3770
3750 LCOUNT% = LCOUNT% + 1 : GOSUB 10730
3760 PRINT #2, "** ERROR ** ELEMENTS module missing"   : ERRORS%=ERRORS%+1
3770 IF NPROP% > 0 THEN GOTO 3800
3780 LCOUNT% = LCOUNT% + 1 : GOSUB 10730
3790 PRINT #2, "** ERROR ** PROPERTIES module missing" : ERRORS%=ERRORS%+1
3800 IF NREST% > 0 THEN GOTO 3830
```

```
3810 LCOUNT% = LCOUNT% + 1 : GOSUB 10730
3820 PRINT #2, "** ERROR ** RESTRAINTS module missing" : ERRORS%=ERRORS%+1
3830 IF (PTYPE$="PFDYNAM") OR (NNLOAD%>0) OR (NELOAD%>0) THEN GOTO 3860
3840 LCOUNT% = LCOUNT% + 1 : GOSUB 10730
3850 PRINT #2, "** ERROR ** LOADING module missing"    : ERRORS%=ERRORS%+1
3860 GOSUB 7280                                    ' Check for unconnected nodes
3870 GOSUB 7450                                    ' Check element nodes & length
3880 GOSUB 8000                                    ' Check restrained nodes
3890 GOSUB 8180                                    ' Check local axes nodes
3900 GOSUB 8400                                    ' Check loaded nodes
3910 GOSUB 8580                                    ' Check loaded elements
3920 GOSUB 8980                                    ' Check added mass loads
3930 GOSUB 9200                                    ' Check bandwidth
3940 PRINT #2,
3950 PRINT #2, "Note: lines terminated with '~~' have been ignored"
3960 CLOSE #1
3970 RETURN
3980 '
3990 '*****************   READ TITLE AND UNITS STRINGS   ******************
4000 '
4010 NLINES% = 2 : GOSUB 10550 : IF DATAEND% THEN GOTO 4100
4020 TITLE$ = WRKSTR$ : I% = 1
4030 WHILE (MID$(TITLE$, I%, 1) = " ")
4040   I% = I% + 1
4050 WEND
4060 TITLE$ = MID$(TITLE$, I%)
4070 '
4080 TARGET$ = "Units" : GOSUB 10380
4090 IF STRFOUND% THEN GOTO 4130
4100 PRINT #2, "** ERROR ** Error reading Title/Units strings"
4110 LCOUNT% = LCOUNT% + 1 : GOSUB 10730
4120 ERRORS% = ERRORS% + 1 : GOTO 4140
4130 UNITS$ = MID$(WRKSTR$,INSTR(WRKSTR$,":- ")+3,32)
4140 RETURN
4150 '
4160 '****************   READ AND CHECK NODAL COORDINATES   ***************
4170 '
4180 GOSUB 10040
4190 WHILE NNOS% > 0
4200   NNODE% = NNODE% + 1
4210   IF NNODE% <> MAXNODE% + 1 THEN GOTO 4260
4220   LCOUNT% = LCOUNT% + 1 : GOSUB 10730
4230   PRINT #2, "** ERROR ** Maximum number of nodes ("; MAXNODE%;
4240   PRINT #2, ") exceeded - checking discontinued"
4250   ERRORS% = ERRORS% + 1
4260   IF NNODE% > MAXNODE% THEN GOTO 4390
4270   IF (NNOS% = NNODENO%) THEN GOTO 4320
4280   LCOUNT% = LCOUNT% + 1 : GOSUB 10730
4290   PRINT #2, "** ERROR ** There should be "; NNODENO%;
4300   PRINT #2, " numbers in this line"
4310   ERRORS% = ERRORS% + 1
4320   FOR J% = 1 TO NNOS%
4330     NODE(NNODE%, J%) = VAL(WKSP$(J%))
4340   NEXT J%
4350   IF NODE(NNODE%,1) > 0 THEN GOTO 4390
4360   LCOUNT% = LCOUNT% + 1 : GOSUB 10730
4370   PRINT #2, "** ERROR **  Node numbers should be positive"
4380   ERRORS% = ERRORS%+1
4390   GOSUB 10040
4400 WEND
4410 FOR I% = 1 TO NNODE%
4420   '
4430   ' ...  Check for missing nodes
4440   '
4450   CHECK% = FAIL%
4460   FOR J% = 1 TO NNODE%
4470     IF NODE(J%,1) = I% THEN CHECK% = OK% : J% = NNODE%
4480   NEXT J%
4490   IF CHECK% = OK% THEN GOTO 4560
4500   LCOUNT% = LCOUNT% + 1 : GOSUB 10730
4510   PRINT #2, "** ERROR **  Node number "; I%; " missing"
4520   ERRORS% = ERRORS% + 1 : GOTO 4660
4530   '
4540   ' ...  Check for coincident nodes
4550   '
4560   FOR J% = I%+1 TO NNODE%
4570     IF I% = J% THEN GOTO 4650
```

```
4580      TEMP = (NODE(J%,2)-NODE(I%,2))^2 + (NODE(J%,3)-NODE(I%,3))^2
4590      IF NNODENO%=4 THEN TEMP = TEMP + (NODE(J%,4)-NODE(I%,4))^2
4600      IF TEMP > .000001 THEN GOTO 4650
4610    LCOUNT% = LCOUNT% + 1 : GOSUB 10730
4620      PRINT #2, "** WARNING ** Nodes ";I%;" and ";J%;
4630      PRINT #2, " are coincident or almost coincident"
4640      WARNINGS% = WARNINGS%+1
4650    NEXT J%
4660 NEXT I%
4670 RETURN
4680 '
4690 '*********************   READ AND CHECK ELEMENT DATA   *****************
4700 '
4710 GOSUB 10040
4720 WHILE NNOS% > 0
4730    NELEM% = NELEM% + 1
4740    IF NELEM% <> MAXELEM% + 1 THEN GOTO 4790
4750    LCOUNT% = LCOUNT% + 1 : GOSUB 10730
4760    PRINT #2, "** ERROR ** Maximum number of elements ("; MAXELEM%;
4770    PRINT #2, ") exceeded - checking discontinued"
4780    ERRORS% = ERRORS% + 1
4790    IF NELEM% > MAXELEM% THEN GOTO 4880
4800    IF (NNOS% = NELEMNO%) THEN GOTO 4850
4810    LCOUNT% = LCOUNT% + 1 : GOSUB 10730
4820    PRINT #2, "** ERROR ** There should be "; NELEMNO%;
4830    PRINT #2, " numbers in this line"
4840    ERRORS% = ERRORS% + 1
4850    FOR J% = 1 TO NNOS%
4860      ELEM(NELEM%, J%) = VAL(WKSP$(J%))
4870    NEXT J%
4880    GOSUB 10040
4890 WEND
4900 RETURN
4910 '
4920 '*******************    READ AND CHECK PROPERTIES    *********************
4930 '
4940 GOSUB 10040
4950 WHILE NNOS% > 0
4960    NPROP% = NPROP% + 1
4970    IF NPROP% <> MAXPROP% + 1 THEN GOTO 5020
4980    LCOUNT% = LCOUNT% + 1 : GOSUB 10730
4990    PRINT #2, "** ERROR ** Maximum number of properties ("; MAXPROP%;
5000    PRINT #2, ") exceeded - checking discontinued"
5010    ERRORS% = ERRORS% + 1
5020    IF NPROP% > MAXPROP% THEN GOTO 5350
5030    IF NNOS% = NPROPNO%  THEN GOTO 5080
5040    LCOUNT% = LCOUNT% + 1 : GOSUB 10730
5050    PRINT #2, "** ERROR ** There should be "; NPROPNO%;
5060    PRINT #2, " numbers in this line"
5070    ERRORS% = ERRORS% + 1
5080    FOR J% = 1 TO NNOS%
5090      PROP(NPROP%, J%) = VAL(WKSP$(J%))
5100    NEXT J%
5110    '
5120    ' ... Check for negative properties
5130    '
5140    FOR I% = 1 TO NPROP%
5150      WRKSTR$ = "property number"   : J%=1 :      GOSUB 5390
5160      IF (PTYPE$ <> "PTRUSS") AND (PTYPE$ <> "STRUSS") THEN GOTO 5190
5170      WRKSTR$ = "A"   : J%=2 :      GOSUB 5390
5180      WRKSTR$ = "E"   : J%=3 :      GOSUB 5390
5190      IF (PTYPE$ <> "PFRAME") THEN GOTO  5230
5200      WRKSTR$ = "A"   : J%=2 :      GOSUB 5390
5210      WRKSTR$ = "I"   : J%=3 :      GOSUB 5390
5220      WRKSTR$ = "E"   : J%=4 :      GOSUB 5390
5230      IF (PTYPE$ <> "GRID")   THEN GOTO  5280
5240      WRKSTR$ = "I"   : J%=2 :      GOSUB 5390
5250      WRKSTR$ = "J"   : J%=3 :      GOSUB 5390
5260      WRKSTR$ = "G"   : J%=4 :      GOSUB 5390
5270      WRKSTR$ = "E"   : J%=5 :      GOSUB 5390
5280      IF (PTYPE$ <> "SFRAME") THEN GOTO  5340
5290      WRKSTR$ = "A"   : J%=2 :      GOSUB 5390
5300      WRKSTR$ = "IYY" : J%=3 :      GOSUB 5390
5310      WRKSTR$ = "IZZ" : J%=4 :      GOSUB 5390
5320      WRKSTR$ = "J"   : J%=5 :      GOSUB 5390
5330      WRKSTR$ = "E"   : J%=7 :      GOSUB 5390
5340    NEXT I%
```

```
5350   WRKSTR$ = "" : GOSUB 10040
5360 WEND
5370 RETURN
5380 '
5390 '--------------------   CHECK FOR NEGATIVE PROPERTY    -----------------
5400 '
5410 IF PROP(NPROP%,J%) > 0 THEN GOTO 5450
5420 LCOUNT% = LCOUNT% + 1 : GOSUB 10730
5430 PRINT #2, "** ERROR **  Negative "; WRKSTR$; " value"
5440 ERRORS% = ERRORS%+1
5450 RETURN
5460 '
5470 '*******************    READ AND CHECK RESTRAINTS    ********************
5480 '
5490 GOSUB 10040
5500 OLDERRORS% = ERRORS% : OLDWARNINGS% = WARNINGS%
5510 IF (PTYPE$ = "PTRUSS")  THEN DIRS$ = "12"     : MINREST% = 2
5520 IF (PTYPE$ = "STRUSS")  THEN DIRS$ = "123"    : MINREST% = 2
5530 IF (PTYPE$ = "PFRAME")  THEN DIRS$ = "123"    : MINREST% = 1
5540 IF (PTYPE$ = "PFDYNAM") THEN DIRS$ = "123"    : MINREST% = 1
5550 IF (PTYPE$ = "GRID")    THEN DIRS$ = "123"    : MINREST% = 1
5560 IF (PTYPE$ = "SFRAME")  THEN DIRS$ = "123456" : MINREST% = 1
5570 WHILE NNOS% > 0
5580   NREST% = NREST% + 1
5590   IF NREST% <> MAXREST% + 1 THEN GOTO 5640
5600   LCOUNT% = LCOUNT% + 1 : GOSUB 10730
5610   PRINT #2, "** ERROR ** Maximum number of restraints ("; MAXREST%;
5620   PRINT #2, ") exceeded - checking discontinued"
5630   ERRORS% = ERRORS% + 1
5640   IF NREST% > MAXREST%  THEN GOTO 5870
5650   IF (NNOS% = NRESTNO%) THEN GOTO 5700
5660   LCOUNT% = LCOUNT% + 1 : GOSUB 10730
5670   PRINT #2, "** ERROR ** There should be "; NRESTNO%;
5680   PRINT #2, " numbers in this line"
5690   ERRORS% = ERRORS% + 1
5700   IF NREST% > MAXREST%  THEN GOTO 5870
5710   FOR J% = 1 TO NNOS%
5720     REST(NREST%, J%) = VAL(WKSP$(J%))
5730   NEXT J%
5740   '
5750   ' ... Check for illegal restraint directions
5760   '
5770   WKSP$(2) = STR$(REST(NREST%,2))      ' Note this string has a leading
blank
5780   WKSP$(2) = RIGHT$(WKSP$(2), LEN(WKSP$(2))-1)
5790   FOR J% = 1 TO LEN(WKSP$(2))
5800     IF INSTR(DIRS$, LEFT$(WKSP$(2),1)) > 0 THEN GOTO 5850
5810     LCOUNT% = LCOUNT% + 1 : GOSUB 10730
5820     PRINT #2, "** ERROR **  Illegal character "; LEFT$(WKSP$(2),1);
5830     PRINT #2, " in restraint direction(s);
5840     ERRORS% = ERRORS% + 1
5850     WKSP$(2) = RIGHT$(WKSP$(2), LEN(WKSP$(2))-1)
5860   NEXT J%
5870   GOSUB 10040
5880 WEND
5890 IF NREST% >= MINREST% THEN GOTO 5940
5900 LCOUNT% = LCOUNT% + 1 : GOSUB 10730
5910 PRINT #2, "** ERROR **  At least ";
5920 PRINT #2, MINREST%; " node(s) must be restrained"
5930 ERRORS% = ERRORS%+1
5940 RETURN
5950 '
5960 '******************    READ AND CHECK LOCAL AXES    ********************
5970 '
5980 GOSUB 10040
5990 WHILE NNOS% > 0
6000   NLAXES% = NLAXES% + 1
6010   IF NLAXES% <> MAXLAXES% + 1 THEN GOTO 6060
6020   LCOUNT% = LCOUNT% + 1 : GOSUB 10730
6030   PRINT #2, "** ERROR ** Maximum number of local axes (";
6040   PRINT #2, ") exceeded - checking discontinued"
6050   ERRORS% = ERRORS% + 1
6060   IF NLAXES% > MAXLAXES% THEN GOTO 6230
6070   IF (NNOS% = NLAXESNO%) THEN GOTO 6120
6080   LCOUNT% = LCOUNT% + 1 : GOSUB 10730
6090   PRINT #2, "** ERROR ** There should be "; NLAXESNO%;
6100   PRINT #2, " numbers in this line"
```

```
6110    ERRORS% = ERRORS% + 1
6120    FOR J% = 1 TO NNOS%
6130      LAXES(NLAXES%, J%) = VAL(WKSP$(J%))
6140    NEXT J%
6150    '
6160    ' ... Check angle
6170    '
6180    IF LAXES(NLAXES%,2) >= -360 OR LAXES(NLAXES%,2) <= 360 THEN GOTO 6230
6190    LCOUNT% = LCOUNT% + 1 : GOSUB 10730
6200    PRINT #2, "** WARNING ** Angle for local axes ";
6210    PRINT #2, "not in range -360 to 360 degrees"
6220    WARNINGS% = WARNINGS%+1
6230    GOSUB 10040
6240 WEND
6250 RETURN
6260 '
6270 '******************    READ AND CHECK NODAL LOADS    *******************
6280 '
6290 IF PTYPE$<>"PFDYNAM" THEN GOTO 6340
6300   LCOUNT% = LCOUNT% + 1 : GOSUB 10730
6310 PRINT #2, "** ERROR ** Illegal module for dynamic analysis"
6320 ERRORS%=ERRORS%+1 : GOTO 6650
6330 '
6340 GOSUB 10040
6350 WHILE NNOS% > 0
6360   NNLOAD% = NNLOAD% + 1
6370   IF NNLOAD% <> MAXNLOAD% + 1 THEN GOTO 6420
6380   LCOUNT% = LCOUNT% + 1 : GOSUB 10730
6390   PRINT #2, "** ERROR ** Maximum number of nodal loads ("; MAXNLOAD%;
6400   PRINT #2, ") exceeded - checking discontinued"
6410   ERRORS% = ERRORS% + 1
6420   IF NNLOAD% > MAXNLOAD% THEN GOTO 6630
6430   IF (NNOS% = NNLOADNO%) THEN GOTO 6480
6440   LCOUNT% = LCOUNT% + 1 : GOSUB 10730
6450   PRINT #2, "** ERROR ** There should be "; NNLOADNO%;
6460   PRINT #2, " numbers in this line"
6470   ERRORS% = ERRORS% + 1
6480   FOR J% = 1 TO NNOS%
6490     NLOAD(NNLOAD%, J%) = VAL(WKSP$(J%))
6500   NEXT J%
6510   '
6520   ' ... Check directions
6530   '
6540   IF (PTYPE$ = "PTRUSS") THEN : J% = 2
6550   IF (PTYPE$ = "STRUSS") THEN : J% = 3
6560   IF (PTYPE$ = "PFRAME") THEN : J% = 3
6570   IF (PTYPE$ = "GRID")   THEN : J% = 3
6580   IF (PTYPE$ = "SFRAME") THEN : J% = 6
6590   IF (NLOAD(NNLOAD%,2) > 0) AND (NLOAD(NNLOAD%,2) <= J%) THEN GOTO 6630
6600   LCOUNT% = LCOUNT% + 1 : GOSUB 10730
6610   PRINT #2, "** ERROR **  Illegal load direction"
6620   ERRORS% = ERRORS%+1
6630   GOSUB 10040
6640 WEND
6650 RETURN
6660 '
6670 '*******************    READ AND CHECK ELEMENT LOADS    ****************
6680 '
6690 IF ((PTYPE$="PFRAME") OR (PTYPE$="GRID")) THEN GOTO 6730
6700 PRINT #2, "** ERROR ** Illegal module for program "; PTYPE$
6710 ERRORS%=ERRORS%+1 : GOTO 6990
6720 '
6730 GOSUB 10040
6740 WHILE NNOS% > 0
6750   NELOAD% = NELOAD% + 1
6760   IF NELOAD% <> MAXELOAD% + 1 THEN GOTO 6810
6770   LCOUNT% = LCOUNT% + 1 : GOSUB 10730
6780   PRINT #2, "** ERROR ** Maximum number of element loads (";MAXELOAD%;
6790   PRINT #2, ") exceeded - checking discontinued"
6800   ERRORS% = ERRORS% + 1
6810   IF NELOAD% > MAXELOAD% THEN GOTO 6970
6820   IF (NNOS% = NELOADNO%) THEN GOTO 6870
6830   LCOUNT% = LCOUNT% + 1 : GOSUB 10730
6840   PRINT #2, "** ERROR ** There should be "; NELOADNO%;
6850   PRINT #2, " numbers in this line"
6860   ERRORS% = ERRORS% + 1
6870   FOR J% = 1 TO NNOS%
```

```
6880      ELOAD(NELOAD%, J%) = VAL(WKSP$(J%))
6890    NEXT J%
6900    '
6910    ' ... Check load type
6920    '
6930    IF (ELOAD(NELOAD%,3)>0) AND (ELOAD(NELOAD%,3)<3) THEN GOTO 6970
6940    LCOUNT% = LCOUNT% + 1 : GOSUB 10730
6950    PRINT #2, "** ERROR **  Illegal element load type"
6960    ERRORS% = ERRORS%+1
6970    GOSUB 10040
6980 WEND
6990 RETURN
7000 '
7010 '*******************    READ AND CHECK ADDED MASSES    ******************
7020 '
7030 IF PTYPE$="PFDYNAM" THEN GOTO 7070
7040 PRINT #2, "** ERROR ** Illegal module for program "; PTYPE$
7050 ERRORS%=ERRORS%+1 : GOTO 7260
7060 '
7070 GOSUB 10040
7080 WHILE NNOS% > 0
7090   NAMASS% = NAMASS% + 1
7100   IF NAMASS% <> MAXAMASS% + 1 THEN GOTO 7150
7110   LCOUNT% = LCOUNT% + 1 : GOSUB 10730
7120   PRINT #2, "** ERROR ** Maximum number of added masses ("; MAXAMASS%;
7130   PRINT #2, ") exceeded - checking discontinued"
7140   ERRORS% = ERRORS% + 1
7150   IF NAMASS% > MAXAMASS% THEN GOTO 7240
7160   IF (NNOS% = NAMASSNO%) THEN GOTO 7210
7170   LCOUNT% = LCOUNT% + 1 : GOSUB 10730
7180   PRINT #2, "** ERROR ** There should be "; NAMASSNO%;
7190   PRINT #2, " numbers in this line"
7200   ERRORS% = ERRORS% + 1
7210   FOR J% = 1 TO NNOS%
7220     AMASS(NAMASS%, J%) = VAL(WKSP$(J%))
7230   NEXT J%
7240   GOSUB 10040
7250 WEND
7260 RETURN
7270 '
7280 '****************    CHECK FOR UNCONNECTED NODES    ********************
7290 '
7300 LCOUNT% = LCOUNT% + 1 : GOSUB 10730
7310 PRINT #2,
7320 FOR I% = 1 TO NNODE%
7330   CHECK% = FAIL%
7340   FOR J% = 1 TO NELEM%
7350     IF ELEM(J%,1)=NODE(I%,1) THEN CHECK%=OK%
7360     IF ELEM(J%,2)=NODE(I%,1) THEN CHECK%=OK%
7370   NEXT J%
7380   IF CHECK% = OK% THEN GOTO 7420
7390   LCOUNT% = LCOUNT% + 1 : GOSUB 10730
7400   PRINT #2, "** ERROR **  Node number "; I%; " is unconnected"
7410   ERRORS% = ERRORS% + 1
7420 NEXT I%
7430 RETURN
7440 '
7450 '*******   CHECK ELEMENT NODE NUMBERS, PROPERTIES AND LENGTHS   ********
7460 '
7470 OLDERRORS%=ERRORS%
7480 FOR I% = 1 TO NELEM%
7490   IN%  = ELEM(I%,1)
7500   JN%  = ELEM(I%,2)
7510   '
7520   ' ...  Check for non-existent nodes
7530   '
7540   CHECK% = FAIL%
7550   FOR J% = 1 TO NNODE%
7560     IF NODE(J%,1) = IN% THEN CHECK% = OK% : J% = NNODE%
7570   NEXT J%
7580   IF CHECK% = OK% THEN GOTO 7630
7590   LCOUNT% = LCOUNT% + 1 : GOSUB 10730
7600   PRINT #2, "** ERROR **  Node "; IN%;
7610   PRINT #2, "referenced in ELEMENTS module does not exist"
7620   ERRORS% = ERRORS%+1
7630   CHECK% = FAIL%
7640   FOR J% = 1 TO NNODE%
```

```
7650      IF NODE(J%,1) = JN% THEN CHECK% = OK% : J% = NNODE%
7660    NEXT J%
7670    IF CHECK% = OK% THEN GOTO 7750
7680    LCOUNT% = LCOUNT% + 1 : GOSUB 10730
7690    PRINT #2, "** ERROR **  Node "; JN%;
7700    PRINT #2, "referenced in ELEMENTS module does not exist"
7710    ERRORS% = ERRORS%+1
7720    '
7730    ' ... Calculate the element length
7740    '
7750    IF OLDERROS%<>ERRORS% THEN GOTO 7980
7760    TEMP = (NODE(JN%,2)-NODE(IN%,2))^2 + (NODE(JN%,3)-NODE(IN%,3))^2
7770    IF (PTYPE$<>"STRUSS") AND (PTYPE$<>"SFRAME") THEN GOTO 7790
7780    TEMP = TEMP + (NODE(JN%,4)-NODE(IN%,4))^2
7790    TEMP = SQR(TEMP)
7800    IF TEMP > .000001 THEN GOTO 7850
7810    LCOUNT% = LCOUNT% + 1 : GOSUB 10730
7820    PRINT #2, "** ERROR **  Element ";IN%;",";JN%;
7830    PRINT #2, " has zero or almost zero length ("; TEMP; ")"
7840    ERRORS% = ERRORS%+1
7850    '
7860    ' ... Check that all properties are defined
7870    '
7880    CHECK% = FAIL%
7890    FOR J% = 1 TO NPROP%
7900      IF ELEM(I%,3) = PROP(J%,1) THEN CHECK% = OK% : J% = NPROP%
7910    NEXT J%
7920    IF CHECK% = OK% THEN GOTO 7970
7930    LCOUNT% = LCOUNT% + 1 : GOSUB 10730
7940    PRINT #2, "** ERROR **  Property number"; ELEM(I%,3);
7950    PRINT #2, "for element "; ELEM(I%,1);",";ELEM(I%,2) " is undefined"
7960    ERRORS% = ERRORS%+1
7970 NEXT I%
7980 RETURN
7990 '
8000 '********************    CHECK RESTRAINED NODES    **********************
8010 '
8020 FOR I% = 1 TO NREST%
8030    '
8040    ' ...  Check for non-existent nodes
8050    '
8060    CHECK% = FAIL%
8070    FOR J% = 1 TO NNODE%
8080      IF NODE(J%,1) = REST(I%,1) THEN CHECK% = OK% : J% = NNODE%
8090    NEXT J%
8100    IF CHECK% = OK% THEN GOTO 8150
8110    LCOUNT% = LCOUNT% + 1 : GOSUB 10730
8120    PRINT #2, "** ERROR **  Node "; REST(I%,1);
8130    PRINT #2, "referenced in RESTRAINTS module does not exist"
8140    ERRORS% = ERRORS%+1
8150 NEXT I%
8160 RETURN
8170 '
8180 '******************    CHECK LOCAL AXES NODES    ************************
8190 '
8200 FOR I% = 1 TO NLAXES%
8210    '
8220    ' ...  Check for non-existent nodes
8230    '
8240    IF LAXES(I%,1) > 0 THEN GOTO 8280
8250    LCOUNT% = LCOUNT% + 1 : GOSUB 10730
8260    PRINT #2, "** ERROR **  Illegal node referenced for local axes"
8270    ERRORS% = ERRORS%+1
8280    CHECK% = FAIL%
8290    FOR J% = 1 TO NNODE%
8300      IF NODE(J%,1) = LAXES(I%,1) THEN CHECK% = OK% : J% = NNODE%
8310    NEXT J%
8320    IF CHECK% = OK% THEN GOTO 8370
8330    LCOUNT% = LCOUNT% + 1 : GOSUB 10730
8340    PRINT #2, "** ERROR **  Node "; LAXES(I%,1);
8350    PRINT #2, "referenced in LOCAL AXES module does not exist"
8360    ERRORS% = ERRORS%+1
8370 NEXT I%
8380 RETURN
8390 '
8400 '********************    CHECK LOADED NODES    ************************
8410 '
```

```
8420 FOR I% = 1 TO NNLOAD%
8430   '
8440   ' ...  Check for non-existent nodes
8450   '
8460   CHECK% = FAIL%
8470   FOR J% = 1 TO NNODE%
8480     IF NODE(J%,1) = NLOAD(I%,1) THEN CHECK% = OK% : J% = NNODE%
8490   NEXT J%
8500   IF CHECK% = OK% THEN GOTO 8550
8510   LCOUNT% = LCOUNT% + 1 : GOSUB 10730
8520   PRINT #2, "** ERROR **  Node "; NLOAD(I%,1);
8530   PRINT #2, "referenced in NODAL LOADS module does not exist"
8540   ERRORS% = ERRORS%+1
8550 NEXT I%
8560 RETURN
8570 '
8580 '********************   CHECK LOADED ELEMENTS    **********************
8590 '
8600 FOR I% = 1 TO NELOAD%
8610   '
8620   ' ...  Check for non-existent nodes
8630   '
8640   CHECK% = FAIL%
8650   FOR J% = 1 TO NNODE%
8660     IF NODE(J%,1) = ELOAD(I%,1) THEN CHECK% = OK% : J% = NNODE%
8670   NEXT J%
8680   IF CHECK% = OK% THEN GOTO 8730
8690   LCOUNT% = LCOUNT% + 1 : GOSUB 10730
8700   PRINT #2, "** ERROR **  Node "; ELOAD(I%,1);
8710   PRINT #2, "referenced in ELEMENT LOADS module does not exist"
8720   ERRORS% = ERRORS%+1
8730   CHECK% = FAIL%
8740   FOR J% = 1 TO NNODE%
8750     IF NODE(J%,1) = ELOAD(I%,2) THEN CHECK% = OK% : J% = NNODE%
8760   NEXT J%
8770   IF CHECK% = OK% THEN GOTO 8850
8780   LCOUNT% = LCOUNT% + 1 : GOSUB 10730
8790   PRINT #2, "** ERROR **  Node "; ELOAD(I%,2);
8800   PRINT #2, "referenced in ELEMENT LOADS  module does not exist""
8810   ERRORS% = ERRORS%+1
8820   '
8830   ' ...  Check for non-existent elements
8840   '
8850   CHECK% = FAIL%
8860   FOR J% = 1 TO NELEM%
8870     IF ELEM(J%,1) <> ELOAD(I%,1) THEN GOTO 8890
8880     IF ELEM(J%,2) =  ELOAD(I%,2) THEN CHECK% = OK% : J% = NELEM%
8890   NEXT J%
8900   IF CHECK% = OK% THEN GOTO 8950
8910   LCOUNT% = LCOUNT% + 1 : GOSUB 10730
8920   PRINT #2, "** ERROR **  Element "; ELOAD(I%,1); ","; ELOAD(I%,2);
8930   PRINT #2, "referenced in ELEMENT LOADS module does not exist"
8940   ERRORS% = ERRORS%+1
8950 NEXT I%
8960 RETURN
8970 '
8980 '********************   CHECK ADDED MASS NODES    *********************
8990 '
9000 FOR I% = 1 TO NAMASS%
9010   '
9020   ' ...  Check for non-existent nodes
9030   '
9040   IF NLOAD(NNLOAD%,1) > 0 THEN GOTO 9080
9050   LCOUNT% = LCOUNT% + 1 : GOSUB 10730
9060   PRINT #2, "** ERROR **  Illegal node referenced for added mass"
9070   ERRORS% = ERRORS%+1
9080   CHECK% = FAIL%
9090   FOR J% = 1 TO NNODE%
9100     IF NODE(J%,1) = AMASS(I%,1) THEN CHECK% = OK% : J% = NNODE%
9110   NEXT J%
9120   IF CHECK% = OK% THEN GOTO 9170
9130   LCOUNT% = LCOUNT% + 1 : GOSUB 10730
9140   PRINT #2, "** ERROR **  Node "; AMASS%(I%,1);
9150   PRINT #2, "referenced in ADDED MASS module does not exist"
9160   ERRORS% = ERRORS%+1
9170 NEXT I%
9180 RETURN
```

```
9190 '
9200 '*********************   CHECK BANDWIDTH   ***************************
9210 '
9220 ' ... No check required for plane and space frames as full matrices
9230 '     stored
9240 '
9250 IF PTYPE$="PTRUSS" OR PTYPE$="STRUSS" THEN GOTO 9760
9260 '
9270 ' ... Zero the freedom vector
9280 '
9290 FOR I% = 1 TO NFREE%*NNODE%
9300   FREE%(I%) = 0
9310 NEXT I%
9320 '
9330 ' ... Loop for all restrained nodes
9340 '
9350 FOR I% = 1 TO NREST%
9360   K% = REST(I%,2)
9370   '
9380   ' ... Evaluate the freedom to be restrained from K%
9390   '
9400   J% = NFREE%*REST(I%,1) - NFREE% + (K% MOD 10)
9410   FREE%(J%) = 1
9420   K% = INT(K%/10)
9430   IF K% > 0 THEN GOTO 9400
9440 NEXT I%
9450 '
9460 ' ... Number the freedoms (restraints set to zero)
9470 '
9480 NDOF% = 0
9490 FOR I% = 1 TO NFREE%*NNODE%
9500   IF FREE%(I%) = 1 THEN GOTO 9540
9510   NDOF% = NDOF% + 1
9520   FREE%(I%) = NDOF%
9530   GOTO 9550
9540   FREE%(I%) = 0
9550 NEXT I%
9560 '
9570 ' ... Calculate the bandwidth
9580 '
9590 FOR K% = 1 TO NELEM%
9600   I% = NFREE%*(ELEM(K%,1)-1)
9610   J% = NFREE%*(ELEM(K%,2)-1)
9620   FOR L% = 1 TO NFREE%
9630     EFREE%(L%)        = FREE%(I%+L%)
9640     EFREE%(L%+NFREE%) = FREE%(J%+L%)
9650   NEXT L%
9660   I% = 0 : J% = 0 : M% = 2*NFREE%+1
9670   FOR L% = 1 TO 2*NFREE%
9680     IF (EFREE%(L%)   <>0) AND (I%=0) THEN I% = EFREE%(L%)
9690     IF (EFREE%(M%-L%)<>0) AND (J%=0) THEN J% = EFREE%(M%-L%)
9700   NEXT L%
9710   IF (J%-I%+1)> BAND% THEN BAND% = J%-I%+1
9720 NEXT K%
9730 LCOUNT% = LCOUNT% + 1 : GOSUB 10730
9740 PRINT #2,
9750 PRINT #2, "Bandwidth = "; BAND%
9760 RETURN
9770 '
9780 '*********************    GET KEYBOARD RESPONSE    *********************
9790 '
9800 ROW% = CSRLIN: COL% = POS(0)
9810 REPLY% = 0
9820 WHILE REPLY% = 0
9830   LOCATE ROW%, COL%: INPUT "", WRKSTR$: REPLY% = VAL(WRKSTR$)
9840   IF (REPLY% < 1)  OR  (REPLY% > NOPTIONS%) THEN GOTO 9870
9850   IF (REPLY% < 10) AND (LEN(WRKSTR$) = 1)   THEN GOTO 9880
9860   IF (REPLY% > 9)  AND (LEN(WRKSTR$) = 2)   THEN GOTO 9880
9870   SOUND 700, 4: REPLY% = 0: LOCATE ROW%, COL%
9880 WEND
9890 PRINT
9900 RETURN
9910 '
9920 '*******************    PROCESS KEYBOARD INPUT    **********************
9930 '
9940 ROW% = CSRLIN: COL% = POS(0)
9950 INPUT ; "", T$
```

```
9960 IF LEN(T$) = 0 THEN GOTO 10000
9970 WKSP$(J%) = T$
9980 PRINT
9990 GOTO 10020
10000 LOCATE ROW%, COL%
10010 PRINT WKSP$(J%)
10020 RETURN
10030 '
10040 '************************   RECORD NUMBERS   *****************************
10050 '
10060 NNOS% = 0 : DONE% = TRUE%
10070 IF EOF(1)<>0 THEN DATAEND% = TRUE% : GOTO 10360
10080 '
10090 ' ... Read next line and check if it is blank
10100 '
10110 LINE INPUT #1, WRKSTR$
10120 FOR I% = 1 TO LEN(WRKSTR$)
10130   IF MID$(WRKSTR$,I%,1)<> " " THEN DONE% = FALSE% : I% = LEN(WRKSTR$)
10140 NEXT I%
10150 LCOUNT% = LCOUNT% + 1 : GOSUB 10730
10160 IF DONE%=TRUE% THEN PRINT #2, WRKSTR$; "~~" : GOTO 10360
10170 PRINT #2, WRKSTR$
10180 FOR I% = 1 TO 10
10190   WKSP$(I%) = ""
10200 NEXT I%
10210 '
10220 ' ... Search for numbers
10230 '
10240 FOR I% = 1 TO LEN(WRKSTR$)
10250   WHILE (MID$(WRKSTR$, I%, 1) = " ") AND I% <= LEN(WRKSTR$)
10260     I% = I% + 1
10270   WEND
10280   IF (I% > LEN(WRKSTR$)) THEN GOTO 10340
10290   IF NNOS% = 10 THEN GOTO 10360
10300   NNOS% = NNOS% + 1: WKSP$(NNOS%) = ""
10310   WHILE (MID$(WRKSTR$, I%, 1) <> " ") AND I% <= LEN(WRKSTR$)
10320     WKSP$(NNOS%) = WKSP$(NNOS%) + MID$(WRKSTR$,I%,1): I% = I% + 1
10330   WEND
10340 NEXT I%
10350 WRKSTR$ = ""
10360 RETURN
10370 '
10380 '***************************   FIND STRING   ****************************
10390 '
10400 STRFOUND% = FALSE% : DONE% = FALSE%
10410 WHILE DONE% = FALSE%
10420   IF EOF(1)<>0 THEN DATAEND% = TRUE% : DONE% = TRUE% : GOTO 10520
10430   LINE INPUT #1, WRKSTR$
10440   IF INSTR(WRKSTR$,TARGET$)>0 THEN STRFOUND% = TRUE% : DONE% = TRUE%
10450   IF(TARGET$="PROGRAM") AND LCOUNT% > 4 THEN DONE% = TRUE%
10460   LCOUNT% = LCOUNT% + 1 : GOSUB 10730
10470   IF DONE%=FALSE% THEN PRINT #2, WRKSTR$; "~~" ELSE PRINT #2, WRKSTR$
10480   '
10490   ' ... Check for new module
10500   '
10510   IF (LEFT$(WRKSTR$,4)="+   ") THEN DONE% = TRUE%
10520 WEND
10530 RETURN
10540 '
10550 '***************************   READ LINES   *****************************
10560 '
10570 FOR I% = 1 TO NLINES%
10580   IF EOF(1)<>0 THEN DATAEND% = TRUE% : I% = NLINES% : GOTO 10610
10590   LCOUNT% = LCOUNT% + 1 : GOSUB 10730
10600   LINE INPUT #1, WRKSTR$ : PRINT #2, WRKSTR$; "~~"
10610 NEXT I%
10620 RETURN
10630 '
10640 '********************   ANY KEY TO CONTINUE   ***********************
10650 '
10660 IF DEVICE$<>"SCRN:" THEN GOTO 10710
10670 LOCATE PAGELEN%+1, 1
10680 PRINT SPACE$(45); "Press any key to continue";
10690 WHILE INKEY$ = "" : WEND
10700 CLS : LCOUNT%=0
10710 RETURN
10720 '
```

```
10730 '**********************    CHECK FOR PAGE END    *********************
10740 '
10750 IF LCOUNT%>=PAGELEN% THEN GOSUB 10640
10760 RETURN
10770 '
10780 '========================    DCHECK.BAS    ==========================
```

Sample Run

There follows a sample run which shows ouput from DCHECK.BAS while checking a PTRUSS file that is supposed to describe the structure analysed in example 5.3. As can be seen DCHECK.BAS finds a number of obvious errors. Note that DCHECK terminates lines that it ignores with ~~ and what is listed here is a report file written by DCHECK.

```
============================================================================~~
~~
    PROGRAM PTRUSS                                 Copyright(c) J.A.D.Balfour
~~
                                DCHECK.BAS check no. 1~~
~~
    File Name :- PT5.DAT                                     Date :- 28-11-91~~
    Units     :- kN & mm                                     Time :- 08:59:44
~~
============================================================================~~
~~
~~
+ + + + + + + + + + + + +~~
+   NODAL COORDINATES   +~~
+ + + + + + + + + + + + +~~
~~
NODE           X              Y
  1            0              0
  2            3000           3000
  3            6000           0
  4            6000           0
~~
** WARNING ** Nodes  3  and  4  are coincident or almost coincident
+ + + + + + + + + +~~
+   ELEMENTS      +~~
+ + + + + + + + + +~~
~~
ELEMENT    PROPERTY
  1  3         1
  1  8         5
~~
+ + + + + +  + + +~~
+   PROPERTIES   +~~
+ + + + + +  + + +~~
~~
PROPERTY        A             E
   -5      -.3000E+04    -.2000E+03
** ERROR **  Negative property number value
** ERROR **  Negative A value
** ERROR **  Negative E value
   2      0.3000E+04    0.2000E+03
~~
+ + + + + + + + + +~~
+   RESTRAINTS    +~~
+ + + + + + + + + +~~
~~
NODE    DIRECTION(S)
  1          26
** ERROR **  Illegal character 6 in restraint direction(s);
~~
** ERROR **  At least  2  node(s) must be restrained
+ + + + + + + + + +~~
+   NODAL LOADS   +~~
+ + + + + + + + + +~~
~~
NODE    DIRECTION     VALUE
```

```
  2         2    -.2000E+03
  6         0    0.1000E+03
** ERROR **  Illegal load direction
--
+ + + + + + + + + + + +--
+   ELEMENT LOADS   +   -     UDL = 1,     POINT LOAD = 2
+ + + + + + + + + + + +--
--
ELEMENT    LOAD      MAGNITUDE     DISTANCE FROM--
NUMBER     TYPE                    LOWER NODE
** ERROR ** Illegal module for program PTRUSS
   0  0     0     0.0000E+00          N/A--

** ERROR **  Node number  2  is unconnected
** ERROR **  Node number  4  is unconnected
** ERROR **  Node  6 referenced in NODAL LOADS module does not exist

Note: lines terminated with '--' have been ignored

DATA CHECKED:-    12  Errors     1  Warnings
```

Chapter 6

Space Trusses

6.1 Introduction

The previous chapter considered only plane trusses. This chapter shows how the theory for plane trusses is easily extended to deal with three-dimensional trusses. The principles underlying the theory developed for plane trusses are equally applicable to space trusses. The implementation is, however, complicated by the introduction of an additional freedom at each node. Consequently the nodal force and displacement vectors now have three components and the transformation matrix and element stiffness submatrices are 3 x 3 arrays. This chapter has been written assuming the reader to be familiar with the material presented in the previous chapter because, for compactness, only the essential differences between the analysis of space trusses and plane trusses are covered.

6.2 The Element Stiffness Matrix in Element Axes

Figure 6.1 shows the element axes and freedoms for a space truss element.

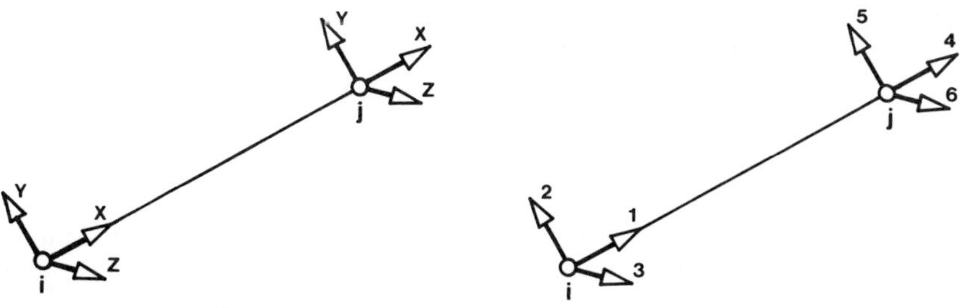

Figure 6.1 Element axes and freedoms for a space truss element

In Chapter 5 the relationship between the element end forces and the element end displacements for a plane truss element was shown to be

$$\begin{bmatrix} f_{ijx} \\ f_{ijy} \\ f_{jix} \\ f_{jiy} \end{bmatrix} = \begin{bmatrix} \frac{EA}{L} & 0 & \frac{-EA}{L} & 0 \\ 0 & 0 & 0 & 0 \\ \frac{-EA}{L} & 0 & \frac{EA}{L} & 0 \\ 0 & 0 & 0 & 0 \end{bmatrix} \begin{bmatrix} \delta_{ijx} \\ \delta_{ijy} \\ \delta_{jix} \\ \delta_{jiy} \end{bmatrix}$$

The principal difference between space trusses and plane trusses is that space trusses have an additional freedom at each node (in the "z" direction). Displacements in the "z" direction (element axes) generate no force in the element (on the basis of small displacement theory). Hence the element stiffness equation for a space truss element is as follows

$$\begin{bmatrix} f_{ijx} \\ f_{ijy} \\ f_{ijz} \\ \hline f_{jix} \\ f_{jiy} \\ f_{jiz} \end{bmatrix} = \left[\begin{array}{ccc|ccc} \frac{EA}{L} & 0 & 0 & \frac{-EA}{L} & 0 & 0 \\ 0 & 0 & 0 & 0 & 0 & 0 \\ 0 & 0 & 0 & 0 & 0 & 0 \\ \hline \frac{-EA}{L} & 0 & 0 & \frac{EA}{L} & 0 & 0 \\ 0 & 0 & 0 & 0 & 0 & 0 \\ 0 & 0 & 0 & 0 & 0 & 0 \end{array}\right] \begin{bmatrix} \delta_{ijx} \\ \delta_{ijy} \\ \delta_{ijz} \\ \hline \delta_{jix} \\ \delta_{jiy} \\ \delta_{jiz} \end{bmatrix}$$

or, in terms of the submatrices indicated by the broken lines

$$f_{ij} = k_{ii}^{j}\,\delta_{ij} + k_{ij}\,\delta_{ji} \tag{6.1}$$

and

$$f_{ji} = k_{ji}\,\delta_{ij} + k_{jj}^{i}\,\delta_{ji} \tag{6.2}$$

The reader should note that equations (6.1) and (6.2) are exactly the same as the corresponding equations for a plane truss element.

6.3 Transformation of Force and Displacement

In general the element axes will not coincide with the nodal axes, making necessary the transformation of forces and displacements from one axes system to another. In section 3.8 transformations in three dimensions were considered and the following transformation matrix was shown to transform force and displacement vectors from the global to the element axes systems.

$$T = \begin{bmatrix} l_x & m_x & n_x \\ l_y & m_y & n_y \\ l_z & m_z & n_z \end{bmatrix}$$

where

l_x is the cosine of the angle between the element "*x*" axis and the global "*x*" axis.

m_x is the cosine of the angle between the element "*x*" axis and the global "*y*" axis.

etc.

for example

$$\delta_{ij} = T_{ij}\,\Delta_i$$

$$\begin{bmatrix} \delta_{ijx} \\ \delta_{ijy} \\ \delta_{ijz} \end{bmatrix} = \begin{bmatrix} l_x & m_x & n_x \\ l_y & m_y & n_y \\ l_z & m_z & n_z \end{bmatrix} \begin{bmatrix} \Delta_{ix} \\ \Delta_{iy} \\ \Delta_{iz} \end{bmatrix}$$

The inverse of the above matrix effects the reverse transformation (i.e. from element to global axes), for example

$$\Delta_i = T_{ij}^{-1}\,\delta_{ij}$$

$$\begin{bmatrix} \Delta_{ix} \\ \Delta_{iy} \\ \Delta_{iz} \end{bmatrix} = \begin{bmatrix} l_x & l_y & l_z \\ m_x & m_y & m_z \\ n_x & n_y & n_z \end{bmatrix} \begin{bmatrix} \delta_{ijx} \\ \delta_{ijy} \\ \delta_{ijz} \end{bmatrix}$$

6.4 The Element Stiffness Matrix in the Global Axes

The following equation was developed in the previous chapter. It expresses, in matrix form, the equilibrium of node "*i*" of a plane truss structure.

$$P_i = K_{ii}\,\Delta_i + K_{ia}\,\Delta_a + K_{ib}\,\Delta_b + \ldots + K_{in}\,\Delta_n$$

where

$$K_{ii} = \sum_{j=a}^{n} K_{ii}^{j}$$

and

$$K_{ii}^{j} = T_{ij}^{-1}\,k_{ii}^{j}\,T_{ij}$$

$$K_{ij} = T_{ij}^{-1}\,k_{ij}\,T_{ij}$$

The reader should note that the above equations have general applicability as they simply state that each node of a structure must be in equilibrium under the action of the external applied forces (which may include reactive forces) and the internal element end forces. In the previous chapter the vectors of nodal force and displacement for a plane truss were seen to have two components and consequently the element stiffness submatrices had dimension 2 x 2. Three-dimensional trusses have an additional freedom at each node. Hence the vectors of nodal force and displacement have three components and the element stiffness submatrices are 3 x 3. Writing the equation of nodal equilibrium for each node of the structure in turn produces a system of simultaneous equations relating the nodal displacements to the nodal loads. That system of equations is, as was seen in the previous chapter, the initial structure stiffness equation.

In practice it is convenient to explicitly evaluate the global element stiffness submatrices as follows

$$\begin{aligned} \boldsymbol{K}_{ii}^{j} &= \boldsymbol{T}_{ij}^{-1}\, \boldsymbol{k}_{ii}^{j}\, \boldsymbol{T}_{ij} \\ &= \begin{bmatrix} l_x & l_y & l_z \\ m_x & m_y & m_z \\ n_x & n_y & n_z \end{bmatrix} \begin{bmatrix} EA/L & 0 & 0 \\ 0 & 0 & 0 \\ 0 & 0 & 0 \end{bmatrix} \begin{bmatrix} l_x & m_x & n_x \\ l_y & m_y & n_y \\ l_z & m_z & n_z \end{bmatrix} \\ &= \frac{EA}{L} \begin{bmatrix} l_x^2 & l_x m_x & l_x n_x \\ l_x m_x & m_x^2 & m_x n_x \\ l_x n_x & m_x n_x & n_x^2 \end{bmatrix} \end{aligned}$$

similarly

$$\begin{aligned} \boldsymbol{K}_{ij} &= \boldsymbol{T}_{ij}^{-1}\, \boldsymbol{k}_{ij}\, \boldsymbol{T}_{ij} \\ &= -\frac{EA}{L} \begin{bmatrix} l_x^2 & l_x m_x & l_x n_x \\ l_x m_x & m_x^2 & m_x n_x \\ l_x n_x & m_x n_x & n_x^2 \end{bmatrix} \end{aligned}$$

Evaluation of the $\boldsymbol{K}_{jj}^{i}$ and the $\boldsymbol{K}_{ji}$ submatrices shows that

$$\boldsymbol{K}_{jj}^{i} = \boldsymbol{K}_{ii}^{j}$$

and

$$\boldsymbol{K}_{ji} = \boldsymbol{K}_{ij}$$

To evaluate the element stiffness submatrices developed above requires a knowledge of the direction cosines of the element "x" axis. These are easily obtained from the coordinates of the nodes at the ends of the element. If the length of element "i,j" is L_{ij} and the coordinates of nodes "i" and "j" are (x_i, y_i, z_i) and (x_j, y_j, z_j) as illustrated in fig 6.2, then

$$l_x = \frac{x_j - x_i}{L_{ij}}$$

$$m_x = \frac{y_j - y_i}{L_{ij}}$$

$$n_x = \frac{z_j - z_i}{L_{ij}}$$

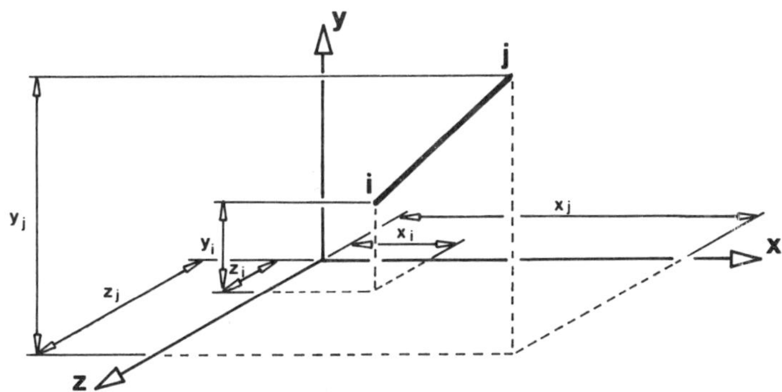

Figure 6.2 Calculation of direction cosines for element "i,j"

6.5 The Structure Stiffness Matrix

Chapter 5 showed that, prior to the application of the boundary conditions, the equations of nodal equilibrium form a mixed set of simultaneous equations. The coefficient matrix of these equations is known as the initial structure stiffness matrix. The procedure to assemble the initial structure stiffness matrix for a three-dimensional truss is identical to that used for a two-dimensional truss. However, because the global element stiffness submatrices are now 3 x 3 they must be added to the initial structure stiffness matrix such that element (1,1) is added to locations ($3i$-2, $3i$-2), ($3i$-2, $3j$-2), ($3j$-2, $3i$-2), and ($3j$-2, $3j$-2) respectively. Figure 6.3 shows the locations where the four global element stiffness submatrices for element 2,5 would be added to the initial structure stiffness matrix. The initial structure stiffness matrix is 18 x 18, indicating that the space truss has 6 nodes.

Chapter 5 also showed that it is more efficient to assemble the final structure stiffness matrix directly, rather than via the initial structure stiffness matrix.

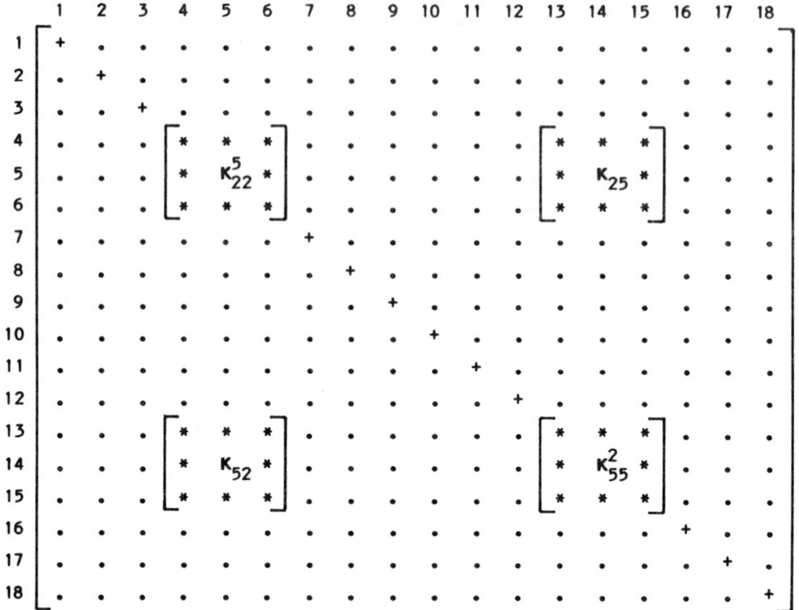

Figure 6.3 Adding element 2,5 to the initial structure stiffness matrix

To assemble the final structure stiffness matrix requires a knowledge of the boundary conditions. Restraints are conveniently described by the number of the restrained node, plus the direction of the restraint. Programming is simplified if the directions of restraint are given numbers. For space trusses the number system that follows will be adopted

1. translation restrained in the global "*x*" direction.
2. translation restrained in the global "*y*" direction.
3. translation restrained in the global "*z*" direction.

The boundary conditions for the truss shown in fig 6.4 would be input as follows.

NODE	DIRECTION
1	1
1	2
1	3
2	1
2	2
2	3
4	2

Input can be reduced by entering the restraint directions as a composite number (see section 5.8). Using composite numbers the above restraints become

NODE	DIRECTIONS
1	123
2	123
4	2

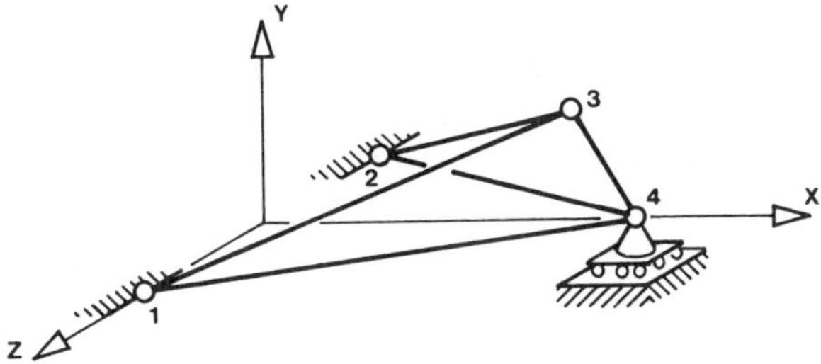

Figure 6.4 A simple three-dimensional truss

Zero the freedom vector	Apply the boundary conditions	Number the freedoms
0	1	0
0	1	0
0	1	0
0	1	0
0	1	0
0	1	0
0	0	1
0	0	2
0	0	3
0	0	4
0	1	0
0	0	5

The procedure to assemble the final stiffness matrix for a space truss is identical to that used to assemble the final stiffness matrix for a plane truss. Hence the freedom vector for the truss shown in fig 6.4 is assembled as follows. Note that the length of the freedom vector is equal to the initial degree of freedom (i.e. 3 times the number of nodes).

The code number for element "i,j" comprises of the elements ($3i$-2), ($3i$-1), ($3i$), ($3j$-2), ($3j$-1) and ($3j$) of the freedom vector.

Consider again the truss shown in fig 6.4. The freedom vector has already been shown to be

$$\{0\ 0\ 0\ 0\ 0\ 0\ 1\ 2\ 3\ 4\ 0\ 5\}$$

Hence the code number for element 1,3 is [0 0 0 1 2 3] which is applied to the global element stiffness matrix as follows.

$$\begin{array}{c} \\ 0 \\ 0 \\ 0 \\ 1 \\ 2 \\ 3 \end{array} \begin{array}{c} \begin{array}{cccccc} 0 & 0 & 0 & 1 & 2 & 3 \end{array} \\ \begin{bmatrix} K_{11} & K_{12} & K_{13} & K_{14} & K_{15} & K_{16} \\ K_{21} & K_{22} & K_{23} & K_{24} & K_{25} & K_{26} \\ K_{31} & K_{32} & K_{33} & K_{34} & K_{35} & K_{36} \\ K_{41} & K_{42} & K_{43} & K_{44} & K_{45} & K_{46} \\ K_{51} & K_{52} & K_{53} & K_{54} & K_{55} & K_{56} \\ K_{61} & K_{62} & K_{63} & K_{64} & K_{65} & K_{66} \end{bmatrix} \end{array}$$

The code number indicates that stiffness term K_{44} should be added to the final structure stiffness matrix at location (1,1), the stiffness term K_{45} should be added at location (1,2), ..., the stiffness term K_{66} should be added at location (3,3). Note that if a zero element appears in either the "horizontal" or the "vertical" code number then that stiffness term is ignored. The final stiffness matrix is complete when the stiffness contributions of all elements have been added using this technique.

6.6 Element End Forces and Reactions

Once the final structure stiffness matrix has been assembled and the equations solved to find the unknown nodal displacements, equation (6.1) is used to find the forces in the elements.

As the nodal displacements are found in the global axes system they must first be transformed into the element axes system.

Equation (6.1) is

$$f_{ij} = k_{ii}^{j} \, \delta_{ij} + k_{ij} \, \delta_{ji}$$

Expressing the element end displacements as transformed nodal displacements gives

$$f_{ij} = k_{ii}^{j} \, T_{ij} \, \Delta_i + k_{ij} \, T_{ij} \, \Delta_j$$

$$\begin{bmatrix} f_{ijx} \\ f_{ijy} \\ f_{ijz} \end{bmatrix} = \begin{bmatrix} EA/L & 0 & 0 \\ 0 & 0 & 0 \\ 0 & 0 & 0 \end{bmatrix} \begin{bmatrix} l_x & m_x & n_x \\ l_y & m_y & n_y \\ l_z & m_z & n_z \end{bmatrix} \begin{bmatrix} \Delta_{ix} \\ \Delta_{iy} \\ \Delta_{iz} \end{bmatrix} +$$

$$\begin{bmatrix} -EA/L & 0 & 0 \\ 0 & 0 & 0 \\ 0 & 0 & 0 \end{bmatrix} \begin{bmatrix} l_x & m_x & n_x \\ l_y & m_y & n_y \\ l_z & m_z & n_z \end{bmatrix} \begin{bmatrix} \Delta_{jx} \\ \Delta_{jy} \\ \Delta_{jz} \end{bmatrix}$$

or

$$f_{ijx} = \frac{EA}{L}[l_x(\Delta_{ix} - \Delta_{jx}) + m_x(\Delta_{iy} - \Delta_{jy}) + n_x(\Delta_{iz} - \Delta_{jz})]$$

$$f_{ijy} = 0$$

$$f_{ijz} = 0$$

Consideration of the equilibrium of the element (or a similar matrix expansion) shows that the forces at end "j" of the element are as follows

$$f_{jix} = -f_{ijx}$$

$$f_{jiy} = 0$$

$$f_{jiz} = 0$$

Note that if f_{ijx} is positive then the element is in *compression*.

Reactions

If node "i" is restrained, the reactive forces (in the global axes system) are found using equation (5.23) which was developed in the previous chapter.

$$\boldsymbol{R}_i = \sum_{j=a}^{n} \boldsymbol{T}_{ij}^{-1} f_{ij} - \boldsymbol{P}_i$$

hence

$$\begin{bmatrix} R_{ix} \\ R_{iy} \\ R_{iz} \end{bmatrix} = \sum \begin{bmatrix} l_x & l_y & l_z \\ m_x & m_y & m_z \\ n_x & n_y & n_z \end{bmatrix} \begin{bmatrix} f_{ijx} \\ f_{ijy} \\ f_{ijz} \end{bmatrix} - \begin{bmatrix} P_{ix} \\ P_{iy} \\ P_{iz} \end{bmatrix}$$

$$= \sum \begin{bmatrix} l_x f_{ijx} \\ m_x f_{ijx} \\ n_x f_{ijx} \end{bmatrix} - \begin{bmatrix} P_{ix} \\ P_{iy} \\ P_{iz} \end{bmatrix}$$

6.7 Space Truss Program - STRUSS.BAS

This program is essentially the plane truss analysis program (PTRUSS.BAS) modified to deal with three-dimensional trusses. The reader should note how little the code has had to be changed during the conversion, which illustrates, once again, the generality of the stiffness method.

Input

Unless STRUSS.BAS has been chained from PRE.BAS or DCHECK.BAS the user is prompted for the name of the data file to be read by the program. STRUSS.BAS then offers the user the following menu which drives the program.

Do you wish to:
(1) Analyse with output to the screen
(2) Analyse with output to the printer
(3) Analyse with output to a file
(4) Edit the data
(5) Check the data
(6) Stop

It is recommeded that the data file should be constructed using the data preprocessor program PRE.BAS which is described in section 5.11 (it can be built using a text editor). The program DCHECK.BAS can be used to check the data for obvious errors. DCHECK.BAS is presented in section 5.12.

Comments on the Algorithm

The program first uses the boundary conditions to establish the freedom vector. The stiffness of each element is then added to the structure stiffness matrix in turn. The upper triangle only of the final structure stiffness matrix is assembled and its symmetry is later used to complete the matrix. No attempt is made to minimise storage by taking advantage of either the symmetry or the bandwidth of the structure stiffness matrix. The loading vector is developed and then the equations are solved using Gauss-Jordan elimination. Once the nodal displacements have been found they are used to calculate the element forces and reactions.

Output

The data from the data file is echoed and messages are output to indicate the beginning of each solution phase. After solution of the stiffness equation nodal displacements are output, followed by the axial forces in the elements and the reactions.

Listing

```
1000 '=======================      STRUSS.BAS      ==============================
1010 '
1020 COMMON DATAFILE$
1030 OPTION BASE 1 : KEY OFF
1040 TRUE% = -1 : FALSE% = 0 : DEVICE$ = "SCRN:" : PAGELEN% = 23
1050 OUTPUTFILE$="SCREEN"
1060 CLS
```

```
1070 PRINT
1080 PRINT "==================================================================="
1090 PRINT
1100 PRINT " PROGRAM STRUSS                         Copyright (c) J.Balfour 1991"
1110 PRINT
1120 PRINT "       Program for the automatic analysis of space trusses"
1130 PRINT "        Data generated with the data preprocessor PRE"
1140 PRINT
1150 PRINT "                 For further information contact"
1160 PRINT " James A.D.Balfour, Heriot-Watt University, Riccarton, Edinburgh"
1170 PRINT "               Tel 031-449-5111, Fax 031-451-3170"
1180 PRINT
1190 PRINT "==================================================================="
1200 '
1210 ' ... Note that variables are defined in Appendix A and the
1220 '     following statements set the maximum problem size
1230 '
1240 MAXNODE%  = 30                     '... Max no of nodes
1250 MAXELEM%  = 50                     '... Max no of elements
1260 MAXPROP%  = 20                     '... Max no of element properties
1270 MAXREST%  = 20                     '... Max no of restrained nodes
1280 MAXNLOAD% = 50                     '... Max no of nodal loads
1290 '
1300 DIM NODE(MAXNODE%,4), ELEM(MAXELEM%,7),   PROP(MAXPROP%,3)
1310 DIM REST(MAXREST%,5), NLOAD(MAXNLOAD%,3), FREE%(3*MAXNODE%), WKSP$(8)
1320 DIM K(3*MAXNODE%,3*MAXNODE%+1),           ESTIFF(6,6),       EFREE%(6)
1330 '
1340 ' ... Check if this program has been chained from PRE
1350 '
1360 IF DATAFILE$ <> "" THEN GOSUB 7970 : GOTO 1390
1370 PRINT
1380 INPUT "Name of the data file  =  ", DATAFILE$
1390 GOSUB 2020
1400 NOPTIONS% = 7
1410 WHILE (REPLY% <> 7)
1420   CLS : PRINT
1430   PRINT "Do you wish to:"
1440   PRINT "  (1) Analyse with output to the screen"
1450   PRINT "  (2) Analyse with output to the printer"
1460   PRINT "  (3) Analyse with output to a file"
1470   PRINT "  (4) Edit the data"
1480   PRINT "  (5) Check the data"
1490   PRINT "  (6) Read a new file"
1500   PRINT "  (7) Stop"
1510   GOSUB 7260
1520   ON REPLY% GOTO 1540, 1570, 1650, 1720, 1750, 1780, 1810
1530     '
1540     OPEN "SCRN:" FOR OUTPUT AS #2                      ' Reply = 1
1550     GOSUB 1860 : GOTO 1810
1560     '
1570     ON ERROR GOTO 1620                                 ' Reply = 2
1580     OPEN "LPT1:" FOR OUTPUT AS #2
1590     PRINT : PRINT "Wait - looking for printer" : PRINT #2,
1600     ON ERROR GOTO 0 : DEVICE$ = "LPT1:" : GOSUB 1860
1610     PRINT #2, CHR$(12); : GOTO 1800
1620     CLS : PRINT : PRINT "** ERROR ** Failed to find the printer"
1630     GOSUB 7970 : RESUME 1810
1640     '
1650     ON ERROR GOTO 1690                                 ' Reply = 3
1660     INPUT "Name of file for output - ", DEVICE$
1670     OPEN DEVICE$ FOR OUTPUT AS #2
1680     ON ERROR GOTO 0 : GOSUB 1860 : GOTO 1810
1690     CLS : PRINT : PRINT "** ERROR ** Failed to open output file"
1700     GOSUB 7970 : RESUME 1810
1710     '
1720     PRINT "Wait - chaining PRE"                          ' Reply = 4
1730     CHAIN "PRE"
1740     '
1750     PRINT "Wait - chaining DCHECK"                       ' Reply = 5
1760     CHAIN "DCHECK"
1770     '
1780     PRINT
1790     INPUT "Name of the data file  =  ", DATAFILE$     ' Reply = 6
1800     GOSUB 2020
1810     CLOSE #2 : DEVICE$="SCRN:" : ON ERROR GOTO 0
1820 WEND
1830 KEY ON
```

```
1840 END
1850 '
1860 ' ************************      ANALYSE     ******************************
1870 '
1880 GOSUB 3340            'Print data
1890 PRINT : PRINT "Generating the freedom vector                  "
1900 GOSUB 4540
1910 PRINT : PRINT "Assembling the stiffness matrix, adding element :- ";
1920 ROW% = CSRLIN: COL% = POS(0)
1930 GOSUB 4870
1940 PRINT : PRINT "Generating the loading vector                  "
1950 GOSUB 5760
1960 PRINT : PRINT "Solving equations - no. of equations = "; NDOF%
1970 GOSUB 5870
1980 GOSUB 6190            'Output the nodal displacements
1990 GOSUB 6420            'Output the element forces and reactions
2000 RETURN
2010 '
2020 '***********************    READ DATA FILE     ***************************
2030 '
2040 ' ...  This subroutine reads the data from a data file
2050 '
2060 PRINT
2070 PRINT "Wait - Reading file "; DATAFILE$; " line";
2080 ROW% = CSRLIN : COL% = POS(0) : LCOUNT% = 0
2090 DATAFILEOK% = FALSE% : DATAEND% = FALSE%
2100 NNODE% = 0 : NELEM% = 0 : NPROP% = 0 : NREST% = 0 : NNLOAD% = 0
2110 ON ERROR GOTO 2130 : OPEN DATAFILE$ FOR INPUT AS #1
2120 ON ERROR GOTO 0 : GOTO 2190
2130 CLS : PRINT
2140 PRINT "** ERROR ** Failed to open file: "; DATAFILE$; " for input"
2150 GOSUB 7970 : RESUME 3310
2160 '
2170 ' ... Check that this is a data file for STRUSS
2180 '
2190 TARGET$ = "STRUSS" : GOSUB 7730
2200 IF STRFOUND% THEN GOTO 2270
2210 CLS : PRINT
2220 PRINT "** ERROR ** Not a file for STRUSS"
2230 GOSUB 7970 : GOTO 3310
2240 '
2250 ' ... Read title and units strings
2260 '
2270 NLINES% = 2 : GOSUB 7880 : IF DATAEND% THEN GOTO 2340
2280 WHILE (MID$(WRKSTR$,I%,1) = " ")
2290   I% = I% + 1
2300 WEND
2310 TITLE$  = MID$(WRKSTR$,I%)
2320 TARGET$ = "Units" : GOSUB 7730
2330 IF STRFOUND% THEN GOTO 2370
2340 CLS : PRINT
2350 PRINT "** ERROR ** Error reading Title/Units strings"
2360 GOSUB 7970 : GOTO 2390
2370 UNITS$ = MID$(WRKSTR$,INSTR(WRKSTR$,":- ")+3,24)
2380 DATAFILEOK% = TRUE%
2390 WHILE NOT DATAEND%
2400   '
2410   ' ... Get module string
2420   '
2430   IF LEFT$(WRKSTR$,4)="+   " THEN GOTO 2450
2440   TARGET$ = "+   " : GOSUB 7730 : IF NOT STRFOUND% THEN GOTO 3300
2450   MODSTR$ = WRKSTR$
2460   '
2470   ' ...  Identify module
2480   '
2490   IF INSTR(MODSTR$,"NODAL COORDINATES")=0 THEN GOTO 2640
2500   '
2510   ' ... Read nodal coordinates
2520   '
2530   TARGET$ = "NODE" : GOSUB 7730 : IF NOT STRFOUND% THEN GOTO 3240
2540   GOSUB 7400
2550   WHILE NNOS% > 0
2560     NNODE% = NNODE% + 1
2570     FOR J% = 1 TO NNOS%
2580       NODE(NNODE%,J%) = VAL(WKSP$(J%))
2590     NEXT J%
2600     GOSUB 7400
```

```
2610    WEND
2620    GOTO 3300
2630    '
2640    IF INSTR(MODSTR$,"ELEMENTS")=0 THEN GOTO 2790
2650    '
2660    ' ... Read elements
2670    '
2680    TARGET$ = "ELEMENT" : GOSUB 7730 : IF NOT STRFOUND% THEN GOTO 3240
2690    GOSUB 7400
2700    WHILE NNOS% > 0
2710      NELEM% = NELEM% + 1
2720      FOR J% = 1 TO NNOS%
2730        ELEM(NELEM%,J%) = VAL(WKSP$(J%))
2740      NEXT J%
2750      GOSUB 7400
2760    WEND
2770    GOTO 3300
2780    '
2790    IF INSTR(MODSTR$,"PROPERTIES")=0 THEN GOTO 2940
2800    '
2810    ' ... Read properties
2820    '
2830    TARGET$ = "PROPERTY" : GOSUB 7730 : IF NOT STRFOUND% THEN GOTO 3240
2840    GOSUB 7400
2850    WHILE NNOS% > 0
2860      NPROP% = NPROP% + 1
2870      FOR J% = 1 TO NNOS%
2880        PROP(NPROP%,J%) = VAL(WKSP$(J%))
2890      NEXT J%
2900      GOSUB 7400
2910    WEND
2920    GOTO 3300
2930    '
2940    IF INSTR(MODSTR$,"RESTRAINTS")=0 THEN GOTO 3090
2950    '
2960    ' ... Read restraints
2970    '
2980    TARGET$ = "NODE" : GOSUB 7730 : IF NOT STRFOUND% THEN GOTO 3240
2990    GOSUB 7400
3000    WHILE NNOS% > 0
3010      NREST% = NREST% + 1
3020      FOR J% = 1 TO NNOS%
3030        REST(NREST%,J%) = VAL(WKSP$(J%))
3040      NEXT J%
3050      GOSUB 7400
3060    WEND
3070    GOTO 3300
3080    '
3090    IF INSTR(MODSTR$,"NODAL LOADS")=0 THEN GOTO 3240
3100    '
3110    ' ... Read nodal loads
3120    '
3130    TARGET$ = "NODE" : GOSUB 7730 : IF NOT STRFOUND% THEN GOTO 3240
3140    GOSUB 7400
3150    WHILE NNOS% > 0
3160      NNLOAD% = NNLOAD% + 1
3170      FOR J% = 1 TO NNOS%
3180        NLOAD(NNLOAD%,J%) = VAL(WKSP$(J%))
3190      NEXT J%
3200      GOSUB 7400
3210    WEND
3220    GOTO 3300
3230    '
3240    CLS : PRINT
3250    PRINT "** ERROR ** in module "; MODSTR$ "run DCHECK for more info";
3260    GOSUB 7970
3270    PRINT
3280    PRINT "Wait - Reading file "; DATAFILE$; " line";
3290    ROW% = CSRLIN : COL% = POS(0)
3300 WEND
3310 CLOSE #1 : ON ERROR GOTO 0
3320 RETURN
3330 '
3340 ' ***********************     PRINT DATA     *****************************
3350 '
3360 GOSUB 3440                          '... List header
3370 GOSUB 3770                          '... List nodal coordinates
```

```
3380 GOSUB 3920                          '... List element data
3390 GOSUB 4070                          '... List list element properties
3400 GOSUB 4230                          '... List restraint data
3410 GOSUB 4380                          '... List nodal loads
3420 RETURN
3430 '
3440 ' ++++++++++++++++++++++   LIST HEADER   +++++++++++++++++++++++++
3450 '
3460 CLS
3470 PRINT #2, : PRINT #2,
3480 PRINT #2, "======================================";
3490 PRINT #2, "======================================"
3500 PRINT #2,
3510 PRINT #2, "    PROGRAM STRUSS"; STRING$(24,32);
3520 PRINT #2, "Copyright(c) J.A.D.Balfour 1991"
3530 PRINT #2,
3540 FOR I% = 1 TO (76-LEN(TITLE$))/2
3550   PRINT #2, " ";
3560 NEXT I%
3570 PRINT #2, TITLE$
3580 PRINT #2,
3590 PRINT #2, "    File Name :- "; DATAFILE$;
3600 FOR I% = 1 TO (40-LEN(DATAFILE$))
3610   PRINT #2, " ";
3620 NEXT I%
3630 PRINT #2, "Date :- "; MID$(DATE$,4,3); MID$(DATE$,1,3);
3640 PRINT #2, MID$(DATE$, 9, 2)
3650 PRINT #2, "    Units     :- "; UNITS$;
3660 FOR I% = 1 TO (40-LEN(UNITS$))
3670   PRINT #2, " ";
3680 NEXT I%
3690 PRINT #2, "Time :- "; TIME$
3700 PRINT #2,
3710 PRINT #2, "======================================";
3720 PRINT #2, "======================================"
3730 PRINT #2,
3740 GOSUB 7970
3750 RETURN
3760 '
3770 ' ++++++++++++++++   LIST NODAL COORDINATES   +++++++++++++++++++++
3780 '
3790 PRINT #2,
3800 PRINT #2, "+ + + + + + + + + + + + + + +"
3810 PRINT #2, "+   NODAL COORDINATES   +"
3820 PRINT #2, "+ + + + + + + + + + + + + + +"
3830 PRINT #2,
3840 PRINT #2, "NODE           X              Y              Z"
3850 FOR I% = 1 TO NNODE%
3860   GOSUB 8060
3870   PRINT #2, " "; NODE(I%,1), NODE(I%,2), NODE(I%,3), NODE(I%,4)
3880 NEXT I%
3890 GOSUB 7970
3900 RETURN
3910 '
3920 ' +++++++++++++++++++++   LIST ELEMENTS   +++++++++++++++++++++++++
3930 '
3940 PRINT #2,
3950 PRINT #2, "+ + + + + + + + + +"
3960 PRINT #2, "+   ELEMENTS    +"
3970 PRINT #2, "+ + + + + + + + + +"
3980 PRINT #2,
3990 PRINT #2, "ELEMENT    PROPERTY"
4000 FOR I% = 1 TO NELEM%
4010   GOSUB 8060
4020   PRINT #2, USING " ## ##        ##"; ELEM(I%,1); ELEM(I%,2); ELEM(I%,3)
4030 NEXT I%
4040 GOSUB 7970
4050 RETURN
4060 '
4070 ' ++++++++++++++++++++   LIST PROPERTIES   ++++++++++++++++++++++++
4080 '
4090 PRINT #2,
4100 PRINT #2, "+ + + + + +  + + +"
4110 PRINT #2, "+   PROPERTIES   +"
4120 PRINT #2, "+ + + + + +  + + +"
4130 PRINT #2,
4140 PRINT #2, "PROPERTY        A              E"
```

```
4150 FOR I% = 1 TO NPROP%
4160   GOSUB 8060
4170   PRINT #2, USING "   ##      #.####^^^^"; PROP(I%,1); PROP(I%,2);
4180   PRINT #2, USING "     #.####^^^^  "; PROP(I%,3)
4190 NEXT I%
4200 GOSUB 7970
4210 RETURN
4220 '
4230 ' +++++++++++++++++++++++   LIST RESTRAINTS    +++++++++++++++++++++++
4240 '
4250 PRINT #2,
4260 PRINT #2, "+ + + + + + + + + + + "
4270 PRINT #2, "+   RESTRAINTS     +"
4280 PRINT #2, "+ + + + + + + + + + + "
4290 PRINT #2,
4300 PRINT #2, "NODE    DIRECTION(S)"
4310 FOR I% = 1 TO NREST%
4320   GOSUB 8060
4330   PRINT #2, USING " ##      ######"; REST(I%,1); REST(I%,2)
4340 NEXT I%
4350 GOSUB 7970
4360 RETURN
4370 '
4380 ' +++++++++++++++++++    LIST NODAL LOADS    ++++++++++++++++++++++++
4390 '
4400 PRINT #2,
4410 PRINT #2, "+ + + + + + + + + + +"
4420 PRINT #2, "+   NODAL LOADS    +"
4430 PRINT #2, "+ + + + + + + + + + +"
4440 PRINT #2,
4450 PRINT #2, "NODE    DIRECTION     VALUE"
4460 FOR I% = 1 TO NNLOAD%
4470   GOSUB 8060
4480   PRINT #2, USING " ##           #";  NLOAD(I%,1); NLOAD(I%,2);
4490   PRINT #2, USING "     #.###^^^^"; NLOAD(I%,3)
4500 NEXT I%
4510 GOSUB 7970
4520 RETURN
4530 '
4540 ' *****************    GENERATE THE FREEDOM VECTOR    *******************
4550 '
4560 ' ... Zero the freedom vector
4570 '
4580 FOR I% = 1 TO 3*NNODE%
4590   FREE%(I%) = 0
4600 NEXT I%
4610 '
4620 ' ... Loop for all restrained nodes
4630 '
4640 FOR I% = 1 TO NREST%
4650   K% = REST(I%,2)
4660   '
4670   ' ... Evaluate the freedom to be restrained from K%
4680   '
4690   J% = 3 * REST(I%,1) - 3 + (K% MOD 10)
4700   FREE%(J%) = 1
4710   K% = INT(K%/10)
4720   IF K% > 0 THEN GOTO 4690
4730 NEXT I%
4740 '
4750 ' ... Number the freedoms (restraints set to zero)
4760 '
4770 NDOF% = 0
4780 FOR I% = 1 TO 3*NNODE%
4790   IF FREE%(I%) = 1 THEN GOTO 4830
4800   NDOF% = NDOF% + 1
4810   FREE%(I%) = NDOF%
4820   GOTO 4840
4830   FREE%(I%) = 0
4840 NEXT I%
4850 RETURN
4860 '
4870 ' ***************    ASSEMBLE THE STIFFNESS MATRIX    ******************
4880 '
4890 ' ... Zero the augmented matrix
4900 '
4910 FOR I% = 1 TO NDOF%
```

```
4920   FOR J% = 1 TO NDOF% + 1
4930     K(I%,J%) = 0
4940   NEXT J%
4950 NEXT I%
4960 '
4970 ' ... Loop for each element
4980 '
4990 FOR K% = 1 TO NELEM%
5000   LOCATE ROW%, COL%: PRINT K%;
5010   IN% = ELEM(K%,1)
5020   JN% = ELEM(K%,2)
5030   PN% = ELEM(K%,3)
5040   '
5050   ' ... Calculate the element length (store in ELEM(K%,4))
5060   '
5070   TEMP(1)     = (NODE(JN%,2)-NODE(IN%,2))^2 + (NODE(JN%,3)-NODE(IN%,3))^2
5080   ELEM(K%,4) = SQR(TEMP(1) + (NODE(JN%,4)-NODE(IN%,4))^2)
5090   EAL        = PROP(PN%,3)*PROP(PN%,2)/ELEM(K%,4)
5100   '
5110   ' ... Calculate the the direction cosines of the member x-axes
5120   '     Store in (store in ELEM(K%,5), ELEM(K%,6) and ELEM(K%,7)
5130   '
5140   ELEM(K%,5) = (NODE(JN%,2) - NODE(IN%,2)) / ELEM(K%,4)
5150   ELEM(K%,6) = (NODE(JN%,3) - NODE(IN%,3)) / ELEM(K%,4)
5160   ELEM(K%,7) = (NODE(JN%,4) - NODE(IN%,4)) / ELEM(K%,4)
5170   '
5180   ' ... Assemble the upper triangle of the element stiffness matrix
5190   '
5200   ESTIFF(1,1) =   EAL * ELEM(K%,5) * ELEM(K%,5)
5210   ESTIFF(1,2) =   EAL * ELEM(K%,5) * ELEM(K%,6)
5220   ESTIFF(1,3) =   EAL * ELEM(K%,5) * ELEM(K%,7)
5230   ESTIFF(1,4) = - ESTIFF(1,1)
5240   ESTIFF(1,5) = - ESTIFF(1,2)
5250   ESTIFF(1,6) = - ESTIFF(1,3)
5260   ESTIFF(2,2) =   EAL * ELEM(K%,6) * ELEM(K%,6)
5270   ESTIFF(2,3) =   EAL * ELEM(K%,6) * ELEM(K%,7)
5280   ESTIFF(2,4) = - ESTIFF(1,2)
5290   ESTIFF(2,5) = - ESTIFF(2,2)
5300   ESTIFF(2,6) = - ESTIFF(2,3)
5310   ESTIFF(3,3) =   EAL * ELEM(K%,7) * ELEM(K%,7)
5320   ESTIFF(3,4) = - ESTIFF(1,3)
5330   ESTIFF(3,5) = - ESTIFF(2,3)
5340   ESTIFF(3,6) = - ESTIFF(3,3)
5350   ESTIFF(4,4) =   ESTIFF(1,1)
5360   ESTIFF(4,5) =   ESTIFF(1,2)
5370   ESTIFF(4,6) =   ESTIFF(1,3)
5380   ESTIFF(5,5) =   ESTIFF(2,2)
5390   ESTIFF(5,6) =   ESTIFF(2,3)
5400   ESTIFF(6,6) =   ESTIFF(3,3)
5410   '
5420   ' ... Set up the element code number
5430   '
5440   '
5450   EFREE%(1) = FREE%(3*IN%-2)
5460   EFREE%(2) = FREE%(3*IN%-1)
5470   EFREE%(3) = FREE%(3*IN%)
5480   EFREE%(4) = FREE%(3*JN%-2)
5490   EFREE%(5) = FREE%(3*JN%-1)
5500   EFREE%(6) = FREE%(3*JN%)
5510   '
5520   ' ... Add the element stiffness to the structure stiffness matrix
5530   '
5540   FOR I% = 1 TO 6
5550     IF EFREE%(I%) = 0 THEN GOTO 5620
5560     FOR J% = I% TO 6
5570       IF EFREE%(J%) = 0 THEN GOTO 5610
5580       L% = EFREE%(I%)
5590       M% = EFREE%(J%)
5600       K(L%,M%) = K(L%,M%) + ESTIFF(I%,J%)
5610     NEXT J%
5620   NEXT I%
5630 NEXT K%
5640 PRINT
5650 '
5660 ' ... Use symmetry to fill out the stiffness matrix
5670 '
5680 FOR I% = 1 TO NDOF% - 1
```

```
5690    FOR J% = I%+1 TO NDOF%
5700      K(J%,I%) = K(I%,J%)
5710    NEXT J%
5720 NEXT I%
5730 '
5740 RETURN
5750 '
5760 '*****************    SET UP THE LOADING VECTOR    ********************
5770 '
5780 ' ... Loop for all loads
5790 '
5800 FOR I% = 1 TO NNLOAD%
5810   J% = 3*NLOAD(I%,1) - 3 + NLOAD(I%,2)
5820   J% = FREE%(J%)
5830   IF J% > 0 THEN K(J%,NDOF%+1) = NLOAD(I%,3)
5840 NEXT I%
5850 RETURN
5860 '
5870 ' **********************    SOLVE EQUATIONS    **************************
5880 '
5890 ' ... Loop for all pivots
5900 '
5910 PRINT "                         - current pivot    = ";
5920 ROW% = CSRLIN: COL% = POS(0):
5930 FOR I% = 1 TO NDOF%
5940   LOCATE ROW%, COL% : PRINT I%;
5950   TEMP(1) = K(I%,I%)
5960   '
5970   ' ... Normalise
5980   '
5990   FOR J% = 1 TO NDOF% + 1
6000     K(I%,J%) = K(I%,J%) / TEMP(1)
6010   NEXT J%
6020   '
6030   ' ... Eliminate for all rows using the current pivot
6040   '
6050   FOR J% = 1 TO NDOF%
6060     IF J% = I% THEN GOTO 6140
6070     TEMP(1) = K(J%,I%)
6080     '
6090     ' ... Loop for all elements in the current row
6100     '
6110     FOR K% = 1 TO NDOF% + 1
6120       K(J%,K%) = K(J%,K%) - TEMP(1)*K(I%,K%)
6130     NEXT K%
6140   NEXT J%
6150 NEXT I%
6160 PRINT
6170 RETURN
6180 '
6190 '********************    OUTPUT THE DISPLACEMENTS  ********************
6200 '
6210 CLS
6220 PRINT #2,
6230 PRINT #2, "* * * * * * * * * * *"
6240 PRINT #2, "*   DISPLACEMENTS   *              * :-  RESTRAINT"
6250 PRINT #2, "* * * * * * * * * * *"
6260 PRINT #2,
6270 PRINT #2, "NODE     X-DISPLACEMENT  Y DISPLACEMENT  Z-DISPLACEMENT"
6280 FOR I% = 1 TO NNODE%
6290   PRINT #2, USING "###    "; I%;
6300   FOR J% = 1 TO 3
6310     K% = FREE%(3*I%+J%-3)
6320     IF K% = 0 THEN GOTO 6350
6330     PRINT #2, USING "      #.####^^^^"; K(K%,NDOF%+1);
6340     GOTO 6360
6350     PRINT #2, "          *      ";
6360   NEXT J%
6370   PRINT #2,
6380 NEXT I%
6390 GOSUB 7970
6400 RETURN
6410 '
6420 ' ****** CALCULATE AND OUTPUT THE ELEMENT FORCES AND REACTIONS    ******
6430 '
6440 ' ... Zero the reaction components
6450 '
```

```
6460 FOR I% = 1 TO NREST%
6470   REST(I%,3) = 0
6480   REST(I%,4) = 0
6490   REST(I%,5) = 0
6500 NEXT I%
6510 CLS
6520 PRINT #2,
6530 PRINT #2, "* * * * * * * * * * * *"
6540 PRINT #2, "*   ELEMENT FORCES    *          TENSION POSITIVE"
6550 PRINT #2, "* * * * * * * * * * * *"
6560 PRINT #2,
6570 PRINT #2, "ELEMENT              FORCE"
6580 FOR I% = 1 TO NELEM%
6590   II% = 3*ELEM(I%,1) - 2
6600   JJ% = 3*ELEM(I%,2) - 2
6610   PN% = ELEM(I%,3)
6620   K% = FREE%(II%)
6630   IF K% = 0 THEN XI = 0 ELSE XI = K(K%,NDOF%+1)
6640   K% = FREE%(II%+1)
6650   IF K% = 0 THEN YI = 0 ELSE YI = K(K%,NDOF%+1)
6660   K% = FREE%(II%+2)
6670   IF K% = 0 THEN ZI = 0 ELSE ZI = K(K%,NDOF%+1)
6680   K% = FREE%(JJ%)
6690   IF K% = 0 THEN XJ = 0 ELSE XJ = K(K%,NDOF%+1)
6700   K% = FREE%(JJ%+1)
6710   IF K% = 0 THEN YJ = 0 ELSE YJ = K(K%, NDOF% + 1)
6720   K% = FREE%(JJ%+2)
6730   IF K% = 0 THEN ZJ = 0 ELSE ZJ = K(K%, NDOF% + 1)
6740   FORCE = ELEM(I%,5)*(XJ-XI) + ELEM(I%,6)*(YJ-YI) + ELEM(I%,7)*(ZJ-ZI)
6750   FORCE = FORCE * PROP(PN%,3) * PROP(PN%,2) / ELEM(I%,4)
6760   PRINT #2, USING "## ##         "; ELEM(I%,1); ELEM(I%,2);
6770   PRINT #2, USING "#.####^^^^"; FORCE
6780   '
6790   ' ... Add any contribution from the element forces to the reactions
6800   '
6810   FOR J% = 1 TO NREST%
6820     K% = 0
6830     IF ELEM(I%,1) = REST(J%,1) THEN K% = -1
6840     IF ELEM(I%,2) = REST(J%,1) THEN K% =  1
6850     IF K% = 0 THEN GOTO 6890
6860     REST(J%,3) = REST(J%,3) + K%*ELEM(I%,5)*FORCE
6870     REST(J%,4) = REST(J%,4) + K%*ELEM(I%,6)*FORCE
6880     REST(J%,5) = REST(J%,5) + K%*ELEM(I%,7)*FORCE
6890   NEXT J%
6900 NEXT I%
6910 GOSUB 7970
6920 '
6930 ' ... Add any applied loads to reactions
6940 '
6950 FOR I% = 1 TO NREST%
6960   FOR J% = 1 TO NNLOAD%
6970     IF REST(I%,1) <> NLOAD(J%,1) THEN GOTO 7000
6980     K% = NLOAD(J%,2)
6990     REST(I%,2+K%) = REST(I%,2+K%) - NLOAD(J%,3)
7000   NEXT J%
7010 NEXT I%
7020 '
7030 ' ... Output the reactions
7040 '
7050 CLS
7060 PRINT #2,
7070 PRINT #2, "* * * * * * * * *"
7080 PRINT #2, "*   REACTIONS   *"
7090 PRINT #2, "* * * * * * * * *"
7100 PRINT #2,
7110 PRINT #2, "                X-FORCE        Y-FORCE        Z-FORCE"
7120 TEMP(1) = 0 : TEMP(2) = 0 : TEMP(3) = 0
7130 FOR I% = 1 TO NREST%
7140   PRINT #2, USING "NODE ##"; REST(I%,1);
7150   PRINT #2, USING "      #.####^^^^"; REST(I%,3); REST(I%,4); REST(I%,5)
7160   TEMP(1) = TEMP(1) + REST(I%,3)
7170   TEMP(2) = TEMP(2) + REST(I%,4)
7180   TEMP(3) = TEMP(3) + REST(I%,5)
7190 NEXT I%
7200 PRINT #2, "           ----------     ----------     ----------"
7210 PRINT #2, "TOTAL  ";
7220 PRINT #2, USING "     #.####^^^^"; TEMP(1); TEMP(2); TEMP(3)
```

```
7230 GOSUB 7970
7240 RETURN
7250 '
7260 '********************     GET KEYBOARD RESPONSE     ********************
7270 '
7280 ROW% = CSRLIN: COL% = POS(0)
7290 REPLY% = 0
7300 WHILE REPLY% = 0
7310   LOCATE ROW%, COL%: INPUT "", WRKSTR$: REPLY% = VAL(WRKSTR$)
7320   IF (REPLY% < 1) OR (REPLY% > NOPTIONS%) THEN GOTO 7350
7330   IF (REPLY% < 10) AND (LEN(WRKSTR$) = 1) THEN GOTO 7360
7340   IF (REPLY% > 9)  AND (LEN(WRKSTR$) = 2) THEN GOTO 7360
7350   SOUND 700, 4: REPLY% = 0: LOCATE ROW%, COL%
7360 WEND
7370 PRINT
7380 RETURN
7390 '
7400 '***********************     RECORD NUMBERS     ***************************
7410 '
7420 NNOS% = 0 : DONE% = TRUE%
7430 IF EOF(1)<>0 THEN DATAEND% = TRUE% : GOTO 7700
7440 '
7450 ' ... Read next line and check if it is blank
7460 '
7470 LINE INPUT #1, WRKSTR$
7480 LCOUNT% = LCOUNT% + 1 : LOCATE ROW%, COL% : PRINT LCOUNT%;
7490 FOR I% = 1 TO LEN(WRKSTR$)
7500   IF MID$(WRKSTR$,I%,1)<> " " THEN DONE% = FALSE% : I% = LEN(WRKSTR$)
7510 NEXT I%
7520 IF DONE% = TRUE% THEN GOTO 7700
7530 FOR I% = 1 TO 8
7540   WKSP$(I%) = ""
7550 NEXT I%
7560 '
7570 ' ... Search for numbers
7580 '
7590 FOR I% = 1 TO LEN(WRKSTR$)
7600   WHILE (MID$(WRKSTR$, I%, 1) = " ") AND I% <= LEN(WRKSTR$)
7610     I% = I% + 1
7620   WEND
7630   IF (I% > LEN(WRKSTR$)) THEN GOTO 7690
7640   IF(TARGET$="STRUSS") AND LCOUNT% > 4 THEN DONE% = TRUE%
7650   NNOS% = NNOS% + 1: WKSP$(NNOS%) = ""
7660   WHILE (MID$(WRKSTR$, I%, 1) <> " ") AND I% <= LEN(WRKSTR$)
7670     WKSP$(NNOS%) = WKSP$(NNOS%) + MID$(WRKSTR$,I%,1): I% = I% + 1
7680   WEND
7690 NEXT I%
7700 WRKSTR$ = ""
7710 RETURN
7720 '
7730 '**************************     FIND STRING     **************************
7740 '
7750 STRFOUND% = FALSE% : DONE% = FALSE%
7760 WHILE DONE% = FALSE%
7770   IF EOF(1)<>0 THEN DATAEND% = TRUE% : DONE% = TRUE% : GOTO 7850
7780   LINE INPUT #1, WRKSTR$
7790   LCOUNT% = LCOUNT% + 1 : LOCATE ROW%, COL% : PRINT LCOUNT%;
7800   IF INSTR(WRKSTR$,TARGET$)>0 THEN STRFOUND% = TRUE% : DONE% = TRUE%
7810   '
7820   ' ... Check for new module
7830   '
7840   IF LEFT$(WRKSTR$,4) ="+   " THEN DONE% = TRUE%
7850 WEND
7860 RETURN
7870 '
7880 '**************************     READ LINES     ***************************
7890 '
7900 FOR I% = 1 TO NLINES%
7910   IF EOF(1)<>0 THEN DATAEND% = TRUE% : I% = NLINES% : GOTO 7940
7920   LINE INPUT #1, WRKSTR$
7930   LCOUNT% = LCOUNT% + 1 : LOCATE ROW%, COL% : PRINT LCOUNT%;
7940 NEXT I%
7950 RETURN
7960 '
7970 '********************     ANY KEY TO CONTINUE     **********************
7980 '
7990 IF DEVICE$<>"SCRN:" THEN GOTO 8040
```

```
8000 LOCATE PAGELEN%+1, 1
8010 PRINT SPACE$(40); "Press any key to continue";
8020 WHILE INKEY$ = "" : WEND
8030 CLS
8040 RETURN
8050 '
8060 '**********************    CHECK FOR PAGE END    **********************
8070 '
8080 IF CSRLIN>=PAGELEN% THEN GOSUB 7970
8090 LCOUNT% = LCOUNT% + 1
8100 RETURN
8110 '
8120 ' ======================    END - STRUSS.BAS    ======================
```

Sample Run

This sample run show STRUSS.BAS being used to analyse the space truss structure shown in the following diagram. Note that $E = 200$ kN/mm^2 and $A = 5000$ mm^2 for all elements. Note this output was produced by selecting the option to analyse with output to file. While output is being written to file STRUSS.BAS displays a series of messages to the screen to indicate the progress of the analysis. The screen output is not shown here.

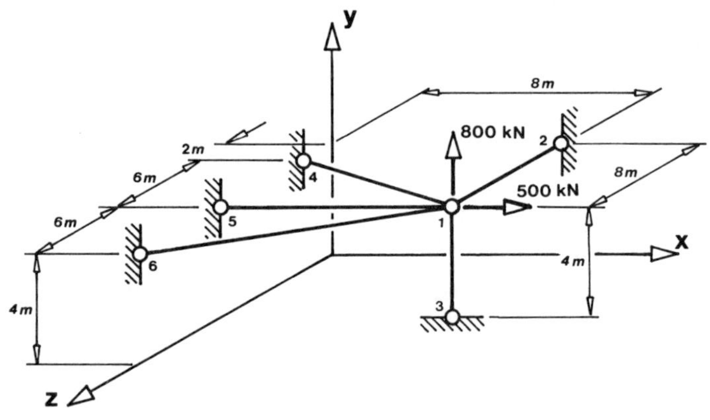


```
==========================================================================

     PROGRAM STRUSS                          Copyright(c) J.A.D.Balfour 1991

                       CASF 2nd Edition - STRUSS.BAS example

     File Name :- ST1.DAT                                 Date :- 27-02-92
     Units     :- kN mm                                   Time :- 19:34:44

==========================================================================

+ + + + + + + + + + + + + +
+   NODAL COORDINATES     +
+ + + + + + + + + + + + + +

NODE         X            Y           Z
  1          8000         4000        8000
  2          8000         4000        0
  3          8000         0           8000
  4          0            4000        2000
  5          0            4000        8000
  6          0            4000        14000
```

```
  3            8000         0            8000
  4            0            4000         2000
  5            0            4000         8000
  6            0            4000         14000

+ + + + + + + + +
+   ELEMENTS    +
+ + + + + + + + +

ELEMENT     PROPERTY
  1  2         1
  1  3         1
  1  4         1
  1  5         1
  1  6         1

+ + + + + +  + + +
+   PROPERTIES   +
+ + + + + +  + + +

PROPERTY         A             E
    1      0.5000E+04     0.2000E+03

+ + + + + + + + + +
+   RESTRAINTS    +
+ + + + + + + + + +

NODE    DIRECTION(S)
  2        123
  3        123
  4        123
  5        123
  6        123

+ + + + + + + + + +
+   NODAL LOADS   +
+ + + + + + + + + +

NODE    DIRECTION     VALUE
  1         1       0.500E+03
  1         2       0.800E+03

* * * * * * * * * * *
*   DISPLACEMENTS   *              * :-  RESTRAINT
* * * * * * * * * * *

NODE      X-DISPLACEMENT  Y-DISPLACEMENT  Z-DISPLACEMENT
  1           0.1976E+01      0.3200E+01      0.0000E+00
  2               *               *               *
  3               *               *               *
  4               *               *               *
  5               *               *               *
  6               *               *               *

* * * * * * * * * * * * *
*   ELEMENT FORCES      *          TENSION POSITIVE
* * * * * * * * * * * * *

ELEMENT            FORCE
 1  2         0.0000E+00
 1  3         0.8000E+03
 1  4         0.1581E+03
 1  5         0.2470E+03
 1  6         0.1581E+03

* * * * * * * * *
*   REACTIONS   *
* * * * * * * * *

                  X-FORCE          Y-FORCE          Z-FORCE
NODE  2       0.0000E+00       0.0000E+00       0.0000E+00
NODE  3       0.0000E+00       -.8000E+03       0.0000E+00
NODE  4       -.1265E+03       0.0000E+00       -.9486E+02
NODE  5       -.2470E+03       0.0000E+00       0.0000E+00
NODE  6       -.1265E+03       0.0000E+00       0.9486E+02
              ----------       ----------       ----------
TOTAL         -.5000E+03       -.8000E+03       0.0000E+00
```

Chapter 7

Plane Frames

7.1 Introduction

Rigidly jointed frames are extensively used in civil engineering structures. Examples include roof trusses, electricity pylons, steel offshore platforms and all types of building frameworks. As the elements are assumed to be rigidly connected at the joints the angles between elements meeting at a joint remain unchanged as the structure deforms under load. Consequently the elements of rigidly jointed frames transmit load, not only axially, but also by bending and shear. This contrasts with pin jointed structures where the inability of the joints to transmit bending ensures that all loads are transmitted by axial force only. Further, rigidly jointed frames are often designed to carry loads both at joints and along the lengths of the elements, whereas pin jointed frames are usually designed to carry loads that are applied only at the joints. This chapter shows that the equations governing the behaviour of trusses also govern the behaviour of rigidly jointed plane frameworks, and that the techniques developed for the analysis of trusses can easily be adapted for the analysis of rigidly jointed frames.

7.2 The Element Stiffness Matrix in Element Axes

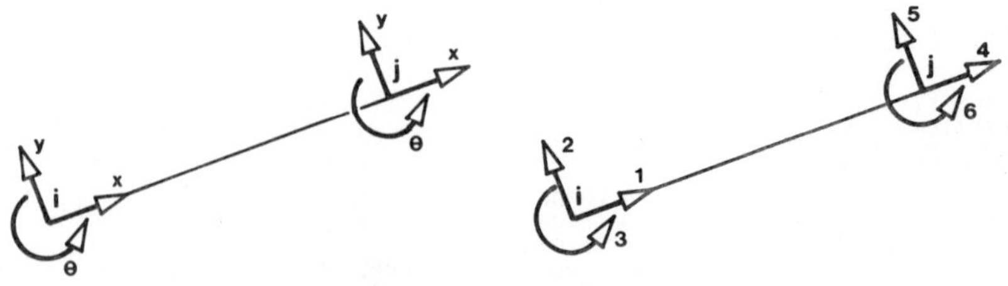

Figure 7.1 Element axes and freedom numbers

The forces at the ends of a plane frame element are related to the displacements at the ends by the element stiffness matrix as follows.

$$\begin{bmatrix} f_{ijx} \\ f_{ijy} \\ m_{ij} \\ f_{jix} \\ f_{jiy} \\ m_{ji} \end{bmatrix} = \begin{bmatrix} k_{11} & k_{12} & k_{13} & k_{14} & k_{15} & k_{16} \\ k_{21} & k_{22} & k_{23} & k_{24} & k_{25} & k_{26} \\ k_{31} & k_{32} & k_{33} & k_{34} & k_{35} & k_{36} \\ k_{41} & k_{42} & k_{43} & k_{44} & k_{45} & k_{46} \\ k_{51} & k_{52} & k_{53} & k_{54} & k_{55} & k_{56} \\ k_{61} & k_{62} & k_{63} & k_{64} & k_{65} & k_{66} \end{bmatrix} \begin{bmatrix} \delta_{ijx} \\ \delta_{ijy} \\ \theta_{ij} \\ \delta_{jix} \\ \delta_{jiy} \\ \theta_{ji} \end{bmatrix}$$

Figure 7.1 shows a typical plane frame element "i,j", and the element axes and freedom numbers that will be adopted in this text. The reader should note the introduction of rotational components of force and displacement at "i" and "j". The element stiffness matrix can be found by treating the element as an unrestrained structure with six freedoms.

In section 3.4 the column "j" of the structure stiffness matrix was shown to be the force system required to maintain unit displacement in the direction of freedom "j" (displacements in the directions of the other freedoms being held at zero). The relationship between the end displacements and the end forces for a line element was investigated in section 2.12. Consider, for example, the force system associated with rotation at end "i" of element "i,j". That force system is shown in fig 2.32, and it follows that the force system associated with unit rotation is as shown in fig 7.2.

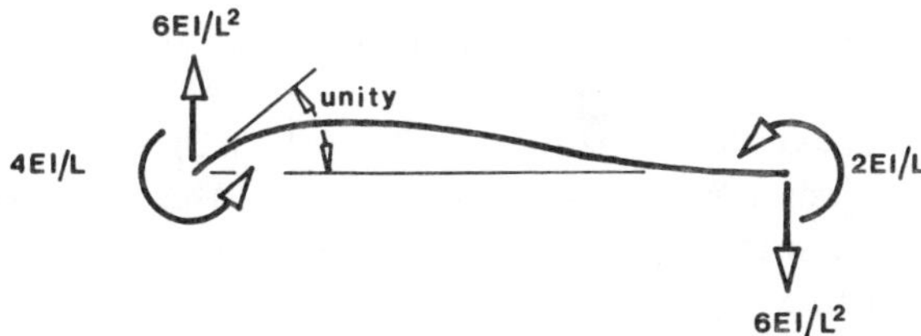

Figure 7.2 Force system associated with freedom 3

By inspection of fig 7.2 the stiffness coefficients associated with freedom 3 are

$$k_{13} = 0 \qquad k_{23} = \frac{6EI}{L^2} \qquad k_{33} = \frac{4EI}{L}$$

$$k_{43} = 0 \qquad k_{53} = -\frac{6EI}{L^2} \qquad k_{63} = \frac{2EI}{L}$$

Treating the other freedoms similarly, and using the stiffness properties developed in section 2.12, the element stiffness relationship can be shown to be

$$\begin{bmatrix} f_{ijx} \\ f_{ijy} \\ m_{ij} \\ \hdashline f_{jix} \\ f_{jiy} \\ m_{ji} \end{bmatrix} = \left[\begin{array}{ccc:ccc} \frac{EA}{L} & 0 & 0 & -\frac{EA}{L} & 0 & 0 \\ 0 & \frac{12EI}{L^3} & \frac{6EI}{L^2} & 0 & -\frac{12EI}{L^3} & \frac{6EI}{L^2} \\ 0 & \frac{6EI}{L^2} & \frac{4EI}{L} & 0 & -\frac{6EI}{L^2} & \frac{2EI}{L} \\ \hdashline -\frac{EA}{L} & 0 & 0 & \frac{EA}{L} & 0 & 0 \\ 0 & -\frac{12EI}{L^3} & -\frac{6EI}{L^2} & 0 & \frac{12EI}{L^3} & -\frac{6EI}{L^2} \\ 0 & \frac{6EI}{L^2} & \frac{2EI}{L} & 0 & -\frac{6EI}{L^2} & \frac{4EI}{L} \end{array}\right] \begin{bmatrix} \delta_{ijx} \\ \delta_{ijy} \\ \theta_{ij} \\ \hdashline \delta_{jix} \\ \delta_{jiy} \\ \theta_{ji} \end{bmatrix} \tag{7.1}$$

When written in terms of the submatrices indicated by the broken lines, the above equation reduces to the well-known expressions

$$\begin{bmatrix} f_{ij} \\ \hdashline f_{ji} \end{bmatrix} = \left[\begin{array}{c:c} k_{ii}^{j} & k_{ij} \\ \hdashline k_{ji} & k_{jj}^{i} \end{array}\right] \begin{bmatrix} \delta_{ij} \\ \hdashline \delta_{ji} \end{bmatrix}$$

hence

$$f_{ij} = k_{ii}^{j}\, \delta_{ij} + k_{ij}\, \delta_{ji} \tag{7.2}$$

and

$$f_{ij} = k_{ji}\, \delta_{ij} + k_{jj}^{i}\, \delta_{ji} \tag{7.3}$$

where the matrix product $k_{ii}^{j}\, \delta_{ij}$ yields the forces at end "i" of element "i,j" due to displacements at end "i", and $k_{ij}\, \delta_{ji}$ gives the forces at end "i" due to the displacements at end "j". Of course the resultant force vector f_{ij} is the matrix sum of these products.

It should be noted that equations (7.2) and (7.3) are identical to the corresponding equations for plane and space trusses. This is because the same matrix equations govern the behaviour of all types of structural frameworks. Of course while the matrix equations are identical, the matrices themselves are quite different. For example, the element end force vector f_{ij} will contain two components of force for a plane truss, three components of force for a space truss, and two components of force and a moment for a plane frame (see section 3.2).

7.3 Transformation of Force and Displacement

Equations (7.2) and (7.3) show how the element end forces (in the element axes system) are related to the element end displacements (also in the element axes system). Previous chapters have shown how nodal equilibrium can be used to develop the stiffness method. Element axes are useful when considering element forces, but they cannot be used when considering nodal equilibrium. This is because elements meeting at a node will, in general, lie at different angles. Consequently the element axes will be inconsistently orientated, which precludes their use when summing nodal forces. Before the forces that a element exerts on a node can be included in the equation of nodal equilibrium they must be transformed into the nodal axes system.

In section 5.3 the matrix T shown below was used to effect the transformation of displacements from global to element axes for plane truss structures.

$$\begin{bmatrix} \delta_{ijx} \\ \delta_{ijy} \end{bmatrix} = \begin{bmatrix} \cos\alpha & \sin\alpha \\ -\sin\alpha & \cos\alpha \end{bmatrix} \begin{bmatrix} \Delta_{ix} \\ \Delta_{iy} \end{bmatrix}$$

or

$$\delta_{ij} = T_{ij}\,\Delta_i$$

where "α" is the clockwise rotation of the element about "i" that will make the element axes coincide with the global axes. Δ_i is the vector of displacement at node "i" described in the nodal axes system and δ_{ij} is the same vector expressed in the element axes system.

The only difference between the nodal displacement vector for a plane truss and the nodal displacement vector for a plane frame is that the plane frame vector has a rotational component of displacement. As the element "x,y" plane lies in the global "x,y" plane rotation about the "z" axis will have the same magnitude whether expressed in the global or in the element axes system. Further, if the sense of the rotation (in this text a right-hand axes systems are used throughout) is consistent between the axes systems then the rotational component of displacement will have the same sign in both axes systems, and the required transformation becomes

$$\begin{bmatrix} \delta_{ijx} \\ \delta_{ijy} \\ \theta_{ij} \end{bmatrix} = \begin{bmatrix} \cos\alpha & \sin\alpha & 0 \\ -\sin\alpha & \cos\alpha & 0 \\ 0 & 0 & 1 \end{bmatrix} \begin{bmatrix} \Delta_{ix} \\ \Delta_{iy} \\ \Theta_i \end{bmatrix}$$

i.e.

$$\delta_{ij} = T_{ij}\,\Delta_i \tag{7.4}$$

and similarly at end "j" of the element

$$\begin{bmatrix} \delta_{jix} \\ \delta_{jiy} \\ \theta_{ji} \end{bmatrix} = \begin{bmatrix} \cos\alpha & \sin\alpha & 0 \\ -\sin\alpha & \cos\alpha & 0 \\ 0 & 0 & 1 \end{bmatrix} \begin{bmatrix} \Delta_{jx} \\ \Delta_{jy} \\ \Theta_j \end{bmatrix}$$

i.e.

$$\boldsymbol{\delta}_{ji} = \boldsymbol{T}_{ij} \, \boldsymbol{\Delta}_j \qquad (7.5)$$

where

$\boldsymbol{T}_{ij}$ is the transformation matrix for element "*i,j*".

$\boldsymbol{\Delta}_i$ is the nodal displacement vector at node "*i*" in the global axes system.

$\boldsymbol{\Delta}_j$ is the nodal displacement vector at node "*j*" in the global axes system.

$\boldsymbol{\delta}_{ij}$ is the nodal displacement vector at node "*i*" in the element axes system for element "*i,j*".

$\boldsymbol{\delta}_{ji}$ is the nodal displacement vector at node "*j*" in the element axes system for element "*j,i*".

The transformation matrix $\boldsymbol{T}_{ij}$ can also be used to transform forces from the global to the element axes system

$$\begin{bmatrix} f_{ijx} \\ f_{ijy} \\ m_{ij} \end{bmatrix} = \begin{bmatrix} cos\alpha & sin\alpha & 0 \\ -sin\alpha & cos\alpha & 0 \\ 0 & 0 & 1 \end{bmatrix} \begin{bmatrix} F_{ijx} \\ F_{ijy} \\ M_{ij} \end{bmatrix}$$

i.e.

$$\boldsymbol{f}_{ij} = \boldsymbol{T}_{ij} \, \boldsymbol{F}_{ij} \qquad (7.6)$$

and similarly at end "*j*" of the element

$$\begin{bmatrix} f_{jix} \\ f_{jiy} \\ m_{ji} \end{bmatrix} = \begin{bmatrix} cos\alpha & sin\alpha & 0 \\ -sin\alpha & cos\alpha & 0 \\ 0 & 0 & 1 \end{bmatrix} \begin{bmatrix} F_{jix} \\ F_{jiy} \\ M_{ji} \end{bmatrix}$$

i.e.

$$\boldsymbol{f}_{ji} = \boldsymbol{T}_{ij} \, \boldsymbol{F}_{ji} \qquad (7.7)$$

where

$\boldsymbol{T}_{ij}$ is the transformation matrix for element "*i,j*".

$\boldsymbol{F}_{ij}$ is the element end force vector at end "*i*" of element "*i,j*" in the global axes system.

$\boldsymbol{F}_{ji}$ is the element end force vector at end "*j*" of element "*i,j*" in the global axes system.

$\boldsymbol{f}_{ij}$ is the element end force vector at end "*i*" of element "*i,j*" in the element axes system.

$\boldsymbol{f}_{ji}$ is the element end force vector at end "*j*" of element "*i,j*" in the element axes system.

Section 5.3 showed how the reverse transformation (from element to global axes) is effected by the inverse of the transformation matrix, i.e.

$$\Delta_i = T^{-1}_{ij}\,\delta_{ij} \qquad \text{and} \qquad \Delta_j = T^{-1}_{ji}\,\delta_{ji}$$

$$F_{ij} = T^{-1}_{ij}\,f_{ij} \qquad \text{and} \qquad F_{ji} = T^{-1}_{ij}\,f_{ji}$$

Summary

The transformation matrix T_{ij} for plane frame element *"i,j"* transforms vectors of force and displacement from the global to the element axes system, and hence its inverse transforms vectors from the element to the global axes system, where

$$T_{ij} = \begin{bmatrix} \cos\alpha & \sin\alpha & 0 \\ -\sin\alpha & \cos\alpha & 0 \\ 0 & 0 & 1 \end{bmatrix}$$

and

$$T^{-1}_{ij} = \begin{bmatrix} \cos\alpha & -\sin\alpha & 0 \\ \sin\alpha & \cos\alpha & 0 \\ 0 & 0 & 1 \end{bmatrix}$$

7.4 The Element Stiffness Matrix in Global Axes

In section 7.2 the relationship between the forces at end *"i"* of element *"i,j"* and the displacements at ends *"i"* and *"j"* was shown to be

$$f_{ij} = k^{j}_{ii}\,\delta_{ij} + k_{ij}\,\delta_{ji}$$

This equation is expressed in terms of element axes and it has already been seen to have general applicability to all types of structural frameworks. In section 5.4 the transformation matrix T enabled the above equation to be rewritten in terms of the global axes system as follows.

$$F_{ij} = T^{-1}_{ij}\,k^{j}_{ii}\,T_{ij}\,\Delta_i + T^{-1}_{ij}\,k_{ij}\,T_{ij}\,\Delta_j$$

or

$$F_{ij} = K^{j}_{ii}\,\Delta_i + K_{ij}\,\Delta_j \qquad (7.8)$$

where

$$K^{j}_{ii} = T^{-1}_{ij}\,k^{j}_{ii}\,T_{ij}$$

and

$$K_{ij} = T^{-1}_{ij}\,k_{ij}\,T_{ij}$$

The matrix product $K^j_{ii}\ \Delta_i$ gives the forces (in the global axes system) at end "i" of element "i,j" associated with displacements at end "i" (also in the global axes system). Similarly the matrix product $K_{ij}\ \Delta_j$ gives the element end forces at end "i" of element "i,j" associated with displacements at end "j". The matrices K^j_{ii} and K_{ij} are known as *global element stiffness submatrices*. It is computationally advantageous to expand these matrices as follows

$$K^j_{ii} = T^{-1}_{ij}\ k^j_{ii}\ T_{ij}$$

$$= \begin{bmatrix} \cos\alpha & -\sin\alpha & 0 \\ \sin\alpha & \cos\alpha & 0 \\ 0 & 0 & 1 \end{bmatrix} \begin{bmatrix} \frac{EA}{L} & 0 & 0 \\ 0 & \frac{12EI}{L^3} & \frac{6EI}{L^2} \\ 0 & \frac{6EI}{L^2} & \frac{4EI}{L} \end{bmatrix} \begin{bmatrix} \cos\alpha & \sin\alpha & 0 \\ -\sin\alpha & \cos\alpha & 0 \\ 0 & 0 & 1 \end{bmatrix}$$

$$= \begin{bmatrix} \left(\frac{EA}{L}\cos^2\alpha + \frac{12EI}{L^3}\sin^2\alpha\right) & \left(\frac{EA}{L} - \frac{12EI}{L^3}\right)\cos\alpha\sin\alpha & -\frac{6EI}{L^2}\sin\alpha \\ \left(\frac{EA}{L} - \frac{12EI}{L^3}\right)\cos\alpha\sin\alpha & \left(\frac{EA}{L}\sin^2\alpha + \frac{12EI}{L^3}\cos^2\alpha\right) & \frac{6EI}{L^2}\cos\alpha \\ -\frac{6EI}{L^2}\sin\alpha & \frac{6EI}{L^2}\cos\alpha & \frac{4EI}{L} \end{bmatrix} \quad (7.9)$$

and $K_{ij} = T^{-1}_{ij}\ k_{ij}\ T_{ij}$

$$= \begin{bmatrix} -\left(\frac{EA}{L}\cos^2\alpha + \frac{12EI}{L^3}\sin^2\alpha\right) & -\left(\frac{EA}{L} - \frac{12EI}{L^3}\right)\cos\alpha\sin\alpha & -\frac{6EI}{L^2}\sin\alpha \\ -\left(\frac{EA}{L} - \frac{12EI}{L^2}\right)\cos\alpha\sin\alpha & -\left(\frac{EA}{L}\sin^2\alpha + \frac{12EI}{L^2}\cos^2\alpha\right) & \frac{6EI}{L^2}\cos\alpha \\ \frac{6EI}{L^2}\sin\alpha & -\frac{6EI}{L^2}\cos\alpha & \frac{2EI}{L} \end{bmatrix}$$

The K_{ji} and the K^i_{jj} submatrices can be found by similar expansions. In practice only the K^j_{ii} submatrix need be evaluated as the other submatrices can be obtained by multiplying the corresponding elements of K^j_{ii} as follows

Find K_{ij} by multiplying corresponding elements of K^j_{ii} by

$$\begin{bmatrix} -1 & -1 & 1 \\ -1 & -1 & 1 \\ -1 & -1 & 0.5 \end{bmatrix} \quad (7.10)$$

Find K_{ji} by multiplying corresponding elements of K_{ii}^{j} by

$$\begin{bmatrix} -1 & -1 & -1 \\ -1 & -1 & -1 \\ 1 & 1 & 0.5 \end{bmatrix} \tag{7.11}$$

Find K_{jj}^{i} by multiplying corresponding elements K_{ii}^{j} of by

$$\begin{bmatrix} 1 & 1 & -1 \\ 1 & 1 & -1 \\ -1 & -1 & 1 \end{bmatrix} \tag{7.12}$$

Note that this is not a matrix multiplication, but simply the multiplication of corresponding elements.

7.5 Nodal Equilibrium

Figure 7.3(a) shows a pitched roof portal, and fig 7.3(b) shows how it might be idealised.

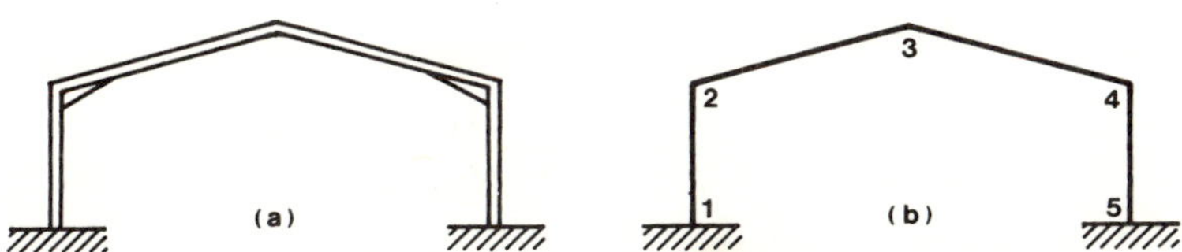

Figure 7.3 A pitched roof portal

In general, each element of a plane frame carries axial force, shear, and bending. Hence the freebody diagram for node 2 will be as shown in fig 7.4(a). As the node is part of a structure in equilibrium then it too must be in equilibrium under the action of the internal element forces (f_{21x}, f_{21y}, m_{21}, f_{23x} etc.) and the external applied loads (P_{2x}, P_{2y}, M_2). Hence if the element end forces are transformed into the nodal axes system as indicated in fig 7.4(b), then the horizontal forces, the vertical forces, and the moments must sum to zero, i.e.

$$\begin{aligned} P_{2x} + F'_{21x} + F'_{23x} &= 0 \\ P_{2y} + F'_{21y} + F'_{23y} &= 0 \\ M_2 + M'_{21} + M'_{23} &= 0 \end{aligned}$$

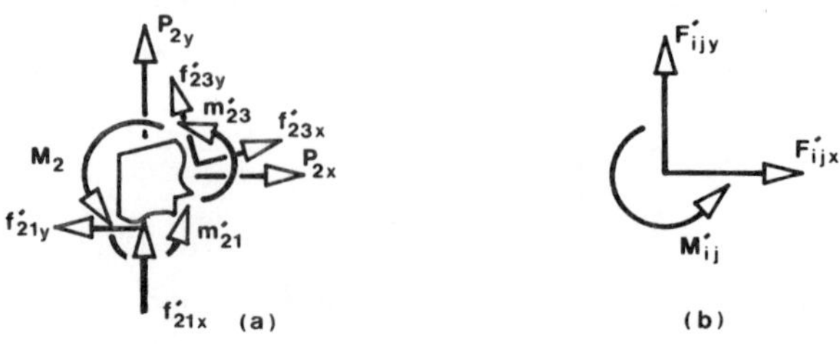

Figure 7.4 Freebody diagram for node 2

In section 5.5 the forces exerted on a node by a element were seen to be equal and opposite to the element end forces. Hence

$$
\begin{aligned}
P_{2x} - F_{21x} - F_{23x} &= 0 \\
P_{2y} - F_{21y} - F_{23y} &= 0 \\
M_2 - M_{21} - M_{23} &= 0
\end{aligned}
$$

or, in matrix notation

$$\boldsymbol{P}_2 = \boldsymbol{F}_{21} + \boldsymbol{F}_{23}$$

where

$$\boldsymbol{P}_2 = \begin{bmatrix} P_{2x} \\ P_{2y} \\ M_2 \end{bmatrix} \qquad \boldsymbol{F}_{21} = \begin{bmatrix} F_{21x} \\ F_{21y} \\ M_{21} \end{bmatrix} \qquad \boldsymbol{F}_{23} = \begin{bmatrix} F_{23x} \\ F_{23y} \\ M_{23} \end{bmatrix}$$

Hence, the equilibrium equation for node "i", which has elements "i,a", "i,b", "i,c", ... "i,n" framing into it, is

$$\boldsymbol{P}_i = \boldsymbol{F}_{ia} + \boldsymbol{F}_{ib} + \boldsymbol{F}_{ic} + \ldots + \boldsymbol{F}_{in}$$

This equation (which is identical to the corresponding equation for plane trusses) simply states that the external applied load vector $\boldsymbol{P}_i$ at node "i" must be balanced by the vectors of internal element end forces $\boldsymbol{F}_{ia}$, $\boldsymbol{F}_{ib}$, $\boldsymbol{F}_{ic}$, ... $\boldsymbol{F}_{in}$. Obviously for the summation to be meaningful the force components must be described in a consistent axes system (see section 3.6).

Equation (7.8) gives the relationship between element end forces and the nodal displacements in the gobal axes system

$$\boldsymbol{F}_{ij} = \boldsymbol{K}_{ii}^{j}\, \boldsymbol{\Delta}_i + \boldsymbol{K}_{ij}\, \boldsymbol{\Delta}_j$$

where

$$\boldsymbol{K}_{ii}^{j} = \boldsymbol{T}_{ij}^{-1}\, \boldsymbol{k}_{ii}^{j}\, \boldsymbol{T}_{ij} \qquad \text{and} \qquad \boldsymbol{K}_{ij} = \boldsymbol{T}_{ij}^{-1}\, \boldsymbol{k}_{ij}\, \boldsymbol{T}_{ij}$$

Substituting for the element end forces in the equation of nodal equilibrium gives

$$P_i = K_{ii}^a \Delta_i + K_{ia} \Delta_a + K_{ii}^b \Delta_i + K_{ib} \Delta_b + \dots + K_{ii}^n \Delta_i + K_{in} \Delta_n$$

or

$$P_i = K_{ii} \Delta_i + K_{ia} \Delta_a + K_{ib} \Delta_b + \dots + K_{in} \Delta_n$$

where

$$K_{ii} = \sum_{j=a}^{n} K_{ii}^j$$

7.6 The Initial Structure Stiffness Matrix

Chapter 5 (section 5.7) showed that, prior to the application of the boundary conditions, the equations of nodal equilibrium form a mixed set of simultaneous equations.

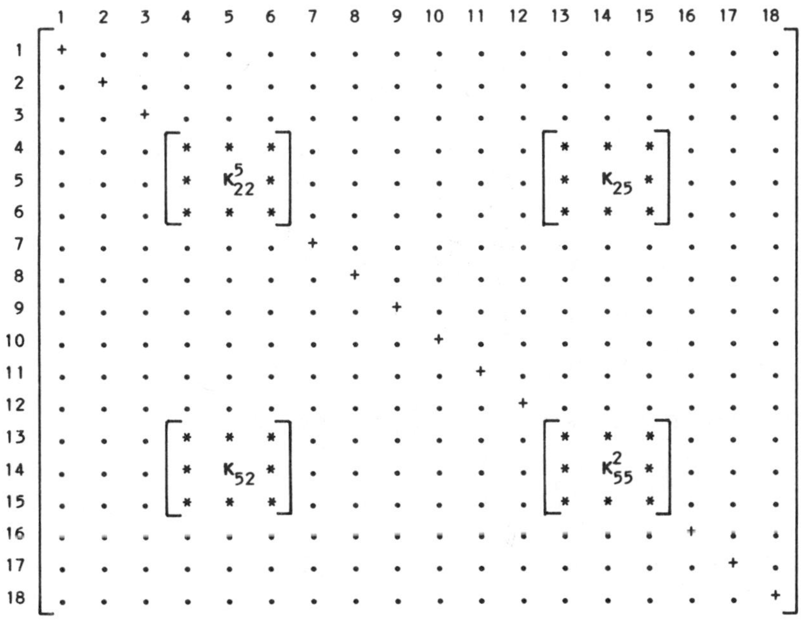

Figure 7.5 Adding element 2,5 to the initial structure stiffness matrix

The coefficient matrix of these equations is known as the initial structure stiffness matrix. The procedure to assemble the initial structure stiffness matrix for a plane frame is identical to that used for two and three-dimensional trusses.

The global element stiffness submatrices K_{ii}^j, K_{ij}, K_{ji} and K_{jj}^i for a plane frame are 3 x 3 square matrices. Hence they must be added to the initial structure stiffness matrix so that their (1,1) elements are added to locations ($3i$-2,$3i$-2), ($3i$-2,$3j$-2), ($3j$-2,$3i$-2), and ($3j$-2, $3j$-2) respectively. Figure 7.5 shows where the global element stiffness submatrices for element 2,5 would be added to the initial structure stiffness matrix. The initial structure matrix is 18 x 18, indicating that the frame has six nodes.

Example 7.1

Assemble the initial stiffness equation, in terms of submatrices, for the portal frame shown in the following diagram.

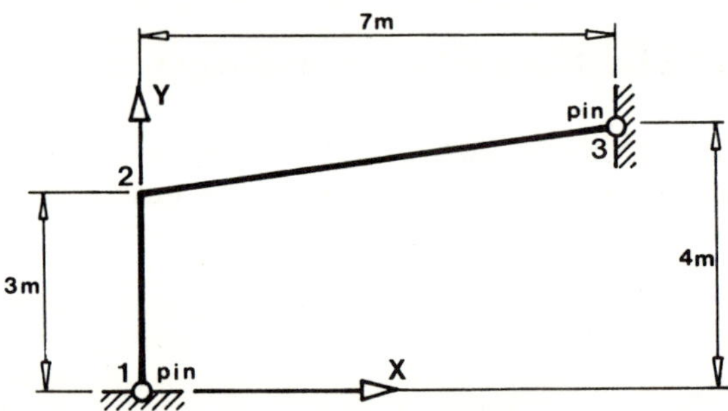

First remove the boundary conditions, leaving the structure completely unrestrained. Now zero the initial structure stiffness matrix (note that each zero represents a 3 x 3 submatrix of zeros).

$$\begin{bmatrix} 0 & 0 & 0 \\ 0 & 0 & 0 \\ 0 & 0 & 0 \end{bmatrix}$$

Add the element stiffness submatrices for element 1,2

$$\begin{bmatrix} K_{11} & K_{12} & 0 \\ K_{21} & K_{22} & 0 \\ 0 & 0 & 0 \end{bmatrix}$$

where $K_{11} = K_{11}^{2}$ and $K_{22} = K_{22}^{1}$

Add the element stiffness submatrices for element 2,3

$$\begin{bmatrix} K_{11} & K_{12} & 0 \\ K_{21} & K_{22} & K_{23} \\ 0 & K_{32} & K_{33} \end{bmatrix}$$

where
$$K_{11} = K_{11}^{2}$$
$$K_{22} = K_{22}^{1} + K_{22}^{3}$$
$$K_{33} = K_{33}^{2}$$

Hence the initial structure stiffness equation, in terms of submatrices, is

$$\begin{bmatrix} K_{11} & K_{12} & 0 \\ K_{21} & K_{22} & K_{23} \\ 0 & K_{32} & K_{33} \end{bmatrix} \begin{bmatrix} \Delta_1 \\ \Delta_2 \\ \Delta_3 \end{bmatrix} = \begin{bmatrix} P_1 \\ P_2 \\ P_3 \end{bmatrix}$$

The reader should note the similarity with example 5.1. Indeed, assembly of the initial structure stiffness matrix using submatrices depends only upon how the structure is connected and is independent of the type of framework.

Example 7.2

If the frame considered in example 7.1 is constructed from steel beam sections having the following properties, $I = 0.25 \times 10^9 \text{ mm}^4$, $A = 9000 \text{ mm}^2$, $E = 200 \text{ kN/mm}^2$ assemble the initial structure stiffness matrix.

The structure has 3 nodes; hence the initial structure stiffness matrix will be 9 x 9. As in example 7.1 the initial structure stiffness matrix is first zeroed, and then the global element stiffness submatrices are added for each element in turn.

Element 1,2 $A = 9000 \text{ mm}^2$, $I = 0.25 \times 10^9 \text{ mm}^4$, $L = 3000 \text{ mm}$,
$E = 200 \text{ kN/mm}^2$, $\alpha = 90°$

$$K_{11}^{2} = \begin{bmatrix} \left(\frac{EA}{L}\cos^2\alpha + \frac{12EI}{L^3}\sin^2\alpha\right) & \left(\frac{EA}{L} - \frac{12EI}{L^3}\right)\cos\alpha\sin\alpha & -\frac{6EI}{L^2}\sin\alpha \\ \left(\frac{EA}{L} - \frac{12EI}{L^3}\right)\cos\alpha\sin\alpha & \left(\frac{EA}{L}\sin^2\alpha + \frac{12EI}{L^3}\cos^2\alpha\right) & \frac{6EI}{L^2}\cos\alpha \\ -\frac{6EI}{L^2}\sin\alpha & \frac{6EI}{L^2}\cos\alpha & \frac{4EI}{L} \end{bmatrix}$$

$$= \begin{bmatrix} 22.2 & 0 & -33333 \\ 0 & 600.0 & 0 \\ -33333 & 0 & 66.7 \times 10^6 \end{bmatrix}$$

The K_{12} and K_{22}^{1} matrices are found from the K_{11}^{2} using the relationships shown in equations (7.10) and (7.12) and then added to the initial structure stiffness matrix (note that the K_{22}^{1} does not have to be added due to the symmetry of the initial structure stiffness matrix).

$$\left[\begin{array}{ccc|ccc|ccc}
22.2 & 0 & -33333 & -22.2 & 0 & -33333 & 0 & 0 & 0 \\
 & 600.0 & 0 & 0 & -600.0 & 0 & 0 & 0 & 0 \\
 & & 66.7E6 & 33333 & 0 & 33.3E6 & 0 & 0 & 0 \\
\hline
 & & & 22.2 & 0 & 33333 & 0 & 0 & 0 \\
 & & & & 600.0 & 0 & 0 & 0 & 0 \\
 & \textit{symmetric} & & & & 66.7E6 & 0 & 0 & 0 \\
\hline
 & & & & & & 0 & 0 & 0 \\
 & & & & & & & 0 & 0 \\
 & & & & & & & & 0
\end{array}\right]$$

Note that, in the preceding matrix, E6 is equivalent to x 10^6.

Element 2,3 $A = 9000\ \text{mm}^2$, $I = 0.25 \times 10^9\ \text{mm}^4$, $L = 7071$ mm,
$E = 200\ \text{kN/mm}^2$, $\alpha = 8.13°$

$$K^3_{22} = \begin{bmatrix} 249.5 & 35.40 & -848.5 \\ 35.40 & 6.754 & 5940 \\ -848.5 & 5940 & 28.3E6 \end{bmatrix}$$

Adding the K^3_{22}, K_{23} and K^2_{33} submatrices completes the assembly of the initial structure stiffness matrix.

$$\left[\begin{array}{ccc|ccc|ccc}
 & \textit{unchanged} & & & \textit{unchanged} & & & \textit{unchanged} & \\
\hline
 & & & 271.7 & 35.40 & 32485 & -249.4 & -35.40 & -848.5 \\
 & & & & 606.8 & 5940 & -35.40 & -6.754 & 5940 \\
 & \textit{symmetric} & & & & 95.0E6 & 848.5 & -5940 & 14.1E6 \\
\hline
 & & & & & & 249.5 & 35.40 & 848.5 \\
 & & & & & & & 6.754 & -5940 \\
 & & & & & & & & 28.3E6
\end{array}\right]$$

The calculation of element stiffness submatrices is a tedious and error-prone procedure that is best done with the aid of a computer. There follows a description of a computer program that calculates the element stiffness matrix for plane frame elements.

Program ESTIFFPF.BAS

This program calculates the element stiffness matrix for plane frame elements.

Input

The following data is input for each element

- A the cross-sectional area of the element.
- I the second moment of area of the element.
- L the length of the element.
- E the elastic constant for the element.
- α the angle the element "*i,j*" makes with the global "*x*" axis. Defined as shown in the following diagram.

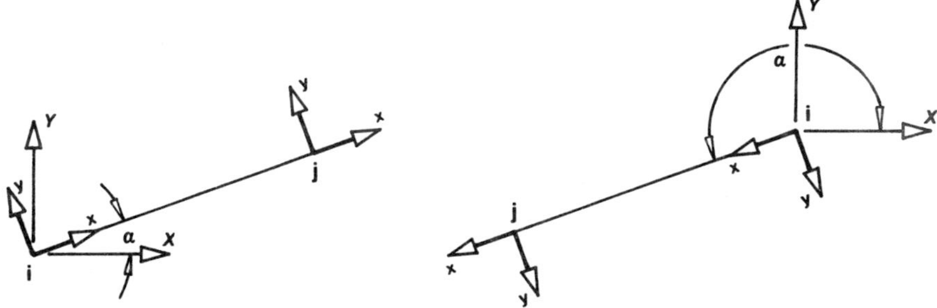

A, I, L, and E must be in consistent units, the unit of force used in E must correspond to the unit of force used for the applied loading, and α is in degrees.

Output

The input is echoed and the element stiffness matrix is output - first in the element axes system and then in the global axes system.

Listing

```
1000 '========================      ESTIFFPF.BAS      ============================
1010 '
1020 OPTION BASE 1
1030 DIM ESTIFF(6,6)
1040 CLS
1050 PRINT
1060 PRINT "=================================================================="
1070 PRINT
1080 PRINT "PROGRAM ESTIFFPF.BAS                    Copyright (c) J.Balfour 1991"
1090 PRINT
1100 PRINT "    Program for the calculation of element stiffness matrix"
1110 PRINT "      in element and global axes for a plane truss elements"
1120 PRINT
1130 PRINT "                  For further information contact"
1140 PRINT " James A.D.Balfour, Heriot-Watt University, Riccarton, Edinburgh"
1150 PRINT "                Tel 031-449-5111, Fax 031-451-3170"
1160 PRINT
1170 PRINT "=================================================================="
1180 '
1190 IN% = 1 : K% = 1
1200 WHILE IN% <> 0
```

```
1210   PRINT
1220   PRINT "ELEMENT "; K%
1230   PRINT
1240   INPUT "Lower node number                = ", IN%
1250   IF IN%=0 THEN GOTO 1460
1260   INPUT "Higher node number               = ", JN%
1270   PRINT
1280   PRINT        "+ + + + + + + + + + + +"
1290   PRINT USING "+   ELEMENT ## ##    +"; IN%, JN%
1300   PRINT        "+ + + + + + + + + + + +"
1310   PRINT
1320   INPUT "Area (A)                         = ", A
1330   INPUT "Second moment of area (I)        = ", A2
1340   INPUT "Length (L)                       = ", L
1350   INPUT "Elastic constant (E)             = ", E
1360   INPUT "Angle of with the global axes    = ", ALPHA
1370   '
1380   ' ... Calculate the element stiffness in the element axes system
1390   '
1400   GOSUB 1490
1410   '
1420   ' ... Calculate the element stiffness in the global axes system
1430   '
1440   GOSUB 1920
1450   K% = K% + 1
1460 WEND
1470 END
1480 '
1490 '********      ELEMENT STIFFNESS IN THE ELEMENT AXES SYSTEM     **********
1500 '
1510 FOR I% = 1 TO 6
1520   FOR J% = 1 TO 6
1530     ESTIFF(I%,J%) = 0
1540   NEXT J%
1550 NEXT I%
1560 EAL = E * A / L
1570 EIL = E * A2 / L
1580 '
1590 ' ... Calculate the element stiffness terms (upper triangle only)
1600 '
1610 ESTIFF(1, 1) =  EAL
1620 ESTIFF(1, 4) = -ESTIFF(1, 1)
1630 ESTIFF(2, 2) = 12 * EIL / L ^ 2
1640 ESTIFF(2, 3) =  6 * EIL / L
1650 ESTIFF(2, 5) = -ESTIFF(2, 2)
1660 ESTIFF(2, 6) =  6 * EIL / L
1670 ESTIFF(3, 3) =  4 * EIL
1680 ESTIFF(3, 5) = -6 * EIL / L
1690 ESTIFF(3, 6) =  2 * EIL
1700 ESTIFF(4, 4) =  ESTIFF(1, 1)
1710 ESTIFF(5, 5) =  ESTIFF(2, 2)
1720 ESTIFF(5, 6) = -6 * EIL / L
1730 ESTIFF(6, 6) =  4 * EIL
1740 '
1750 ' ... Output the element stiffness matrix
1760 '
1770 PRINT
1780 PRINT "* * * * * * * * * * * * * * * * * * * * *"
1790 PRINT "*    ELEMENT STIFFNESS MATRIX IN      *"
1800 PRINT "*    THE ELEMENT AXES SYSTEM          *"
1810 PRINT "* * * * * * * * * * * * * * * * * * * * *"
1820 PRINT
1830 FOR I% = 1 TO 6
1840   FOR J% = 1 TO 6
1850     IF I% <= J% THEN PRINT USING "#.####^^^^    "; ESTIFF(I%,J%);
1860     IF I% >  J% THEN PRINT USING "#.####^^^^    "; ESTIFF(J%,I%);
1870   NEXT J%
1880   PRINT
1890 NEXT I%
1900 RETURN
1910 '
1920 '**********     ELEMENT STIFFNESS IN THE GLOBAL AXES SYSTEM      **********
1930 '
1940 ' ... Convert the angle to radians
1950 '
1960 AR = 4 * ATN(1) * ALPHA / 180
1970 C  = COS(AR)
```

```
1980 S  = SIN(AR)
1990 '
2000 ' ... Zero the stiffness matrix
2010 '
2020 FOR I% = 1 TO 6
2030   FOR J% = 1 TO 6
2040     ESTIFF(I%,J%) = 0
2050   NEXT J%
2060 NEXT I%
2070 '
2080 ' ... Calculate the element stiffness terms (upper triangle only)
2090 '
2100 ESTIFF(1,1) =   EAL*C*C + 12*EIL*S*S/L^2
2110 ESTIFF(1,2) =   EAL*S*C - 12*EIL*S*C/L^2
2120 ESTIFF(1,3) = - 6*EIL*S/L
2130 ESTIFF(1,4) = - EAL*C*C - 12*EIL*S*S/L^2
2140 ESTIFF(1,5) = - EAL*C*S + 12*EIL*S*C/L^2
2150 ESTIFF(1,6) = - 6*EIL*S/L
2160 ESTIFF(2,2) =   EAL*S*S + 12*EIL*C*C/L^2
2170 ESTIFF(2,3) =   6*EIL*C/L
2180 ESTIFF(2,4) = - EAL*S*C + 12*EIL*C*S/L^2
2190 ESTIFF(2,5) = - EAL*S*S - 12*EIL*C*C/L^2
2200 ESTIFF(2,6) =   6*EIL*C/L
2210 ESTIFF(3,3) =   4*EIL
2220 ESTIFF(3,4) =   6*EIL*S/L
2230 ESTIFF(3,5) = - 6*EIL*C/L
2240 ESTIFF(3,6) =   2*EIL
2250 ESTIFF(4,4) =   EAL*C*C + 12*EIL*S*S/L^2
2260 ESTIFF(4,5) =   EAL*S*C - 12*EIL*S*C/L^2
2270 ESTIFF(4,6) =   6*EIL*S/L
2280 ESTIFF(5,5) =   EAL*S*S + K1*EIL*C*C/L^2
2290 ESTIFF(5,6) = - 6*EIL*C/L
2300 ESTIFF(6,6) =   4*EIL
2310 '
2320 ' ... Output the element stiffness matrix
2330 '
2340 PRINT
2350 PRINT "* * * * * * * * * * * * * * * * * * * * **"
2360 PRINT "*   ELEMENT STIFFNESS MATRIX IN       *"
2370 PRINT "*   THE GLOBAL AXES SYSTEM            *"
2380 PRINT "* * * * * * * * * * * * * * * * * * * * **"
2390 PRINT
2400 FOR I% = 1 TO 6
2410   FOR J% = 1 TO 6
2420     IF I% <= J% THEN PRINT USING "#.####^^^^   "; ESTIFF(I%,J%);
2430     IF I% >  J% THEN PRINT USING "#.####^^^^   "; ESTIFF(J%,I%);
2440   NEXT J%
2450   PRINT
2460 NEXT I%
2470 RETURN
2480 '
2490 '========================   ESTIFFPF.BAS    ============================
```

Sample Run

The following sample run shows program ESTIFFPF.BAS being used to evaluate the element stiffness matrices for the structure analysed in example 7.2.

```
=================================================================

PROGRAM ESTIFFPF.BAS                  Copyright (c) J.Balfour 1991

    Program for the calculation of element stiffness matrix
      in element and global axes for a plane truss elements

                 For further information contact
 James A.D.Balfour, Heriot-Watt University, Riccarton, Edinburgh
               Tel 031-449-5111, Fax 031-451-3170

=================================================================
```

```
ELEMENT  1

Lower node number              = 1
Higher node number             = 2

+ + + + + + + + + + + +
+   ELEMENT  1  2   +
+ + + + + + + + + + + +

Area (A)                        = 9000
Second moment of area (I)       = .25E9
Length (L)                      = 3000
Elastic constant (E)            = 200
Angle of with the global axes   = 90

* * * * * * * * * * * * * * * * * * * * *
*   ELEMENT STIFFNESS MATRIX IN        *
*   THE ELEMENT AXES SYSTEM            *
* * * * * * * * * * * * * * * * * * * * *

0.6000E+03   0.0000E+00   0.0000E+00   -.6000E+03   0.0000E+00   0.0000E+00
0.0000E+00   0.2222E+02   0.3333E+05   0.0000E+00   -.2222E+02   0.3333E+05
0.0000E+00   0.3333E+05   0.6667E+08   0.0000E+00   -.3333E+05   0.3333E+08
-.6000E+03   0.0000E+00   0.0000E+00   0.6000E+03   0.0000E+00   0.0000E+00
0.0000E+00   -.2222E+02   -.3333E+05   0.0000E+00   0.2222E+02   -.3333E+05
0.0000E+00   0.3333E+05   0.3333E+08   0.0000E+00   -.3333E+05   0.6667E+08

* * * * * * * * * * * * * * * * * * * * *
*   ELEMENT STIFFNESS MATRIX IN        *
*   THE GLOBAL AXES SYSTEM             *
* * * * * * * * * * * * * * * * * * * * *

0.2222E+02   -.9413E-04   -.3333E+05   -.2222E+02   0.9413E-04   -.3333E+05
-.9413E-04   0.6000E+03   -.5431E-02   0.9413E-04   -.6000E+03   -.5431E-02
-.3333E+05   -.5431E-02   0.6667E+08   0.3333E+05   0.5431E-02   0.3333E+08
-.2222E+02   0.9413E-04   0.3333E+05   0.2222E+02   -.9413E-04   0.3333E+05
0.9413E-04   -.6000E+03   0.5431E-02   -.9413E-04   0.6000E+03   0.5431E-02
-.3333E+05   -.5431E-02   0.3333E+08   0.3333E+05   0.5431E-02   0.6667E+08

ELEMENT  2

Lower node number              = 2
Higher node number             = 3

+ + + + + + + + + + + +
+   ELEMENT  2  3   +
+ + + + + + + + + + + +

Area (A)                        = 9000
Second moment of area (I)       = .25E9
Length (L)                      = 7071
Elastic constant (E)            = 200
Angle of with the global axes   = 8.13

* * * * * * * * * * * * * * * * * * * * *
*   ELEMENT STIFFNESS MATRIX IN        *
*   THE ELEMENT AXES SYSTEM            *
* * * * * * * * * * * * * * * * * * * * *

0.2546E+03   0.0000E+00   0.0000E+00   -.2546E+03   0.0000E+00   0.0000E+00
0.0000E+00   0.1697E+01   0.6000E+04   0.0000E+00   -.1697E+01   0.6000E+04
0.0000E+00   0.6000E+04   0.2828E+08   0.0000E+00   -.6000E+04   0.1414E+08
-.2546E+03   0.0000E+00   0.0000E+00   0.2546E+03   0.0000E+00   0.0000E+00
0.0000E+00   -.1697E+01   -.6000E+04   0.0000E+00   0.1697E+01   -.6000E+04
0.0000E+00   0.6000E+04   0.1414E+08   0.0000E+00   -.6000E+04   0.2828E+08

* * * * * * * * * * * * * * * * * * * * *
*   ELEMENT STIFFNESS MATRIX IN        *
*   THE GLOBAL AXES SYSTEM             *
* * * * * * * * * * * * * * * * * * * * *

0.2495E+03   0.3540E+02   -.8485E+03   -.2495E+03   -.3540E+02   -.8485E+03
0.3540E+02   0.6754E+01   0.5940E+04   -.3540E+02   -.6754E+01   0.5940E+04
-.8485E+03   0.5940E+04   0.2828E+08   0.8485E+03   -.5940E+04   0.1414E+08
-.2495E+03   -.3540E+02   0.8485E+03   0.2495E+03   0.3540E+02   0.8485E+03
-.3540E+02   -.6754E+01   -.5940E+04   0.3540E+02   0.5091E+01   -.5940E+04
-.8485E+03   0.5940E+04   0.1414E+08   0.8485E+03   -.5940E+04   0.2828E+08
```

```
ELEMENT  3

Lower node number           = 0
```

Program ISTIFFPF.BAS

This program illustrates how ESTIFFPF.BAS can be developed to automatically assemble the initial structure stiffness matrix for a plane frame structure.

Input

The program loops for each element in turn requesting the element data. The element data is used to calculate the element stiffness matrix which is then added to the initial structure stiffness matrix. Input is terminated be entering a lower node number of zero.

Comments on the Algorithm

Only the upper triangle of the initial structure stiffness matrix is assembled. The symmetry of the stiffness matrix is used to complete the matrix.

Output

The initial structure stiffness matrix is output, with advantage being taken of the symmetry of the matrix to output the sub-diagonal elements.

Listing

```
1000 '========================      ISTIFFPF.BAS      ===========================
1010 '
1020 ' ... Set maximum number of nodes
1030 '
1040 NNODE% = 20 : TRUE% = 1 : FALSE% = 0
1050 OPTION BASE 1
1060 DIM EFREE%(6), K(3*NNODE%,3*NNODE%), ESTIFF(6,6)
1070 CLS
1080 PRINT
1090 PRINT "=================================================================="
1100 PRINT
1110 PRINT "PROGRAM ISTIFFPF.BAS                    Copyright (c) J.Balfour 1991"
1120 PRINT
1130 PRINT "      Program for the calculation of element stiffness matrices"
1140 PRINT "         in element and global axes for a plane truss element"
1150 PRINT
1160 PRINT "                     For further information contact"
1170 PRINT " James A.D.Balfour, Heriot-Watt University, Riccarton, Edinburgh"
1180 PRINT "                  Tel 031-449-5111, Fax 031-451-3170"
1190 PRINT
1200 PRINT "=================================================================="
1210 PRINT
1220 PRINT "Maximum number of nodes = "NNODE%
1230 '
1240 ' ... Zero the initial stiffness matrix
1250 '
1260 NDOF%  = 3 * NNODE%
1270 FOR I% = 1 TO NDOF%
1280   FOR J% = 1 TO NDOF%
```

```
1290      K(I%,J%) = 0
1300    NEXT J%
1310 NEXT I%
1320 '
1330 ' ... Input elements and add stiffness to the initial stiffness matrix
1340 '
1350 GOSUB 1430
1360 '
1370 ' ... Output the initial structure stiffness matrix
1380 '
1390 GOSUB 2200
1400 '
1410 END
1420 '
1430 '*********************     INPUT THE ELEMENT DATA     *********************
1440 '
1450 K% = 1 : NDOF% = 0 : DONE% = FALSE%
1460 WHILE DONE% = FALSE%
1470   PRINT
1480   PRINT "ELEMENT "; K%
1490   PRINT
1500   INPUT "Lower node number (0 to stop) = ", IN%
1510   IF IN% = 0 THEN DONE% = TRUE% : GOTO 2170
1520   INPUT "Higher node number            = ", JN%
1530   PRINT
1540   PRINT "+ + + + + + + + + + + +"
1550   PRINT USING "+    ELEMENT ## ##    +"; IN%; JN%
1560   PRINT "+ + + + + + + + + + + +"
1570   PRINT
1580   INPUT "Area (A)                      = ", A
1590   INPUT "Second moment of area (I)     = ", A2
1600   INPUT "Length (L)                    = ", L
1610   INPUT "Elastic constant (E)          = ", E
1620   INPUT "Angle with the global x-axis  = ", ALPHA
1630   PRINT
1640   IF 2*JN% > NDOF% THEN NDOF% = 2*JN%
1650   '
1660   ' ... Convert the angle to radians
1670   '
1680   AR = 4 * ATN(1) * ALPHA / 180
1690   C  = COS(AR)
1700   S  = SIN(AR)
1710   '
1720   ' ... Calculate the upper triangle of the element stiffness matrix
1730   '
1740   ESTIFF(1,1) =   E * A * C * C / L  +  12 * E * A2 * S * S / L^3
1750   ESTIFF(1,2) =   E * A * S * C / L  -  12 * E * A2 * S * C / L^3
1760   ESTIFF(1,3) = - 6 * E * A2 * S / L^2
1770   ESTIFF(1,4) = - ESTIFF(1,1)
1780   ESTIFF(1,5) = - ESTIFF(1,2)
1790   ESTIFF(1,6) =   ESTIFF(1,3)
1800   ESTIFF(2,2) =   E * A * S * S / L  +  12 * E * A2 * C * C / L^3
1810   ESTIFF(2,3) =   6 * E * A2 * C / L^2
1820   ESTIFF(2,4) = - ESTIFF(1,2)
1830   ESTIFF(2,5) = - ESTIFF(2,2)
1840   ESTIFF(2,6) =   ESTIFF(2,3)
1850   ESTIFF(3,3) =   4 * E * A2 / L
1860   ESTIFF(3,4) = - ESTIFF(1,3)
1870   ESTIFF(3,5) = - ESTIFF(2,3)
1880   ESTIFF(3,6) =   .5 * ESTIFF(3,3)
1890   ESTIFF(4,4) =   ESTIFF(1,1)
1900   ESTIFF(4,5) =   ESTIFF(1,2)
1910   ESTIFF(4,6) = - ESTIFF(1,3)
1920   ESTIFF(5,5) =   ESTIFF(2,2)
1930   ESTIFF(5,6) = - ESTIFF(2,3)
1940   ESTIFF(6,6) =   ESTIFF(3,3)
1950   '
1960   ' ... Set up the element code number (initial)
1970   '
1980   EFREE%(1) = 3*IN%-2
1990   EFREE%(2) = 3*IN%-1
2000   EFREE%(3) = 3*IN%
2010   EFREE%(4) = 3*JN%-2
2020   EFREE%(5) = 3*JN%-1
2030   EFREE%(6) = 3*JN%
2040   IF 3*JN% > NDOF% THEN NDOF% = 3*JN%
2050   '
```

```
2060    ' ... Add the element stiffness terms to the upper triangle of the
2070    '      of the initial structure stiffness matrix
2080    '
2090    FOR I% = 1 TO 6
2100      FOR J% = I% TO 6
2110        L% = EFREE%(I%)
2120        M% = EFREE%(J%)
2130        K(L%,M%) = K(L%,M%) + ESTIFF(I%,J%)
2140      NEXT J%
2150    NEXT I%
2160    K% = K% + 1
2170 WEND
2180 RETURN
2190 '
2200 '**********    OUTPUT THE INITIAL STRUCTURE STIFFNESS MATRIX    ********
2210 '
2220 PRINT : PRINT
2230 PRINT "* * * * * * * * * * * * * * *"
2240 PRINT "*   INITIAL STRUCTURE    *"
2250 PRINT "*   STIFFNESS MATRIX     *"
2260 PRINT "* * * * * * * * * * * * * * *"
2270 PRINT
2280 FOR I% = 1 TO NDOF%
2290   FOR J% = 1 TO NDOF%
2300     IF I% <= J% THEN PRINT USING "#.####^^^^    "; K(I%,J%);
2310     IF I% >  J% THEN PRINT USING "#.####^^^^    "; K(J%,I%);
2320     IF J% MOD 6 = 0 THEN PRINT
2330   NEXT J%
2340   IF J% MOD 6 <> 0 THEN PRINT
2350   PRINT
2360 NEXT I%
2370 RETURN
2380 '
2390 '========================    ISTIFFPF.BAS    ============================
```

Sample Run

The following sample run shows program ISTIFFPF.BAS being used to evaluate the initial stiffness matrices for the structure analysed in example 7.2.

```
=================================================================

PROGRAM ISTIFFPF.BAS                 Copyright (c) J.Balfour 1991

     Program for the calculation of element stiffness matrices
       in element and global axes for a plane truss element

                 For further information contact
 James A.D.Balfour, Heriot-Watt University, Riccarton, Edinburgh
            Tel 031-449-5111, Fax 031-451-3170

=================================================================

Maximum number of nodes =  20


ELEMENT  1

Lower node number (0 to stop) = 1
Higher node number            = 2


+ + + + + + + + + + +
+   ELEMENT  1  2   +
+ + + + + + + + + + +

Area (A)                      = 9000
Second moment of area (I)     = .25E9
Length (L)                    = 3000
Elastic constant (E)          = 200
Angle with the global x-axis  = 90
```

```
ELEMENT  2

Lower node number (0 to stop) = 2
Higher node number            = 3


+ + + + + + + + + + + +
+   ELEMENT  2  3    +
+ + + + + + + + + + + +

Area (A)                      = 9000
Second moment of area (I)     = .25E9
Length (L)                    = 7071
Elastic constant (E)          = 200
Angle with the global x-axis  = 8.13

ELEMENT  3

Lower node number (0 to stop) = 0


* * * * * * * * * * * * * *
*   INITIAL STRUCTURE     *
*   STIFFNESS MATRIX      *
* * * * * * * * * * * * * *

0.2222E+02   -.9413E-04   -.3333E+05   -.2222E+02   0.9413E-04   -.3333E+05
0.0000E+00   0.0000E+00   0.0000E+00

-.9413E-04   0.6000E+03   -.5431E-02   0.9413E-04   -.6000E+03   -.5431E-02
0.0000E+00   0.0000E+00   0.0000E+00

-.3333E+05   -.5431E-02   0.6667E+08   0.3333E+05   0.5431E-02   0.3333E+08
0.0000E+00   0.0000E+00   0.0000E+00

-.2222E+02   0.9413E-04   0.3333E+05   0.2717E+03   0.3540E+02   0.3248E+05
-.2495E+03   -.3540E+02   -.8485E+03

0.9413E-04   -.6000E+03   0.5431E-02   0.3540E+02   0.6068E+03   0.5940E+04
-.3540E+02   -.6754E+01   0.5940E+04

-.3333E+05   -.5431E-02   0.3333E+08   0.3248E+05   0.5940E+04   0.9495E+08
0.8485E+03   -.5940E+04   0.1414E+08

0.0000E+00   0.0000E+00   0.0000E+00   -.2495E+03   -.3540E+02   0.8485E+03
0.2495E+03   0.3540E+02   0.8485E+03

0.0000E+00   0.0000E+00   0.0000E+00   -.3540E+02   -.6754E+01   -.5940E+04
0.3540E+02   0.6754E+01   -.5940E+04

0.0000E+00   0.0000E+00   0.0000E+00   -.8485E+03   0.5940E+04   0.1414E+08
0.8485E+03   -.5940E+04   0.2828E+08
```

7.7 Application of the Boundary Conditions

In section 5.7 the initial structure stiffness equation was shown to be a mixed set of simultaneous equations containing known displacements (boundary conditions) and unknown displacements in the displacement vector, and known forces (applied loads) and unknown forces (reactions) in the force vector. Section 5.7 also showed that the final structure stiffness equation can be obtained from the initial structure stiffness equation by deleting the rows and columns associated with the boundary conditions. If freedom "*m*" is restrained (i.e. has zero displacement) then row "*m*" and column "*m*" are deleted from the initial stiffness equation.

Example 7.3

Find the displacements for the frame shown in the following diagram given that $I = 0.25 \times 10^9\ \text{mm}^4$, $A = 9000\ \text{mm}^2$, $E = 200\ \text{kN/mm}^2$ for both elements.

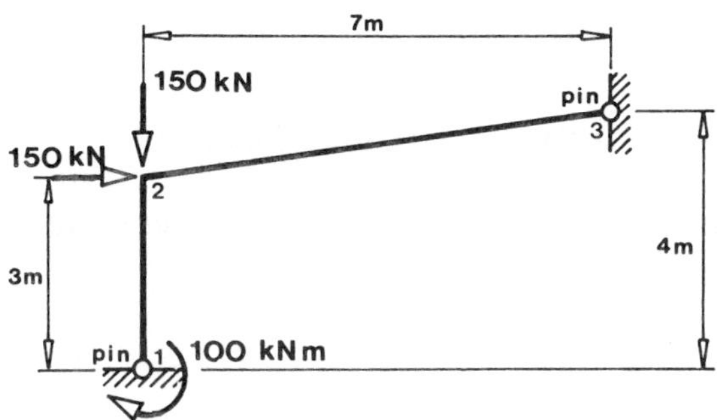

Nodes 1 and 3 are restrained against translation in the global "x" and "y" directions, therefore $\Delta_{1x} = \Delta_{1y} = \Delta_{3x} = \Delta_{3y} = 0$. Hence the final stiffness equation is obtained by deleting rows and columns centred on diagonal elements 1, 2, 7, and 8 of the initial stiffness equation. The frame in this example is, in fact, identical to the frame analysed in example 7.2, making unnecessary the evaluation of the initial stiffness matrix, and allowing the initial structure stiffness equation to be written directly as follows

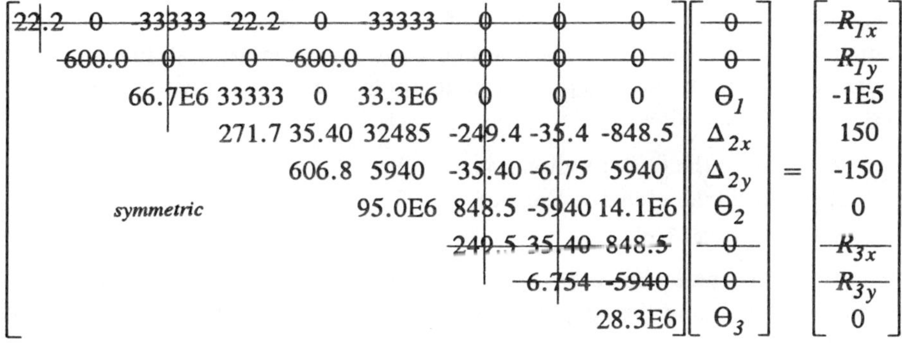

$$\begin{bmatrix} 22.2 & 0 & 33333 & -22.2 & 0 & 33333 & 0 & 0 & 0 \\ & 600.0 & 0 & 0 & -600.0 & 0 & 0 & 0 & 0 \\ & & 66.7E6 & 33333 & 0 & 33.3E6 & 0 & 0 & 0 \\ & & & 271.7 & 35.40 & 32485 & -249.4 & -35.4 & -848.5 \\ & & & & 606.8 & 5940 & -35.40 & -6.75 & 5940 \\ & \textit{symmetric} & & & & 95.0E6 & 848.5 & -5940 & 14.1E6 \\ & & & & & & 249.5 & 35.40 & 848.5 \\ & & & & & & & 6.754 & -5940 \\ & & & & & & & & 28.3E6 \end{bmatrix} \begin{bmatrix} 0 \\ 0 \\ \theta_1 \\ \Delta_{2x} \\ \Delta_{2y} \\ \theta_2 \\ 0 \\ 0 \\ \theta_3 \end{bmatrix} = \begin{bmatrix} R_{1x} \\ R_{1y} \\ -1E5 \\ 150 \\ -150 \\ 0 \\ R_{3x} \\ R_{3y} \\ 0 \end{bmatrix}$$

Note that this matrix is symmetric and, for simplicity, only the upper triangle has been shown.

The boundary conditions are applied by striking out rows and columns as shown. Solution of the remaining equations (using any of the equation solving programs in Chapter 4) yields

$$\begin{bmatrix} \theta_1 \\ \Delta_{2x} \\ \Delta_{2y} \\ \theta_2 \\ \theta_3 \end{bmatrix} = \begin{bmatrix} -2.16 \times 10^{-3} \text{ rads} \\ 0.792 \text{ mm} \\ -0.297 \text{ mm} \\ 0.532 \times 10^{-3} \text{ rads} \\ -0.179 \times 10^{-3} \text{ rads} \end{bmatrix}$$

7.8 The Final Structure Stiffness Matrix

In practice it is uneconomical to assemble the initial structure stiffness matrix and apply the boundary conditions afterwards.

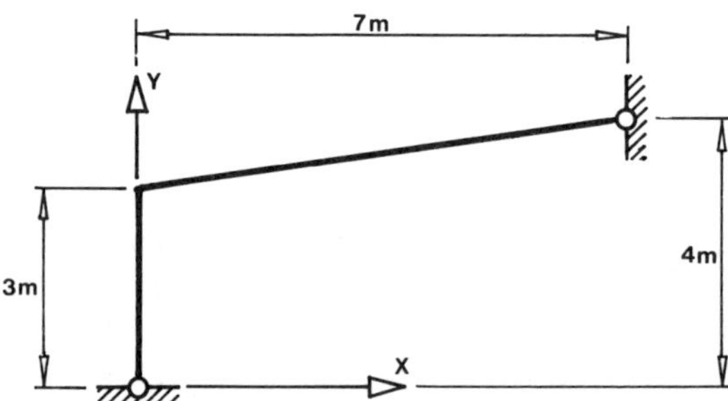

Figure 7.6 The frame analysed in example 7.3

When dealing with plane and space trusses the final stiffness matrix was assembled directly through the use of a freedom vector and element code numbers. In the case of a plane frame each node can be restrained against translational or rotational displacement and in this text the directions of restraint are numbered as follows

Restraint in the global x-direction	-	1
Restraint in the global y-direction	-	2
Rotational restraint	-	3

Consider the frame shown in fig 7.6 (note that this frame was analysed in example 7.3). Using the adopted restraint numbering convention the boundary conditions would be input as follows.

NODE	DIRECTION(S)	
1	12	(node 1 restrained in the "*x*" and "*y*" directions)
3	12	(node 3 restrained in the "*x*" and "*y*" directions)

Note that the restraint directions have been described using composite numbers. If a node is restrained in all directions then the composite restraint number would be 123 or 213 or 231, etc. From the boundary conditions a vector of the freedom numbers can be produced. The procedure used is exactly the same as was employed in the analysis of plane and space trusses; the freedom vector for the frame in example 7.3 being found as shown below.

$$\textit{Zero the freedom vector} \begin{bmatrix} 0 \\ 0 \\ 0 \\ 0 \\ 0 \\ 0 \\ 0 \\ 0 \\ 0 \end{bmatrix} \quad \textit{Apply the boundary conditions} \begin{bmatrix} 1 \\ 1 \\ 0 \\ 0 \\ 0 \\ 0 \\ 1 \\ 1 \\ 0 \end{bmatrix} \quad \textit{Number the freedoms} \begin{bmatrix} 0 \\ 0 \\ 1 \\ 2 \\ 3 \\ 4 \\ 0 \\ 0 \\ 5 \end{bmatrix}$$

To add the stiffness of element *"i,j"* to the final structure stiffness matrix the element code number is constructed from elements *(3i-2)*, *(3i-1)*, *(3i)*, *(3j-2)*, *(3j-1)* and *(3j)* of the freedom vector. The code number, when applied to the global element stiffness matrix, gives the destinations of the element stiffness coefficients in the final structure stiffness matrix.

For example, the code number of element 2,3 in example 7.3 is [2 3 4 0 0 5] which, when applied to the global element stiffness matrix as shown, indicates that element K_{11} should be added to location (2,2) in the final structure stiffness matrix, element K_{12} should be added to location (2,3) in the final structure stiffness matrix, ... element K_{66} should be added to location (5,5) in the final structure stiffness matrix.

$$\begin{array}{c} \\ 2 \\ 3 \\ 4 \\ 0 \\ 4 \\ 0 \end{array} \begin{array}{c} \begin{array}{cccccc} 2 & 3 & 4 & 0 & 0 & 5 \end{array} \\ \begin{bmatrix} K_{11} & K_{12} & K_{13} & K_{14} & K_{15} & K_{16} \\ K_{21} & K_{22} & K_{23} & K_{24} & K_{25} & K_{26} \\ K_{31} & K_{32} & K_{33} & K_{34} & K_{35} & K_{36} \\ K_{41} & K_{42} & K_{43} & K_{44} & K_{45} & K_{46} \\ K_{51} & K_{52} & K_{53} & K_{54} & K_{55} & K_{56} \\ K_{61} & K_{62} & K_{63} & K_{64} & K_{65} & K_{66} \end{bmatrix} \end{array}$$

Note that if a zero element appears in either the "horizontal" or the "vertical" code number then that stiffness term is ignored. The final structure stiffness matrix is complete when the stiffness contributions of all elements have been added using this technique.

Example 7.4

Assemble the final stiffness matrix for the frame shown below using the structure freedom vector and element code numbers. Note that the frame is identical to that analysed in example 7.2, and the freedom vector has already been found to be { 0 0 1 2 3 4 0 0 5 }.

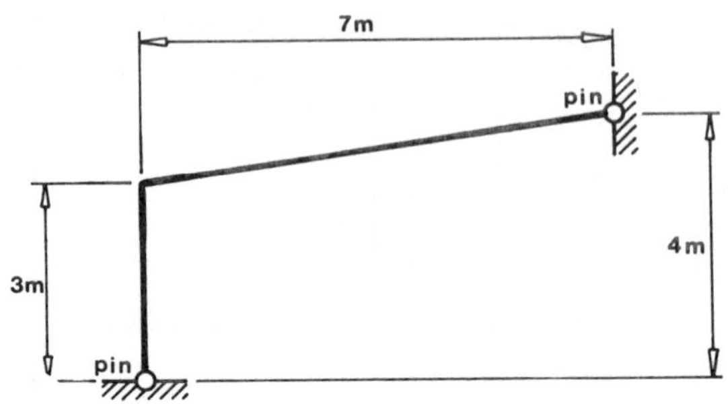

Element 1,2 - code number [0 0 1 2 3 4]

Apply the code number to the global element stiffness matrix (from example 7.2).

	0	0	1	2	3	4
0	22.2	0	-33333	-22.2	0	-33333
0	0	600.0	0	0	-600.0	0
1	-33333	0	66.7E6	33333	0	33.3E6
2	-22.2	0	33333	22.2	0	33333
3	0	-600.0	0	0	600.0	0
4	-33333	0	33.3E6	33333	0	66.6E6

Adding the stiffness terms to the structure stiffness matrix gives

$$\begin{bmatrix} 66.7E6 & 33333 & 0 & 33.3E6 & 0 \\ 33333 & 22.2 & 0 & 33333 & 0 \\ 0 & 0 & 600.0 & 0 & 0 \\ 33.3E6 & 33333 & 0 & 66.7E6 & 0 \\ 0 & 0 & 0 & 0 & 0 \end{bmatrix}$$

Element 2,3 - code number [2 3 4 0 0 5]

Apply the code number to the element stiffness matrix.

	2	3	4	0	0	5
2	249.5	35.40	-848.5	-249.5	-35.40	-848.5
3	35.40	6.754	5940	-35.40	-6.754	5940
4	-848.5	5940	28.3E6	848.5	-5940	14.1E6
0	-249.5	-35.40	848.5	249.5	35.40	848.5
0	-35.40	-6.754	-5940	35.40	6.754	-5940
5	-848.5	5940	14.1E6	848.5	-5940	28.3E6

Adding the stiffness terms completes the final structure stiffness matrix.

$$\begin{bmatrix} 66.7E6 & 33333 & 0 & 33.3E6 & 0 \\ 33333 & 271.7 & 35.40 & 32484 & -848.5 \\ 0 & 35.40 & 606.8 & 5940 & 5940 \\ 33.3E6 & 32484 & 5940 & 95.0E6 & 14.1E6 \\ 0 & -848.5 & 5940 & 14.1E6 & 28.3E6 \end{bmatrix}$$

Program FSTIFFPF.BAS

This program shows how the freedom vector is set up and then used to assemble the final structure stiffness matrix from the global element stiffness matrices for a plane frame. Note that advantage has been taken of the symmetry and the bandwidth of the structure stiffness matrix to save space by assembling only its upper semi-band (i.e. elements that are on or above the diagonal and within the bandwidth).

Input

The program requests details of the boundary conditions, and then the data for each element in turn.

Comments on the Algorithm

Once the boundary conditions have been established the freedom vector is developed. The program then loops for each element in turn; first generating the global element stiffness matrix and then the element code number which is used to add the element stiffness to the structure stiffness matrix. To conserve space only the upper semi-band of the stiffness matrix is stored.

Output

The upper semi-band of the final structure stiffness matrix is output. The reader should observe that the matrix is in the correct form to be used directly by the equation solving programs GAUSS.BAS and UDU.BAS described in Chapter 4.

Listing

```
1000 '=======================     FSTIFFPF.BAS     ===========================
1010 '
1020 OPTION BASE 1
1030 '
1040 ' ... Problem size set by the following dimension statements
1050 '
1060 MAXNNODE% = 20 : MAXBAND% = 40
1070 DIM ESTIFF(6,6), K(3*MAXNNODE%,MAXBAND%), EFREE%(6), FREE%(3*MAXNNODE%)
1080 CLS
1090 PRINT
```

```
1100 PRINT "=================================================================="
1110 PRINT
1120 PRINT "PROGRAM FSTIFFPF.BAS                   Copyright (c) J.Balfour 1991"
1130 PRINT
1140 PRINT "         Program for the assembly of the final stiffness"
1150 PRINT "        matrix for rigidly jointed plane frame structures"
1160 PRINT
1170 PRINT "                For further information contact"
1180 PRINT " James A.D.Balfour, Heriot-Watt University, Riccarton, Edinburgh"
1190 PRINT "              Tel 031-449-5111, Fax 031-451-3170"
1200 PRINT
1210 PRINT "=================================================================="
1220 PRINT
1230 PRINT "Maximum number of nodes = "; MAXNNODE%
1240 PRINT "Maximum semi-bandwidth  = "; MAXBAND%
1250 INPUT "Actual number of nodes  =  ", NNODE%
1260 '
1270 ' ... Input the boundary conditions and generate the freedom vector
1280 '
1290 GOSUB 1410
1300 '
1310 ' ... Input element data and add the element stiffness
1320 '
1330 GOSUB 1860
1340 '
1350 ' ... Output the final structure stiffness matrix
1360 '
1370 GOSUB 2740
1380 '
1390 END
1400 '
1410 '**********************   GENERATE THE FREEDOM VECTOR    ****************
1420 '
1430 ' ... Zero the freedom vector
1440 '
1450 FOR I% = 1 TO 3*NNODE%
1460   FREE%(I%) = 0
1470 NEXT I%
1480 PRINT : PRINT
1490 PRINT "+ + + + + + + + + + + + + +"
1500 PRINT "+   BOUNDARY CONDITIONS   +"
1510 PRINT "+ + + + + + + + + + + + + +"
1520 PRINT
1530 PRINT "INPUT BOUNDARY CONDITIONS"
1540 PRINT "Restraint in the X-direction  =  1"
1550 PRINT "Restraint in the Y-direction  =  2"
1560 PRINT "Rotational restraint          =  3"
1570 PRINT "Enter restraints as a composite number"
1580 PRINT "(e.g. if restrained in the X- and Y-directions enter 12)"
1590 IN% = 1
1600 WHILE IN% <> 0
1610   PRINT
1620   INPUT "Node number (0 to finish)   =  ", IN%
1630   IF IN%=0 THEN GOTO 1720
1640   INPUT "Direction(s)                =  ", DIRN%
1650   '
1660   ' ... Identify the freedom to be restrained from the RH digit
1670   '
1680   J% = 3*IN% - 3 + (DIRN% MOD 10)
1690   FREE%(J%) = 1
1700   DIRN% = INT(DIRN%/10)
1710   IF DIRN% > 0 THEN GOTO 1680
1720 WEND
1730 '
1740 ' ... Number the freedoms (restraints set to zero)
1750 '
1760 NDOF% = 0
1770 FOR I% = 1 TO 3*NNODE%
1780   IF FREE%(I%) = 1 THEN GOTO 1820
1790   NDOF% = NDOF% + 1
1800   FREE%(I%) = NDOF%
1810   GOTO 1830
1820   FREE%(I%) = 0
1830 NEXT I%
1840 RETURN
1850 '
1860 '************   ASSEMBLE THE STRUCTURE STIFFNESS MATRIX    **************
```

```
1870 '
1880 ' ... Zero the structure stiffness matrix
1890 '
1900 FOR I% = 1 TO NDOF%
1910   FOR J% = 1 TO NDOF%
1920     K(I%,J%) = 0
1930   NEXT J%
1940 NEXT I%
1950 '
1960 ' ...  Loop for all elements
1970 '
1980 IN% = 1 : K% = 1 : NNODE% = 0 : BAND% = 0
1990 WHILE IN% <> 0
2000   PRINT : PRINT
2010   PRINT "ELEMENT "; K% : PRINT
2020   INPUT "Lower  node number (zero to stop) =  ", IN%
2030   IF IN% = 0 THEN GOTO 2710
2040   INPUT "Higher node number                =  ", JN%
2050   PRINT
2060   PRINT        "+ + + + + + + + + + + +"
2070   PRINT USING "+    ELEMENT ## ##    +"; IN%, JN%
2080   PRINT        "+ + + + + + + + + + + +"
2090   PRINT
2100   INPUT "Area                            =  ", A
2110   INPUT "Second moment of area           =  ", A2
2120   INPUT "Length                          =  ", L
2130   INPUT "Elastic constant                =  ", E
2140   INPUT "Angle with the global X-axis    =  ", ALPHA
2150   '
2160   ' ... Convert the angle to radians
2170   '
2180   AR = 4 * ATN(1) * ALPHA / 180
2190   C  = COS(AR)
2200   S  = SIN(AR)
2210   EAL = E * A / L
2220   EIL = E * A2 / L
2230   '
2240   ' ... Generate the upper triangle of the element stiffness matrix
2250   '
2260   ESTIFF(1,1) =   EAL*C*C + 12*EIL*S*S/L^2
2270   ESTIFF(1,2) =   EAL*S*C - 12*EIL*S*C/L^2
2280   ESTIFF(1,3) = - 6*EIL*S/L
2290   ESTIFF(1,4) = - EAL*C*C - 12*EIL*S*S/L^2
2300   ESTIFF(1,5) = - EAL*C*S + 12*EIL*S*C/L^2
2310   ESTIFF(1,6) = - 6*EIL*S/L
2320   ESTIFF(2,2) =   EAL*S*S + 12*EIL*C*C/L^2
2330   ESTIFF(2,3) =   6*EIL*C/L
2340   ESTIFF(2,4) = - EAL*S*C + 12*EIL*C*S/L^2
2350   ESTIFF(2,5) = - EAL*S*S - 12*EIL*C*C/L^2
2360   ESTIFF(2,6) =   6*EIL*C/L
2370   ESTIFF(3,3) =   4*EIL
2380   ESTIFF(3,4) =   6*EIL*S/L
2390   ESTIFF(3,5) = - 6*EIL*C/L
2400   ESTIFF(3,6) =   2*EIL
2410   ESTIFF(4,4) =   EAL*C*C + 12*EIL*S*S/L^2
2420   ESTIFF(4,5) =   EAL*S*C - 12*EIL*S*C/L^2
2430   ESTIFF(4,6) =   6*EIL*S/L
2440   ESTIFF(5,5) =   EAL*S*S + 12*EIL*C*C/L^2
2450   ESTIFF(5,6) = - 6*EIL*C/L
2460   ESTIFF(6,6) =   4*EIL
2470   '
2480   ' ... Set up the element code number
2490   '
2500   EFREE%(1) = FREE%(3*IN%-2)
2510   EFREE%(2) = FREE%(3*IN%-1)
2520   EFREE%(3) = FREE%(3*IN%)
2530   EFREE%(4) = FREE%(3*JN%-2)
2540   EFREE%(5) = FREE%(3*JN%-1)
2550   EFREE%(6) = FREE%(3*JN%)
2560   '
2570   ' ... Add the element stiffness terms
2580   '
2590   FOR I% = 1 TO 6
2600     IF EFREE%(I%) = 0 THEN GOTO 2680
2610     FOR J% = I% TO 6
2620       IF EFREE%(J%) = 0 THEN GOTO 2670
2630       L% = EFREE%(I%)
```

```
2640        M% = EFREE%(J%) - L% + 1
2650        IF M% > BAND% THEN BAND% = M%
2660        K(L%,M%) = K(L%,M%) + ESTIFF(I%,J%)
2670      NEXT J%
2680    NEXT I%
2690    IF JN% > NNODE% THEN NNODE% = JN%
2700    K% = K% + 1
2710 WEND
2720 RETURN
2730 '
2740 '***************    OUTPUT THE FINAL STIFFNESS MATRIX    ****************
2750 '
2760 PRINT : PRINT
2770 PRINT "* * * * * * * * * * * * * * * * * **"
2780 PRINT "**   UPPER SEMI-BAND OF THE      *"
2790 PRINT "**   FINAL STIFFNESS MATRIX      *"
2800 PRINT "* * * * * * * * * * * * * * * * * **"
2810 PRINT
2820 FOR I% = 1 TO NDOF%
2830   FOR J% = 1 TO BAND%
2840     PRINT USING "#.####^^^^     "; K(I%,J%);
2850     IF (J% MOD 6) = 0 THEN PRINT
2860   NEXT J%
2870   IF (BAND% MOD 6) <> 0 THEN PRINT
2880   PRINT
2890 NEXT I%
2900 RETURN
2910 '
2920 '=======================    FSTIFFPF.BAS    ===========================
```

Sample Run

The following sample run shows program FSTIFFPF.BAS being used to evaluate the final stiffness matrix for the structure analysed in example 7.4.

```
=================================================================

PROGRAM FSTIFFPF.BAS                   Copyright (c) J.Balfour 1991

        Program for the assembly of the final stiffness
       matrix for rigidly jointed plane frame structures

               For further information contact
 James A.D.Balfour, Heriot-Watt University, Riccarton, Edinburgh
          Tel 031-449-5111, Fax 031-451-3170

=================================================================


Maximum number of nodes =  20
Maximum semi-bandwidth  =  40
Actual number of nodes  =  3


+ + + + + + + + + + + + + + +
+   BOUNDARY CONDITIONS    +
+ + + + + + + + + + + + + + +

INPUT BOUNDARY CONDITIONS
Restraint in the X-direction  =  1
Restraint in the Y-direction  =  2
Rotational restraint          =  3
Enter restraints as a composite number
(e.g. if restrained in the X- and Y-directions enter 12)

Node number (0 to finish)    =  1
Direction(s)                 =  12

Node number (0 to finish)    =  3
Direction(s)                 =  12

Node number (0 to finish)    =  0
```

```
ELEMENT  1

Lower  node number (zero to stop) =  1
Higher node number                =  2

+ + + + + + + + + + +
+   ELEMENT  1  2   +
+ + + + + + + + + + +

Area                          =  9000
Second moment of area         =  .25E9
Length                        =  3000
Elastic constant              =  200
Angle with the global X-axis  =  90

ELEMENT  2

Lower  node number (zero to stop) =  2
Higher node number                =  3

+ + + + + + + + + + +
+   ELEMENT  2  3   +
+ + + + + + + + + + +

Area                          =  9000
Second moment of area         =  .25E9
Length                        =  7071
Elastic constant              =  200
Angle with the global X-axis  =  8.13

ELEMENT  3

Lower  node number (zero to stop) =  0

* * * * * * * * * * * * * * * *
*   UPPER SEMI-BAND OF THE     *
*   FINAL STIFFNESS MATRIX     *
* * * * * * * * * * * * * * * *

0.6667E+08    0.3333E+05    0.5431E-02    0.3333E+08

0.2717E+03    0.3540E+02    0.3248E+05    -.8485E+03

0.6068E+03    0.5940E+04    0.5940E+04    0.0000E+00

0.9495E+08    0.1414E+08    0.0000E+00    0.0000E+00

0.2828E+08    0.0000E+00    0.0000E+00    0.0000E+00
```

7.9 Element End Forces and Reactions

The element stiffness submatrices in the element axes system are used to calculate the element end forces from the element end displacements using the following equation (see section 7.2)

$$f_{ij} \;=\; k_{ii}^{j}\,\delta_{ij} \;+\; k_{ij}\,\delta_{ji}$$

Solution of the final stiffness equation yields the nodal displacements in terms of the global axes systems. Before the above equation can be used to find the element end forces the nodal displacements must first be transformed into the element axes system.

$$f_{ij} \;=\; k_{ii}^{j}\,T_{ij}\,\Delta_i \;+\; k_{ij}\,T_{ij}\,\Delta_j$$

hence

$$\begin{bmatrix} f_{ijx} \\ f_{ijy} \\ m_{ij} \end{bmatrix} = \begin{bmatrix} EA/L & 0 & 0 \\ 0 & 12EI/L^3 & 6EI/L^2 \\ 0 & 6EI/L^2 & 4EI/L \end{bmatrix} \begin{bmatrix} cos\alpha & sin\alpha & 0 \\ -sin\alpha & cos\alpha & 0 \\ 0 & 0 & 1 \end{bmatrix} \begin{bmatrix} \Delta_{ix} \\ \Delta_{iy} \\ \theta_i \end{bmatrix} +$$

$$\begin{bmatrix} -EA/L & 0 & 0 \\ 0 & -12EI/L^3 & 6EI/L^2 \\ 0 & -6EI/L^2 & 2EI/L \end{bmatrix} \begin{bmatrix} cos\alpha & sin\alpha & 0 \\ -sin\alpha & cos\alpha & 0 \\ 0 & 0 & 1 \end{bmatrix} \begin{bmatrix} \Delta_{jx} \\ \Delta_{jy} \\ \theta_j \end{bmatrix}$$

The forces at end "j" can be found in a similar manner by evaluation of the following matrix equation

$$f_{ji} = k_{ji}\, T_{ij}\, \Delta_i + k^i_{jj}\, T_{ij}\, \Delta_j$$

Example 7.5

Find the element end forces for element 2,3 of the structure shown in the following diagram.

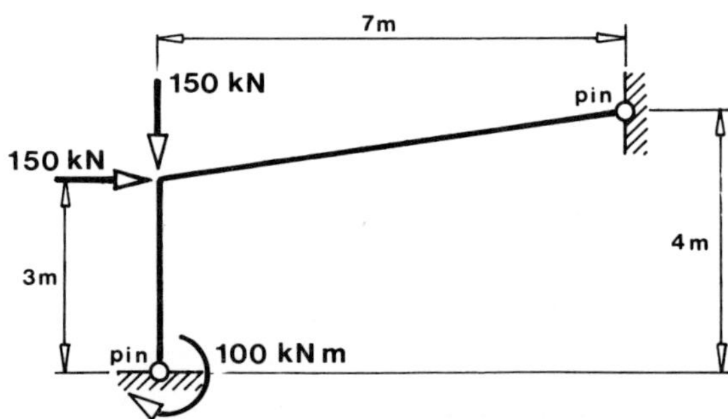

Note that the final stiffness matrix for this structure was found in example 7.4 and the final stiffness equation is

$$\begin{bmatrix} 66.7E6 & 33333 & 0 & 33.3E6 & 0 \\ 33333 & 271.7 & 35.40 & 32484 & -848.5 \\ 0 & 35.40 & 606.8 & 5940 & 5940 \\ 33.3E6 & 32484 & 5940 & 95.0E6 & 14.1E6 \\ 0 & -848.5 & 5940 & 14.1E6 & 28.3E6 \end{bmatrix} \begin{bmatrix} \theta_1 \\ \Delta_{2x} \\ \Delta_{2y} \\ \theta_2 \\ \theta_3 \end{bmatrix} = \begin{bmatrix} -100000 \\ 150 \\ -150 \\ 0 \\ 0 \end{bmatrix}$$

Solving the above equation using any of the equation solvers listed in Chapter 4 yields

$$\begin{bmatrix} \theta_1 \\ \Delta_{2x} \\ \Delta_{2y} \\ \theta_2 \\ \theta_3 \end{bmatrix} = \begin{bmatrix} -2.16 \text{ x } 10^{-3} \text{ rads} \\ 0.792 \text{ mm} \\ -0.297 \text{ mm} \\ 0.532 \text{ x } 10^{-3} \text{ rads} \\ -0.179 \text{ x } 10^{-3} \text{ rads} \end{bmatrix}$$

Element forces for element 2,3 are found using the following equation

$$f_{23} = k_{22}^{3} T_{23} \Delta_2 + k_{23} T_{23} \Delta_3$$

$$\begin{bmatrix} f_{23x} \\ f_{23y} \\ m_{23} \end{bmatrix} = \begin{bmatrix} 255 & 0 & 0 \\ 0 & 1.7 & 6000 \\ 0 & 6000 & 28.3E6 \end{bmatrix} \begin{bmatrix} 0.990 & 0.141 & 0 \\ -0.141 & 0.990 & 0 \\ 0 & 0 & 1 \end{bmatrix} \begin{bmatrix} 0.792 \\ -0.297 \\ 0.532E-3 \end{bmatrix} +$$

$$\begin{bmatrix} -255 & 0 & 0 \\ 0 & -1.7 & 6000 \\ 0 & -6000 & 14.1E6 \end{bmatrix} \begin{bmatrix} 0.990 & 0.141 & 0 \\ -0.141 & 0.990 & 0 \\ 0 & 0 & 1 \end{bmatrix} \begin{bmatrix} 0 \\ 0 \\ -0.179E-3 \end{bmatrix}$$

$$= \begin{bmatrix} 189 \\ 2.50 \\ 12621 \end{bmatrix} + \begin{bmatrix} 0 \\ -1.07 \\ -2524 \end{bmatrix} = \begin{bmatrix} 189 \text{ kN} \\ 1.43 \text{ kN} \\ 10097 \text{ kN mm} \end{bmatrix}$$

Similarly

$$\begin{bmatrix} f_{32x} \\ f_{32y} \\ m_{32} \end{bmatrix} = \begin{bmatrix} -189 \\ -2.05 \\ 5067 \end{bmatrix} + \begin{bmatrix} 0 \\ 1.07 \\ -5066 \end{bmatrix} = \begin{bmatrix} -189 \text{ kN} \\ -1.43 \text{ kN} \\ 1.00 \text{ kN mm} \end{bmatrix}$$

Note that at end 3, where the moment should be zero, there is a small moment of 1.0 kN mm. This error arises from the limited number of significant figures used in the calculations.

It is computationally advantageous to expand the expressions for the element end forces at end "i", i.e.

$$f_{ijx} = \frac{EA}{L} [(\Delta_{ix} - \Delta_{jx}) \cos\alpha + (\Delta_{iy} - \Delta_{jy}) \sin\alpha] \tag{7.13}$$

$$f_{ijy} = \frac{12EI}{L^3}[(-\Delta_{ix} + \Delta_{jx})\, sin\alpha + (\Delta_{iy} - \Delta_{jy})\, cos\alpha] + \frac{6EI}{L^2}[(\Theta_i + \Theta_j)] \tag{7.15}$$

$$m_{ij} = \frac{6EI}{L^2}[(-\Delta_{ix} + \Delta_{jx})\, sin\alpha + (\Delta_{iy} - \Delta_{jy})\, cos\alpha] + \frac{2EI}{L}[(2\Theta_i + \Theta_j)] \tag{7.16}$$

By considering the equilibrium of the element it is a simple matter to show that the forces at end "*j*" of the element are as follows

$$f_{jix} = -f_{ijx} \tag{7.17}$$

$$f_{jiy} = -f_{ijy} \tag{7.18}$$

$$m_{ji} = -m_{ij} + L f_{ijy} \tag{7.19}$$

Reactions

If node "*i*" is restrained then the reactive forces (in the global axes system) can be found using equation (5.22) which, although developed for plane trusses, is generally applicable.

$$R_i = \sum_{j=a}^{n} T_{ij}^{-1} f_{ij} - P_i$$

i.e.

$$\begin{bmatrix} R_{ix} \\ R_{iy} \\ RM_i \end{bmatrix} = \sum_{j=a}^{n} \begin{bmatrix} cos\alpha & -sin\alpha & 0 \\ sin\alpha & cos\alpha & 0 \\ 0 & 0 & 1 \end{bmatrix} \begin{bmatrix} f_{ijx} \\ f_{ijy} \\ m_{ij} \end{bmatrix} + \begin{bmatrix} P_{ix} \\ P_{iy} \\ M_i \end{bmatrix}$$

where RM_i is the reactive moment at node "*i*" and the summation is for all elements framing into node "*i*".

Program EFORCEPF.BAS

This program calculates the axial force, shear force, and bending moment at the ends of a plane frame element from a knowledge of the end displacements.

Input

The program prompts for the element properties plus end displacements for each element.

Output

Axial force, shear force and bending moment at each end of the element are output.

Listing

```
1000 '=======================    EFORCEPF.BAS    ===========================
1010 '
1020 TRUE% = 1 : FALSE% = 0
1030 CLS
1040 PRINT
1050 PRINT "================================================================="
1060 PRINT
1070 PRINT "PROGRAM EFORCEPF.BAS                     Copyright (c) J.Balfour 1991"
1080 PRINT
1090 PRINT "         Program for the calculation of element end forces"
1100 PRINT "                    for a plane frame element"
1110 PRINT
1120 PRINT "               For further information contact"
1130 PRINT " James A.D.Balfour, Heriot-Watt University, Riccarton, Edinburgh"
1140 PRINT "              Tel 031-449-5111, Fax 031-451-3170"
1150 PRINT
1160 PRINT "================================================================="
1170 '
1180 ' ... Loop for all members
1190 '
1200 DONE% = FALSE% : K% = 1
1210 WHILE DONE% = FALSE%
1220   PRINT
1230   PRINT "ELEMENT "; K%
1240   PRINT
1250   INPUT "Lower node number (0 to stop) = ", IN%
1260   IF IN%=0 THEN DONE% = TRUE% : GOTO 1810
1270   INPUT "Higher node number            = ", JN%
1280   PRINT
1290   PRINT        "+ + + + + + + + + + + +"
1300   PRINT USING "+   ELEMENT ## ##    +"; IN%, JN%
1310   PRINT        "+ + + + + + + + + + + +"
1320   PRINT
1330   INPUT "Area (A)                      = ", A
1340   INPUT "Second moment of area (I)     = ", A2
1350   INPUT "Length (L)                    = ", L
1360   INPUT "Elastic constant (E)          = ", E
1370   INPUT "Angle with the global x-axis  = ", ALPHA
1380   PRINT
1390   PRINT "Enter the nodal displacements (global axes)"
1400   PRINT
1410   PRINT USING "X-displacement at node ##     =  "; IN%;
1420   INPUT "", XI
1430   PRINT USING "Y-displacement at node ##     =  "; IN%;
1440   INPUT "", YI
1450   PRINT USING "Rotation at node ##           =  "; IN%;
1460   INPUT "", RI
1470   PRINT USING "X-displacement at node ##     =  "; JN%;
1480   INPUT "", XJ
1490   PRINT USING "Y-displacement at node ##     =  "; JN%;
1500   INPUT "", YJ
1510   PRINT USING "Rotation at node ##           =  "; JN%;
1520   INPUT "", RJ
1530   '
1540   ' ... Convert the angle to radians
1550   '
1560   AR = 4 * ATN(1) * ALPHA / 180
1570   C  = COS(AR)
1580   S  = SIN(AR)
1590   '
1600   ' ... Calculate and output the element end forces
1610   '
1620   FXI =     A*E* ( (XI-XJ)*C + (YI-YJ)*S )/L
1630   FYI =  12*E*A2*( (XJ-XI)*S + (YI-YJ)*C )/L^3 + 6*E*A2*(RI+RJ)/L^2
1640   FRI =          (-6*XI + 6*XJ)*S + (6*YI - 6*YJ)*C
```

```
1650    FRI =     E*A2*(FRI/L^2 + (4*RI + 2*RJ)/L)
1660    FXJ = - FXI
1670    FYJ = - FYI
1680    FRJ = - FRI + L*FYI
1690    PRINT : PRINT
1700    PRINT "* * * * * * * * * * * * * *"
1710    PRINT "*   ELEMENT END FORCES   *"
1720    PRINT "* * * * * * * * * * * * * *"
1730    PRINT
1740    PRINT "                    AXIAL        SHEAR       BENDING"
1750    PRINT "                    FORCE        FORCE        MOMENT"
1760    PRINT USING "NODE ##      "; IN%;
1770    PRINT USING "#.####^^^^   #.####^^^^    #.####^^^^"; FXI, FYI, FRI
1780    PRINT USING "NODE ##      "; JN%;
1790    PRINT USING "#.####^^^^   #.####^^^^    #.####^^^^"; FXJ, FYJ, FRJ
1800    K% = K% + 1
1810 WEND
1820 END
1830 '
1840 '========================    EFORCEPF.BAS    ============================
```

Sample Run

In the following sample run the program EFORCEPF.BAS is used to evaluate the element end forces for the structure analysed in example 7.5.

```
==================================================================

PROGRAM EFORCEPF.BAS                    Copyright (c) J.Balfour 1991

         Program for the calculation of element end forces
                    for a plane frame element

               For further information contact
 James A.D.Balfour, Heriot-Watt University, Riccarton, Edinburgh
            Tel 031-449-5111, Fax 031-451-3170

==================================================================

ELEMENT  1

Lower node number (0 to stop) = 1
Higher node number            = 2

+ + + + + + + + + + + +
+   ELEMENT  1  2    +
+ + + + + + + + + + + +

Area (A)                      = 9000
Second moment of area (I)     = .25E9
Length (L)                    = 3000
Elastic constant (E)          = 200
Angle with the global x-axis  = 90

Enter the nodal displacements (global axes)

X-displacement at node  1     =  0
Y-displacement at node  1     =  0
Rotation at node  1           =  -2.16E-3
X-displacement at node  2     =  .792
Y-displacement at node  2     =  -.297
Rotation at node  2           =  .532E-3


* * * * * * * * * * * * * *
*   ELEMENT END FORCES   *
* * * * * * * * * * * * * *

                   AXIAL         SHEAR        BENDING
                   FORCE         FORCE         MOMENT
NODE  1      0.1782E+03    -.3667E+02    -.9987E+05
NODE  2      -.1782E+03    0.3667E+02    -.1013E+05
```

```
ELEMENT  2

Lower node number (0 to stop) = 2
Higher node number            = 3

+ + + + + + + + + + +
+   ELEMENT  2  3   +
+ + + + + + + + + + +

Area (A)                       = 9000
Second moment of area (I)      = .25E9
Length (L)                     = 7071
Elastic constant (E)           = 200
Angle with the global x-axis   = 8.13

Enter the nodal displacements (global axes)

X-displacement at node  2      =  .792
Y-displacement at node  2      =  -.297
Rotation at node  2            =  .532E-3
X-displacement at node  3      =  0
Y-displacement at node  3      =  0
Rotation at node  3            =  -.179E-3


* * * * * * * * * * * * *
*   ELEMENT END FORCES   *
* * * * * * * * * * * * *

                AXIAL         SHEAR        BENDING
                FORCE         FORCE         MOMENT
NODE  2    0.1889E+03    0.1429E+01    0.1008E+05
NODE  3    -.1889E+03    -.1429E+01    0.2459E+02

ELEMENT  3

Lower node number (0 to stop) = 0
```

Chapter 8

Plane Frames - Further Topics

8.1 Introduction

The previous chapter dealt with the analysis of plane frames having rigidly jointed prismatic elements, simple boundary conditions, and loads applied only at the nodes. In practice, however, plane frames are often subjected to more complex loading regimens involving element loads, temperature effects, settlement, and lack of fit. The analysis is likely to be futher complicated by inclined supports, elastic foundations, elements with pins, and non-prismatic elements.

This chapter shows how these effects can easily be incorporated into the stiffness method. The last section of this chapter presents a program for the automatic analysis of plane frames. This program caters for element loads, elements with pins and local axes; thus demonstrating the practical implementation of some of the theory presented in this chapter.

8.2 Element Loads, Temperature, Settlement and Lack of Fit

So far, loads have been applied only at nodes. In practice, however, framed structures often carry loads along the length of their elements. Such forces can be dealt with using the concepts of fixed end force, equivalent joint force, and superposition. The theory developed for element loads is easily extended to allow temperature effects, settlement, and lack of fit to be included in the analysis. To illustrate, consider the structure shown in fig 8.1 which carries a uniformly distributed load along the length of element 2,3. Although a specific problem is considered here, the theory is presented in general terms to allow the same strategy to be adopted for element loads, temperature effects, settlement, and lack of fit.

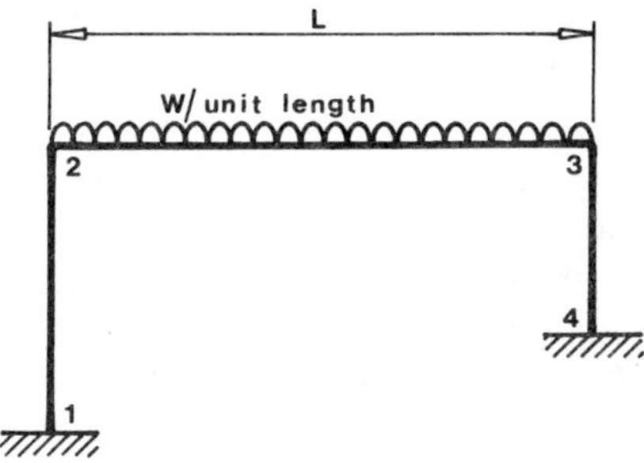

Figure 8.1 A portal frame carrying element loads

The procedure is as follows.

1. Lock all joints against displacement (rotation and translation).
2. Apply element loads, temperature change, settlement, and force fit ill-fitting elements.
3. Calculate the forces developed at the locked ends of the elements (the *fixed end forces*).
4. Transform the fixed end forces into the nodal axes system.
5. Add the negative of the transformed fixed end forces (the *equivalent joint forces*) to the nodal loads and then analyse the structure.
6. The final element forces and displacements are found by superimposing the fixed end force system on the results of the analysis described in 5.

In the case of the portal shown in fig 8.1 the fixed end forces are as shown in fig 8.2.

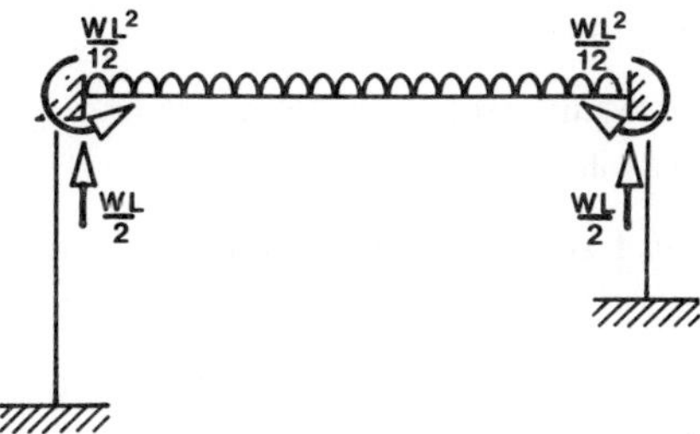

Figure 8.2 Fixed end forces

It is convenient, in the first instance, to express the fixed end forces for each element in terms of the element axes system. For an element "i,j" affected by

$$f_{ij}^{f} = \begin{bmatrix} f_{ijx}^{f} \\ f_{ijy}^{f} \\ m_{ij}^{f} \end{bmatrix} \quad \text{and} \quad f_{ji}^{f} = \begin{bmatrix} f_{jix}^{f} \\ f_{jiy}^{f} \\ m_{ji}^{f} \end{bmatrix}$$

The fixed end forces are an artificial system of applied loads that serve to hold the nodal displacements at zero. To return to the true structural behaviour it is necessary to superimpose the negative of these forces (i.e. the equivalent joint forces), as illustrated in fig 8.3.

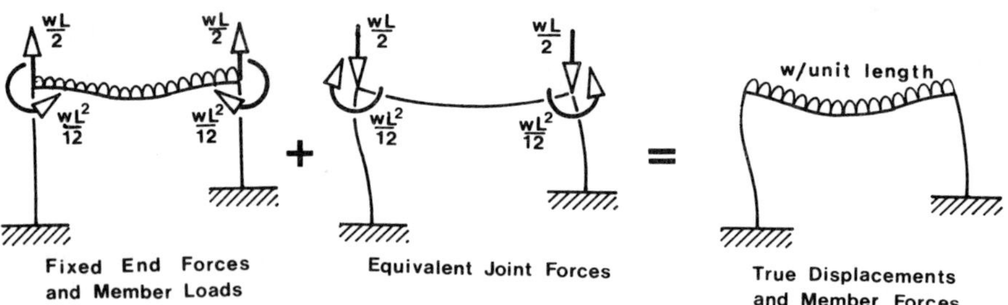

Figure 8.3 Superposition of the fixed forces and the equivalent joint forces

Before the equivalent joint forces can be included in the equations of nodal equilibrium they must be expressed in the global axes system. Hence the vectors of equivalent joint force are obtained by transforming the negative of the fixed end force vectors into the global axes system, i.e.

$$P_{ij}^{e} = -T_{ij}^{-1} f_{ij}^{f} \quad \text{and} \quad P_{ji}^{e} = -T_{ij}^{-1} f_{ji}^{f}$$

where

P_{ij}^{e} is the vector of equivalent joint forces at end "i" of element "i,j" (in the global axes system).

P_{ji}^{e} is the vector of equivalent joint forces at end "j" of element "i,j" (in the global axes system).

f_{ij}^{f} is the vector of fixed end forces at end "i" of element "i,j" (in the element axes system).

f_{ji}^{f} is the vector of fixed end forces at end "j" of element "i,j" (in the element axes system).

T_{ij}^{-1} is the inverse of the transformation matrix for element "i,j".

At each node the vectors of equivalent joint force are added to the vectors of nodal loads to produce the final nodal force vectors. Hence the final nodal force

At each node the vectors of equivalent joint force are added to the vectors of nodal loads to produce the final nodal force vectors. Hence the final nodal force vector at node "*i*" with elements "*i,a*", "*i,b*", "*i,c*" ...... "*i,n*", framing into it takes the form

$$P_i = P_i^n + \sum_{j=a}^{n} P_{ij}^e$$

where

P_i is the final force vector at node "*i*".

P_i^n is the vector of nodal loads at node "*i*".

$\sum P_{ij}^e$ is the sum of the vectors of equivalent joint force at end "*i*" of all of the elements framing into joint "*i*".

The loading vector for the complete structure P is now produced from the final nodal force vectors P_i, and the stiffness equation $K\,\Delta = P$ is solved to find the unknown nodal displacements.

The element end forces and the structural displacements are obtained by the superposition of the effects of the fixed end forces and the final nodal forces (which will include any nodal loads). Hence when calculating the element end forces for element "*i,j*", the fixed end forces (if any) must be added to the forces calculated from the nodal displacements.

$$f_{ij} = f_{ij}^n + f_{ij}^f$$

$$f_{ij} = k_{ii}^j\, T_{ij}\, \Delta_i + k_{ij}\, T_{ij}\, \Delta_j + f_{ij}^f \qquad (8.1)$$

The solution of the structure stiffness equation using the final nodal loads yields the true nodal displacements. Away from the nodes the displacements caused by the fixed end force system (see fig 8.3) are not necessarily zero, and must be included if displacements are to be calculated away from the nodes. If, for example, the vertical displacement is required half way along the horizontal element of the portal shown in fig 8.3, then $wL^4/384EI$ (the displacement due to the fixed end force system) must be added to the displacement due to the final force system.

Element Loads

In practice the elements of framed structures are often loaded along their length. If the element loads consist of point loads only, then by locating a node under each load the problem can be solved using the theory developed in Chapter 7. However, the introduction of additional nodes increases the problem size and hence the cost of the analysis. Furthermore, distributed loads cannot be satisfactorily catered for by the introduction of additional nodes, and the use of equivalent joint forces should be used whenever element loads are present.

Lack of Fit

Lack of fit is a problem associated with steel structures which usually arises from tolerance on the element sizes or inaccuracy in the position of column footings. During construction the structure is forced together causing a small change in the design geometry and an unintentional stressing of the elements. Usually the lack of fit is small; the force fit being achieved during the tightening of bolted connections, and the resulting change in geometry is usually neglected for design purposes.

Temperature Effects

Whereas lack of fit normally affects only steel structures, temperature changes can affect all types of structure. In general, temperature changes can cause extremely complex stress patterns. This chapter only considers the simple case where one or more of the elements of a structure changes temperature uniformly relative to the rest of the structure. In such cases the temperature effects can be considered as a special case of lack of fit. When the temperature of an element changes, the element tries to change its length. If this change of length is resisted by the other elements of the structure (or the restraints) then stress will be developed. To the structural analyst the problem is identical to that presented by lack of fit, only the "lack of fit" is due to temperature change, rather than structural tolerance. The "lack of fit" caused by temperature change is calculated using the coefficient of thermal expansion "c", which is, in effect, the strain induced per unit change in temperature. If an unrestrained element undergoes a change in temperature of Δt, then its length will change according to the following relationship.

$$\delta L = c \, L \, \Delta t$$

If, however, the ends of the element were fully restrained during a Δt change in temperature, then the axial force induced in the element would be

$$\begin{aligned} F &= \textit{cross sectional area x stress} = A \, \sigma \\ &= A \, \varepsilon \, E = A \, \delta L \, E / L = A \, c \, \Delta t \, E \qquad (8.2) \end{aligned}$$

Equation (8.2) is, of course, the fixed end force for a change of temperature Δt in the element.

Finally it should be noted that statically determinate structures are not stressed by lack of fit or temperature changes of this type. For example, if element 2,4 of the structure shown in fig 8.4 is short (or cools relative to the other elements) then the geometry of the structure changes, but there is no associated change of stress (if small displacement theory is assumed to be valid).

other elements) then the geometry of the structure changes, but there is no associated change of stress (if small displacement theory is assumed to be valid).

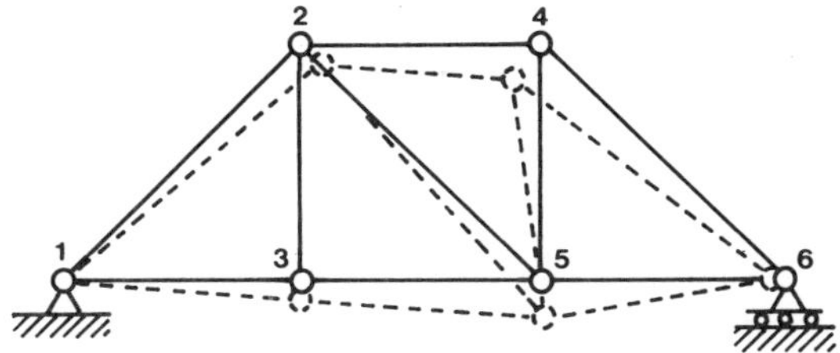

Figure 8.4 Lack of fit in a statically determinate structure

If the structure is made statically indeterminate by the introduction of a new element as shown in fig 8.5 then any lack of fit (or temperature change) of element 2,4 will stress the structure.

Note that the force in some of the elements of the structure shown in fig 8.5 is statically determinate. Consequently lack of fit of these elements (e.g. element 1,2) will not stress the structure. Conversely, any lack of fit of the elements that are statically indeterminate (e.g. element 2,4) will cause stressing, but only in the statically indeterminate elements of the structure.

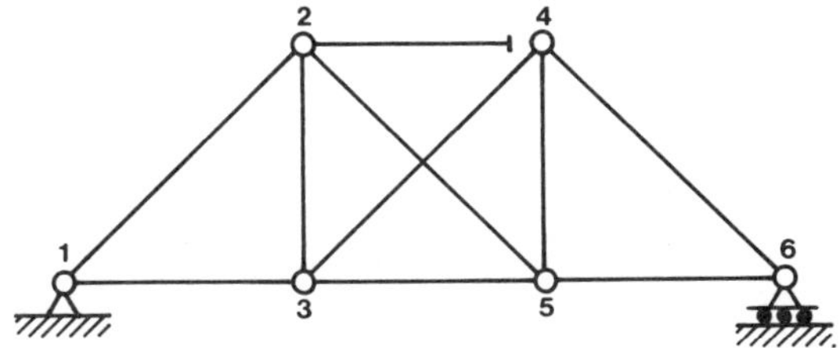

Figure 8.5 Lack of fit in a statically indeterminate structure

Settlement

Before a civil engineering structure is designed a site investigation is usually conducted to determine the nature of the ground upon which it will be built. Often the site investigation will reveal soil conditions that indicate that the structure will continue to settle long after the construction is finished. If all of the foundations settle by the same amount then the settlement amounts to rigid body motion of the structure, and the designer has little need for concern as no stress is induced. However, because soil conditions often vary over the plan area of a structure and the footings usually carry different loadings, differential settlement frequently occurs. Calculation of settlements is a complex and inexact branch of civil engineering that is not within the scope of this book, but once the amount of settlement has been estimated the problem can, like temperature change, be regarded as a special case of lack of fit. Here the lack of fit applies not to the elements, but to the structural boundaries.

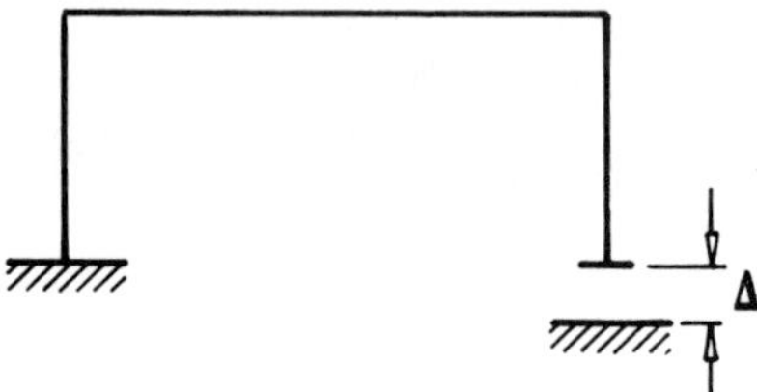

Figure 8.6 Settlement to a footing of a portal

Figure 8.6 shows a portal suffering from vertical settlement of the right-hand footing. The right-hand column has been severed to unstress the structure. In common with the other topics in this section the solution procedure involves finding the set of forces that will hold the unknown nodal displacements at zero (note that the displacements at boundaries that have settled are not unknown). Figure 8.7 shows the fixed end force system due to settlement of the right-hand footing of the portal frame shown in fig 8.6.

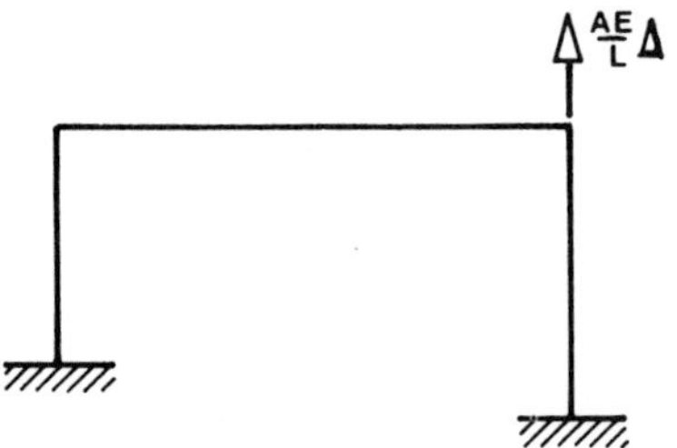

Figure 8.7 Fixed end forces

The structure is then analysed for the negative of the fixed end forces (the equivalent joint forces) as shown in fig 8.8.

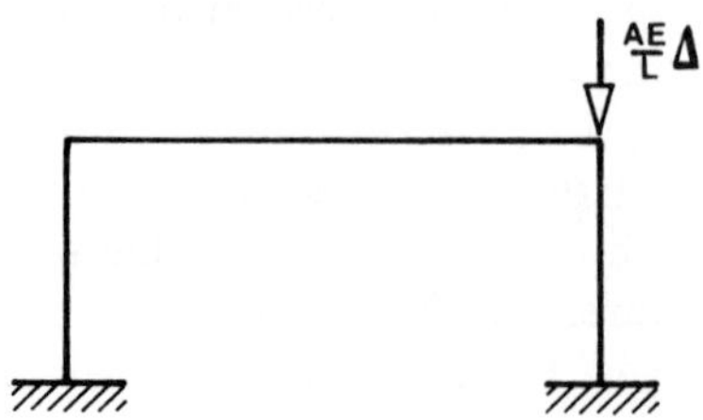

Figure 8.8 Equivalent joint forces

The final displacements and stresses are found by the superposition of the effects of the fixed end force system and the equivalent joint force system.

Fixed End Forces

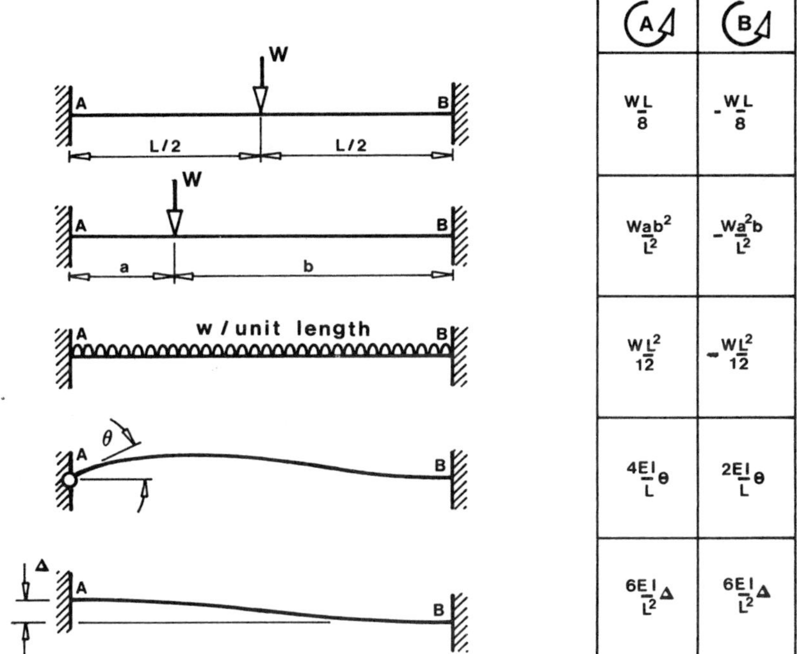

Figure 8.9 Fixed end moments for some common loading cases

A variety of methods can be used to calculate fixed end forces, and a selection of the more popular methods is presented in most structural analysis textbooks. For convenience fixed end moments for some common loading cases are given in fig 8.9.

Example 8.1

Find the element end forces for element 1,2 of the structure shown below if

(i) A uniformly distributed load of 10 kN/m is applied to element 2,3, and a horizontal force of 50 kN (to the right) is applied to node 2.

(ii) Element 1,2 undergoes a 20° centigrade rise in temperature. The structure is constructed from a material having a coefficient of thermal expansion of 1.1×10^{-5} per degree centigrade.

(iv) Element 2,3 is 20 mm too short.

(v) And finally, find the horizontal displacement at mid-height of element 1,2 under loading (iii).

Given that E = 200 kN/mm^2, A = 9000 mm^2 I = 0.25 x 10^9 mm^4 for both elements.

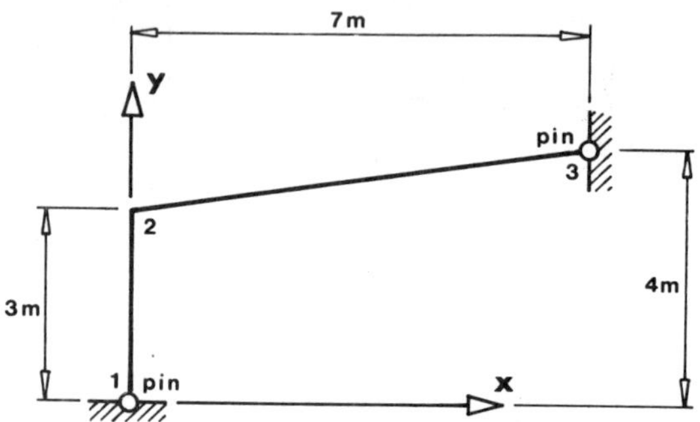

(i) A uniformly distributed load of 10 kN/m is applied to element 2,3, and a horizontal force of 50 kN is applied to node 2.

Fixed end forces

$$\frac{wL}{2} = 35.4 \text{ kN} \quad \text{and} \quad \frac{wL^2}{12} = 41.7 \text{ kN m}$$

hence

$$f^f_{23} = \begin{bmatrix} 0.0 \text{ kN} \\ 35.4 \text{ kN} \\ 41.7 \text{ kN m} \end{bmatrix} \qquad f^f_{32} = \begin{bmatrix} 0.0 \text{ kN} \\ 35.4 \text{ kN} \\ -41.7 \text{ kN m} \end{bmatrix}$$

Transform the fixed end force vectors into equivalent joint force vectors (α_{23} = 8.13°).

$$P^e_{23} = -T^{-1}_{23} f^f_{23}$$

$$P^e_{23} = -T^{-1}_{23} f^f_{23}$$

$$= -\begin{bmatrix} 0.990 & -0.141 & 0 \\ 0.141 & 0.990 & 0 \\ 0 & 0 & 1 \end{bmatrix}\begin{bmatrix} 0 \\ 35.4 \\ 41.7 \end{bmatrix} = \begin{bmatrix} 4.99 \text{ kN} \\ -35.0 \text{ kN} \\ -41.7 \text{ kN m} \end{bmatrix}$$

Similarly

$$P^e_{32} = -T^{-1}_{23} f^f_{32} = \begin{bmatrix} 4.99 \text{ kN} \\ -35.0 \text{ kN} \\ 41.7 \text{ kN m} \end{bmatrix}$$

Calculate the vectors of final nodal force.

$$P_1 = P^n_1 + P^e_{12}$$

$$= \begin{bmatrix} 0 \\ 0 \\ 0 \end{bmatrix} + \begin{bmatrix} 0 \\ 0 \\ 0 \end{bmatrix} = \begin{bmatrix} 0 \text{ kN} \\ 0 \text{ kN} \\ 0 \text{ kN m} \end{bmatrix}$$

$$P_2 = P^n_2 + P^e_{21} + P^e_{23}$$

$$= \begin{bmatrix} 50 \\ 0 \\ 0 \end{bmatrix} + \begin{bmatrix} 0 \\ 0 \\ 0 \end{bmatrix} + \begin{bmatrix} 4.99 \\ -35.0 \\ -41.7 \end{bmatrix} = \begin{bmatrix} 5.0 \text{ kN} \\ -35.0 \text{ kN} \\ -41.7 \text{ kN m} \end{bmatrix}$$

$$P_3 = P^n_3 + P^e_{32}$$

$$= \begin{bmatrix} 0 \\ 0 \\ 0 \end{bmatrix} + \begin{bmatrix} 4.99 \\ -35.0 \\ 41.7 \end{bmatrix} = \begin{bmatrix} 5.0 \text{ kN} \\ -35.0 \text{ kN} \\ 41.7 \text{ kN m} \end{bmatrix}$$

Discarding the forces associated with the boundary conditions (i.e. P_{1x}, P_{1y}, P_{3x} and P_{3y}) and employing the structure stiffness found previously (this structure is identical to the one analysed in example 7.3) produces the final stiffness equation. Note that the units are kN and mm.

$$\begin{bmatrix} 66.7E6 & 33333 & 0 & 33.3E6 & 0 \\ 33333 & 271.7 & 35.40 & 32484 & -848.5 \\ 0 & 35.40 & 606.8 & 5940 & 5940 \\ 33.3E6 & 32484 & 5940 & 95.0E6 & 14.1E6 \\ 0 & -848.5 & 5940 & 14.1E6 & 28.3E6 \end{bmatrix}\begin{bmatrix} \theta_1 \\ \Delta_{2x} \\ \Delta_{2y} \\ \theta_2 \\ \theta_3 \end{bmatrix} = \begin{bmatrix} 0.0 \\ 55.0 \\ -35.1 \\ -41700 \\ 41700 \end{bmatrix}$$

which yields

$$\begin{bmatrix} \theta_1 \\ \Delta_{2x} \\ \Delta_{2y} \\ \theta_2 \\ \theta_3 \end{bmatrix} = \begin{bmatrix} 3.22 \times 10^{-4} \text{ rads} \\ 0.292 \text{ mm} \\ -0.085 \text{ mm} \\ -9.39 \times 10^{-4} \text{ rads} \\ 1.97 \times 10^{-3} \text{ rads} \end{bmatrix}$$

The element end forces for element 1,2 are found by adding the forces associated with the above displacements to the fixed end forces.

$$\begin{aligned} f_{12} &= f_{12}^n + f_{12}^f \\ &= k_{11}^2 \, T_{12} \, \Delta_1 + k_{12} \, T_{12} \, \Delta_2 + f_{12}^f \end{aligned}$$

hence

$$\begin{bmatrix} f_{12x} \\ f_{12y} \\ m_{12} \end{bmatrix} = \begin{bmatrix} EA/L & 0 & 0 \\ 0 & 12EI/L^3 & 6EI/L^2 \\ 0 & 6EI/L^2 & 4EI/L \end{bmatrix} \begin{bmatrix} \cos\alpha & \sin\alpha & 0 \\ -\sin\alpha & \cos\alpha & 0 \\ 0 & 0 & 1 \end{bmatrix} \begin{bmatrix} \Delta_{1x} \\ \Delta_{1y} \\ \theta_1 \end{bmatrix} +$$

$$\begin{bmatrix} -EA/L & 0 & 0 \\ 0 & -12EI/L^3 & 6EI/L^2 \\ 0 & -6EI/L^2 & 2EI/L \end{bmatrix} \begin{bmatrix} \cos\alpha & \sin\alpha & 0 \\ -\sin\alpha & \cos\alpha & 0 \\ 0 & 0 & 1 \end{bmatrix} \begin{bmatrix} \Delta_{2x} \\ \Delta_{2y} \\ \theta_2 \end{bmatrix} +$$

$$\begin{bmatrix} f_{12x}^f \\ f_{12y}^f \\ m_{12}^f \end{bmatrix}$$

$$= \begin{bmatrix} 600 & 0 & 0 \\ 0 & 22.2 & 33333 \\ 0 & 33333 & 66.7E6 \end{bmatrix} \begin{bmatrix} 0 & 1 & 0 \\ -1 & 0 & 0 \\ 0 & 0 & 1 \end{bmatrix} \begin{bmatrix} 0 \\ 0 \\ 3.22E\text{-}4 \end{bmatrix} +$$

$$\begin{bmatrix} -600 & 0 & 0 \\ 0 & -22.2 & 33333 \\ 0 & -33333 & 33.3E6 \end{bmatrix} \begin{bmatrix} 0 & 1 & 0 \\ -1 & 0 & 0 \\ 0 & 0 & 1 \end{bmatrix} \begin{bmatrix} 0.292 \\ -0.085 \\ -9.39E\text{-}4 \end{bmatrix} +$$

$$\begin{bmatrix} 0 \\ 0 \\ 0 \end{bmatrix}$$

$$= \begin{bmatrix} 0.0 \\ 10.7 \\ 21480 \end{bmatrix} + \begin{bmatrix} 51.0 \\ -24.8 \\ -21540 \end{bmatrix} + \begin{bmatrix} 0 \\ 0 \\ 0 \end{bmatrix} = \begin{bmatrix} 51.0 \text{ kN} \\ -14.1 \text{ kN} \\ -60.0 \text{ kN m} \end{bmatrix}$$

Similarly

$$f_{21} = f_{21}^{n} + f_{21}^{f}$$

$$= k_{21}\, T_{12}\, \Delta_1 + k_{22}^{1}\, T_{12}\, \Delta_2 + f_{21}^{f}$$

$$= \begin{bmatrix} 0.0 \\ -10.7 \\ 10720 \end{bmatrix} + \begin{bmatrix} -51.0 \\ 24.8 \\ -52900 \end{bmatrix} + \begin{bmatrix} 0 \\ 0 \\ 0 \end{bmatrix} = \begin{bmatrix} -51.0 \text{ kN} \\ 14.1 \text{ kN} \\ -42180 \text{ kN m} \end{bmatrix}$$

(ii) Element 1,2 undergoes a 20°C rise in temperature.

Fixed end forces

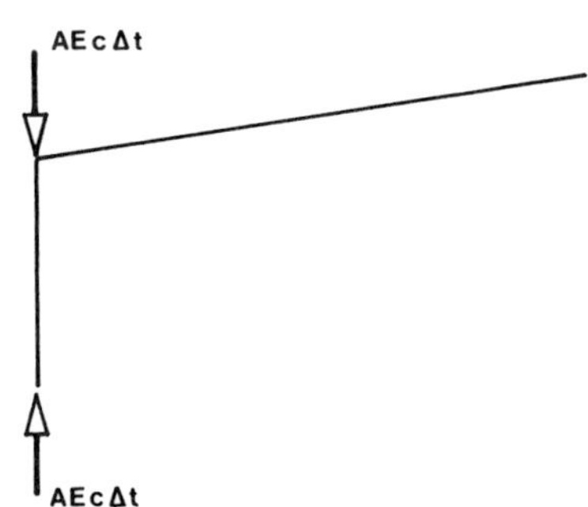

$$c\,\Delta t\,E\,A = 1.1 \text{ x } 10^{-5} \text{ x } 20 \text{ x } 9000 \text{ x } 200 = 396 \text{ kN}$$

Hence

$$f_{12}^{f} = \begin{bmatrix} 396 \\ 0 \\ 0 \end{bmatrix} \qquad f_{32}^{f} = \begin{bmatrix} -396 \\ 0 \\ 0 \end{bmatrix}$$

Transform the fixed end force vectors into equivalent joint force vectors, note that ($\alpha_{12} = 90°$).

$$P_{12}^{e} = -T_{ij}^{-1}\, f_{12}^{f}$$

$$= -\begin{bmatrix} 0 & -1 & 0 \\ 1 & 0 & 0 \\ 0 & 0 & 1 \end{bmatrix}\begin{bmatrix} 396 \\ 0 \\ 0 \end{bmatrix} = \begin{bmatrix} 0 \\ -396 \\ 0 \end{bmatrix}$$

Similarly

$$P^e_{21} = \begin{bmatrix} 0 \\ 396 \\ 0 \end{bmatrix}$$

After calculating the nodal force vectors and applying the boundary conditions, the final stiffness equation is

$$\begin{bmatrix} 66.7E6 & 33333 & 0 & 33.3E6 & 0 \\ 33333 & 271.7 & 35.40 & 32484 & -848.5 \\ 0 & 35.40 & 606.8 & 5940 & 5940 \\ 33.3E6 & 32484 & 5940 & 95.0E6 & 14.1E6 \\ 0 & -848.5 & 5940 & 14.1E6 & 28.3E6 \end{bmatrix} \begin{bmatrix} \theta_1 \\ \Delta_{2x} \\ \Delta_{2y} \\ \theta_2 \\ \theta_3 \end{bmatrix} = \begin{bmatrix} 0 \\ 0 \\ 396 \\ 0 \\ 0 \end{bmatrix}$$

which yields

$$\begin{bmatrix} \theta_1 \\ \Delta_{2x} \\ \Delta_{2y} \\ \theta_2 \\ \theta_3 \end{bmatrix} = \begin{bmatrix} 0.490 \text{ x } 10^{-4} \text{ rads} \\ -0.0916 \text{ mm} \\ 0.659 \text{ mm} \\ -6.67 \text{ x } 10^{-6} \text{ rads} \\ -1.38 \text{ x } 10^{-4} \text{ rads} \end{bmatrix}$$

and the element end forces for element 1,2 are given by

$$\begin{aligned} f_{12} &= f^n_{12} + f^f_{12} \\ &= k^2_{11} T_{12} \Delta_1 + k_{12} T_{12} \Delta_2 + f^f_{12} \\ &= \begin{bmatrix} 0 \\ 1.633 \\ 3268 \end{bmatrix} + \begin{bmatrix} -395.4 \\ -2.256 \\ -3275 \end{bmatrix} + \begin{bmatrix} 396 \\ 0 \\ 0 \end{bmatrix} = \begin{bmatrix} 0.6 \text{ kN} \\ -0.623 \text{ kN} \\ -7 \text{ kN mm} \end{bmatrix} \end{aligned}$$

Similarly

$$\begin{aligned} f_{21} &= f^n_{21} + f^f_{21} \\ &= k_{21} T_{12} \Delta_1 + k^1_{22} T_{12} \Delta_2 + f^f_{21} \end{aligned}$$

$$= \begin{bmatrix} 0.0 \\ -1.633 \\ 1632 \end{bmatrix} + \begin{bmatrix} 395.4 \\ 2.256 \\ -3498 \end{bmatrix} + \begin{bmatrix} -396 \\ 0 \\ 0 \end{bmatrix}$$

$$= \begin{bmatrix} -0.6 \text{ kN} \\ 0.623 \text{ kN} \\ -1866 \text{ kN mm} \end{bmatrix}$$

(iii) Support 1 settles 15 mm vertically (downwards) and 10 mm horizontally (to the right).

Fixed end forces

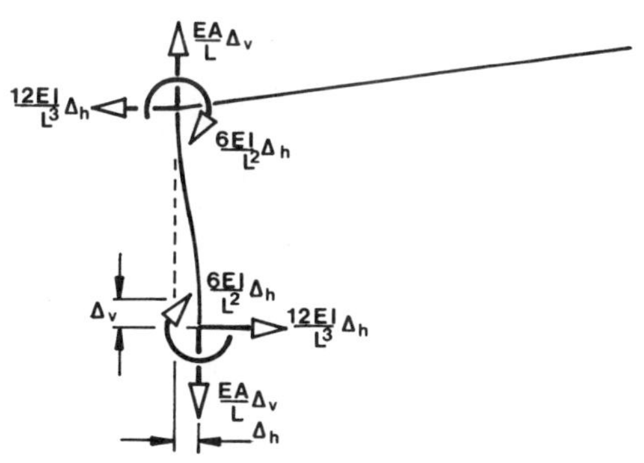

$$\frac{6EI}{L^2}\Delta_h = 333 \text{ kN m} \qquad \frac{12EI}{L^3}\Delta_h = 222 \text{ kN} \qquad \frac{AE}{L}\Delta_v = 9000 \text{ kN}$$

$$f_{12}^f = \begin{bmatrix} -9000 \text{ kN} \\ -222 \text{ kN} \\ -333000 \text{ kN mm} \end{bmatrix} \qquad f_{32}^f = \begin{bmatrix} 9000 \text{ kN} \\ 222 \text{ kN} \\ -333000 \text{ kN mm} \end{bmatrix}$$

After calculating the final force vectors and applying the boundary conditions the final stiffness equation is

$$\begin{bmatrix} 66.7E6 & 33333 & 0 & 33.3E6 & 0 \\ 33333 & 271.7 & 35.40 & 32484 & -848.5 \\ 0 & 35.40 & 606.8 & 5940 & 5940 \\ 33.3E6 & 32484 & 5940 & 95.0E6 & 14.1E6 \\ 0 & -848.5 & 5940 & 14.1E6 & 28.3E6 \end{bmatrix} \begin{bmatrix} \theta_1 \\ \Delta_{2x} \\ \Delta_{2y} \\ \theta_2 \\ \theta_3 \end{bmatrix} = \begin{bmatrix} 333000 \\ 222 \\ -9000 \\ 333000 \\ 0 \end{bmatrix}$$

which yields

$$\begin{bmatrix} \theta_1 \\ \Delta_{2x} \\ \Delta_{2y} \\ \theta_2 \\ \theta_3 \end{bmatrix} = \begin{bmatrix} 2.68 \times 10^{-3} \text{ rads} \\ 2.15 \text{ mm} \\ -15.0 \text{ mm} \\ 2.47 \times 10^{-3} \text{ rads} \\ 1.98 \times 10^{-3} \text{ rads} \end{bmatrix}$$

and the element end forces for element 1,2 are given by

$$\begin{aligned} f_{12} &= f_{12}^n + f_{12}^f \\ &= k_{11}^2 T_{12} \Delta_1 + k_{12} T_{12} \Delta_2 + f_{12}^f \\ &= \begin{bmatrix} 0.0 \\ 89.33 \\ 178800 \end{bmatrix} + \begin{bmatrix} 9000 \\ 130.1 \\ 153900 \end{bmatrix} + \begin{bmatrix} -9000 \\ -222 \\ -333000 \end{bmatrix} \\ &= \begin{bmatrix} 0.0 \text{ kN} \\ -2.57 \text{ kN} \\ -300 \text{ kN mm} \end{bmatrix} \end{aligned}$$

Similarly

$$\begin{aligned} f_{21} &= f_{21}^n + f_{21}^f \\ &= k_{21} T_{12} \Delta_1 + k_{22}^1 T_{12} \Delta_2 + f_{21}^f \\ &= \begin{bmatrix} 0.0 \\ -89.33 \\ 89240 \end{bmatrix} + \begin{bmatrix} -9000 \\ -130.1 \\ 236400 \end{bmatrix} + \begin{bmatrix} 9000 \\ 222 \\ -333000 \end{bmatrix} \\ &= \begin{bmatrix} 0.0 \text{ kN} \\ 2.57 \text{ kN} \\ -7360 \text{ kN mm} \end{bmatrix} \end{aligned}$$

Note that the element end forces due to nodal displacements and the fixed end forces both contain large numbers of approximately equal magnitude. In these conditions more significant figures could be usefully employed in the calculations.

(iv) Element 2,3 is 20 mm too short, hence the fixed end forces are

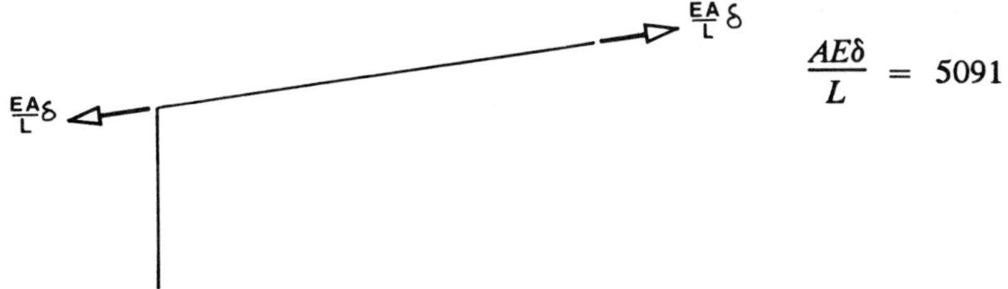

$$\frac{AE\delta}{L} = 5091$$

From here the solution follows exactly the same procedures as were used in (i), (ii) and (iii), yielding the following equivalent joint force and solution vectors

$$P_{23}^{e} = \begin{bmatrix} 5040 \\ 720 \\ 0 \end{bmatrix} \quad \text{and} \quad P_{32}^{e} = \begin{bmatrix} -5040 \\ -720 \\ 0 \end{bmatrix}$$

and the final stiffness equation yields

$$\begin{bmatrix} \theta_1 \\ \Delta_{2x} \\ \Delta_{2y} \\ \theta_2 \\ \theta_3 \end{bmatrix} = \begin{bmatrix} -0.7737 \text{ x } 10^{-2} \text{ rads} \\ 20.05 \text{ mm} \\ -0.03345 \text{ mm} \\ -0.4574 \text{ x } 10^{-2} \text{ rads} \\ 0.2882 \text{ x } 10^{-2} \text{ rads} \end{bmatrix}$$

Finding the element forces as before produces

$$\begin{aligned} f_{12} &= f_{12}^{n} + f_{12}^{f} \\ &= k_{11}^{2} \, T_{12} \, \Delta_1 + k_{12} \, T_{12} \, \Delta_2 + f_{12}^{f} \\ &= \begin{bmatrix} 0.0 \\ -257.9 \\ -515800 \end{bmatrix} + \begin{bmatrix} -20.07 \\ 293.1 \\ 515900 \end{bmatrix} + \begin{bmatrix} 0.0 \\ 0.0 \\ 0.0 \end{bmatrix} = \begin{bmatrix} -20.1 \\ 35.2 \\ 100.0 \end{bmatrix} \end{aligned}$$

Similarly

$$\begin{aligned} f_{21} &= f_{21}^{n} + f_{21}^{f} \\ &= k_{21} \, T_{12} \, \Delta_1 + k_{22}^{1} \, T_{12} \, \Delta_2 + f_{21}^{f} \\ &= \begin{bmatrix} 0.0 \\ 257.9 \\ -257900 \end{bmatrix} + \begin{bmatrix} 20.07 \\ -293.1 \\ 363400 \end{bmatrix} + \begin{bmatrix} 0.0 \\ 0.0 \\ 0.0 \end{bmatrix} = \begin{bmatrix} 20.07 \\ -35.2 \\ 105500 \end{bmatrix} \end{aligned}$$

(v) The displacement at mid-height of element 1,2 under loading (iii) can be found as follows from the fact that the deformed shape of a prismatic beam loaded only at its ends is a cubic polynomial (see section 2.11).

$$y = ax^3 + bx^2 + cx + d$$

The boundary conditions are (from the solution vector found in (iii)) are

at $x = 0$	$y = 0$	$dy/dx = 0.00268$
at $x = 3000$	$y = -2.15$	$dy/dx = 0.00247$

Substitution into the expressions for y and dy/dx yields

$$a = 0.729\text{E-}9 \quad b = -3.32\text{E-}6 \quad c = 2.680\text{E-}3 \quad d = 0.0$$

hence

$$y = 0.729\text{E-}9\,x^3 - 3.32\text{E-}6\,x^2 + 2.680\text{E-}3\,x$$

which yields , $y = -0.99$ mm when $x = 1500$ mm.

Note that this is the displacement at mid-height of element 1,2 in terms of the shape function axes under the action of the equivalent joint forces. As the shape function "y" axis lies at 180 degrees to the global "x" axis, the above displacement represents a horizontal displacement of 0.99 mm to the right. The final displacement is found by adding the displacement from the fixed end forces (i.e. half of the horizontal displacement at node 1) to the above displacement.

Final displacement $= 0.99 + 10.0/2 = 5.99$ mm

Program FIX.BAS

This program calculates the fixed end forces and the equivalent joint forces for a number of common loading conditions on plane frame elements with a variety of end connections.

Input

The program loops for each load case in turn. Once the user has responded to prompts for details of the loading and the element the fixed end and equivalent joint forces are calculated.

Comments on the Algorithm

The fixed end forces are first evaluated (in the element axes system). The equivalent joint forces are then found by transforming the fixed end forces into the global axes system. Note that elements with pins are dealt with in sections 8.5 and 8.7.

Output

The fixed end forces (in the element axes system) and the equivalent joint forces (in the global axes system) are output.

Listing

```
1000 '===========================     FIX.BAS     =============================
1010 '
1020 TRUE% = 1 : FALSE% = 0
1030 CLS
1040 PRINT
1050 PRINT "================================================================="
1060 PRINT
1070 PRINT "PROGRAM FIX.BAS                           Copyright (c) J.Balfour 1991"
1080 PRINT
1090 PRINT "      Program for the calculation of fixed end and equivalent"
1100 PRINT "              end forces for a plane frame element"
1110 PRINT
1120 PRINT "               For further information contact"
1130 PRINT " James A.D.Balfour, Heriot-Watt University, Riccarton, Edinburgh"
1140 PRINT "              Tel 031-449-5111, Fax 031-451-3170"
1150 PRINT
1160 PRINT "================================================================="
1170 '
1180 ' ... Loop for all elements
1190 '
1200 PRINT
1210 PRINT "ELEMENT      LOWER        HIGHER"
1220 PRINT " TYPE        NODE         NODE"
1230 PRINT "   1         Rigid        Rigid"
1240 PRINT "   2         Rigid        Pinned"
1250 PRINT "   3         Pinned       Rigid"
1260 PRINT "   4         Pinned       Pinned"
1270 '
1280 ' ... Loop for all elements
1290 '
1300 DONE% = FALSE% : K% = 1
1310 WHILE DONE%=FALSE%
1320   '
1330   ' ... Input the element data
1340   '
1350   GOSUB 1490
1360   IF DONE% = TRUE% THEN GOTO 1460
1370   '
1380   ' ... Calculate the fixed end forces and equivalent joint forces
1390   '
1400   GOSUB 1980
1410   '
1420   ' ... Output results
1430   '
1440   GOSUB 3070
1450   K% = K% + 1
1460 WEND
1470 END
1480 '
1490 '*******************     INPUT THE ELEMENT DATA     ********************
1500 '
1510 PRINT : PRINT
1520 PRINT "LOAD CASE"; K%
1530 PRINT
1540 INPUT "Lower load number (0 to stop) = ", IN%
1550 IF IN% = 0 THEN DONE% = TRUE% : GOTO 1960
1560 INPUT "Higher node number            = ", JN%
1570 PRINT
1580 PRINT       "+ + + + + + + + + + + + +"
1590 PRINT USING "+   ELEMENT ## ##   +"; IN%, JN%
1600 PRINT       "+ + + + + + + + + + + + +"
1610 PRINT
1620 PRINT "SELECT LOADING TYPE"
1630 PRINT "(1) Uniformly distributed"
1640 PRINT "(2) Point load"
```

```
1650 PRINT "(3) Lack of fit"
1660 PRINT "(4) Thermal expansion"
1670 PRINT "(5) Settlement"
1680 INPUT LTYPE%
1690 '
1700 ' ... Input the necessary element data
1710 '
1720 IF LTYPE% > 2 THEN GOTO 1760
1730 PRINT
1740 PRINT "Note that a positive element load load acts"
1750 PRINT "in the direction of the element Y-axis"
1760 PRINT
1770 IF LTYPE% = 1 THEN GOTO 1840
1780 IF LTYPE% = 2 THEN GOTO 1840
1790 INPUT "Area (A)                          = ", A
1800 IF LTYPE% = 3 THEN GOTO 1840
1810 IF LTYPE% = 4 THEN L = 1
1820 IF LTYPE% = 4 THEN GOTO 1870
1830 INPUT "Second moment of area (I)         = ", A2
1840 INPUT "Length (L)                        = ", L
1850 IF LTYPE% = 1 THEN GOTO 1880
1860 IF LTYPE% = 2 THEN GOTO 1880
1870 INPUT "Elastic constant (E)              = ", E
1880 INPUT "Angle with the global x-axis      = ", ALPHA
1890 INPUT "Element type                      = ", MTYPE%
1900 '
1910 ' ... Convert the angle to radians
1920 '
1930 AR = 4 * ATN(1) * ALPHA / 180
1940 C  = COS(AR)
1950 S  = SIN(AR)
1960 RETURN
1970 '
1980 '*******    CALCULATE THE FIXED END AND EQUIVALENT JOINT FORCES    ******
1990 '
2000 ON LTYPE% GOTO 2040, 2150, 2290, 2400, 2520
2010 '
2020 ' ... Uniformly distributed load
2030 '
2040 INPUT "Load intensity                    = ", W
2050 FEF(1) =   0
2060 FEF(2) = - W * L / 2
2070 FEF(3) = - W * L^2 / 12
2080 FEF(4) =   0
2090 FEF(5) =   FEF(2)
2100 FEF(6) = - FEF(3)
2110 GOTO 2820
2120 '
2130 ' ... Point load
2140 '
2150 PRINT "Distance  -  Load to node"; IN; "   =  ";
2160 INPUT "", X
2170 INPUT "Magnitude of the load             = ", P
2180 Y      =   L - X
2190 FEF(1) =   0
2200 FEF(2) = - P * (Y/L)^2 * (1 + 2*X/L)
2210 FEF(3) = - P * X * Y^2 / L^2
2220 FEF(4) =   0
2230 FEF(5) = - P * (X/L)^2 * (1 + 2*Y/L)
2240 FEF(6) =   P * Y * X^2 / L^2
2250 GOTO 2820
2260 '
2270 ' ... Lack of fit
2280 '
2290 INPUT "Lack of fit (short is +ve)        = ", X
2300 FEF(1) = - X * A * E / L
2310 FEF(2) =   0
2320 FEF(3) =   0
2330 FEF(4) = - FEF(1)
2340 FEF(5) =   0
2350 FEF(6) =   0
2360 GOTO 2820
2370 '
2380 ' ... Thermal expansion
2390 '
2400 INPUT "Coeff. of thermal expansion       = ", CO
2410 INPUT "Temperature rise                  = ", X
```

```
2420 FEF(1) =   A * CO * X * E
2430 FEF(2) =   0
2440 FEF(3) =   0
2450 FEF(4) = - FEF(1)
2460 FEF(5) =   0
2470 FEF(6) =   0
2480 GOTO 2820
2490 '
2500 ' ... Settlement
2510 '
2520 PRINT "Does the settlement take place at"
2530 PRINT "    (1) Node", IN%
2540 PRINT "    (2) Node", JN%
2550 INPUT NS%
2560 PRINT
2570 INPUT "Settlement Global X-direction  =  ", X
2580 INPUT "Settlement Global Y-direction  =  ", Y
2590 INPUT "Rotation (degrees)             =  ", R
2600 '
2610 ' ... Convert settlements to local axes then calc F.E.F.S
2620 '
2630 R      =   4 * ATN(1) * R / 180
2640 CC     =   COS(AR)
2650 SS     =   SIN(AR)
2660 IF NS% = 1 THEN GOTO 2690
2670 CC     =   COS(AR + 4*ATN(1))
2680 SS     =   SIN(AR + 4*ATN(1))
2690 FEF(1) =        E * A  * ( X*CC + Y*SS) / L
2700 FEF(2) =   12 * E * A2 * (-X*SS + Y*CC) / L^3  +  6 * E * A2 * R / L^2
2710 FEF(3) =    6 * E * A2 * (-X*SS + Y*CC) / L^2  +  4 * E * A2 * R / L
2720 FEF(4) = - FEF(1)
2730 FEF(5) = - FEF(2)
2740 FEF(6) = - FEF(3) + L*FEF(2)
2750 IF NS  = 1 THEN GOTO 2820
2760 X      = FEF(3)
2770 FEF(3) = FEF(6)
2780 FEF(6) = X
2790 '
2800 ' ... Take account of any pins
2810 '
2820 ON MTYPE% GOTO 2990, 2880, 2830, 2830
2830 FEF(2) = FEF(2) - 1.5*FEF(3)/L
2840 FEF(5) = FEF(5) + 1.5*FEF(3)/L
2850 FEF(6) = FEF(6) - FEF(3)/2
2860 FEF(3) = 0
2870 IF MTYPE = 3 THEN GOTO 2990 ELSE GOTO 2930
2880 FEF(2) = FEF(2) - 1.5*FEF(6)/L
2890 FEF(3) = FEF(3) - FEF(6)/2
2900 FEF(5) = FEF(5) + 1.5*FEF(6)/L
2910 FEF(6) = 0
2920 GOTO 2990
2930 FEF(2) = FEF(2) - FEF(6)/L
2940 FEF(5) = FEF(5) + FEF(6)/L
2950 FEF(6) = 0
2960 '
2970 ' ... Convert fixed end forces to equivalent joint forces
2980 '
2990 EJF(1) = - (FEF(1)*C - FEF(2)*S)
3000 EJF(2) = - (FEF(1)*S + FEF(2)*C)
3010 EJF(3) = -  FEF(3)
3020 EJF(4) = - (FEF(4)*C - FEF(5)*S)
3030 EJF(5) = - (FEF(4)*S + FEF(5)*C)
3040 EJF(6) = -  FEF(6)
3050 RETURN
3060 '
3070 '*********    OUTPUT THE FIXED END AND EQUIVALENT JOINT FORCES    ******
3080 '
3090 PRINT
3100 PRINT "* * * * * * * * * * * * * * * * *"
3110 PRINT "*   FIXED END FORCES            *"
3120 PRINT "*   (ELEMENT AXES SYSTEM)       *"
3130 PRINT "* * * * * * * * * * * * * * * * *"
3140 PRINT
3150 PRINT USING "                   NODE ##         NODE ##"; IN%, JN%
3160 IF MTYPE = 1 THEN PRINT "                   (Rigid)         (Rigid)"
3170 IF MTYPE = 2 THEN PRINT "                   (Rigid)         (Pinned)"
3180 IF MTYPE = 3 THEN PRINT "                   (Pinned)        (Rigid)"
```

```
3190 IF MTYPE = 4 THEN PRINT "                    (Pinned)           (Pinned)"
3200 PRINT USING "X-FORCE      #.####^^^^        #.####^^^^"; FEF(1), FEF(4)
3210 PRINT USING "Y-FORCE      #.####^^^^        #.####^^^^"; FEF(2), FEF(5)
3220 PRINT USING "MOMENT       #.####^^^^        #.####^^^^"; FEF(3), FEF(6)
3230 PRINT
3240 PRINT "* * * * * * * * * * * * * * * * * *"
3250 PRINT "*   EQUIVALENT JOINT FORCES      *"
3260 PRINT "*   (GLOBAL AXES SYSTEM)         *"
3270 PRINT "* * * * * * * * * * * * * * * * * *"
3280 PRINT
3290 PRINT USING "                 NODE ##          NODE ##"; IN%, JN%
3300 IF MTYPE = 1 THEN PRINT "                    (Rigid)            (Rigid)"
3310 IF MTYPE = 2 THEN PRINT "                    (Rigid)            (Pinned)"
3320 IF MTYPE = 3 THEN PRINT "                    (Pinned)           (Rigid)"
3330 IF MTYPE = 4 THEN PRINT "                    (Pinned)           (Pinned)"
3340 PRINT USING "X-FORCE      #.####^^^^        #.####^^^^"; EJF(1), EJF(4)
3350 PRINT USING "Y-FORCE      #.####^^^^        #.####^^^^"; EJF(2), EJF(5)
3360 PRINT USING "MOMENT       #.####^^^^        #.####^^^^"; EJF(3), EJF(6)
3370 RETURN
3380 '
3390 '===========================    FIX.BAS    ==============================
```

Sample Run

The following sample shows program FIX.BAS being used to evaluate the fixed end forces and the equivalent joint forces for example 8.1.

```
=================================================================

PROGRAM FIX.BAS                          Copyright (c) J.Balfour 1991

      Program for the calculation of fixed end and equivalent
              end forces for a plane frame element

               For further information contact
 James A.D.Balfour, Heriot-Watt University, Riccarton, Edinburgh
            Tel 031-449-5111, Fax 031-451-3170

=================================================================


ELEMENT      LOWER        HIGHER
 TYPE        NODE         NODE
   1         Rigid        Rigid
   2         Rigid        Pinned
   3         Pinned       Rigid
   4         Pinned       Pinned


LOAD CASE 1

Lower load number (0 to stop) = 2
Higher node number            = 3

+ + + + + + + + + + + +
+   ELEMENT  2  3   +
+ + + + + + + + + + + +

SELECT LOADING TYPE
(1) Uniformly distributed
(2) Point load
(3) Lack of fit
(4) Thermal expansion
(5) Settlement
? 1

Note that a positive element load load acts
in the direction of the element Y-axis

Length (L)                      =  7071
Angle with the global x-axis    =  8.13
Element type                    =  1
Load intensity                  =  -0.01
```

```
* * * * * * * * * * * * * * *
*   FIXED END FORCES        *
*   (ELEMENT AXES SYSTEM)   *
* * * * * * * * * * * * * * *

              NODE  2         NODE  3
X-FORCE     0.0000E+00      0.0000E+00
Y-FORCE     0.3536E+02      0.3536E+02
MOMENT      0.4167E+05      -.4167E+05

* * * * * * * * * * * * * * * *
*   EQUIVALENT JOINT FORCES    *
*   (GLOBAL AXES SYSTEM)       *
* * * * * * * * * * * * * * * *

              NODE  2         NODE  3
X-FORCE     0.5000E+01      0.5000E+01
Y-FORCE     -.3500E+02      -.3500E+02
MOMENT      -.4167E+05      0.4167E+05

LOAD CASE 2

Lower load number (0 to stop) = 1
Higher node number            = 2

+ + + + + + + + + + +
+   ELEMENT  1  2   +
+ + + + + + + + + + +

SELECT LOADING TYPE
(1) Uniformly distributed
(2) Point load
(3) Lack of fit
(4) Thermal expansion
(5) Settlement
? 4

Area (A)                        =  9000
Elastic constant (E)            =  200
Angle with the global x-axis    =  90
Element type                    =  1
Coeff. of thermal expansion     =  1.1E-5
Temperature rise                =  20

* * * * * * * * * * * * * * *
*   FIXED END FORCES        *
*   (ELEMENT AXES SYSTEM)   *
* * * * * * * * * * * * * * *

              NODE  1         NODE  2
X-FORCE     0.3960E+03      -.3960E+03
Y-FORCE     0.0000E+00      0.0000E+00
MOMENT      0.0000E+00      0.0000E+00

* * * * * * * * * * * * * * * *
*   EQUIVALENT JOINT FORCES    *
*   (GLOBAL AXES SYSTEM)       *
* * * * * * * * * * * * * * * *

              NODE  1         NODE  2
X-FORCE     0.6452E-04      -.6452E-04
Y-FORCE     -.3960E+03      0.3960E+03
MOMENT      0.0000E+00      0.0000E+00

LOAD CASE 3

Lower load number (0 to stop) = 1
Higher node number            = 2

+ + + + + + + + + + +
+   ELEMENT  1  2   +
+ + + + + + + + + + +

SELECT LOADING TYPE
(1) Uniformly distributed
(2) Point load
(3) Lack of fit
```

```
(4) Thermal expansion
(5) Settlement
? 5

Area (A)                          =  9000
Second moment of area (I)         =  .25E9
Length (L)                        =  3000
Elastic constant (E)              =  200
Angle with the global x-axis      =  90
Element type                      =  1
Does the settlement take place at
   (1) Node    1
   (2) Node    2
? 1

Settlement Global X-direction     =  10
Settlement Global Y-direction     =  -15
Rotation (degrees)                =  0

* * * * * * * * * * * * * * *
*   FIXED END FORCES        *
*   (ELEMENT AXES SYSTEM)   *
* * * * * * * * * * * * * * *

               NODE  1         NODE  2
X-FORCE      -.9000E+04      0.9000E+04
Y-FORCE      -.2222E+03      0.2222E+03
MOMENT       -.3333E+06      -.3333E+06

* * * * * * * * * * * * * * * * *
*   EQUIVALENT JOINT FORCES     *
*   (GLOBAL AXES SYSTEM)        *
* * * * * * * * * * * * * * * * *

               NODE  1         NODE  2
X-FORCE      -.2222E+03      0.2222E+03
Y-FORCE      0.9000E+04      -.9000E+04
MOMENT       0.3333E+06      0.3333E+06

LOAD CASE 4

Lower load number (0 to stop) = 2
Higher node number            = 3

+ + + + + + + + + + +
+   ELEMENT  2  3   +
+ + + + + + + + + + +

SELECT LOADING TYPE
(1) Uniformly distributed
(2) Point load
(3) Lack of fit
(4) Thermal expansion
(5) Settlement
? 3

Area (A)                          =  9000
Length (L)                        =  7071
Elastic constant (E)              =  200
Angle with the global x-axis      =  8.13
Element type                      =  1
Lack of fit (short is +ve)        =  5

* * * * * * * * * * * * * * *
*   FIXED END FORCES        *
*   (ELEMENT AXES SYSTEM)   *
* * * * * * * * * * * * * * *

               NODE  2         NODE  3
X-FORCE      -.1273E+04      0.1273E+04
Y-FORCE      0.0000E+00      0.0000E+00
MOMENT       0.0000E+00      0.0000E+00

* * * * * * * * * * * * * * * * *
*   EQUIVALENT JOINT FORCES     *
*   (GLOBAL AXES SYSTEM)        *
* * * * * * * * * * * * * * * * *
```

```
              NODE  2         NODE  3
X-FORCE    0.1260E+04      -.1260E+04
Y-FORCE    0.1800E+03      -.1800E+03
MOMENT     0.0000E+00      0.0000E+00

LOAD CASE 5

Lower load number (0 to stop) = 0
```

8.3 Local Axes

In the section 3.6 two types of nodal axes were described.

Global axes these are nodal axes that point in the directions of the coordinate axes.

Local axes these are arbitrarily orientated nodal axes that are used when it is either inconvenient or impossible to describe the boundary conditions in terms of the global axes.

The solution procedure is greatly simplified by the use of global axes at all nodes.

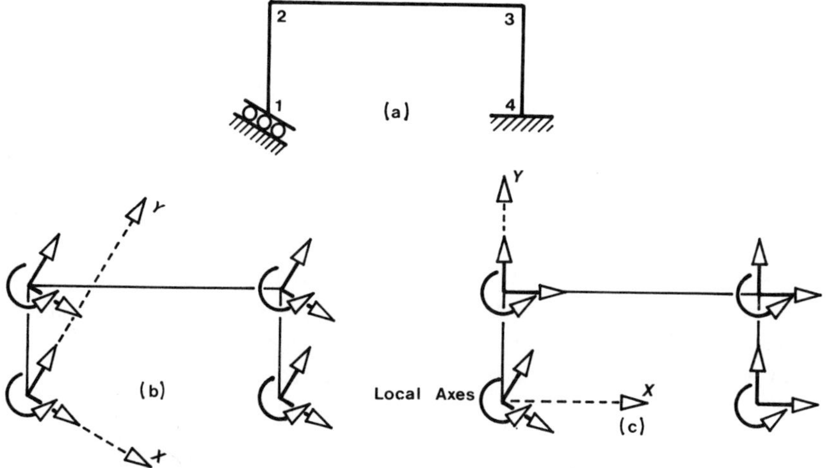

Figure 8.10 Two possible choices of coordinate axes

The advantage of the coordinate axes system shown in fig 8.10(b) is that it allows the analysis to proceed without the complication of local axes. Its disadvantages are that the calculation of nodal coordinates would be more difficult and error-prone and, if the nodal loads are horizontal and vertical, then they would have to be transformed into the inclined global axes systems prior to analysis. In general, if the available computer program allows the use of local axes then the choice of the coordinate axes system should be governed by the ease of data preparation.

Some structures, such as that shown in fig 8.11, can only be analysed through the use of local axes as no single axes can be used to describe the boundary conditions.

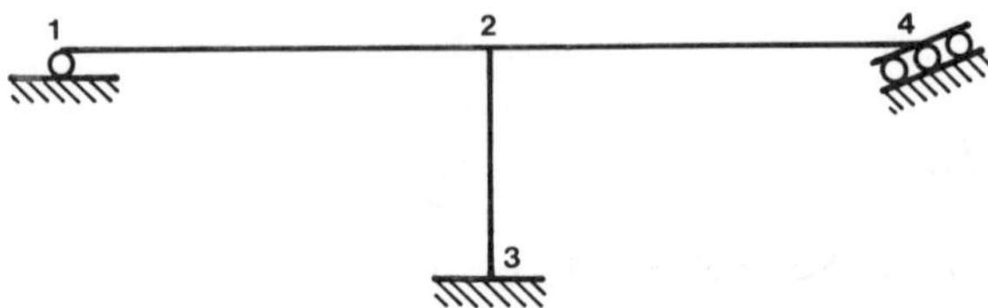

Figure 8.11 Example of a structure that demands the use of local axes

Figure 8.12 shows nodal axes systems that could be used in the analysis of the structure shown in fig 8.11.

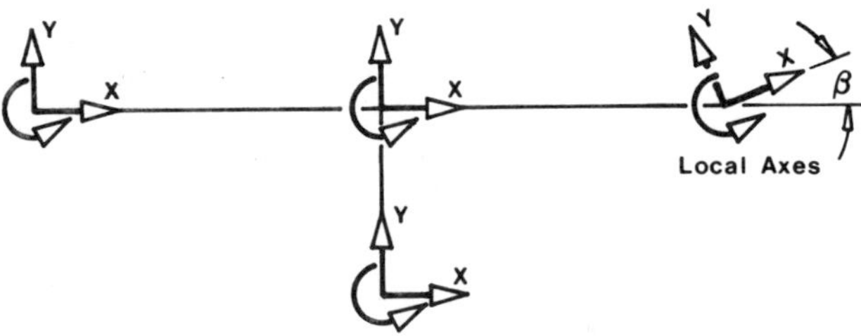

Figure 8.12 An example of the use of local axes

Note that the difference between the nodal axes shown in fig 8.12 and the nodal axes dealt with previously is that one set of axes lies at an angle to the other nodal axes. In principle all the nodal axes might point in different directions. In order to proceed it is necessary to delevope an element stiffness matrix for an element with nodal axes that point in arbitrary directions.

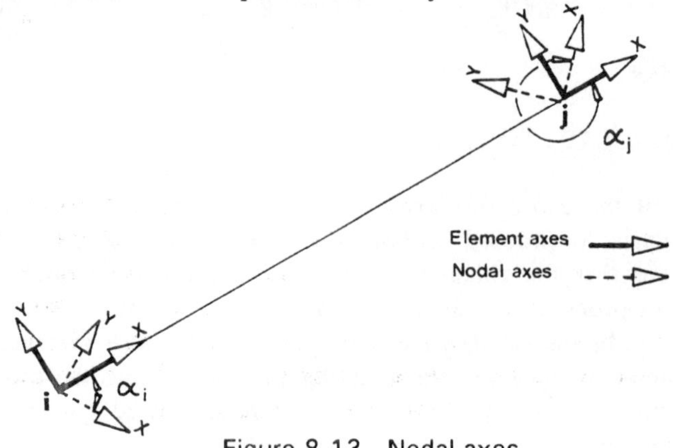

Figure 8.13 Nodal axes

Consider the element shown in fig 8.13. Note that the nodal axes need to be rotated *anticlockwise* by α_i and α_j at nodes "*i*" and "*j*" respectively to make them coincide with the element axes.

The objective here is to express the forces at the ends of the element in terms of the displacements at its ends where both the forces and the displacements are expressed in the nodal axes systems. The element stiffness equation in terms of the element axes has already been shown to take the form.

$$\begin{bmatrix} f_{ij} \\ f_{ji} \end{bmatrix} = \begin{bmatrix} k_{ii}^{j} & k_{ij} \\ k_{ji} & k_{jj}^{i} \end{bmatrix} \begin{bmatrix} \delta_{ij} \\ \delta_{ji} \end{bmatrix}$$

The force and displacement vectors at can be transformed into the nodal axes as follows

$$\delta_{ij} = T_{ij}\,\Delta_i \qquad \text{and} \qquad f_{ij} = T_{ij}\,F_{ij}$$

and similarly at end "*j*" of the element

$$\delta_{ji} = T_{ji}\,\Delta_j \qquad \text{and} \qquad f_{ji} = T_{ji}\,F_{ji}$$

where

$$T_{ij} = \begin{bmatrix} cos\alpha_i & sin\alpha_i & 0 \\ -sin\alpha_i & cos\alpha_i & 0 \\ 0 & 0 & 1 \end{bmatrix}$$

and

$$T_{ji} = \begin{bmatrix} cos\alpha_j & sin\alpha_j & 0 \\ -sin\alpha_j & cos\alpha_j & 0 \\ 0 & 0 & 1 \end{bmatrix}$$

hence the element stiffness equation becomes

$$\begin{bmatrix} T_{ij}F_{ij} \\ T_{ji}F_{ji} \end{bmatrix} = \begin{bmatrix} k_{ii}^{j} & k_{ij} \\ k_{ji} & k_{jj}^{i} \end{bmatrix} \begin{bmatrix} T_{ij}\Delta_i \\ T_{ji}\Delta_j \end{bmatrix} = \begin{bmatrix} k_{ii}^{j}T_{ij} & k_{ij}T_{ji} \\ k_{ji}T_{ij} & k_{jj}^{i}T_{ji} \end{bmatrix} \begin{bmatrix} \Delta_i \\ \Delta_j \end{bmatrix}$$

Now multiply both sides by the inverse of the transformation matrices as shown.

$$\begin{bmatrix} T_{ij}^{-1} & T_{ji}^{-1} \end{bmatrix} \begin{bmatrix} T_{ij}F_{ij} \\ T_{ji}F_{ji} \end{bmatrix} = \begin{bmatrix} T_{ij}^{-1} & T_{ji}^{-1} \end{bmatrix} \begin{bmatrix} k_{ii}^{j}T_{ij} & k_{ij}T_{ji} \\ k_{ji}T_{ij} & k_{jj}^{i}T_{ji} \end{bmatrix} \begin{bmatrix} \Delta_i \\ \Delta_j \end{bmatrix}$$

or

$$\begin{bmatrix} F_{ij} \\ F_{ji} \end{bmatrix} = \begin{bmatrix} T_{ij}^{-1}k_{ii}^{j}T_{ij} & T_{ij}^{-1}k_{ij}T_{ji} \\ T_{ji}^{-1}k_{ji}T_{ij} & T_{ji}^{-1}k_{jj}^{i}T_{ji} \end{bmatrix} \begin{bmatrix} \Delta_i \\ \Delta_j \end{bmatrix}$$

The previous equation can be rewritten as

$$\begin{bmatrix} F_{ij} \\ F_{ji} \end{bmatrix} = \begin{bmatrix} K_{ii}^{j} & K_{ij} \\ K_{ji} & K_{jj}^{i} \end{bmatrix} \begin{bmatrix} \Delta_i \\ \Delta_j \end{bmatrix}$$

$$K_{ii}^{j} = T_{ij}^{-1}\, k_{ii}^{j}\, T_{ij}$$

$$= \begin{bmatrix} \left(\frac{EA}{L}\cos^2\alpha_i + \frac{12EI}{L^3}\sin^2\alpha_i\right) & \left(\frac{EA}{L} - \frac{12EI}{L^3}\right)\cos\alpha_i\,\sin\alpha_i & -\frac{6EI}{L^2}\sin\alpha_i \\ \left(\frac{EA}{L} - \frac{12EI}{L^3}\right)\cos\alpha_i\,\sin\alpha_i & \left(\frac{EA}{L}\sin^2\alpha_i + \frac{12EI}{L^3}\cos^2\alpha_i\right) & \frac{6EI}{L^2}\cos\alpha_i \\ -\frac{6EI}{L^2}\sin\alpha_i & \frac{6EI}{L^2}\cos\alpha_i & \frac{4EI}{L} \end{bmatrix} \tag{8.3}$$

$$K_{ij} = T_{ij}^{-1}\, k_{ij}\, T_{ji}$$

$$= \begin{bmatrix} -\left(\frac{EA}{L}c\alpha_i c\alpha_j + \frac{12EI}{L^3}s\alpha_i s\alpha_j\right) & \left(-\frac{EA}{L}c\alpha_i s\alpha_j + \frac{12EI}{L^3}s\alpha_i c\alpha_j\right) & -\frac{6EI}{L^2}s\alpha_i \\ \left(-\frac{EA}{L}s\alpha_i c\alpha_j + \frac{12EI}{L^3}c\alpha_i s\alpha_j\right) & \left(-\frac{EA}{L}s\alpha_i s\alpha_j - \frac{12EI}{L^3}c\alpha_i c\alpha_j\right) & \frac{6EI}{L^2}c\alpha_i \\ \frac{6EI}{L^2}s\alpha_j & -\frac{6EI}{L^2}c\alpha_j & \frac{2EI}{L} \end{bmatrix} \tag{8.4}$$

where $c\alpha_i = \cos\alpha_i$, $s\alpha_i = \sin\alpha_i$, $c\alpha_j = \cos\alpha_j$ and $s\alpha_j = \sin\alpha_j$

$$K_{jj}^{i} = T_{ji}^{-1}\, k_{jj}^{i}\, T_{ji}$$

$$= \begin{bmatrix} \left(\frac{EA}{L}\cos^2\alpha_j + \frac{12EI}{L^3}\sin^2\alpha_j\right) & \left(\frac{EA}{L} - \frac{12EI}{L^3}\right)\cos\alpha_j\,\sin\alpha_j & \frac{6EI}{L^2}\sin\alpha_j \\ \left(\frac{EA}{L} - \frac{12EI}{L^3}\right)\cos\alpha_j\,\sin\alpha_j & \left(\frac{EA}{L}\sin^2\alpha_j + \frac{12EI}{L^3}\cos^2\alpha_j\right) & -\frac{6EI}{L^2}\cos\alpha_j \\ \frac{6EI}{L^2}\sin\alpha_j & -\frac{6EI}{L^2}\cos\alpha_j & \frac{4EI}{L} \end{bmatrix} \tag{8.5}$$

and to maintain the symmetry of the element stiffness matrix

$$K_{ji} = K_{ij}^{T}$$

The element stiffness equation is now known in terms of arbitrarily orientated nodal axes and the analysis can proceed as before. Solution of the final structure stiffness equation yields the displacements in the nodal axes systems and care must be taken to ensure that the correct transformations are applied to the nodal displacements when calculating the element end forces (in the element axes systems). The required equations are

$$f_{ij} = k_{ii}^{j}\, T_{ij}\, \Delta_i + k_{ij}\, T_{ji}\, \Delta_j + f_{ij}^{f} \tag{8.6}$$

$$f_{ji} = k_{ji}\, T_{ij}\, \Delta_i + k_{jj}^{i}\, T_{ji}\, \Delta_j + f_{ji}^{f} \tag{8.7}$$

Example 8.2

Find the displacements for the structure shown in the following diagram. Find also the element end forces for element 2,4.

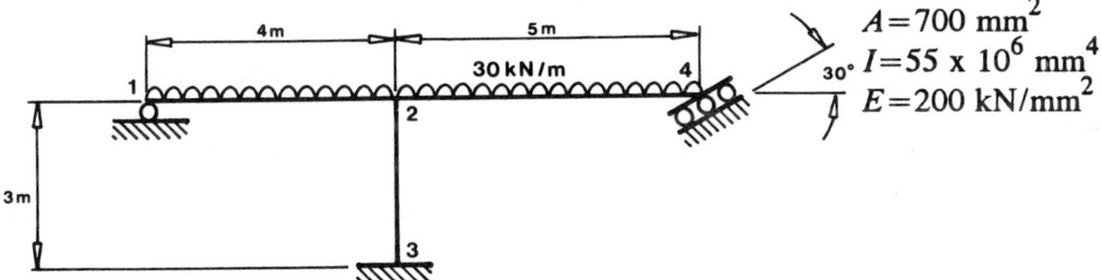

The following diagram shows the unrestrained nodal displacements and the freedom numbers. Note that the nodal 'x' axis at node 4 points in the direction of freedom 6.

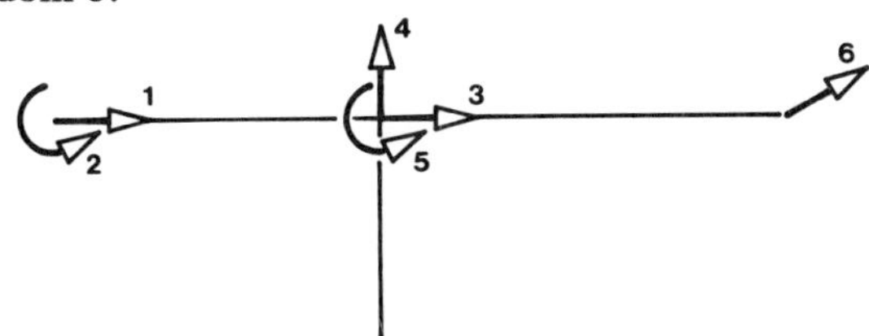

i.e the freedom vector is {1 0 2 3 4 5 0 0 0 6 0 0}

Fixed end forces (in the element axes systems) - units kN and m.

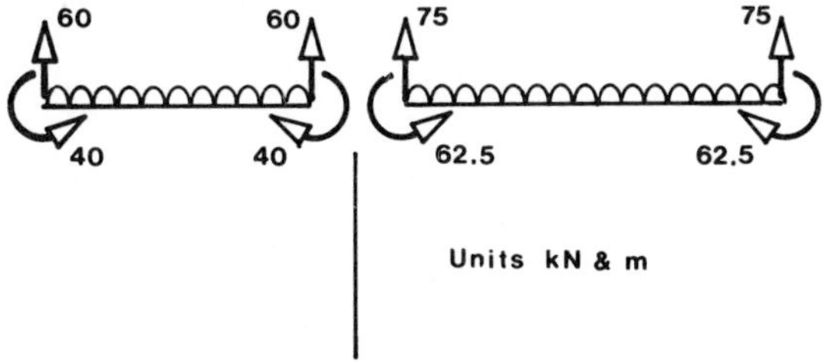

The nodal loading vectors are calculated using the appropriate transformation matrices (note that α_4 has be chosen to be 330°). For example

$$P^e_{12} = -T^{-1}_{12} f^f_{12}$$

$$= -\begin{bmatrix} 1 & 0 & 0 \\ 0 & 1 & 0 \\ 0 & 0 & 1 \end{bmatrix} \begin{bmatrix} 0 \\ 60 \\ 40000 \end{bmatrix} = \begin{bmatrix} 0 \\ -60 \\ -40000 \end{bmatrix}$$

$$P^e_{42} = -T^{-1}_{42} f^f_{42}$$

$$= -\begin{bmatrix} 0.866 & 0.5 & 0 \\ -0.5 & 0.866 & 0 \\ 0 & 0 & 1 \end{bmatrix} \begin{bmatrix} 0 \\ 75 \\ -62500 \end{bmatrix} = \begin{bmatrix} -37.5 \\ -65 \\ 62500 \end{bmatrix}$$

Equivalent joint forces (in the global axes systems) - units kN and m.

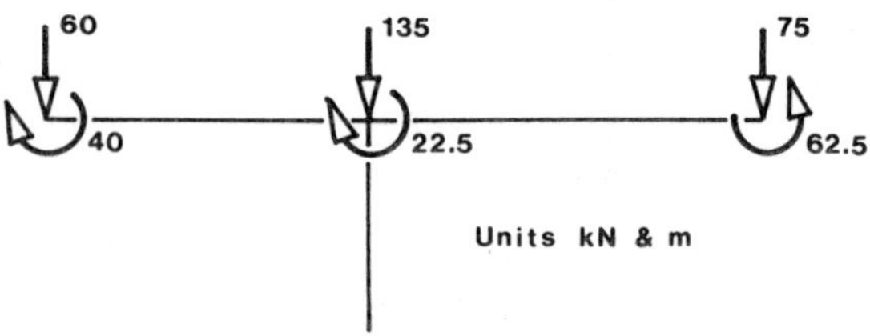

Final loadings (in the nodal axes systems)

Element 1,2 $A = 4700\ \text{mm}^2$, $I = 55 \times 10^6\ \text{mm}^4$, $L = 4000$ mm, $E = 200\ \text{kN/mm}^2$, code no. = [1 0 2 3 4 5], $\alpha_1 = 0°$, $\alpha_2 = 0°$.

	1	0	2	3	4	5
1	235	0	0	-235	0	0
0	0	2.06	4130	0	-2.06	4130
2	0	4130	11E6	0	-4130	5.5E6
3	-235	0	0	235	0	0
4	0	-2.06	-4130	0	2.06	-4130
5	0	4130	5.5E6	0	-4130	11E6

Adding the stiffness terms to the structure stiffness matrix gives

$$\begin{bmatrix} 235 & 0 & -235 & 0 & 0 & 0 \\ 0 & 11E6 & 0 & -4130 & 5.5E6 & 0 \\ -235 & 0 & 235 & 0 & 0 & 0 \\ 0 & -4130 & 0 & 2.06 & -4130 & 0 \\ 0 & 5.5E6 & 0 & -4130 & 11E6 & 0 \\ 0 & 0 & 0 & 0 & 0 & 0 \end{bmatrix}$$

Element 2,3 $A = 4700\ mm^2$, $I = 55 \times 10^6\ mm^4$, $L = 3000$ mm, $E = 200\ kN/mm^2$, code no = [3 4 5 0 0 0], $\alpha_2 = 270°$, $\alpha_3 = 270°$

	3	4	5	0	0	0
3	4.89	0	7330	-4.89	0	7330
4	0	313	0	0	-313	0
5	7330	0	14.7E6	-7330	0	7.4E6
0	-4.89	0	-7330	4.89	0	-7330
0	0	-313	0	0	313	0
0	7330	0	7.4E6	-7330	0	14.7E6

Adding the stiffness terms to the structure stiffness matrix gives

$$\begin{bmatrix} 235 & 0 & -235 & 0 & 0 & 0 \\ 0 & 11E6 & 0 & -4130 & 5.5E6 & 0 \\ -235 & 0 & 240 & 0 & 7330 & 0 \\ 0 & -4130 & 0 & 315 & -4130 & 0 \\ 0 & 5.5E6 & 7330 & -4130 & 25.7E6 & 0 \\ 0 & 0 & 0 & 0 & 0 & 0 \end{bmatrix}$$

Element 2,4 $A = 4700\ mm^2$, $I = 55 \times 10^6\ mm^4$, $L = 5000$ mm, $E = 200\ kN/mm^2$, code no = [3 4 5 6 0 0], $\alpha_2 = 0°$, $\alpha_4 = 330°$

	3	4	5	6	0	0
3	188	0	0	-163	94.0	0
4	0	1.06	2640	-0.53	-0.92	2640
5	0	2640	8.8E6	-1320	-2286	4.4E6
6	-163	-0.53	-1320	141	-80.9	-1320
0	94.0	-0.92	-2286	-80.9	47.8	-2286
0	0	2640	4.4E6	-1320	-2286	8.8E6

Adding the stiffness of this element to the structure stiffness matrix allows the final structure stiffness equation to be written as follows.

$$
\begin{bmatrix} 235 & 0 & -235 & 0 & 0 & 0 \\ 0 & 11E6 & 0 & -4130 & 5.5E6 & 0 \\ -235 & 0 & 428 & 0 & 7330 & -163 \\ 0 & -4130 & 0 & 316 & -1490 & -0.53 \\ 0 & 5.5E6 & 7330 & -1490 & 34.5E6 & -1320 \\ 0 & 0 & -163 & -0.53 & -1320 & 141 \end{bmatrix} \begin{bmatrix} \Delta_{1x} \\ \theta_1 \\ \Delta_{2x} \\ \Delta_{2y} \\ \theta_2 \\ \Delta_{4x} \end{bmatrix} = \begin{bmatrix} 0 \\ -40000 \\ 0 \\ -135 \\ -22500 \\ -37.5 \end{bmatrix}
$$

Solution of these equations yields the following displacements.

$$
\begin{bmatrix} \Delta_{1x} \\ \theta_1 \\ \Delta_{2x} \\ \Delta_{2y} \\ \theta_2 \\ \Delta_{4x} \end{bmatrix} = \begin{bmatrix} -12.32 \text{ mm} \\ -0.00491 \text{ rads} \\ -12.32 \text{ mm} \\ -0.505 \text{ mm} \\ 0.00217 \text{ rads} \\ -14.49 \text{ mm} \end{bmatrix}
$$

The element end forces for element 2,4 are found by adding the forces associated with the above displacements to the fixed end forces. Note that different transformation matrices are required at ends 2 and 4 to transform from nodal to element coordinates.

End 2

$$
\begin{aligned}
f_{24} &= f_{24}^{n} + f_{24}^{f} \\
&= k_{22}^{4}\, T_{24}\, \Delta_2 + k_{24}\, T_{42}\, \Delta_4 + f_{24}^{f} \\
&= \begin{bmatrix} 188 & 0 & 0 \\ 0 & 1.06 & 2640 \\ 0 & 2640 & 8.8E6 \end{bmatrix} \begin{bmatrix} 1 & 0 & 0 \\ 0 & 1 & 0 \\ 0 & 0 & 1 \end{bmatrix} \begin{bmatrix} -12.32 \\ -0.505 \\ 0.00217 \end{bmatrix} + \\
&\quad \begin{bmatrix} -188 & 0 & 0 \\ 0 & -1.06 & 2640 \\ 0 & -2640 & 4.4E6 \end{bmatrix} \begin{bmatrix} 0.866 & -0.500 & 0 \\ 0.500 & 0.866 & 0 \\ 0 & 0 & 1 \end{bmatrix} \begin{bmatrix} -14.49 \\ 0 \\ 0 \end{bmatrix} + \\
&\quad \begin{bmatrix} 0 \\ 75 \\ 62500 \end{bmatrix}
\end{aligned}
$$

$$= \begin{bmatrix} -2322 \\ 5.19 \\ 17600 \end{bmatrix} + \begin{bmatrix} 2358 \\ 7.68 \\ 19140 \end{bmatrix} + \begin{bmatrix} 0 \\ 75 \\ 62500 \end{bmatrix} = \begin{bmatrix} 36.0 \\ 87.9 \\ 99240 \end{bmatrix}$$

End 4

$$f_{42} = f^n_{42} + f^f_{42}$$

$$= k_{42}\, T_{24}\, \Delta_2 + k^2_{44}\, T_{42}\, \Delta_4 + f^f_{42}$$

$$= \begin{bmatrix} 2322 \\ -5.19 \\ 8215 \end{bmatrix} + \begin{bmatrix} -2358 \\ -7.68 \\ 19140 \end{bmatrix} + \begin{bmatrix} 0 \\ 75 \\ -62500 \end{bmatrix} = \begin{bmatrix} -36.0 \\ 62.1 \\ -35140 \end{bmatrix}$$

8.4 Elastic Supports

Elastic supports can be used to model "imperfect" boundary conditions (e.g. when the foundations are expected to undergo significant movement due to the elastic deformation of the subsoil). In some structures elastic supports are deliberately introduced in order to reduce peak reactions and stresses.

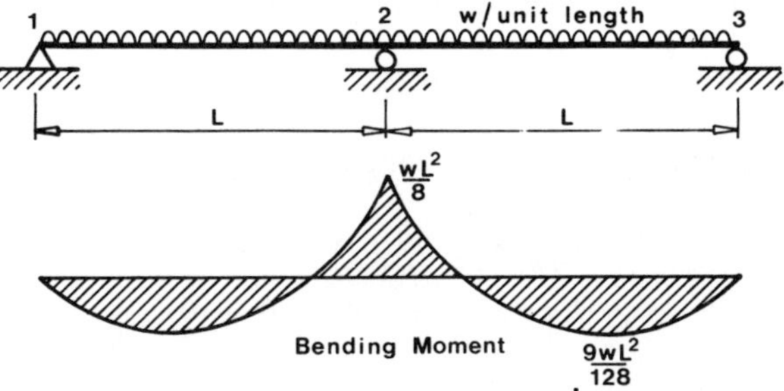

Figure 8.14 An example of rigid supports

For example, if the supports to the continuous beam shown in fig 8.14 are rigid then the hogging moment over the central support will exceed the sagging moment in the span by 78%. Also the magnitude of the central reaction will be more than three times that of the end reactions.

If the rigid central support is replaced by an elastic support of suitable stiffness as shown in fig 8.15, then the hogging and sagging moments can be equalised, and some of the central reaction will be distributed to the end supports.

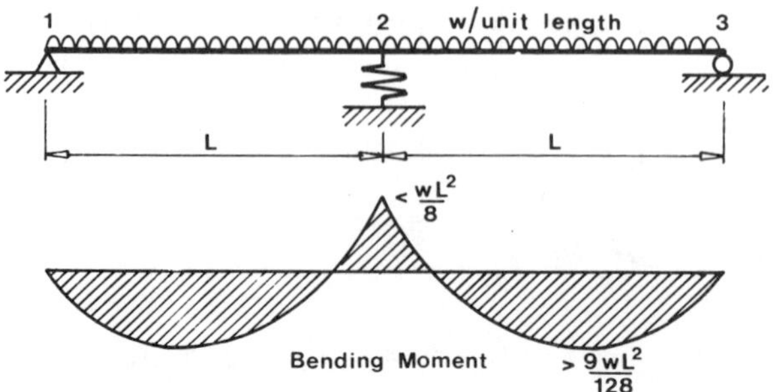

Figure 8.15 Elastic central support

Of course if the spring is too weak then the sagging moment will exceed the hogging moment. It should also be noted that although the ground is usually regarded as rigid for the purposes of structural analysis it will, in fact, always deform under the loads from the foundations. Occasionally the engineer may have to regard the ground as an elastic support in order to represent satisfactorily the true structural behaviour. The structure shown in fig 8.15 will be used to illustrate how elastic supports can be incorporated into the stiffness method. The initial stiffness equation for the structure shown in fig 8.15, in terms of submatrices, is

$$\begin{bmatrix} K_{11} & K_{12} & \mathbf{0} \\ K_{21} & K_{22} & K_{23} \\ \mathbf{0} & K_{32} & K_{33} \end{bmatrix} \begin{bmatrix} \Delta_1 \\ \Delta_2 \\ \Delta_3 \end{bmatrix} = \begin{bmatrix} P_1 \\ P_2 \\ P_3 \end{bmatrix}$$

Expanding and applying the boundary conditions

$$\begin{bmatrix} K_{11} & K_{12} & K_{13} & K_{14} & K_{15} & K_{16} & 0 & 0 & 0 \\ K_{21} & K_{22} & K_{23} & K_{24} & K_{25} & K_{26} & 0 & 0 & 0 \\ K_{31} & K_{32} & K_{33} & K_{34} & K_{35} & K_{36} & 0 & 0 & 0 \\ K_{41} & K_{42} & K_{43} & K_{44} & K_{45} & K_{46} & K_{47} & K_{48} & K_{49} \\ K_{51} & K_{52} & K_{53} & K_{54} & K_{55} & K_{56} & K_{57} & K_{58} & K_{59} \\ K_{61} & K_{62} & K_{63} & K_{64} & K_{65} & K_{66} & K_{67} & K_{68} & K_{69} \\ 0 & 0 & 0 & K_{74} & K_{75} & K_{76} & K_{77} & K_{78} & K_{79} \\ 0 & 0 & 0 & K_{84} & K_{85} & K_{86} & K_{87} & K_{88} & K_{89} \\ 0 & 0 & 0 & K_{94} & K_{95} & K_{96} & K_{97} & K_{98} & K_{99} \end{bmatrix} \begin{bmatrix} \Delta_{1x} \\ \Delta_{1y} \\ \theta_1 \\ \Delta_{2x} \\ \Delta_{2y} \\ \theta_2 \\ \Delta_{3x} \\ \Delta_{3y} \\ \theta_3 \end{bmatrix} = \begin{bmatrix} P_{1x} \\ P_{1y} \\ M_1 \\ P_{2x} \\ P_{2y} \\ M_2 \\ P_{3x} \\ P_{3y} \\ M_3 \end{bmatrix}$$

Note that had the central support been rigid then Δ_{2y} would have been another zero displacement.

If the spring at node 2 did not exist then the "vertical" stiffness of node 2 would be

$$K_{55} = \frac{12EI}{L_{12}^3} + \frac{12EI}{L_{23}^3}$$

The presence of the spring will change the force required to cause unit vertical displacement at node 2 by K_{2y}, where K_{2y} is the stiffness of the spring. The spring will not, however, alter any other stiffness coefficient. Hence elastic supports are dealt with simply by adding the spring stiffness to the appropriate stiffness coefficient on the diagonal of the structure stiffness matrix. If node "*i*" is connected to an elastic support having stiffnesses K_{ix} and K_{iy} against translation and stiffness K_{ir} against rotation, then the elastic support can be regarded as a "special" element whose only contribution to the structure stiffness matrix is the submatrix K_{ii}^s where

$$K_{ii}^s = \begin{bmatrix} K_{ix} & 0 & 0 \\ 0 & K_{iy} & 0 \\ 0 & 0 & K_{ir} \end{bmatrix}$$

These stiffnesses are, of course, expressed in terms of the nodal axes system at node "*i*", and the above submatrix is added to the structure stiffness matrix in exactly the same way as the element stiffness submatrices K_{ii}^j and K_{jj}^i.

Example 8.3

Determine the bending moment diagram for the structure shown below if the spring has a stiffness of 40 kN/mm. Demonstrate the equilibrium of node 2.

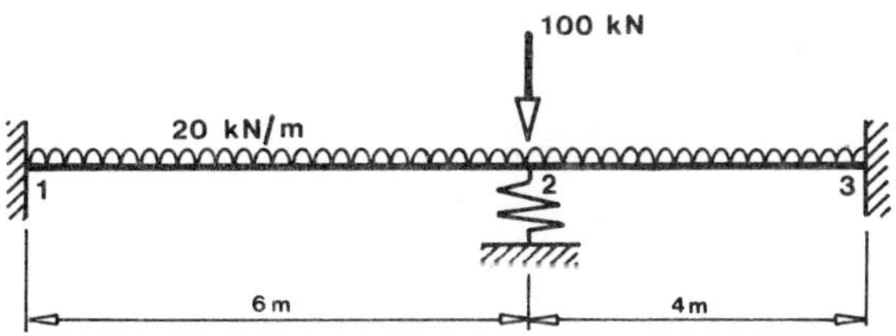

Given that $E = 15$ kN/mm^2, $A = 0.245 \times 10^3$ mm^2, $I = 12 \times 10^9$ mm^4 for both spans. The initial structure stiffness equation, in submatrices, is as follows

$$\begin{bmatrix} K_{11} & K_{12} & 0 \\ K_{21} & K_{22} & K_{23} \\ 0 & K_{32} & K_{33} \end{bmatrix} \begin{bmatrix} \Delta_1 \\ \Delta_2 \\ \Delta_3 \end{bmatrix} = \begin{bmatrix} P_1 \\ P_2 \\ P_3 \end{bmatrix}$$

Applying the boundary conditions $\Delta_1 = \Delta_3 = 0$ gives the final stiffness equation

$$K_{22}\,\Delta_2 = P_2$$

where

$$K_{22} = K_{22}^{1} + K_{22}^{2} + K_{22}^{s}$$

$$= \begin{bmatrix} 613 & 0 & 0 \\ 0 & 10 & -30000 \\ 0 & -30000 & 120E6 \end{bmatrix} + \begin{bmatrix} 919 & 0 & 0 \\ 0 & 33.8 & 67500 \\ 0 & 67500 & 180E6 \end{bmatrix} +$$

$$\begin{bmatrix} 0 & 0 & 0 \\ 0 & 40 & 0 \\ 0 & 0 & 0 \end{bmatrix}$$

Loading

$$P_{21}^{e} = -T_{21}^{-1} f_{21}^{f}$$

$$= -\begin{bmatrix} 1 & 0 & 0 \\ 0 & 1 & 0 \\ 0 & 0 & 1 \end{bmatrix}\begin{bmatrix} 0 \\ 60 \\ -60000 \end{bmatrix} = \begin{bmatrix} 0 \\ -60 \\ 60000 \end{bmatrix}$$

Similarly

$$P_{23}^{e} = \begin{bmatrix} 0 \\ -40 \\ -26667 \end{bmatrix}$$

and the vector of nodal loads at node 2 is

$$P_{2}^{n} = \begin{bmatrix} 0 \\ -100 \\ 0 \end{bmatrix}$$

hence

$$P_2 = P_{2}^{n} + P_{21}^{e} + P_{23}^{e}$$

$$= \begin{bmatrix} 0 \\ -100 \\ 0 \end{bmatrix} + \begin{bmatrix} 0 \\ -60 \\ 60000 \end{bmatrix} + \begin{bmatrix} 0 \\ -40 \\ -26667 \end{bmatrix} = \begin{bmatrix} 0 \\ -200 \\ 33333 \end{bmatrix}$$

The final stiffness equation is

$$\begin{bmatrix} 1532 & 0 & 0 \\ 0 & 83.8 & 37500 \\ 0 & 37500 & 300E6 \end{bmatrix} \begin{bmatrix} \Delta_{2x} \\ \Delta_{2y} \\ \theta_2 \end{bmatrix} = \begin{bmatrix} 0 \\ -200 \\ 33333 \end{bmatrix}$$

which yields

$$\begin{bmatrix} \Delta_{2x} \\ \Delta_{2y} \\ \theta_2 \end{bmatrix} = \begin{bmatrix} 0.0 \text{ mm} \\ -2.58 \text{ mm} \\ 0.434E\text{-}3 \text{ rads} \end{bmatrix}$$

The force at end 1 of element 1,2 is found using the now familar equation

$$f_{12} = k_{11}^{2} T_{12} \Delta_1 + k_{12} T_{12} \Delta_2 + f_{12}^{f}$$

$$\begin{bmatrix} f_{12x} \\ f_{12y} \\ m_{12} \end{bmatrix} = \begin{bmatrix} 613 & 0 & 0 \\ 0 & 10 & 30000 \\ 0 & 30000 & 120E6 \end{bmatrix} \begin{bmatrix} 1 & 0 & 0 \\ 0 & 1 & 0 \\ 0 & 0 & 1 \end{bmatrix} \begin{bmatrix} 0 \\ 0 \\ 0 \end{bmatrix} + \begin{bmatrix} -613 & 0 & 0 \\ 0 & -10 & 30000 \\ 0 & -30000 & 60E6 \end{bmatrix} \begin{bmatrix} 1 & 0 & 0 \\ 0 & 1 & 0 \\ 0 & 0 & 1 \end{bmatrix} \begin{bmatrix} 0 \\ -2.58 \\ 0.434E\text{-}3 \end{bmatrix} + \begin{bmatrix} 0 \\ 60 \\ 60000 \end{bmatrix}$$

$$= \begin{bmatrix} 0 \\ 0 \\ 0 \end{bmatrix} + \begin{bmatrix} 0 \\ 38.8 \\ 103400 \end{bmatrix} + \begin{bmatrix} 0 \\ 60 \\ 60000 \end{bmatrix} = \begin{bmatrix} 0 \\ 98.8 \\ 163400 \end{bmatrix}$$

Similar calculation shows the other element end forces to be

$$\begin{bmatrix} f_{21x} \\ f_{21y} \\ m_{21} \end{bmatrix} = \begin{bmatrix} 0 \\ 21.2 \\ 69500 \end{bmatrix} \qquad \begin{bmatrix} f_{23x} \\ f_{23y} \\ m_{23} \end{bmatrix} = \begin{bmatrix} 0 \\ -17.9 \\ -69500 \end{bmatrix} \qquad \begin{bmatrix} f_{32x} \\ f_{32y} \\ m_{32} \end{bmatrix} = \begin{bmatrix} 0 \\ 97.9 \\ -161700 \end{bmatrix}$$

The freebody diagram for node 2 is shown in the following diagram.

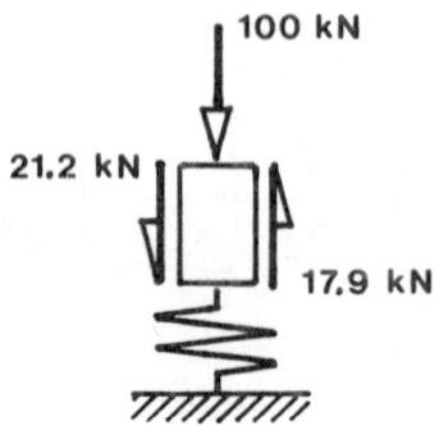

For vertical equilibrium of the element, the force in the spring must be 103.3 kN, and the force in the spring calculated from the vertical deflection of node 2 is 40 x 2.58 = 103.2 kN.

8.5 Elements with Pins

In Chapters 5 and 6 structures consisting exclusively of pin ended elements were considered, and so far this chapter has considered structures consisting entirely of rigidly connected elements. Although most civil engineering structures can be regarded as falling into one of these categories, there are times when the engineer has to deal with hybrid structures containing both rigidly jointed and pinned connections. Figure 8.16 gives three examples of such structures.

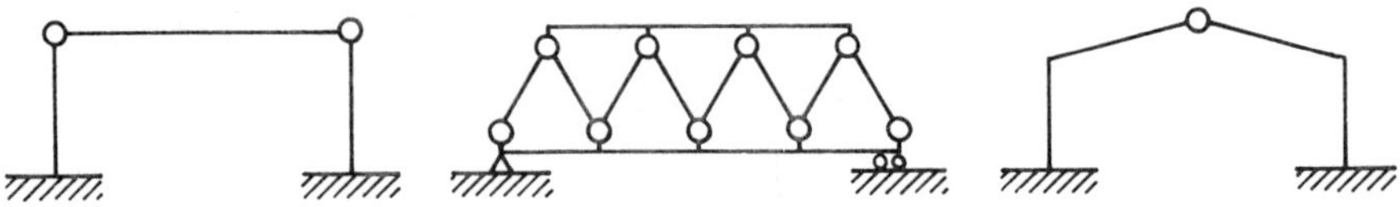

Figure 8.16 Structures containing both rigidly jointed and pin ended elements

Consider the left-hand structure shown in fig 8.16. Obviously the structure consists of three elements. The problem facing the structural analyst is where to locate the nodes.

Figure 8.17 shows three possible idealisations. Where the nodes are located affects the character of the connections at the ends of the elements as shown in the following table.

Idealisation	Element 1,2	Element 2,3	Element 3,4
(a)	rigid/pinned	rigid/rigid	pinned/rigid
(b)	rigid/rigid	pinned/pinned	rigid/rigid
(c)	rigid/rigid	pinned/rigid	pinned/rigid

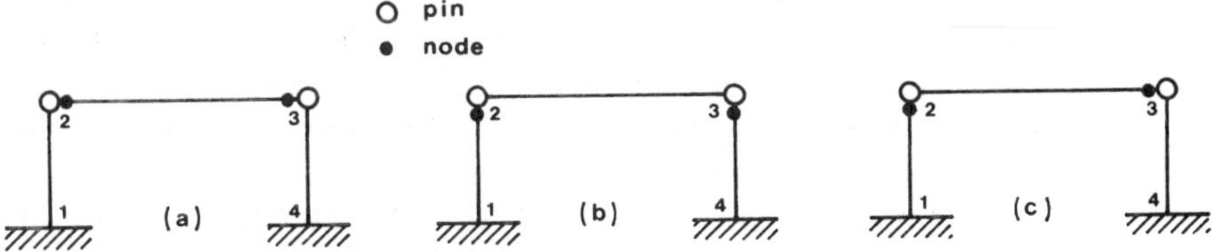

Figure 8.17 Structural idealisations

Note that although the way in which the structure is idealised affects the character of the individual elements, the character of the structure as a whole is unaffected (i.e. the deformed shape of the structure under a given system of loads will be the same for all three idealisations). The nodal rotations found during the analysis would, however, be different because of the discontinuity of rotation at the pins.

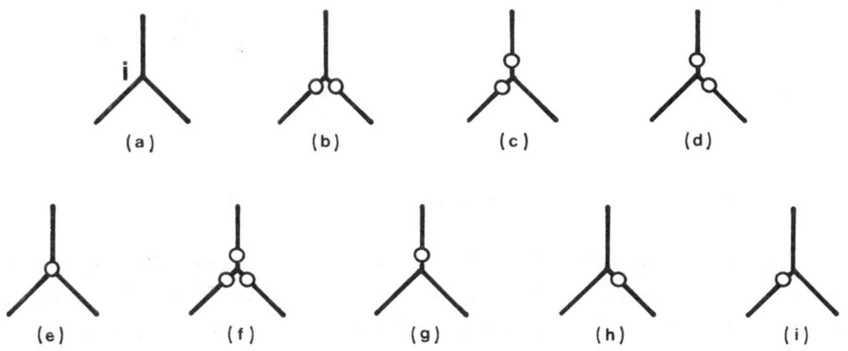

Figure 8.18 Joint idealisations

If the connection between a joint and an element is assumed pinned, then the end of the element connected to that joint can rotate relative to the joint. It is important that the structure is idealised such that the behaviour of the actual, "as built" structure is realistically represented. Consider a joint "*i*" which has three elements framing into it. Figure 8.18 shows ways in which the joint might be idealised.

(a) shows the idealisation that should be used if all three elements are rigidly connected.

(b),(c),(d) show idealisations that could be used to represent a pinned connection. All three connections are structurally equivalent (i.e. all three elements can rotate relative to each other). Note that the use of this idealisation throughout the structure results in a pin jointed frame.

(e),(f) show idealisations that result in the joint having no rotational stiffness (i.e. the joint becomes a mechanism). If the analysis is concerned with joint rotations, e.g. plane frame analysis, then such idealisations lead to a breakdown in the solution procedure as the joint rotations become indeterminate.

Idealisation (e) is frequently used to represent the joints of a pin jointed frame. Pin jointed frame analysis is unconcerned with joint rotations; hence the fact that the joints have no rotational stiffness is unimportant.

(g),(h),(i) show idealisations where one element framing into the joint has a pinned connection, and the other two are rigidly connected. Note that these idealisations are not structurally equivalent.

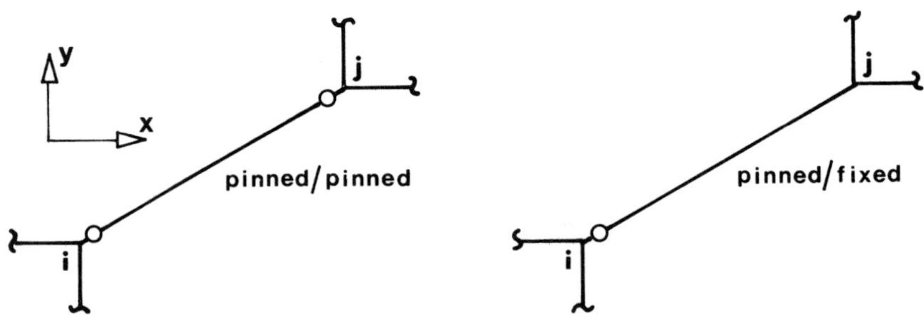

Figure 8.19 Pinned/pinned and pinned/rigid elements

The theory already developed has been based upon elements that are rigidly connected to the joints at their ends. That theory is, however, generally applicable, and all that is required to include pinned/pinned and pinned/rigid elements in the analysis is the element stiffness matrices for these types of elements. At this point it is worthwhile recalling that the element stiffness matrix can be evaluated simply by finding the force system required to produce unit displacement in the direction of each freedom in turn.

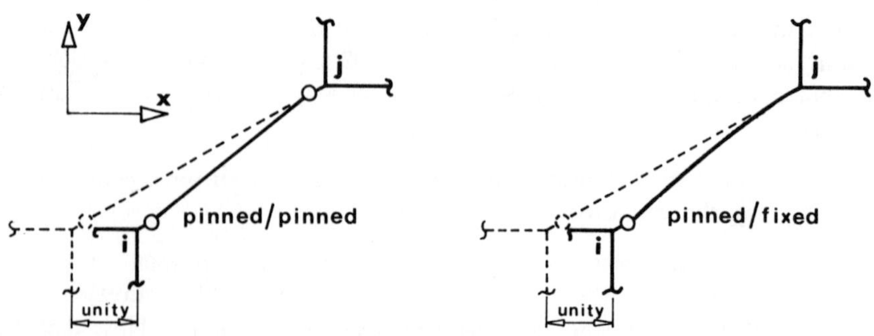

Figure 8.20 Unit displacement in the "x" direction

Figure 8.19 shows a pinned/pinned and a pinned/rigid element running between nodes "i" and "j". It should be obvious that both elements will affect the translational stiffness of joints "i" and "j". Further it should be noted that the pinned/pinned element adds no rotational stiffness to either joint, and that the pinned/rigid element adds rotational stiffness only to joint "j".

Figure 8.20 shows unit displacement in the "x" direction and illustrates how moment will be developed at joint "j" in the pinned/rigid case, but not in the pinned/pinned case.

Element Pinned at Both Ends

The element stiffness matrix in the element axes system for a pinned/pinned element can be written down by inspection and partitioned in the normal way to produce submatrices as shown in the following equation.

$$\begin{bmatrix} f_{ijx} \\ f_{ijy} \\ m_{ij} \\ \hline f_{jix} \\ f_{jiy} \\ m_{ji} \end{bmatrix} = \left[\begin{array}{ccc|ccc} \frac{EA}{L} & 0 & 0 & \frac{-EA}{L} & 0 & 0 \\ 0 & 0 & 0 & 0 & 0 & 0 \\ 0 & 0 & 0 & 0 & 0 & 0 \\ \hline \frac{-EA}{L} & 0 & 0 & \frac{EA}{L} & 0 & 0 \\ 0 & 0 & 0 & 0 & 0 & 0 \\ 0 & 0 & 0 & 0 & 0 & 0 \end{array}\right] \begin{bmatrix} \delta_{ijx} \\ \delta_{ijy} \\ \theta_{ij} \\ \hline \delta_{jix} \\ \delta_{jiy} \\ \theta_{ji} \end{bmatrix}$$

i.e.

$$f_{ij} = k_{ii}^{j}\,\delta_{ij} + k_{ij}\,\delta_{ji}$$

and

$$f_{ji} = k_{ji}\,\delta_{ij} + k_{jj}^{i}\,\delta_{ji}$$

The element stiffness submatrices in the global axes system are found using the familiar triple matrix multiplication

$$K_{ii}^{j} = T_{ij}^{-1}\,k_{ii}^{j}\,T_{ij}$$

where

$$T_{ij} = \begin{bmatrix} \cos\alpha & \sin\alpha & 0 \\ -\sin\alpha & \cos\alpha & 0 \\ 0 & 0 & 1 \end{bmatrix}$$

hence

$$K_{ii}^{j} = \begin{bmatrix} \frac{EA}{L}\cos^2\alpha & \frac{EA}{L}\cos\alpha\,\sin\alpha & 0 \\ \frac{EA}{L}\cos\alpha\,\sin\alpha & \frac{EA}{L}\sin^2\alpha & 0 \\ 0 & 0 & 0 \end{bmatrix} \tag{8.8}$$

The other three element stiffness submatrices are similar in form and can, in fact, be found from the K_{ii}^{j} submatrix as shown below.

Find K_{ij} and K_{ji} by multiplying corresponding elements of K_{ii}^{j} by

$$\begin{bmatrix} -1 & -1 & 1 \\ -1 & -1 & 1 \\ 1 & 1 & 1 \end{bmatrix}$$

Find K_{jj}^{i} by multiplying corresponding elements K_{ii}^{j} of by

$$\begin{bmatrix} 1 & 1 & 1 \\ 1 & 1 & 1 \\ 1 & 1 & 1 \end{bmatrix}$$

As might be expected, these element stiffness submatrices are identical to the submatrices for a standard plane frame element having no flexural stiffness. By using the above submatrices elements with pins at both ends can be included in the analysis using exactly the same procedures as were used for rigidly connected elements.

Element Pinned at End "*i*"

The procedure to deal with elements pinned at only one end is identical to that used for elements pinned at both ends. The first stage is to determine the element stiffness matrix. Here the element will be assumed to be pinned at node "*i*" and rigidly connected at node "*j*", as shown in fig 8.21.

Figure 8.21 Element pinned at end "*i*"

The axial stiffness of the element is unaffected by the pin, hence

$$K_{11} = \frac{EA}{L} \qquad K_{41} = -\frac{EA}{L} \qquad K_{21} = K_{31} = K_{51} = K_{61} = 0$$

There are many ways in which the remaining stiffness terms can be found. Here use will again be made of shape functions. To illustrate, consider the stiffness terms associated with freedom 2 (see fig 8.22).

Section 2.12 showed that the deformed shape of a prismatic line element loaded only at its ends is described exactly by a cubic polynomial.

$$y = ax^3 + bx^2 + cx + d$$

$$\frac{dy}{dx} = 3ax^2 + 2bx + c$$

$$\frac{d^2y}{dx^2} = 6ax + 2b$$

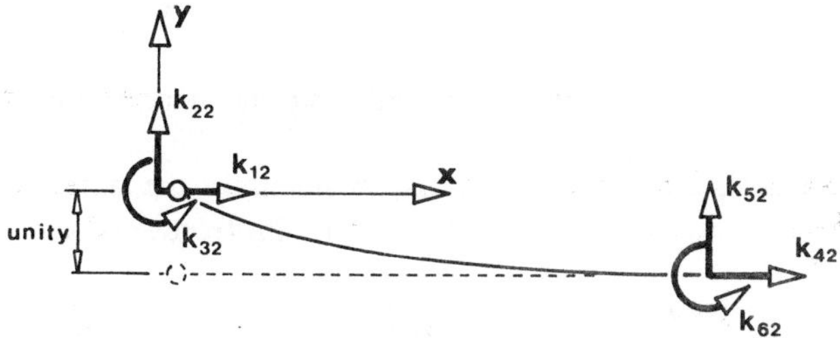

Figure 8.22 Unit displacement at freedom 2

The boundary conditions are

at $x = 0$

$$y = 0 \qquad \text{therefore} \qquad 0 = 0 + 0 + 0 + d$$

$$\frac{d^2y}{dx^2} = 0 \qquad \text{therefore} \qquad 0 = 0 + 2b$$

at $x = L$

$$y = -1 \qquad \text{therefore} \qquad -1 = aL^3 + bL^2 + cL + d$$

$$\frac{dy}{dx} = 0 \qquad \text{therefore} \qquad 0 = 3aL^2 + 2bL + c$$

hence

$$a = \frac{1}{2L^3} \qquad b = 0 \qquad c = -\frac{3}{2L} \qquad d = 0$$

giving

$$y = \frac{x^3}{2L^3} - \frac{3x}{2L}$$

but

$$M = EI\frac{d^2y}{dx^2} = \frac{3EI}{L^3}x$$

therefore

$$M = 0 \qquad \text{when} \qquad x = 0$$

$$M = \frac{3EI}{L^3} \qquad \text{when} \qquad x = L$$

The sign means that the applied bending moment causes positive curvature at the right-hand end of the beam. Hence the force system associated with unit displacement in the direction of freedom 2 is as shown in fig 8.23.

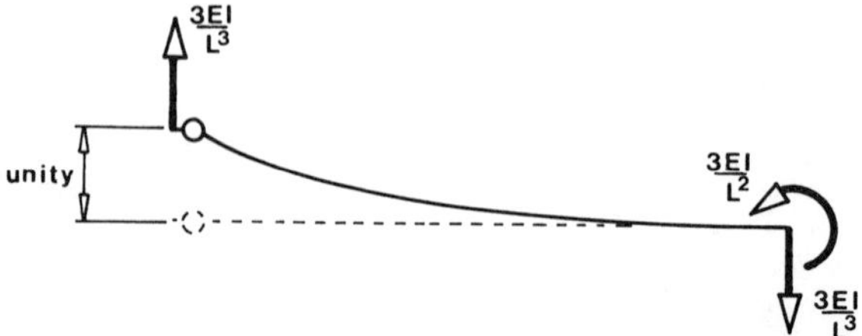

Figure 8.23 Force system associated with unit displacement at freedom 2

Note that the forces $3EI/L^3$ are necessary to maintain the equilibrium of the element, and the stiffness coefficients associated with freedom 2 are

$$K_{22} = \frac{3EI}{L^3} \qquad K_{52} = -\frac{3EI}{L^3} \qquad K_{62} = \frac{3EI}{L^2}$$

$$K_{12} = 0 \qquad K_{32} = 0 \qquad K_{42} = 0$$

Treating the other freedoms similarly the element stiffness relationship can be shown to be

$$\begin{bmatrix} f_{ijx} \\ f_{ijy} \\ m_{ij} \\ f_{jix} \\ f_{jiy} \\ m_{ji} \end{bmatrix} = \begin{bmatrix} \frac{EA}{L} & 0 & 0 & -\frac{EA}{L} & 0 & 0 \\ 0 & \frac{3EI}{L^3} & 0 & 0 & -\frac{3EI}{L^3} & \frac{3EI}{L^2} \\ 0 & 0 & 0 & 0 & 0 & 0 \\ -\frac{EA}{L} & 0 & 0 & \frac{EA}{L} & 0 & 0 \\ 0 & -\frac{3EI}{L^3} & 0 & 0 & \frac{3EI}{L^3} & -\frac{3EI}{L^2} \\ 0 & \frac{3EI}{L^2} & 0 & 0 & -\frac{3EI}{L^2} & \frac{3EI}{L} \end{bmatrix} \begin{bmatrix} \delta_{ijx} \\ \delta_{ijy} \\ \theta_{ij} \\ \delta_{jix} \\ \delta_{jiy} \\ \theta_{ji} \end{bmatrix} \qquad (8.9)$$

The above equation can be written in terms of the submatrices indicated by the broken lines.

$$\begin{bmatrix} \boldsymbol{f}_{ij} \\ \boldsymbol{f}_{ji} \end{bmatrix} = \begin{bmatrix} \boldsymbol{k}_{ii}^{j} & \boldsymbol{k}_{ij} \\ \boldsymbol{k}_{ji} & \boldsymbol{k}_{jj}^{i} \end{bmatrix} \begin{bmatrix} \boldsymbol{\delta}_{ij} \\ \boldsymbol{\delta}_{ji} \end{bmatrix}$$

The global element stiffness submatrices are found as follows.

$$\boldsymbol{K}_{ii}^{j} = \boldsymbol{T}_{ij}^{-1} \boldsymbol{k}_{ii}^{j} \boldsymbol{T}_{ij}$$

$$= \begin{bmatrix} \left(\frac{EA}{L}\cos^2\alpha + \frac{3EI}{L^3}\sin^2\alpha\right) & \left(\frac{EA}{L} - \frac{3EI}{L^3}\right)\cos\alpha\,\sin\alpha & 0 \\ \left(\frac{EA}{L} - \frac{3EI}{L^3}\right)\cos\alpha\,\sin\alpha & \left(\frac{EA}{L}\sin^2\alpha + \frac{3EI}{L^3}\cos^2\alpha\right) & 0 \\ 0 & 0 & 0 \end{bmatrix} \tag{8.10}$$

$$K_{ij} = T^{-1}_{ij}\, k_{ij}\, T_{ij}$$

$$= \begin{bmatrix} -\left(\frac{EA}{L}\cos^2\alpha + \frac{3EI}{L^3}\sin^2\alpha\right) & -\left(\frac{EA}{L} - \frac{3EI}{L^3}\right)\cos\alpha\,\sin\alpha & -\frac{3EI}{L^2}\sin\alpha \\ -\left(\frac{EA}{L} - \frac{3EI}{L^3}\right)\cos\alpha\,\sin\alpha & -\left(\frac{EA}{L}\sin^2\alpha + \frac{3EI}{L^3}\cos^2\alpha\right) & \frac{3EI}{L^2}\cos\alpha \\ 0 & 0 & 0 \end{bmatrix} \tag{8.11}$$

$$K_{ji} = K^T_{ij} \tag{8.12}$$

$$K^i_{jj} = T^{-1}_{ij}\, k^i_{jj}\, T_{ij}$$

$$= \begin{bmatrix} \left(\frac{EA}{L}\cos^2\alpha + \frac{3EI}{L^3}\sin^2\alpha\right) & \left(\frac{EA}{L} - \frac{3EI}{L^3}\right)\cos\alpha\,\sin\alpha & \frac{3EI}{L^2}\sin\alpha \\ \left(\frac{EA}{L} - \frac{3EI}{L^3}\right)\cos\alpha\,\sin\alpha & \left(\frac{EA}{L}\sin^2\alpha + \frac{3EI}{L^3}\cos^2\alpha\right) & -\frac{3EI}{L^2}\cos\alpha \\ \frac{3EI}{L^2}\sin\alpha & -\frac{3EI}{L^2}\cos\alpha & \frac{3EI}{L} \end{bmatrix} \tag{8.13}$$

For hand calculation it is best to first determine the K^i_{jj} submatrix and then obtain the other submatrices using the relationships expressed in the following matrices.

Find K^j_{ii} by multiplying corresponding elements of K^i_{jj} by

$$\begin{bmatrix} 1 & 1 & 0 \\ 1 & 1 & 0 \\ 0 & 0 & 0 \end{bmatrix} \tag{8.14}$$

Find K_{ij} by multiplying corresponding elements of K^i_{jj} by

$$\begin{bmatrix} -1 & -1 & -1 \\ -1 & -1 & -1 \\ 0 & 0 & 0 \end{bmatrix} \tag{8.15}$$

Find K_{ji} by multiplying corresponding elements of K_{jj}^{i} by

$$\begin{bmatrix} -1 & -1 & 0 \\ -1 & -1 & 0 \\ -1 & -1 & 0 \end{bmatrix} \tag{8.16}$$

Note that this is not a matrix multiplication, but simply the multiplication of corresponding elements.

Element Pinned at End "j"

The techniques used for an element with a pin at "i" can be used to show that the element stiffness equation (in the element axes system) for an element with a pin at "j" and rigidly jointed at "i" is as shown in the following equation.

$$\begin{bmatrix} f_{ijx} \\ f_{ijy} \\ m_{ij} \\ f_{jix} \\ f_{jiy} \\ m_{ji} \end{bmatrix} = \begin{bmatrix} \frac{EA}{L} & 0 & 0 & -\frac{EA}{L} & 0 & 0 \\ 0 & \frac{3EI}{L^3} & \frac{3EI}{L^2} & 0 & -\frac{3EI}{L^3} & 0 \\ 0 & \frac{3EI}{L^2} & \frac{3EI}{L} & 0 & -\frac{3EI}{L^2} & 0 \\ -\frac{EA}{L} & 0 & 0 & \frac{EA}{L} & 0 & 0 \\ 0 & -\frac{3EI}{L^3} & -\frac{3EI}{L^2} & 0 & \frac{3EI}{L^3} & 0 \\ 0 & 0 & 0 & 0 & 0 & 0 \end{bmatrix} \begin{bmatrix} \delta_{ijx} \\ \delta_{ijy} \\ \theta_{ij} \\ \delta_{jix} \\ \delta_{jiy} \\ \theta_{ji} \end{bmatrix}$$

The global element stiffness submatrices are developed in the usual way as follows

$$K_{ii}^{j} = T_{ij}^{-1}\, k_{ii}^{j}\, T_{ij}$$

$$= \begin{bmatrix} \left(\frac{EA}{L}\cos^2\alpha + \frac{3EI}{L^3}\sin^2\alpha\right) & \left(\frac{EA}{L} - \frac{3EI}{L^3}\right)\cos\alpha\,\sin\alpha & -\frac{3EI}{L^2}\sin\alpha \\ \left(\frac{EA}{L} - \frac{3EI}{L^3}\right)\cos\alpha\,\sin\alpha & \left(\frac{EA}{L}\sin^2\alpha + \frac{3EI}{L^3}\cos^2\alpha\right) & \frac{3EI}{L^2}\cos\alpha \\ -\frac{3EI}{L^2}\sin\alpha & \frac{3EI}{L^2}\cos\alpha & \frac{3EI}{L} \end{bmatrix} \tag{8.17}$$

$$K_{ij} = T_{ij}^{-1} k_{ij} T_{ij}$$

$$= \begin{bmatrix} -\left(\frac{EA}{L}\cos^2\alpha + \frac{3EI}{L^3}\sin^2\alpha\right) & -\left(\frac{EA}{L} - \frac{3EI}{L^3}\right)\cos\alpha\sin\alpha & 0 \\ -\left(\frac{EA}{L} - \frac{3EI}{L^3}\right)\cos\alpha\sin\alpha & -\left(\frac{EA}{L}\sin^2\alpha + \frac{3EI}{L^3}\cos^2\alpha\right) & 0 \\ \frac{3EI}{L^2}\sin\alpha & -\frac{3EI}{L^2}\cos\alpha & 0 \end{bmatrix} \tag{8.18}$$

$$K_{ji} = K_{ij}^T \tag{8.19}$$

$$K_{jj}^i = T_{ij}^{-1} k_{jj}^i T_{ij}$$

$$= \begin{bmatrix} \left(\frac{EA}{L}\cos^2\alpha + \frac{3EI}{L^3}\sin^2\alpha\right) & \left(\frac{EA}{L} - \frac{3EI}{L^3}\right)\cos\alpha\sin\alpha & 0 \\ \left(\frac{EA}{L} - \frac{3EI}{L^3}\right)\cos\alpha\sin\alpha & \left(\frac{EA}{L}\sin^2\alpha + \frac{3EI}{L^3}\cos^2\alpha\right) & 0 \\ 0 & 0 & 0 \end{bmatrix} \tag{8.20}$$

For hand calculation it is best to first determine the K_{ii}^j submatrix and then obtain the other submatrices as follows.

Find K_{jj}^i by multiplying corresponding elements of K_{ii}^j by

$$\begin{bmatrix} 1 & 1 & 0 \\ 1 & 1 & 0 \\ 0 & 0 & 0 \end{bmatrix} \tag{8.21}$$

Find K_{ij} by multiplying corresponding elements of K_{ii}^j by

$$\begin{bmatrix} -1 & -1 & 0 \\ -1 & -1 & 0 \\ -1 & -1 & 0 \end{bmatrix} \tag{8.22}$$

Find K_{ji} by multiplying corresponding elements of K_{ii}^j by

$$\begin{bmatrix} -1 & -1 & -1 \\ -1 & -1 & -1 \\ 0 & 0 & 0 \end{bmatrix} \tag{8.23}$$

Note again that this is not a matrix multiplication, but simply the multiplication of corresponding elements.

Element Loads and Element End Forces

If an element is pinned at one or more ends then the pin (or pins) must be taken into account when determining the fixed end and the equivalent joint forces.

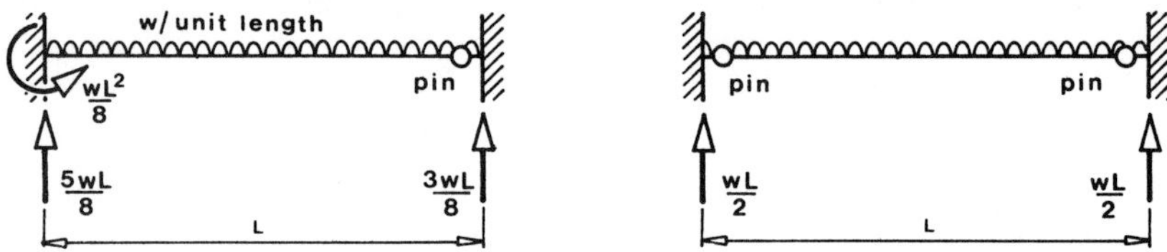

Figure 8.24 Fixed end forces for elements containing pins

Figure 8.24 shows the fixed end forces due to a uniformly distributed load acting on elements containing pins. The fixed end forces for the element pinned at both ends can be obtained by inspection. The element pinned at one end behaves, under transverse loads, as a propped cantilever, and fixed end forces for common loading cases are to be found in many textbooks. It is worthwhile noting that if the fixed end forces for a rigid/rigid element are known, then the fixed end forces for the rigid/pinned case can easily be found by superposition as illustrated in fig 8.25.

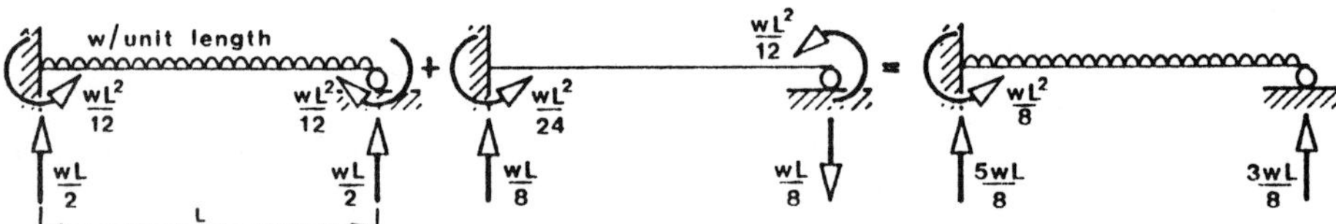

Figure 8.25 Fixed end forces for a rigid/pinned element

Once the nodal displacements have been found the now familiar matrix equations are used to find the element end forces.

$$f_{ij} = k_{ii}^{j}\, T_{ij}\, \Delta_i + k_{ij}\, T_{ij}\, \Delta_j + f_{ij}^{f}$$

$$f_{ji} = k_{ji}\, T_{ij}\, \Delta_i + k_{jj}^{i}\, T_{ij}\, \Delta_j + f_{ji}^{f}$$

The stiffness submatrices used in the above equations must, of course, be consistent with the end conditions of the element (e.g. if the element has a pinned connection at end "*i*" and a rigid connection at end "*j*" the submatrices must be as shown in equation (8.9)).

Example 8.4

For the structure shown prove that the reactive forces at node 1 are independent of which element is assumed to contain the pin. Given that $E = 15\ \text{kN/mm}^2$, $A = 45 \times 10^3\ \text{mm}^2$ $I = 340 \times 10^6\ \text{mm}^4$ for both spans.

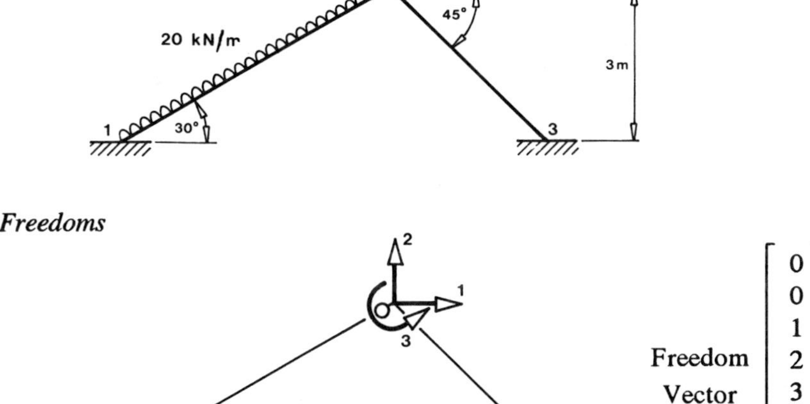

Freedoms

Freedom Vector $\begin{bmatrix} 0 \\ 0 \\ 1 \\ 2 \\ 3 \\ 0 \\ 0 \\ 0 \end{bmatrix}$

(i) If element 1,2 contains the pin, then the fixed end and equivalent joint forces are as shown in the following diagram.

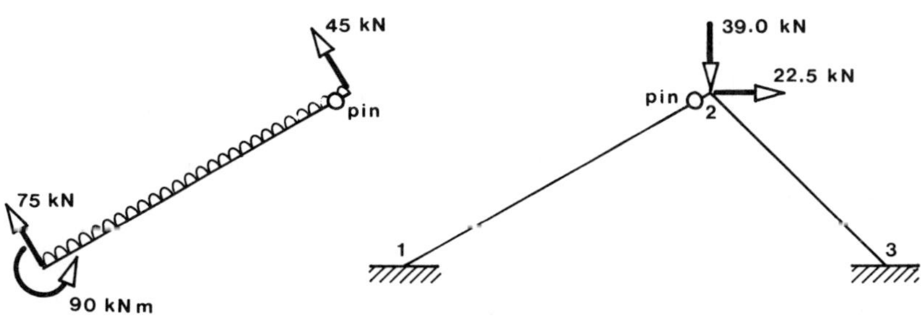

Element 1,2 $A = 45000\ \text{mm}^2$, $I = 340 \times 10^6\ \text{mm}^4$, $L = 6000\ \text{mm}$
$E = 15\ \text{kN/mm}^2$, code no. = [0 0 0 1 2 3], $\alpha = 30°$

This is a rigid/pinned element, therefore equation (8.17) yields the K_{ii}^j global element stiffness submatrix, and the remaining submatrices are found using transformations (8.21)-(8.23). Apply the code number to the global element stiffness matrix.

$$\begin{array}{c|cccccc} & 0 & 0 & 0 & 1 & 2 & 3 \\ \hline 0 & 84.4 & 48.7 & -213 & -84.4 & -48.7 & 0 \\ 0 & 48.7 & 28.2 & 368 & -48.7 & -28.2 & 0 \\ 0 & -213 & 368 & 2.55E6 & 213 & -368 & 0 \\ 1 & -84.4 & -48.7 & 213 & 84.4 & 48.7 & 0 \\ 2 & -48.7 & -28.2 & -368 & 48.7 & 28.2 & 0 \\ 3 & 0 & 0 & 0 & 0 & 0 & 0 \end{array}$$

Adding the stiffness terms to the structure stiffness matrix gives

$$\begin{bmatrix} 84.4 & 48.7 & 0 \\ 48.7 & 28.2 & 0 \\ 0 & 0 & 0 \end{bmatrix}$$

Element 2,3 $A = 45000 \text{ mm}^2$, $I = 340 \times 10^6 \text{ mm}^4$, $L = 4243$ mm
$E = 15 \text{ kN/mm}^2$, code no. = [1 2 3 0 0 0], $\alpha = 315°$

This is a rigid/rigid element, therefore equation (7.9) yields the K^j_{ii} global element stiffness submatrix, and the remaining submatrices are found using transformations (7.10) - (7.12). Apply the code number to the global element stiffness matrix.

$$\begin{array}{c|cccccc} & 1 & 2 & 3 & 0 & 0 & 0 \\ \hline 1 & 80.0 & -79.2 & 1200 & -80.0 & 79.2 & 1200 \\ 2 & -79.2 & 80.0 & 1200 & 79.2 & -80.0 & 1200 \\ 3 & 1200 & 1200 & 4.81E6 & -1200 & -1200 & 2.40E6 \\ 0 & -80.0 & 79.2 & -1200 & 80.0 & -79.2 & -1200 \\ 0 & 79.2 & -80.0 & -1200 & -79.2 & 80.0 & -1200 \\ 0 & 1200 & 1200 & 2.40E6 & -1200 & -1200 & 4.81E6 \end{array}$$

Adding the stiffness terms to the structure stiffness matrix completes its assembly, and the final structure stiffness equation is

$$\begin{bmatrix} 164.4 & -30.5 & 1200 \\ -30.5 & 108.2 & 1200 \\ 1200 & 1200 & 4.81E6 \end{bmatrix} \begin{bmatrix} \Delta_{2x} \\ \Delta_{2y} \\ \theta_2 \end{bmatrix} = \begin{bmatrix} 30.0 \\ -39.0 \\ 0 \end{bmatrix}$$

which yields

$$\begin{bmatrix} \Delta_{2x} \\ \Delta_{2y} \\ \theta_2 \end{bmatrix} = \begin{bmatrix} 0.0732 \text{ mm} \\ -0.341 \text{ mm} \\ 0.667E\text{-}4 \text{ rads} \end{bmatrix}$$

Calculate the reactive forces at node 1 from the element end forces.

Element 1,2 (rigid/pinned)

$$f_{12} = k_{11}^{2}\, T_{12}\, \Delta_1 + k_{12}\, T_{12}\, \Delta_2 + f_{12}^{f}$$

$$= k_{12}\, T_{12}\, \Delta_2 + f_{12}^{f} \qquad \text{as} \qquad \Delta_1 = \mathbf{0}$$

hence

$$\begin{bmatrix} f_{12x} \\ f_{12y} \\ m_{12} \end{bmatrix} = \begin{bmatrix} -EA/L & 0 & 0 \\ 0 & -3EI/L^3 & 0 \\ 0 & -3EI/L^2 & 0 \end{bmatrix} \begin{bmatrix} cos\alpha & sin\alpha & 0 \\ -sin\alpha & cos\alpha & 0 \\ 0 & 0 & 1 \end{bmatrix} \begin{bmatrix} \Delta_{2x} \\ \Delta_{2y} \\ \theta_2 \end{bmatrix} + \begin{bmatrix} f_{12x}^{f} \\ f_{12y}^{f} \\ m_{12}^{f} \end{bmatrix}$$

$$= \begin{bmatrix} -113 & 0 & 0 \\ 0 & -0.071 & 0 \\ 0 & -425 & 0 \end{bmatrix} \begin{bmatrix} 0.866 & 0.5 & 0 \\ -0.5 & 0.866 & 0 \\ 0 & 0 & 1 \end{bmatrix} \begin{bmatrix} 0.792 \\ -0.341 \\ 667E-7 \end{bmatrix} + \begin{bmatrix} 0 \\ 75 \\ 90000 \end{bmatrix}$$

$$= \begin{bmatrix} 12.1 \\ 0.0236 \\ 141 \end{bmatrix} + \begin{bmatrix} 0 \\ 75 \\ 90000 \end{bmatrix} = \begin{bmatrix} 12.1 \\ 75 \\ 90140 \end{bmatrix}$$

$$R_i = \sum_{j=a}^{n} T_{ij}^{-1} f_{ij} - P_i = T_{12}^{-1} f_{12}$$

i.e.

$$\begin{bmatrix} R_{1x} \\ R_{1y} \\ RM_1 \end{bmatrix} = \begin{bmatrix} 0.866 & -0.5 & 0 \\ 0.5 & 0.866 & 0 \\ 0 & 0 & 1 \end{bmatrix} \begin{bmatrix} 12.1 \\ 75 \\ 90140 \end{bmatrix} + \begin{bmatrix} -27.0 \text{ kN} \\ 71.0 \text{ kN} \\ 90140 \text{ kN mm} \end{bmatrix}$$

(ii) If element 2,3 contains the pin, then the fixed end and equivalent joint forces are

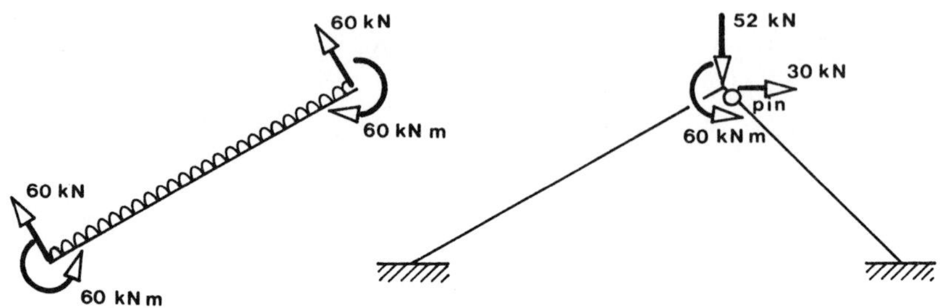

Element 1,2 A = 45000 mm^2, I = 340 x 10^6 mm^4, L= 6000 mm
E = 15 kN/mm^2, code no. = [0 0 0 1 2 3], α = 30°

This is a rigid/rigid element, therefore equation (7.11) yields the K^j_{ii} global element stiffness submatrix, and the remaining submatrices are found using relationships (7.12)-(7.14). Apply the code number to the global element stiffness matrix.

	0	0	0	1	2	3
0	84.4	48.6	-425	-84.4	-48.6	-425
0	48.6	28.3	736	-48.7	-28.3	736
0	-425	736	3.4E6	425	-736	1.7E6
1	-84.4	-48.7	425	84.4	48.6	425
2	-48.6	-28.3	-736	48.6	28.3	-736
3	-425	736	1.7E6	425	-736	3.4E6

Adding the stiffness terms to the structure stiffness matrix gives

$$\begin{bmatrix} 84.4 & 48.6 & 425 \\ 48.6 & 28.3 & -736 \\ 425 & -736 & 3.4E6 \end{bmatrix}$$

Element 2,3 A = 45000 mm^2, I = 340 x 10^6 mm^4, L= 4243 mm
E = 15 kN/mm^2, code no. = [1 2 3 0 0 0], α = 315°

This is a pinned/rigid element, therefore equation (8.13) yields the K^i_{jj} global element stiffness submatrix, and the remaining submatrices are found using transformations (8.14)-(8.16). Apply the code number to the global element stiffness matrix.

$$
\begin{array}{c} \\ 1 \\ 2 \\ 3 \\ 0 \\ 0 \\ 0 \end{array}
\begin{array}{c}
\begin{array}{cccccc} 1 & 2 & 3 & 0 & 0 & 0 \end{array} \\
\begin{bmatrix}
79.6 & -79.4 & 0 & -79.6 & 79.4 & 601 \\
-79.4 & 79.6 & 0 & 79.4 & -79.6 & 601 \\
0 & 0 & 0 & 0 & 0 & 0 \\
-79.6 & 79.4 & 0 & 79.6 & -79.4 & -601 \\
79.4 & -79.6 & 0 & -79.4 & 80.0 & -601 \\
601 & 601 & 0 & -601 & -601 & 3.61E6
\end{bmatrix}
\end{array}
$$

Adding the stiffness terms to the structure stiffness matrix completes its assembly, and the final structure stiffness equation is

$$
\begin{bmatrix} 164.0 & -30.8 & 425 \\ -30.8 & 107.9 & -736 \\ 425 & -736 & 3.4E6 \end{bmatrix}
\begin{bmatrix} \Delta_{2x} \\ \Delta_{2y} \\ \theta_2 \end{bmatrix}
=
\begin{bmatrix} 30.0 \\ -52.0 \\ 60000 \end{bmatrix}
$$

which yields

$$
\begin{bmatrix} \Delta_{2x} \\ \Delta_{2y} \\ \theta_2 \end{bmatrix}
=
\begin{bmatrix} 0.0733 \text{ mm} \\ -0.341 \text{ mm} \\ 0.0176 \text{ rads} \end{bmatrix}
$$

Calculate the reactive forces at node 1 from the element end forces.

Element 1,2 (rigid/rigid)

$$
\begin{aligned}
f_{12} &= k_{11}^{2} T_{12} \Delta_1 + k_{12} T_{12} \Delta_2 + f_{12}^{f} \\
&= k_{12} T_{12} \Delta_2 + f_{12}^{f} \qquad \text{as} \qquad \Delta_1 = \mathbf{0}
\end{aligned}
$$

hence

$$
\begin{bmatrix} f_{12x} \\ f_{12y} \\ m_{12} \end{bmatrix}
=
\begin{bmatrix} -EA/L & 0 & 0 \\ 0 & -12EI/L^3 & 6EI/L^2 \\ 0 & -6EI/L^2 & 2EI/L \end{bmatrix}
\begin{bmatrix} \cos\alpha & \sin\alpha & 0 \\ -\sin\alpha & \cos\alpha & 0 \\ 0 & 0 & 1 \end{bmatrix}
\begin{bmatrix} \Delta_{2x} \\ \Delta_{2y} \\ \theta_2 \end{bmatrix}
+
\begin{bmatrix} f_{12x}^{f} \\ f_{12y}^{f} \\ m_{12}^{f} \end{bmatrix}
$$

$$
=
\begin{bmatrix} -113 & 0 & 0 \\ 0 & -0.283 & 850 \\ 0 & -850 & 1.7E6 \end{bmatrix}
\begin{bmatrix} 0.866 & 0.5 & 0 \\ -0.5 & 0.866 & 0 \\ 0 & 0 & 1 \end{bmatrix}
\begin{bmatrix} 0.0733 \\ -0.341 \\ 0.0176 \end{bmatrix}
+
\begin{bmatrix} 0 \\ 60 \\ 60000 \end{bmatrix}
$$

$$= \begin{bmatrix} 12.1 \\ 15.1 \\ 30200 \end{bmatrix} + \begin{bmatrix} 0 \\ 60 \\ 60000 \end{bmatrix} = \begin{bmatrix} 12.1 \\ 75.1 \\ 90200 \end{bmatrix}$$

$$R_i = \sum_{j=a}^{n} T_{ij}^{-1} f_{ij} - P_i = T_{12}^{-1} f_{12}$$

i.e.

$$\begin{bmatrix} R_{1x} \\ R_{1y} \\ RM_1 \end{bmatrix} = \begin{bmatrix} 0.866 & -0.5 & 0 \\ 0.5 & 0.866 & 0 \\ 0 & 0 & 1 \end{bmatrix} \begin{bmatrix} 12.1 \\ 75.1 \\ 90200 \end{bmatrix} = \begin{bmatrix} -27.1 \text{ kN} \\ 71.1 \text{ kN} \\ 90200 \text{ kN mm} \end{bmatrix}$$

These figures are, given the limited number of significant figures employed, the same as when the pin was assumed to be in element 1,2.

8.6 Non-Prismatic Elements

All of the elements considered so far have been straight and of uniform cross-section, i.e. prismatic elements. If, however, an element is curved or has section properties which vary along its length, then it is known as a non-prismatic element. Figure 8.26 shows four examples of non-prismatic elements.

(a) is a stepped member
(b) is a tapered member
(c) is a curved member
(d) is a curved and tapered member

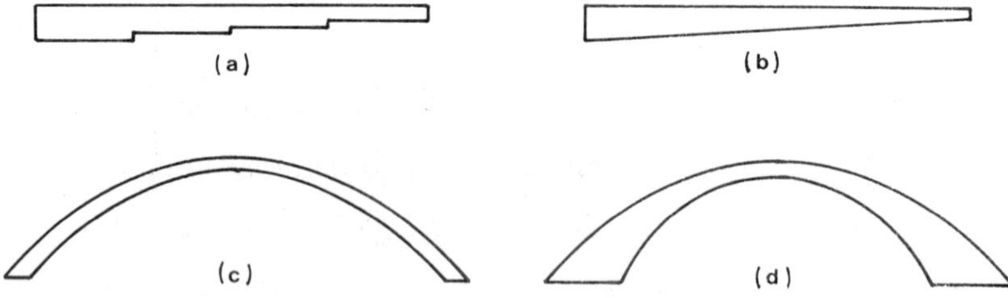

Figure 8.26 Example of non-prismatic elements

One way to deal with such elements is to evaluate their stiffness either analytically or numerically and then treat them as standard elements (as was done with elements containing pins). A variety of techniques can be used to evaluate the stiffness of non-prismatic elements. The majority of problems involving non-prismatic elements are, however, most easily tackled by treating the non-prismatic elements as a series of standard prismatic beam elements as illustrated in fig 8.27.

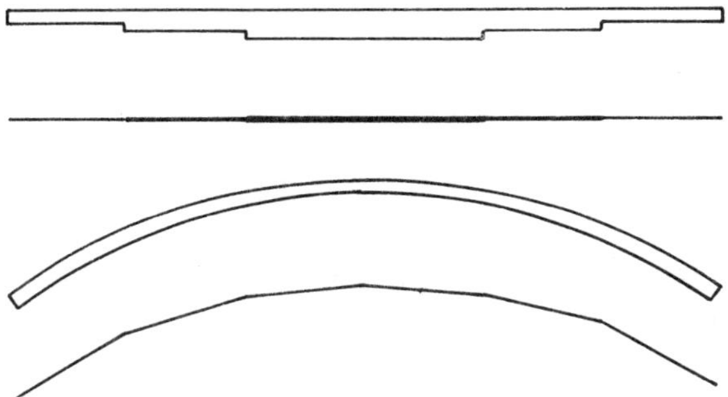

Figure 8.27 Non-prismatic elements treated as a series of prismatic elements

Note that the disadvantage of this approach is that additional nodes are introduced. This increases the size of the problem and hence the cost of the analysis. In the case of the straight non-prismatic element illustrated in fig 8.27 an additional four nodes have been introduced which increases the degree of freedom by 12. Usually, however, the disadvantage of increased problem size is more than outweighed by the ease of analysis. Only when there are many non-prismatic elements, or when the sub-division of non-prismatic elements makes the problem too expensive or too large for the available computer should the engineer consider the alternative method of calculating the actual stiffness of the non-prismatic elements.

Example 8.5

The program PFRAME.BAS presented later in this chapter has been used to compare the end deflection of the tapered element in the following diagram when it is idealised as 2, 4, and 8 prismatic elements.

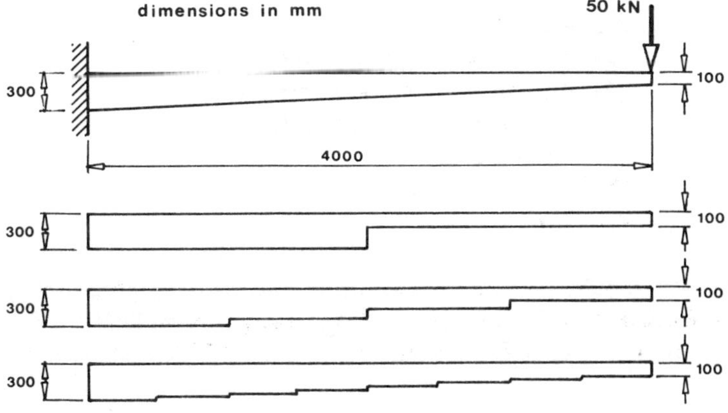

The cross-section is rectangular with a breadth of 400 mm and E = 200 kN/mm^2.

The results are as follows.

Number of Elements	End Deflection (downwards)
2	25.4
4	13.5
8	12.5

Notice how the end deflection quickly approaches the true value of 12.8 mm as the number of elements increases.

8.7 Putting Some of it Together - Program PFRAME.BAS

This chapter and the previous chapter have now covered the procedures necessary for the analysis of plane frames using the stiffness method. In this section the short computer programs to assemble the final stiffness matrix (FSTIFFPF.BAS) and to calculate element forces (EFORCEPF.BAS) are extended and combined with the Gaussian elimination equation solver (GAUSS.BAS) to produce an automatic plane frame analysis program (PFRAME.BAS). To illustrate how the theory covered in this chapter might be implemented in an automatic plane frame analysis program PFRAME.BAS has been given the ability to deal with local axes, element loads and elements containing pins.

Input

Unless PTRUSS.BAS has been chained from PRE.BAS or DCHECK.BAS the user is prompted for the name of the data file to be read by the program. PTRUSS.BAS then offers the user the following menu which drives the program.

Do you wish to:
(1) Analyse with output to the screen
(2) Analyse with output to the printer
(3) Analyse with output to a file
(4) Edit the data
(5) Check the data
(6) Stop

This data file should be constructed using the data preprocessor program PRE.BAS which is described in section 5.11. The program DCHECK.BAS can be used to check the data for obvious errors. DCHECK.BAS is presented in section 5.12.

Comments on the Algorithm

The length of each element and the cosine and sine of the angle it makes with the global "x" axis are calculated from the element node numbers and the nodal coordinates. Only the upper semi-band of the final stiffness matrix is assembled. Elements containing pins are dealt with by using the variables K_1 - K_8 in the element stiffness matrix as shown in the following equation. The values of the K factors are tabulated in Table 8.1.

Table 8.1 *K* Factors

ELEMENT TYPE	NODE "i"	NODE "j"	K_1	K_2	K_3	K_4	K_5	K_6	K_7	K_8
1	Rigid	Rigid	12	6	6	4	-6	2	-6	4
2	Rigid	Pinned	3	3	0	3	-3	0	0	0
3	Pinned	Rigid	3	0	3	0	0	0	-3	-3
4	Pinned	Pinned	0	0	0	0	0	0	0	0

$$\begin{bmatrix} f_{ijx} \\ f_{ijy} \\ m_{ij} \\ f_{jix} \\ f_{jiy} \\ m_{ji} \end{bmatrix} = \begin{bmatrix} \frac{EA}{L} & 0 & 0 & -\frac{EA}{L} & 0 & 0 \\ & K_1\frac{EI}{L^3} & K_2\frac{EI}{L^2} & 0 & -K_1\frac{EI}{L^3} & K_3\frac{EI}{L^2} \\ & & K_4\frac{EI}{L} & 0 & K_5\frac{EI}{L^2} & K_6\frac{EI}{L} \\ & \text{symmetric} & & \frac{EA}{L} & 0 & 0 \\ & & & & K_1\frac{EI}{L^3} & K_7\frac{EI}{L^2} \\ & & & & & K_8\frac{EI}{L} \end{bmatrix} \begin{bmatrix} \delta_{ijx} \\ \delta_{ijy} \\ \theta_{ij} \\ \delta_{jix} \\ \delta_{jiy} \\ \theta_{ji} \end{bmatrix}$$

Local axes are defined by specifing the node number, "i", where the local axes are located, and the anticlockwise rotation of the local axes from the coordinate axes - the "β" angle. When assembling the structure stiffness matrix PFRAME.BAS calculates the angle that the element makes with the cooordinate axes, "α", from the nodal coordinates and then checks if there is a set of local axes at either or both ends of the element. This sets up two angles "β_i" and "β_j" (which are zero if no local axes are defined). From the "α" and "β" angles "α_i" and "α_j" are calculated as follows

$$\alpha_i = \alpha - \beta_i \quad \text{and} \quad \alpha_j = \alpha - \beta_j$$

Uniformly distributed loads and point loads acting perpendicular to the elements are catered for through the use of fixed end forces and equivalent joint forces. Note that positive element loads act in the *positive "y" direction* (element axes). Initially the fixed end forces are calculated on the assumption that the element is rigidly connected to the nodes at its ends. The moments developed at pinned connections are then released to produce fixed end forces that are consistent with the end connections.

Once the unknown nodal displacements have been found they are used to calculate the element end forces. Note that the stiffness factors K_1 - K_8 are used to facilitate the calculation of the element end forces. The reactions are evaluated by summing the contributions from the elements connected to the restrained nodes.

Output

The data from the data file is echoed and messages are output to indicate the beginning of each solution phase. After solution of the stiffness equation nodal displacements are output. The axial force, shear force, and bending moment at each end of each element are then output. Finally the components of reaction (in the global axes system) at each restrained node are output.

Listing

```
1000 '=======================     PFRAME.BAS     ==============================
1010 '
1020 COMMON DATAFILE$
1030 OPTION BASE 1 : KEY OFF
1040 TRUE% = -1 : FALSE% = 0 : DEVICE$ = "SCRN:" : PAGELEN% = 23
1050 OUTPUTFILE$="SCREEN"
1060 CLS
1070 PRINT
1080 PRINT "==============================================================="
1090 PRINT
1100 PRINT " PROGRAM PFRAME                          Copyright (c) J.Balfour 1991"
1110 PRINT
1120 PRINT "      Program for the automatic analysis of plane frames"
1130 PRINT "        Data generated with the data preprocessor PRE"
1140 PRINT
1150 PRINT "                For further information contact"
1160 PRINT " James A.D.Balfour, Heriot-Watt University, Riccarton, Edinburgh"
1170 PRINT "              Tel 031-449-5111, Fax 031-451-3170"
1180 PRINT
1190 PRINT "==============================================================="
1200 '
1210 ' ... Note that variables are defined in Appendix A and the
1220 '     following statements set the maximum problem size
1230 '
1240 MAXNODE%   =  45                       '... Max no of nodes
1250 MAXELEM%   = 100                       '... Max no of elements
1260 MAXPROP%   =  20                       '... Max no of element properties
1270 MAXLAXES% =  20                       '... Max no of local axes
1280 MAXREST%   =  20                       '... Max no of restrained nodes
1290 MAXNLOAD% =  45                       '... Max no of nodal loads
1300 MAXELOAD% =  20                       '... Max no of element loads
1310 MAXBAND%   =  40                       '... Max semi-bandwidth
1320 '
1330 DIM NODE(MAXNODE%,4),          ELEM(MAXELEM%,9),    PROP(MAXPROP%,5)
1340 DIM LAXES(MAXLAXES%,2),        REST(MAXREST%,5),    NLOAD(MAXNLOAD%,3)
1350 DIM ELOAD(MAXELOAD%,9),        FREE%(3*MAXNODE%),  EFREE%(6)
1360 DIM K(3*MAXNODE%,MAXBAND%),  P(3*MAXNODE%),        ESTIFF(6,6)
1370 DIM WKSP$(8),                  TEMP(12)
1380 '
1390 ' ... Check if this program has been chained from PRE
1400 '
```

```
1410 IF DATAFILE$ <> "" THEN GOSUB 10560 : GOTO 1440
1420 PRINT
1430 INPUT "Name of the data file  =  ", DATAFILE$
1440 GOSUB 2160
1450 NOPTIONS% = 7
1460 WHILE (REPLY% <> 7)
1470   CLS : PRINT
1480   PRINT "Do you wish to:"
1490   PRINT "   (1) Analyse with output to the screen"
1500   PRINT "   (2) Analyse with output to the printer"
1510   PRINT "   (3) Analyse with output to a file"
1520   PRINT "   (4) Edit the data"
1530   PRINT "   (5) Check the data"
1540   PRINT "   (6) Read a new data file
1550   PRINT "   (7) Stop"
1560   GOSUB 9840
1570   ON REPLY% GOTO 1590, 1630, 1710, 1780, 1810, 1840, 1870
1580     '
1590     OUTPUTFILE$ = "SCREEN"
1600     OPEN "SCRN:" FOR OUTPUT AS #2                         ' Reply = 1
1610     GOSUB 1920 : GOTO 1870
1620     '
1630     ON ERROR GOTO 1680                                    ' Reply = 2
1640     OPEN "LPT1:" FOR OUTPUT AS #2
1650     PRINT : PRINT "Wait - looking for printer" : PRINT #2,
1660     ON ERROR GOTO 0 : DEVICE$ = "LPT1:"
1670     GOSUB 1920 : PRINT #2, CHR$(12); : GOTO 1870
1680     CLS : PRINT : PRINT "** ERROR ** Failed to find the printer"
1690     GOSUB 10550 : RESUME 1870
1700     '
1710     ON ERROR GOTO 1750                                    ' Reply = 3
1720     INPUT "Name of file for output = ", DEVICE$
1730     OPEN DEVICE$ FOR OUTPUT AS #2
1740     ON ERROR GOTO 0 : GOSUB 1920 : GOTO 1870
1750     CLS : PRINT : PRINT "** ERROR ** Failed to open output file"
1760     GOSUB 10550 : RESUME 1870
1770     '
1780     PRINT "Wait - chaining PRE"                            ' Reply = 4
1790     CHAIN "PRE"
1800     '
1810     PRINT "Wait - chaining DCHECK"                         ' Reply = 5
1820     CHAIN "DCHECK"
1830     '
1840     PRINT                                                 ' Reply = 6
1850     INPUT "Name of the data file  =  ", DATAFILE$
1860     GOSUB 2160
1870     CLOSE #2 : DEVICE$ = "SCRN:" : ON ERROR GOTO 0
1880 WEND
1890 KEY ON
1900 END
1910 '
1920 '*************************     ANALYSE     ******************************
1930 '
1940 IF DATAFILEOK% THEN GOTO 1980
1950 PRINT
1960 PRINT "** ERROR ** Analysis aborted no valid data file loaded"
1970 GOTO 2140
1980 GOSUB 3800           '... Print data
1990 PRINT : PRINT "Generating the freedom vector                         "
2000 GOSUB 5400
2010 PRINT
2020 PRINT "Assembling the stiffness matrix, adding element :- ";
2030 ROW% = CSRLIN : COL% = POS(0)
2040 GOSUB 5730
2050 PRINT : PRINT "Generating the loading vector                         "
2060 GOSUB 6820
2070 PRINT
2080 PRINT "Solving equations - no. of equations = "; NDOF%
2090 PRINT "                  - semi-bandwidth   = "; BAND%
2100 GOSUB 7630           '... Solve the equations
2110 GOSUB 10550
2120 GOSUB 8230           '... Output the nodal displacements
2130 GOSUB 8470           '... Output the element forces and reactions
2140 RETURN
2150 '
2160 '**********************     READ DATA FILE     ***************************
2170 '
```

```
2180 ' ...  This subroutine reads the data from a data file
2190 '
2200 PRINT
2210 PRINT "Wait - Reading file "; DATAFILE$; " line";
2220 ROW% = CSRLIN : COL% = POS(0) : LCOUNT% = 0
2230 DATAFILEOK% = FALSE% : DATAEND% = FALSE%
2240 NNODE%  = 0 : NELEM%  = 0 : NPROP% = 0 : NREST% = 0 : NLAXES% = 0
2250 NNLOAD% = 0 : NELOAD% = 0
2260 ON ERROR GOTO 2280 : OPEN DATAFILE$ FOR INPUT AS #1
2270 ON ERROR GOTO 0 : GOTO 2340
2280 CLS : PRINT
2290 PRINT "** ERROR ** Failed to open file: "; DATAFILE$; " for input"
2300 GOSUB 10550 : RESUME 3770
2310 '
2320 ' ... Check problem type
2330 '
2340 TARGET$ = "PFRAME" : GOSUB 10310
2350 IF STRFOUND% THEN GOTO 2420
2360 CLS : PRINT
2370 PRINT #2, "** ERROR ** Not a file for PFRAME"
2380 GOSUB 10550 : GOTO 3770
2390 '
2400 ' ... Read title and units strings
2410 '
2420 NLINES% = 2 : GOSUB 10460 : IF DATAEND% THEN GOTO 2490
2430 WHILE (MID$(WRKSTR$,I%,1) = " ")
2440   I% = I% + 1
2450 WEND
2460 TITLE$  = MID$(WRKSTR$,I%)
2470 TARGET$ = "Units" : GOSUB 10310
2480 IF STRFOUND% THEN GOTO 2520
2490 CLS : PRINT : PRINT
2500 PRINT "** ERROR ** Error reading Title/Units strings"
2510 GOSUB 10550 : GOTO 2540
2520 UNITS$ = MID$(WRKSTR$,INSTR(WRKSTR$,":- ")+3,24)
2530 DATAFILEOK% = TRUE%
2540 WHILE NOT DATAEND%
2550   '
2560   ' ... Get module string
2570   '
2580   IF LEFT$(WRKSTR$,4)="+   " THEN GOTO 2600
2590   TARGET$ = "+   " : GOSUB 10310 : IF NOT STRFOUND% THEN GOTO 3690
2600   MODSTR$ = WRKSTR$
2610   '
2620   ' ...  Identify module
2630   '
2640   IF INSTR(MODSTR$,"NODAL COORDINATES")=0 THEN GOTO 2790
2650   '
2660   ' ... Read nodal coordinates
2670   '
2680   TARGET$ = "NODE" : GOSUB 10310 : IF NOT STRFOUND% THEN GOTO 3690
2690   GOSUB 9980
2700   WHILE NNOS% > 0
2710     NNODE% = NNODE% + 1
2720     FOR J% = 1 TO NNOS%
2730       NODE(NNODE%,J%) = VAL(WKSP$(J%))
2740     NEXT J%
2750     GOSUB 9980
2760   WEND
2770   GOTO 3760
2780   '
2790   IF INSTR(MODSTR$,"ELEMENTS")=0 THEN GOTO 2940
2800   '
2810   ' ... Read elements
2820   '
2830   TARGET$ = "ELEMENT" : GOSUB 10310 : IF NOT STRFOUND% THEN GOTO 3690
2840   GOSUB 9980
2850   WHILE NNOS% > 0
2860     NELEM% = NELEM% + 1
2870     FOR J% = 1 TO NNOS%
2880       ELEM(NELEM%,J%) = VAL(WKSP$(J%))
2890     NEXT J%
2900     GOSUB 9980
2910   WEND
2920   GOTO 3760
2930   '
2940   IF INSTR(MODSTR$,"PROPERTIES")=0 THEN GOTO 3090
```

```
2950    '
2960    ' ... Read properties
2970    '
2980    TARGET$ = "PROPERTY" : GOSUB 10310 : IF NOT STRFOUND% THEN GOTO 3690
2990    GOSUB 9980
3000    WHILE NNOS% > 0
3010      NPROP% = NPROP% + 1
3020      FOR J% = 1 TO NNOS%
3030        PROP(NPROP%,J%) = VAL(WKSP$(J%))
3040      NEXT J%
3050      GOSUB 9980
3060    WEND
3070    GOTO 3760
3080    '
3090    IF INSTR(MODSTR$,"RESTRAINTS")=0 THEN GOTO 3240
3100    '
3110    ' ... Read restraints
3120    '
3130    TARGET$ = "NODE" : GOSUB 10310 : IF NOT STRFOUND% THEN GOTO 3690
3140    GOSUB 9980
3150    WHILE NNOS% > 0
3160      NREST% = NREST% + 1
3170      FOR J% = 1 TO NNOS%
3180        REST(NREST%,J%) = VAL(WKSP$(J%))
3190      NEXT J%
3200      GOSUB 9980
3210    WEND
3220    GOTO 3760
3230    '
3240    IF INSTR(MODSTR$,"LOCAL AXES")=0 THEN GOTO 3390
3250    '
3260    ' ... Read local axes
3270    '
3280    TARGET$ = "NODE" : GOSUB 10310 : IF NOT STRFOUND% THEN GOTO 3690
3290    GOSUB 9980
3300    WHILE NNOS% > 0
3310      NLAXES% = NLAXES% + 1
3320      FOR J% = 1 TO NNOS%
3330        LAXES(NLAXES%,J%) = VAL(WKSP$(J%))
3340      NEXT J%
3350      GOSUB 9980
3360    WEND
3370    GOTO 3760
3380    '
3390    IF INSTR(MODSTR$,"NODAL LOADS")=0 THEN GOTO 3540
3400    '
3410    ' ... Read nodal loads
3420    '
3430    TARGET$ = "NODE" : GOSUB 10310 : IF NOT STRFOUND% THEN GOTO 3690
3440    GOSUB 9980
3450    WHILE NNOS% > 0
3460      NNLOAD% = NNLOAD% + 1
3470      FOR J% = 1 TO NNOS%
3480        NLOAD(NNLOAD%,J%) = VAL(WKSP$(J%))
3490      NEXT J%
3500      GOSUB 9980
3510    WEND
3520    GOTO 3760
3530    '
3540    IF INSTR(MODSTR$,"ELEMENT LOADS")=0 THEN GOTO 3690
3550    '
3560    ' ... Read element loads
3570    '
3580    TARGET$ = "TYPE" : GOSUB 10310 : IF NOT STRFOUND% THEN GOTO 3690
3590    GOSUB 9980
3600    WHILE NNOS% > 0
3610      NELOAD% = NELOAD% + 1
3620      FOR J% = 1 TO NNOS%
3630        ELOAD(NELOAD%,J%) = VAL(WKSP$(J%))
3640      NEXT J%
3650      GOSUB 9980
3660    WEND
3670    GOTO 3760
3680    '
3690    CLS : PRINT
3700    PRINT "** ERROR ** in module "; MODSTR$ " run DCHECK for more info."
3710    GOSUB 10550
```

```
3720   PRINT
3730   PRINT "Wait - Reading file "; DATAFILE$; " line";
3740   ROW% = CSRLIN : COL% = POS(0)
3750   DATAFILEOK% = FALSE% : WRKSTR$ = ""
3760 WEND
3770 CLOSE #1 : ON ERROR GOTO 0
3780 RETURN
3790 '
3800 ' ***********************    PRINT DATA    ******************************
3810 '
3820 GOSUB 3920                           '... List header
3830 GOSUB 4250                           '... List nodal coordinates
3840 GOSUB 4400                           '... List element data
3850 GOSUB 4560                           '... List list element properties
3860 GOSUB 4720                           '... List local axes data
3870 GOSUB 4880                           '... List restraint data
3880 GOSUB 5030                           '... List nodal loads
3890 GOSUB 5200                           '... List element loads
3900 RETURN
3910 '
3920 ' +++++++++++++++++++++++    LIST HEADER    +++++++++++++++++++++++++++
3930 '
3940 CLS
3950 PRINT #2, : PRINT #2,
3960 PRINT #2, "======================================";
3970 PRINT #2, "======================================"
3980 PRINT #2,
3990 PRINT #2, "    PROGRAM PFRAME"; STRING$(24,32);
4000 PRINT #2, "Copyright(c) J.A.D.Balfour 1991"
4010 PRINT #2,
4020 FOR I% = 1 TO (76-LEN(TITLE$))/2
4030   PRINT #2, " ";
4040 NEXT I%
4050 PRINT #2, TITLE$
4060 PRINT #2,
4070 PRINT #2, "    File Name :- "; DATAFILE$;
4080 FOR I% = 1 TO (40-LEN(DATAFILE$))
4090   PRINT #2, " ";
4100 NEXT I%
4110 PRINT #2, "Date :- "; MID$(DATE$,4,3); MID$(DATE$,1,3);
4120 PRINT #2, MID$(DATE$, 9, 2)
4130 PRINT #2, "    Units     :- "; UNITS$;
4140 FOR I% = 1 TO (40-LEN(UNITS$))
4150   PRINT #2, " ";
4160 NEXT I%
4170 PRINT #2, "Time :- "; TIME$
4180 PRINT #2,
4190 PRINT #2, "======================================";
4200 PRINT #2, "======================================"
4210 PRINT #2,
4220 GOSUB 10550
4230 RETURN
4240 '
4250 ' +++++++++++++++++    LIST NODAL COORDINATES    +++++++++++++++++++++
4260 '
4270 PRINT #2,
4280 PRINT #2, "+ + + + + + + + + + + + + +"
4290 PRINT #2, "+   NODAL COORDINATES   +"
4300 PRINT #2, "+ + + + + + + + + + + + + +"
4310 PRINT #2,
4320 PRINT #2, "NODE           X              Y"
4330 FOR I% = 1 TO NNODE%
4340   GOSUB 10640
4350   PRINT #2, " "; NODE(I%,1), NODE(I%,2), NODE(I%,3)
4360 NEXT I%
4370 GOSUB 10550
4380 RETURN
4390 '
4400 ' ++++++++++++++++++++++    LIST ELEMENTS    +++++++++++++++++++++++++
4410 '
4420 PRINT #2,
4430 PRINT #2, "+ + + + + + + + + +"
4440 PRINT #2, "+   ELEMENTS    +"
4450 PRINT #2, "+ + + + + + + + + +"
4460 PRINT #2,
4470 PRINT #2,         "ELEMENT    PROPERTY     TYPE"
4480 FOR I% = 1 TO NELEM%
```

```
4490    GOSUB 10640
4500    PRINT #2, USING "### ###        ##"; ELEM(I%,1); ELEM(I%,2); ELEM(I%,3);
4510    PRINT #2, USING "          ##"; ELEM(I%,4)
4520 NEXT I%
4530 GOSUB 10550
4540 RETURN
4550 '
4560 ' ++++++++++++++++++++++   LIST PROPERTIES    ++++++++++++++++++++++++
4570 '
4580 PRINT #2,
4590 PRINT #2, "+ + + + + +  + + +"
4600 PRINT #2, "+   PROPERTIES   +"
4610 PRINT #2, "+ + + + + +  + + +"
4620 PRINT #2,
4630 PRINT #2, "PROPERTY         A               I             E"
4640 FOR I% = 1 TO NPROP%
4650    GOSUB 10640
4660    PRINT #2, USING "  ###      #.####^^^^  ";      PROP(I%,1); PROP(I%,2);
4670    PRINT #2, USING "   #.####^^^^    #.####^^^^"; PROP(I%,3); PROP(I%,4)
4680 NEXT I%
4690 GOSUB 10550
4700 RETURN
4710 '
4720 ' ++++++++++++++++++++++    LIST LOCAL AXES    ++++++++++++++++++++++++
4730 '
4740 IF (NLAXES%=0) THEN GOTO 4860
4750 PRINT #2,
4760 PRINT #2, "+ + + + + + + + + + + "
4770 PRINT #2, "+   LOCAL AXES    +"
4780 PRINT #2, "+ + + + + + + + + + + "
4790 PRINT #2,
4800 PRINT #2, "NODE      ANGLE"
4810 FOR I% = 1 TO NLAXES%
4820    GOSUB 10640
4830    PRINT #2, USING " ###      ####.#"; LAXES(I%,1); LAXES(I%,2)
4840 NEXT I%
4850 GOSUB 10550
4860 RETURN
4870 '
4880 ' ++++++++++++++++++++++++   LIST RESTRAINTS    ++++++++++++++++++++++++
4890 '
4900 PRINT #2,
4910 PRINT #2, "+ + + + + + + + + + + "
4920 PRINT #2, "+   RESTRAINTS    +"
4930 PRINT #2, "+ + + + + + + + + + + "
4940 PRINT #2,
4950 PRINT #2, "NODE    DIRECTION(S)"
4960 FOR I% = 1 TO NREST%
4970    GOSUB 10640
4980    PRINT #2, USING "###      ######"; REST(I%,1); REST(I%,2)
4990 NEXT I%
5000 GOSUB 10550
5010 RETURN
5020 '
5030 ' ++++++++++++++++++++    LIST NODAL LOADS    +++++++++++++++++++++++++
5040 '
5050 IF (NNLOAD%=0) THEN GOTO 5180
5060 PRINT #2,
5070 PRINT #2, "+ + + + + + + + + + +"
5080 PRINT #2, "+   NODAL LOADS   +"
5090 PRINT #2, "+ + + + + + + + + + +"
5100 PRINT #2,
5110 PRINT #2, "NODE    DIRECTION        VALUE"
5120 FOR I% = 1 TO NNLOAD%
5130    GOSUB 10640
5140    PRINT #2, USING "###        ###";  NLOAD(I%,1); NLOAD(I%,2);
5150    PRINT #2, USING "          #.####^^^^"; NLOAD(I%,3)
5160 NEXT I%
5170 GOSUB 10550
5180 RETURN
5190 '
5200 ' ++++++++++++++++++++    LIST ELEMENT LOADS    ++++++++++++++++++++++
5210 '
5220 IF (NELOAD%=0) THEN GOTO 5380
5230 PRINT #2,
5240 PRINT #2, "+ + + + + + + + + + + + "
5250 PRINT #2, "+   ELEMENT LOADS   +    -    UDL = 1,    POINT LOAD = 2"
```

```
5260 PRINT #2, "+ + + + + + + + + + + "
5270 PRINT #2,
5280 PRINT #2, "ELEMENT   LOAD     MAGNITUDE     DISTANCE FROM"
5290 PRINT #2, "          TYPE                   LOWER NODE"
5300 FOR I% = 1 TO NELOAD%
5310   GOSUB 10640
5320   PRINT #2, USING "### ###   ###"; ELOAD(I%,1); ELOAD(I%,2); ELOAD(I%,3);
5330   PRINT #2, USING "     #.####^^^^"; ELOAD(I%,4);
5340   IF ELOAD(I%,3) =  2 THEN PRINT #2, USING "    #.####^^^^"; ELOAD(I%,5)
5350   IF ELOAD(I%,3) <> 2 THEN PRINT #2, "          N/A"
5360 NEXT I%
5370 GOSUB 10550
5380 RETURN
5390 '
5400 ' *****************    GENERATE THE FREEDOM VECTOR   *******************
5410 '
5420 ' ... Zero the freedom vector
5430 '
5440 FOR I% = 1 TO 3*NNODE%
5450   FREE%(I%) = 0
5460 NEXT I%
5470 '
5480 ' ... Loop for all restrained nodes
5490 '
5500 FOR I% = 1 TO NREST%
5510   K% = REST(I%,2)
5520   '
5530   ' ... Evaluate the freedom to be restrained from K%
5540   '
5550   J% = 3 * REST(I%,1) - 3 + (K% MOD 10)
5560   FREE%(J%) = 1
5570   K% = INT(K%/10)
5580   IF K% > 0 THEN GOTO 5550
5590 NEXT I%
5600 '
5610 ' ... Number the freedoms (restraints set to zero)
5620 '
5630 NDOF% = 0
5640 FOR I% = 1 TO 3*NNODE%
5650   IF FREE%(I%) = 1 THEN GOTO 5690
5660   NDOF% = NDOF% + 1
5670   FREE%(I%) = NDOF%
5680   GOTO 5700
5690   FREE%(I%) = 0
5700 NEXT I%
5710 RETURN
5720 '
5730 ' ***************    ASSEMBLE THE STIFFNESS MATRIX    ******************
5740 '
5750 ' ... Zero the stiffness matrix and the loading vector
5760 '
5770 FOR I% = 1 TO NDOF%
5780   FOR J% = 1 TO MAXBAND%
5790     K(I%,J%) = 0
5800   NEXT J%
5810   P(I%) = 0
5820 NEXT I%
5830 '
5840 ' ... Loop for each element
5850 '
5860 BAND% = 0
5870 FOR K% = 1 TO NELEM%
5880   LOCATE ROW%, COL% : PRINT K%;
5890   IN% = ELEM(K%,1)
5900   JN% = ELEM(K%,2)
5910   PN% = ELEM(K%,3)
5920   '
5930   ' ... Calculate the element length (store in ELEM(K%,5))
5940   '
5950   ELEM(K%,5) = (NODE(JN%,2)-NODE(IN%,2))^2 + (NODE(JN%,3)-NODE(IN%,3))^2
5960   ELEM(K%,5) = SQR(ELEM(K%,5))
5970   EAL        = PROP(PN%,4)*PROP(PN%,2)/ELEM(K%,5)
5980   EIL        = PROP(PN%,4)*PROP(PN%,3)/ELEM(K%,5)
5990   '
6000   ' ... Calculate the the direction cosines of the member x-axes
6010   '
6020   C = (NODE(JN%,2) - NODE(IN%,2)) / ELEM(K%,5)
```

```
6030    S = (NODE(JN%,3) - NODE(IN%,3)) / ELEM(K%,5)
6040    '
6050    ' ... CHECK FOR LOCAL AXES
6060    '
6070    BETAI = 0 : BETAJ = 0
6080    FOR I% = 1 TO NLAXES%
6090      IF LAXES(I%,1) = IN% THEN BETAI = LAXES(I%,2) * 3.142 / 180
6100      IF LAXES(I%,1) = JN% THEN BETAJ = LAXES(I%,2) * 3.142 / 180
6110    NEXT I%
6120    REM
6130    REM  ...  Store cosine and sine of alpha I and alpha J in ELEM(K%,6-9)
6140    REM
6150    ELEM(K%,6) = C * COS(BETAI) + S * SIN(BETAI)
6160    ELEM(K%,7) = S * COS(BETAI) - C * SIN(BETAI)
6170    ELEM(K%,8) = C * COS(BETAJ) + S * SIN(BETAJ)
6180    ELEM(K%,9) = S * COS(BETAJ) - C * SIN(BETAJ)
6190    GOSUB 9600                                      ' Get K values
6200    '
6210    ' ... Assemble the upper triangle of the element stiffness matrix
6220    '
6230    L   = ELEM(K%,5)
6240    CI  = ELEM(K%,6)
6250    SI  = ELEM(K%,7)
6260    CJ  = ELEM(K%,8)
6270    SJ  = ELEM(K%,9)
6280    EAL = PROP(PN%,4)*PROP(PN%,2)/ELEM(K%,5)
6290    EIL = PROP(PN%,4)*PROP(PN%,3)/ELEM(K%,5)
6300    ESTIFF(1,1) =   EAL*CI*CI + K1*EIL*SI*SI/L^2
6310    ESTIFF(1,2) =   EAL*SI*CI - K1*EIL*SI*CI/L^2
6320    ESTIFF(1,3) = - K2*EIL*SI/L
6330    ESTIFF(1,4) = - EAL*CI*CJ - K1*EIL*SI*SJ/L^2
6340    ESTIFF(1,5) = - EAL*CI*SJ + K1*EIL*SI*CJ/L^2
6350    ESTIFF(1,6) = - K3*EIL*SI/L
6360    ESTIFF(2,2) =   EAL*SI*SI + K1*EIL*CI*CI/L^2
6370    ESTIFF(2,3) =   K2*EIL*CI/L
6380    ESTIFF(2,4) = - EAL*SI*CJ + K1*EIL*CI*SJ/L^2
6390    ESTIFF(2,5) = - EAL*SI*SJ - K1*EIL*CI*CJ/L^2
6400    ESTIFF(2,6) =   K3*EIL*CI/L
6410    ESTIFF(3,3) =   K4*EIL
6420    ESTIFF(3,4) = - K5*EIL*SJ/L
6430    ESTIFF(3,5) =   K5*EIL*CJ/L
6440    ESTIFF(3,6) =   K6*EIL
6450    ESTIFF(4,4) =   EAL*CJ*CJ + K1*EIL*SJ*SJ/L^2
6460    ESTIFF(4,5) =   EAL*SJ*CJ - K1*EIL*SJ*CJ/L^2
6470    ESTIFF(4,6) = - K7*EIL*SJ/L
6480    ESTIFF(5,5) =   EAL*SJ*SJ + K1*EIL*CJ*CJ/L^2
6490    ESTIFF(5,6) =   K7*EIL*CJ/L
6500    ESTIFF(6,6) =   K8*EIL
6510    '
6520    ' ... Set up the element code number
6530    '
6540    '
6550    EFREE%(1) = FREE%(3*IN%-2)
6560    EFREE%(2) = FREE%(3*IN%-1)
6570    EFREE%(3) = FREE%(3*IN%)
6580    EFREE%(4) = FREE%(3*JN%-2)
6590    EFREE%(5) = FREE%(3*JN%-1)
6600    EFREE%(6) = FREE%(3*JN%)
6610    '
6620    ' ... Add the element stiffness to the structure stiffness matrix
6630    '
6640    FOR I% = 1 TO 6
6650      IF EFREE%(I%) = 0 THEN GOTO 6770
6660      FOR J% = I% TO 6
6670        IF EFREE%(J%) = 0 THEN GOTO 6760
6680        L% = EFREE%(I%)
6690        M% = EFREE%(J%) - L% + 1
6700        IF M%<=BAND% THEN GOTO 6750
6710        BAND% = M%
6720        IF BAND% <= MAXBAND% THEN GOTO 6750
6730        PRINT #2, "Max semi-bandwidth ("; STR$(MAXBAND%); ") exceeded"
6740        END
6750        K(L%,M%) = K(L%,M%) + ESTIFF(I%,J%)
6760      NEXT J%
6770    NEXT I%
6780 NEXT K%
6790 PRINT
```

```
6800 RETURN
6810 '
6820 '*****************     SET UP THE LOADING VECTOR     ********************
6830 '
6840 ' ... Loop for all point loads
6850 '
6860 FOR I% = 1 TO NNLOAD%
6870   J% = 3*NLOAD(I%,1) - 3 + NLOAD(I%,2)
6880   J% = FREE%(J%)
6890   IF J% > 0 THEN P(J%) = NLOAD(I%,3)
6900 NEXT I%
6910 '
6920 ' ... Add element loads to the loading vector
6930 '
6940 FOR I% = 1 TO NELOAD%
6950   IN% = ELOAD(I%,1)
6960   JN% = ELOAD(I%,2)
6970   FOR K% =  1 TO NELEM%
6980     IF ELEM(K%,1) <> IN% THEN GOTO 7000
6990     IF ELEM(K%,2) =  JN% THEN J% = K%
7000   NEXT K%
7010   W  = ELOAD(I%,4)
7020   L  = ELEM(J%,5)
7030   X  = ELOAD(I%,5)
7040   Y  = L - X
7050   CI = ELEM(J%,6)
7060   SI = ELEM(J%,7)
7070   CJ = ELEM(J%,8)
7080   SJ = ELEM(J%,9)
7090   '
7100   ' ... Calc and store F.E.F. (initially assume a rigid/rigid element)
7110   '
7120   ON ELOAD(I%,3) GOTO 7160, 7240
7130   '
7140   ' ... Uniformly distributed load
7150   '
7160   ELOAD(I%,6) = - W * L / 2
7170   ELOAD(I%,7) = - W * L^2 / 12
7180   ELOAD(I%,8) =   ELOAD(I%,6)
7190   ELOAD(I%,9) = - ELOAD(I%,7)
7200   GOTO 7310
7210   '
7220   ' ... Point load
7230   '
7240   ELOAD(I%,6) = - W * (Y/L)^2 * (1 + 2*X/L)
7250   ELOAD(I%,7) = - W * X * Y^2 / L^2
7260   ELOAD(I%,8) = - W * (X/L)^2 * (1 + 2*Y/L)
7270   ELOAD(I%,9) =   W * X^2 * Y / L^2
7280   '
7290   ' ... Take account of any pins
7300   '
7310   ON ELEM(J%,4) GOTO 7480, 7370, 7320, 7320
7320   ELOAD(I%,6) = ELOAD(I%,6) - 1.5*ELOAD(I%,7)/L
7330   ELOAD(I%,8) = ELOAD(I%,8) + 1.5*ELOAD(I%,7)/L
7340   ELOAD(I%,9) = ELOAD(I%,9) -     ELOAD(I%,7)/2
7350   ELOAD(I%,7) = 0
7360   IF ELEM(J%,4) = 3 THEN GOTO 7480 ELSE GOTO 7420
7370   ELOAD(I%,6) = ELOAD(I%,6) - 1.5*ELOAD(I%,9)/L
7380   ELOAD(I%,7) = ELOAD(I%,7) -     ELOAD(I%,9)/2
7390   ELOAD(I%,8) = ELOAD(I%,8) + 1.5*ELOAD(I%,9)/L
7400   ELOAD(I%,9) = 0
7410   GOTO 7480
7420   ELOAD(I%,6) = ELOAD(I%,6) - ELOAD(I%,9)/L
7430   ELOAD(I%,8) = ELOAD(I%,8) + ELOAD(I%,9)/L
7440   ELOAD(I%,9) = 0
7450   '
7460   ' ... Add equivalent joint forces to the loading vector
7470   '
7480   J% = FREE%(3*IN% - 2)
7490   IF J% <> 0 THEN P(J%) = P(J%) + ELOAD(I%,6)*SI
7500   J% = FREE%(3*IN% - 1)
7510   IF J% <> 0 THEN P(J%) = P(J%) - ELOAD(I%,6)*CI
7520   J% = FREE%(3*IN%)
7530   IF J% <> 0 THEN P(J%) = P(J%) - ELOAD(I%,7)
7540   J% = FREE%(3*JN% - 2)
7550   IF J% <> 0 THEN P(J%) = P(J%) + ELOAD(I%,8)*SJ
7560   J% = FREE%(3*JN% - 1)
```

```
7570   IF J% <> 0 THEN P(J%) = P(J%) - ELOAD(I%,8)*CJ
7580   J% = FREE%(3*JN%)
7590   IF J% <> 0 THEN P(J%) = P(J%) - ELOAD(I%,9)
7600 NEXT I%
7610 RETURN
7620 '
7630 ' **********************   SOLVE EQUATIONS    **************************
7640 '
7650 ' ... This subroutine solves a system of symmetric, banded equations
7660 '     using Gaussian elimation with no row interchange.  Only the
7670 '     uppper semi-band of the coefficient matrix is stored.
7680 '
7690 ' ... Loop for all pivots
7700 '
7710 BB% = BAND%
7720 PRINT "                    - current pivot    = ";
7730 ROW% = CSRLIN: COL% = POS(0):
7740 FOR I% = 1 TO NDOF%
7750   LOCATE ROW%, COL% : PRINT I%;
7760   '
7770   ' ... Check if in the unused triangle
7780   '
7790   IF I% > NDOF%-BAND%+1 THEN BB% = NDOF%-I%+1
7800   PIVOT = K(I%,1)
7810   '
7820   ' ... Normalise
7830   '
7840   FOR J% = 1 TO BB%
7850     K(I%,J%) = K(I%,J%) / PIVOT
7860   NEXT J%
7870   P(I%) = P(I%) / PIVOT
7880   '
7890   ' ... Check if in the last row
7900   '
7910   IF BB% = 1 THEN GOTO 8100
7920   '
7930   ' ... Eliminate (within band) for all rows above the pivot
7940   '
7950   FOR K% = 2 TO BB%
7960     '
7970     ' ... Calculate the row number and evaluate the multiplier
7980     '
7990     L%   = I% + K% - 1
8000     MULT = K(I%,K%) * PIVOT
8010     '
8020     ' ... Loop for all elements in the elimination row
8030     '
8040     FOR J% = K% TO BB%
8050       M% = J% - K% + 1
8060       K(L%,M%) = K(L%,M%) - MULT * K(I%,J%)
8070     NEXT J%
8080     P(L%) = P(L%) - MULT*P(I%)
8090   NEXT K%
8100 NEXT I%
8110 '
8120 ' ... Back substitute
8130 '
8140 FOR I%  = 1 TO NDOF%-1
8150   IF I% > BAND%-1 THEN BB% = BAND%-1 ELSE BB% = I%
8160   L% = NDOF%-I%
8170   FOR J%  = 1 TO BB%
8180     P(L%) = P(L%) - K(L%,J%+1)*P(L%+J%)
8190   NEXT J%
8200 NEXT I%
8210 RETURN
8220 '
8230 '********************    OUTPUT THE DISPLACEMENTS   ********************
8240 '
8250 CLS
8260 PRINT #2,
8270 PRINT #2, "* * * * * * * * * * * * **"
8280 PRINT #2, "*   DISPLACEMENTS    *            * :-  RESTRAINT"
8290 PRINT #2, "* * * * * * * * * * * * **"
8300 PRINT #2,
8310 PRINT #2, "NODE     X-DISPLACEMENT  Y-DISPLACEMENT        ROTATION"
8320 FOR I% = 1 TO NNODE%
8330   GOSUB 10640
```

```
8340   PRINT #2, USING "###    "; I%;
8350   FOR J% = 1 TO 3
8360     K% = FREE%(3*I%+J%-3)
8370     IF K% = 0 THEN GOTO 8400
8380     PRINT #2, USING "      #.####^^^^"; P(K%);
8390     GOTO 8410
8400     PRINT #2, "          *      ";
8410   NEXT J%
8420   PRINT #2,
8430 NEXT I%
8440 GOSUB 10550          '... Pause
8450 RETURN
8460 '
8470 ' ****** CALCULATE AND OUTPUT THE ELEMENT FORCES AND REACTIONS     ******
8480 '
8490 ' ... Zero the reaction components
8500 '
8510 FOR I% = 1 TO NREST%
8520   REST(I%,3) = 0
8530   REST(I%,4) = 0
8540   REST(I%,5) = 0
8550 NEXT I%
8560 CLS
8570 PRINT #2,
8580 PRINT #2, "* * * * * * * * * * * **"
8590 PRINT #2, "*   ELEMENT FORCES   *"
8600 PRINT #2, "* * * * * * * * * * * **"
8610 PRINT #2,
8620 PRINT #2, "              ---------   END I   ---------";
8630 PRINT #2, "       ---------   END J   ---------"
8640 PRINT #2, "ELEMENT       AXIAL       SHEAR     BENDING";
8650 PRINT #2, "       AXIAL       SHEAR     BENDING"
8660 PRINT #2, "              FORCE       FORCE      MOMENT";
8670 PRINT #2, "       FORCE       FORCE      MOMENT"
8680 '
8690 ' ...  Loop for all elements
8700 '
8710 FOR K% = 1 TO NELEM%
8720   PN% = ELEM(K%,3)
8730   L   = ELEM(K%,5)
8740   CI  = ELEM(K%,6)
8750   SI  = ELEM(K%,7)
8760   CJ  = ELEM(K%,8)
8770   SJ  = ELEM(K%,9)
8780   EAL = PROP(PN%,4)*PROP(PN%,2)/ELEM(K%,5)
8790   EIL = PROP(PN%,4)*PROP(PN%,3)/ELEM(K%,5)
8800   I%  = 3*ELEM(K%,1) - 3
8810   '
8820   '  ...  Extract the nodal displacement - store in TEMP(1)-(6)
8830   '
8840   FOR M% = 1 TO 6
8850     J%    = FREE%(I%+M%)
8860     IF J% = 0 THEN TEMP(M%) = 0 ELSE TEMP(M%) = P(J%)
8870     IF M% = 3 THEN I% = 3*ELEM(K%,2) - 6
8880   NEXT M%
8890   '
8900   '  ...  Get K values, calc element end forces - store in TEMP(7)-(12)
8910   '
8920   GOSUB 9600
8930   TEMP(7)  =   EAL*(TEMP(1)*CI-TEMP(4)*CJ+TEMP(2)*SI-TEMP(5)*SJ)
8940   TEMP(8)  =   K1*EIL*(TEMP(4)*SJ-TEMP(1)*SI+TEMP(2)*CI-TEMP(5)*CJ)/L^2
8950   TEMP(8)  =   TEMP(8) + EIL*(K2*TEMP(3)+K3*TEMP(6))/L
8960   TEMP(9)  = - K2*TEMP(1)*SI-K5*TEMP(4)*SJ+K2*TEMP(2)*CI+K5*TEMP(5)*CJ
8970   TEMP(9)  =   EIL*(TEMP(9)/L+(K4*TEMP(3)+K6*TEMP(6)))
8980   TEMP(10) = - TEMP(7)
8990   TEMP(11) = - TEMP(8)
9000   TEMP(12) = - TEMP(9) + L*TEMP(8)
9010   '
9020   '  ...  Add in fixed end forces
9030   '
9040   IF NELOAD% = 0 THEN GOTO 9140
9050   FOR I%    = 1 TO NELOAD%
9060     IF ELEM(K%,1) <> ELOAD(I%,1) THEN GOTO 9120
9070     IF ELEM(K%,2) <> ELOAD(I%,2) THEN GOTO 9120
9080     TEMP(8)  = TEMP(8)  + ELOAD(I%,6)
9090     TEMP(9)  = TEMP(9)  + ELOAD(I%,7)
9100     TEMP(11) = TEMP(11) + ELOAD(I%,8)
```

```
9110     TEMP(12) = TEMP(12) + ELOAD(I%,9)
9120   NEXT I%
9130   GOSUB 10640
9140   PRINT #2, USING "## ##  "; ELEM(K%,1), ELEM(K%,2);
9150   PRINT #2, USING " ##.####^^^^"; TEMP(7), TEMP(8), TEMP(9);
9160   PRINT #2, USING " ##.####^^^^"; TEMP(10), TEMP(11), TEMP(12)
9170   '
9180   ' ...  Add any contribution from the element forces to the reactions
9190   '
9200   FOR I% = 1 TO NREST%
9210     IF (ELEM(K%,1) <> REST(I%,1)) THEN GOTO 9250
9220     REST(I%,3) = REST(I%,3) + CI*TEMP(7) - SI*TEMP(8)
9230     REST(I%,4) = REST(I%,4) + SI*TEMP(7) + CI*TEMP(8)
9240     REST(I%,5) = REST(I%,5) + TEMP(9)
9250     IF (ELEM(K%,2) <> REST(I%,1)) THEN GOTO 9290
9260     REST(I%,3) = REST(I%,3) + CJ*TEMP(10) - SJ*TEMP(11)
9270     REST(I%,4) = REST(I%,4) + SJ*TEMP(10) + CJ*TEMP(11)
9280     REST(I%,5) = REST(I%,5) + TEMP(12)
9290   NEXT I%
9300 NEXT K%
9310 GOSUB 10550            '... Pause
9320 '
9330 ' ...   Add any applied loads
9340 '
9350 FOR I% = 1 TO NREST%
9360   FOR J% = 1 TO NNLOAD%
9370     IF REST(I%,1) <> NLOAD(J%,1) THEN GOTO 9400
9380     K%      = NLOAD(J%,2)
9390     REST(I%,2+K%) = REST(I%,2+K%) - NLOAD(J%,3)
9400   NEXT J%
9410 NEXT I%
9420 '
9430 ' ...  Output the reactions
9440 '
9450 CLS
9460 PRINT #2,
9470 PRINT #2, "* * * * * * * * * *"
9480 PRINT #2, "*   REACTIONS    *"
9490 PRINT #2, "* * * * * * * * * *"
9500 PRINT #2,
9510 PRINT #2, "                X-COMPONENT    Y-COMPONENT          MOMENT"
9520 FOR I%   = 1 TO NREST%
9530   GOSUB 10640
9540   PRINT #2, USING "NODE ##   "; REST(I%,1);
9550   PRINT #2, USING "    ##.####^^^^"; REST(I%,3), REST(I%,4), REST(I%,5)
9560 NEXT I%
9570 GOSUB 10550            '... Pause
9580 RETURN
9590 '
9600 '********    STIFFNESS FACTORS FOR DIFFERENT END CONNECTIONS    *********
9610 '
9620 ' ... Zero stiffness factore (i.e. pinned/pinned element)
9630 '
9640 K1 =  0 : K2 =  0 : K3 =  0 : K4 =  0
9650 K5 =  0 : K6 =  0 : K7 =  0 : K8 =  0
9660 ON ELEM(K%,4) GOTO 9700, 9760, 9810, 9820
9670 '
9680 ' ... Rigid/rigid element
9690 '
9700 K1 = 12 : K2 =  6 : K3 =  6 : K4 =  4
9710 K5 = -6 : K6 =  2 : K7 = -6 : K8 =  4
9720 GOTO 9820
9730 '
9740 ' ... Rigid/pinned element
9750 '
9760 K1 =  3 : K2 =  3 : K4 =  3 : K5 = -3
9770 GOTO 9820
9780 '
9790 ' ... Pinned/rigid element
9800 '
9810 K1 =  3 : K3 =  3 : K7 = -3 : K8 =  3
9820 RETURN
9830 '
9840 ' ******************    GET KEYBOARD RESPONSE    *********************
9850 '
9860 ROW% = CSRLIN : COL% = POS(0)
9870 REPLY% = 0
```

```
9880 WHILE REPLY% = 0
9890   LOCATE ROW%, COL%: INPUT "", WRKSTR$: REPLY% = VAL(WRKSTR$)
9900   IF (REPLY% <  1) OR  (REPLY% > NOPTIONS%) THEN GOTO 9930
9910   IF (REPLY% < 10) AND (LEN(WRKSTR$) = 1)   THEN GOTO 9940
9920   IF (REPLY% >  9) AND (LEN(WRKSTR$) = 2)   THEN GOTO 9940
9930   SOUND 700, 4 : REPLY% = 0 : LOCATE ROW%, COL%
9940 WEND
9950 PRINT
9960 RETURN
9970 '
9980 '************************   RECORD NUMBERS   ******************************
9990 '
10000 NNOS% = 0 : DONE% = TRUE%
10010 IF EOF(1)<>0 THEN DATAEND% = TRUE% : GOTO 10290
10020 '
10030 ' ... Read next line and check if it is blank
10040 '
10050 LINE INPUT #1, WRKSTR$
10060 LCOUNT% = LCOUNT% + 1 : LOCATE ROW%, COL% : PRINT LCOUNT%;
10070 FOR I% = 1 TO LEN(WRKSTR$)
10080   IF MID$(WRKSTR$,I%,1)<> " " THEN DONE% = FALSE% : I% = LEN(WRKSTR$)
10090 NEXT I%
10100 IF DONE% = TRUE% THEN GOTO 10290
10110 FOR I% = 1 TO 8
10120   WKSP$(I%) = ""
10130 NEXT I%
10140 '
10150 ' ... Search for numbers
10160 '
10170 FOR I% = 1 TO LEN(WRKSTR$)
10180   WHILE (I% <= LEN(WRKSTR$)) AND (MID$(WRKSTR$,I%,1) = " ")
10190     I% = I% + 1
10200   WEND
10210   IF (I% > LEN(WRKSTR$)) THEN GOTO 10270
10220   IF(TARGET$="PFRAME") AND LCOUNT% > 4 THEN DONE% = TRUE%
10230   NNOS% = NNOS% + 1: WKSP$(NNOS%) = ""
10240   WHILE (MID$(WRKSTR$, I%, 1) <> " ") AND I% <= LEN(WRKSTR$)
10250     WKSP$(NNOS%) = WKSP$(NNOS%) + MID$(WRKSTR$,I%,1): I% = I% + 1
10260   WEND
10270 NEXT I%
10280 WRKSTR$ = ""
10290 RETURN
10300 '
10310 '***************************   FIND STRING   ****************************
10320 '
10330 STRFOUND% = FALSE% : DONE% = FALSE%
10340 WHILE NOT DONE%
10350   IF EOF(1)<>0 THEN DATAEND% = TRUE% : DONE% = TRUE% : GOTO 10430
10360   LINE INPUT #1, WRKSTR$
10370   LCOUNT% = LCOUNT% + 1 : LOCATE ROW%, COL% : PRINT LCOUNT%;
10380   IF INSTR(WRKSTR$,TARGET$)>0 THEN STRFOUND% = TRUE% : DONE% = TRUE%
10390   '
10400   ' ... Check for new module
10410   '
10420   IF (LEFT$(WRKSTR$,4) ="+   ") THEN DONE% = TRUE%
10430 WEND
10440 RETURN
10450 '
10460 '***************************   READ LINES   *****************************
10470 '
10480 FOR I% = 1 TO NLINES%
10490   IF EOF(1)<>0 THEN DATAEND% = TRUE% : I% = NLINES% : GOTO 10520
10500   LINE INPUT #1, WRKSTR$
10510   LCOUNT% = LCOUNT% + 1 : LOCATE ROW%, COL% : PRINT LCOUNT%;
10520 NEXT I%
10530 RETURN
10540 '
10550 '*********************   ANY KEY TO CONTINUE   ************************
10560 '
10570 IF DEVICE$<>"SCRN:" THEN GOTO 10620
10580 LOCATE PAGELEN%+1, 1
10590 PRINT SPACE$(40); "Press any key to continue";
10600 WHILE INKEY$ = "" : WEND
10610 CLS
10620 RETURN
10630 '
10640 '************************   CHECK FOR PAGE END   *********************
```

```
10650 '
10660 IF CSRLIN>=PAGELEN% THEN GOSUB 10550
10670 RETURN
10680 '
10690 ' ========================    PFRAME.BAS    =========================
```

Sample Run 1

The following sample run shows sample output from the program PFRAME.BAS. The structure is the one analysed in example 7.5.

```
==========================================================================

    PROGRAM PFRAME                          Copyright(c) J.A.D.Balfour 1991

               CASF 2nd Edition - Example 7.5 - Rigid Elements

    File Name :- PF1.DAT                                   Date :- 16-02-92
    Units     :- kN & mm                                   Time :- 20:24:38

==========================================================================


+ + + + + + + + + + + + +
+   NODAL COORDINATES   +
+ + + + + + + + + + + + +

NODE            X             Y
  1             0             0
  2             0             3000
  3             7000          4000

+ + + + + + + + +
+   ELEMENTS    +
+ + + + + + + + +

ELEMENT     PROPERTY     TYPE
  1   2         1          1
  2   3         1          1

+ + + + + +  + + +
+   PROPERTIES   +
+ + + + + +  + + +

PROPERTY         A              I            E
    1     0.9000E+04     0.2500E+09   0.2000E+03

+ + + + + + + + + +
+   RESTRAINTS    +
+ + + + + + + + + +

NODE    DIRECTION(S)
  1          12
  3          12

+ + + + + + + + + +
+   NODAL LOADS   +
+ + + + + + + + + +

NODE    DIRECTION        VALUE
  1          3         -.1000E+06
  2          1         0.1500E+03
  2          2         -.1500E+03

* * * * * * * * * * *
*   DISPLACEMENTS   *              * :-  RESTRAINT
* * * * * * * * * * *

NODE     X-DISPLACEMENT  Y-DISPLACEMENT        ROTATION
  1             *               *             -.2163E-02
  2        0.7916E+00      -.2969E+00         0.5339E-03
  3             *               *             -.1809E-03
```

```
* * * * * * * * * * * *
*   ELEMENT FORCES   *
* * * * * * * * * * * *

                  ---------   END I    ---------           ---------   END J    ---------
ELEMENT          AXIAL       SHEAR      BENDING          AXIAL       SHEAR       BENDING
                 FORCE       FORCE       MOMENT          FORCE       FORCE        MOMENT
 1  2       1.7812E+02 -3.6703E+01 -1.0000E+05 -1.7812E+02  3.6703E+01 -1.0109E+04
 2  3       1.8880E+02  1.4296E+00  1.0109E+04 -1.8880E+02 -1.4296E+00  1.9531E-03

* * * * * * * * *
*   REACTIONS   *
* * * * * * * * *

               X-COMPONENT      Y-COMPONENT          MOMENT
NODE  1         3.6703E+01       1.7812E+02      1.5625E-02
NODE  3        -1.8670E+02      -2.8116E+01      1.9531E-03
```

Sample Run 2

The following sample run shows the program PFRAME.BAS being used to analyse the structure shown in example 8.4 when the pin is assumed to be in element 1,2. The data was assembled using the program PRE.BAS and was stored on file PF2.DAT.

```
==============================================================================

    PROGRAM PFRAME.BAS -   ANALYSIS        Copyright(c) J.A.D.Balfour

                                      EXAMPLE 8.4

    File Name :- PF2.DAT                                      Date :- 13-11-91
    Units     :- kN and mm                                    Time :- 07:07:17

==============================================================================

+ + + + + + + + + + + + + +
+   NODAL COORDINATES   +
+ + + + + + + + + + + + + +

Node            X           Y
  1             0           0
  2             5196        3000
  3             8196        0

+ + + + + +  + +
+   ELEMENTS   +
+ + + + + +  + +

ELEMENT     PROPERTY      TYPE
  1  2           1          2
  2  3           1          1

+ + + + + +  + + +
+   PROPERTIES   +
+ + + + + +  + + +

PROPERTY         A               I            E
     1      0.4500E+05     0.3400E+09    0.1500E+02

+ + + + + + + + + + +
+   RESTRAINTS     +
+ + + + + + + + + + +

NODE     DIRECTION(S)
  1          123
  3          123

+ + + + + + + + + + + +
+   ELEMENT LOADS    +    -      UDL = 1,     POINT LOAD = 2
+ + + + + + + + + + + +
```

```
ELEMENT    LOAD      MAGNITUDE      DISTANCE FROM
NUMBER     TYPE                     LOWER NODE
   1  2      1      -.2000E-01          N/A

* * * * * * * * * * *
*   DISPLACEMENTS   *                * :-   RESTRAINT
* * * * * * * * * * *

NODE      X-DISPLACEMENT  Y-DISPLACEMENT         ROTATION
  1              *               *                   *
  2         0.7330E-01       -.3405E+00       0.6680E-04
  3              *               *                   *

* * * * * * * * * * *
*   ELEMENT FORCES  *
* * * * * * * * * * *

                ---------      END I    ---------         ---------      END J     ---------
ELEMENT          AXIAL         SHEAR       BENDING          AXIAL         SHEAR       BENDING
                 FORCE         FORCE        MOMENT          FORCE         FORCE        MOMENT
 1  2       1.2011E+01    7.5022E+01   9.0137E+04  -1.2011E+01    4.4976E+01   1.5259E-05
 2  3       4.6552E+01   -3.7851E-02   3.4985E-05  -4.6552E+01    3.7851E-02  -1.6059E+02

* * * * * * * * *
*   REACTIONS   *
* * * * * * * * *

               X-COMPONENT     Y-COMPONENT           MOMENT

NODE  1        -2.7110E+01      7.0976E+01       9.0137E+04
NODE  3        -3.2890E+01      3.2944E+01      -1.6059E+02
```

Chapter 9

Grillages

9.1 Introduction

Grillages, like plane frames, are rigidly jointed planar structures. The two types of structure differ only in the way in which they are loaded; plane frames are loaded in the plane of the structure, whereas grillages (grids) are loaded only perpendicular to the plane of the structure, as illustrated in fig 9.1. Structures such as floor systems, roof systems, and bridge decks are frequently analysed as grillages.

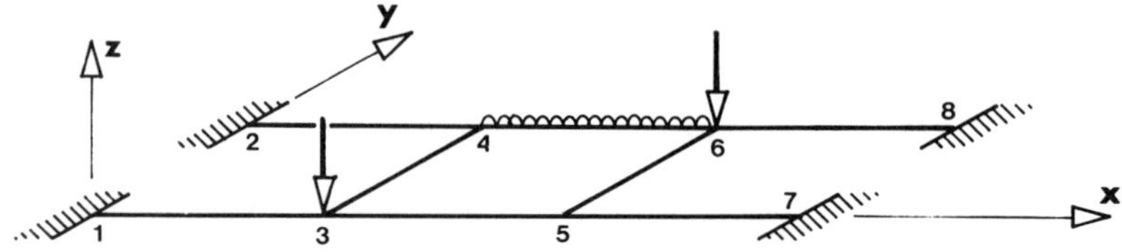

Figure 9.1 A simple grillage

Each unrestrained node of a grillage has three freedoms, and if the global axes are located as shown in fig 9.2, these freedoms are: translation in the "z" direction, rotation about the "x" axis, and rotation about the "y" axis. In other words grillage analysis is concerned with the out-of-plane displacements and rotations caused by out-of-plane loads applied to a plane structure. As rotation can occur out of the plane of the elements, torsion (i.e. twisting along the length of the element) is possible. Take, for example, element 5,6 of the grillage shown in fig 9.1. Under the loading shown, differing degrees of rotation about the "y" axis at nodes 5 and 6 are likely to occur, resulting in torsion of the element.

9.2 The Element Stiffness Matrix in Element Axes

Figure 9.2 shows the element axes and freedoms adopted for grillage elements in this book. Further, the element axes will be orientated such that the element "z" axis always coincides with the global "z" axis, i.e. the load axis.

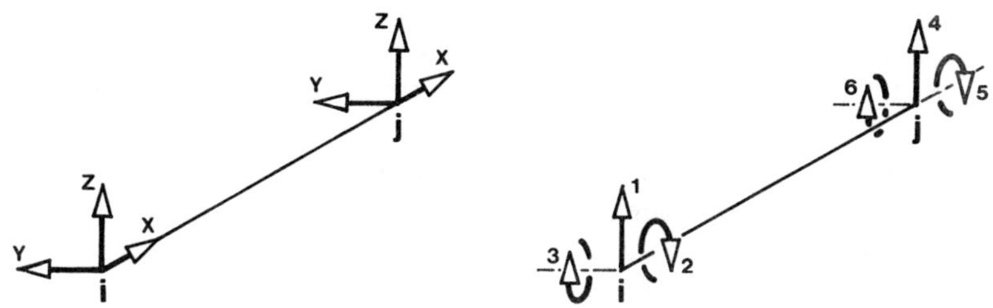

Figure 9.2 Element axes and freedoms for a grillage element

The element stiffness coefficients are found, as before, by evaluation of the force systems required to produce unit displacement in the direction of each freedom in turn. With the exception of the torsional stiffness terms, all other stiffness coefficients have already been encountered. The torsional behaviour of line elements is covered in section 2.12. Using equation (2.4) and the flexural stiffness terms from Chapter 7 the element stiffness matrix for a grillage element can easily be shown to be.

$$\begin{bmatrix} f_{ijz} \\ m_{ijx} \\ m_{ijy} \\ \hdashline f_{jiz} \\ m_{jix} \\ m_{jiy} \end{bmatrix} = \begin{bmatrix} \frac{12EI}{L^3} & 0 & -\frac{6EI}{L^2} & -\frac{12EI}{L^3} & 0 & -\frac{6EI}{L^2} \\ 0 & \frac{GJ}{L} & 0 & 0 & -\frac{GJ}{L} & 0 \\ -\frac{6EI}{L^2} & 0 & \frac{4EI}{L} & \frac{6EI}{L^2} & 0 & \frac{2EI}{L} \\ \hdashline -\frac{12EI}{L^3} & 0 & \frac{6EI}{L^2} & \frac{12EI}{L^3} & 0 & \frac{6EI}{L^2} \\ 0 & -\frac{GJ}{L} & 0 & 0 & \frac{GJ}{L} & 0 \\ -\frac{6EI}{L^2} & 0 & \frac{2EI}{L} & \frac{6EI}{L^2} & 0 & \frac{4EI}{L} \end{bmatrix} \begin{bmatrix} \delta_{ijz} \\ \theta_{ijx} \\ \theta_{ijy} \\ \hdashline \delta_{jiz} \\ \theta_{jix} \\ \theta_{jiy} \end{bmatrix} \tag{9.1}$$

When written in terms of the submatrices indicated by the broken lines, the above equation reduces to the well-known expressions

$$\begin{bmatrix} f_{ij} \\ \hdashline f_{ji} \end{bmatrix} = \begin{bmatrix} k_{ii}^{j} & k_{ij} \\ \hdashline k_{ji} & k_{jj}^{i} \end{bmatrix} \begin{bmatrix} \delta_{ij} \\ \hdashline \delta_{ji} \end{bmatrix}$$

hence

$$f_{ij} = k_{ii}^{j}\,\delta_{ij} + k_{ij}\,\delta_{ji} \tag{9.2}$$

and

$$f_{ji} = k_{ji}\,\delta_{ij} + k^{i}_{jj}\,\delta_{ji} \qquad (9.3)$$

9.3 Transformation of Force and Displacement

The element axes system adopted in the previous section demands that the element "z" axis coincides with the global "z" axis. Hence force and displacement in the "z" direction are the same in both axes systems; rendering transformation unnecessary. However, transformation of rotation and moment between the element "x" and "y" axes and the global "x" and "y" axes is necessary. In general the element "x" axis will lie at an angle to the coordinate "x" axis. Let that angle be "α", measured as shown in fig 9.3 (i.e. "α" is the anticlockwise rotation of global axes that will make them point in the same directions as the element axes.).

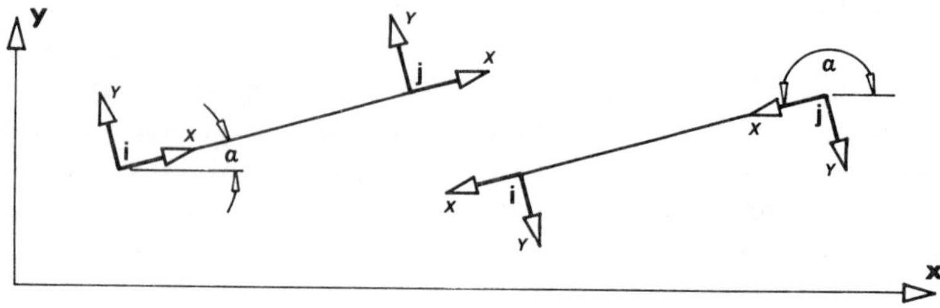

Figure 9.3 Element orientation

Rotations and bending moments can be split into components in the same way as translations and forces. For instance, consider a bending moment about the global "x" axis as shown in fig 9.4(a).

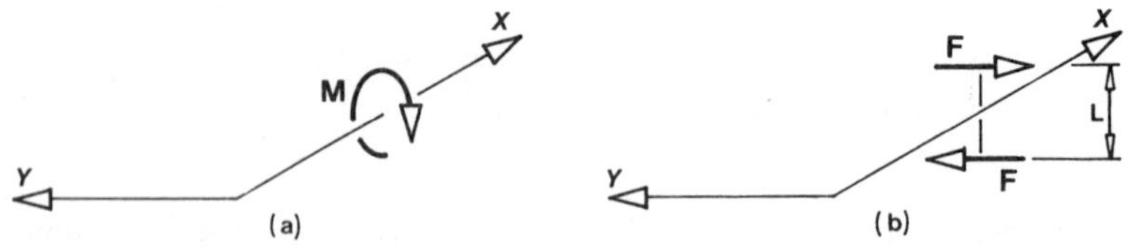

Figure 9.4 Replacement of a moment by a couple

The bending moment can be replaced by two equal and opposite forces equidistant from the "x" axis as shown in fig 9.4(b). Such a pair of forces is usually referred to as a *couple*, and $M = F\,L$ (where "L" is the distance between forces).

Throughout this text moment and rotation about any axis are defined as positive if the sense is anticlockwise when observed looking down the axis towards the origin. Hence the direction of the upper force of positive couples about the "x" and "y" axes is as shown in fig 9.5.

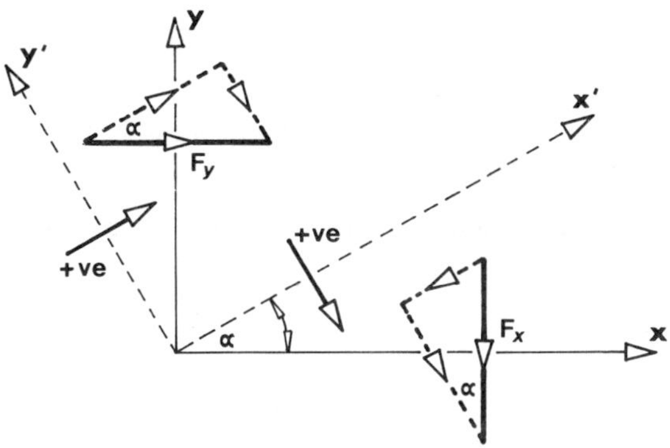

Figure 9.5 Transformation of a couple

The upper forces (F_x and F_y) of the couples about the "x" and "y" axes can be split into components in the directions of the "x'" and "y'" axes as indicated by the broken lines. The lower force in each couple can be treated similarly, and it is evident that moments can be transformed in exactly the same way as forces. If the "x,y" axes system is regarded as the global axes system at node "i", and the "x',y'" axes system is regarded as the element axes system at end "i" of element "i,j", it is a simple matter to show that transformation of a nodal force vector from the global to the element axes system takes the following form.

$$\begin{bmatrix} f_{ijz} \\ m_{ijx} \\ m_{ijy} \end{bmatrix} = \begin{bmatrix} 1 & 0 & 0 \\ 0 & cos\alpha & sin\alpha \\ 0 & -sin\alpha & cos\alpha \end{bmatrix} \begin{bmatrix} F_{ijz} \\ M_{ijx} \\ M_{ijy} \end{bmatrix}$$

or

$$f_{ij} = T_{ij} \; F_{ij}$$

where

f_{ij} is the element end force vector at end "i" of element "i,j" in the element axes system.

F_{ij} is the element end force vector at end "i" of element "i,j" in the global axes system.

T_{ij} is the transformation matrix for element "i,j".

Of course, the same transformation matrix will transform a nodal displacement vector from the global axes system to the element axes system, i.e.

$$\begin{bmatrix} \delta_{ijz} \\ \theta_{ijx} \\ \theta_{ijy} \end{bmatrix} = \begin{bmatrix} 1 & 0 & 0 \\ 0 & cos\alpha & sin\alpha \\ 0 & -sin\alpha & cos\alpha \end{bmatrix} \begin{bmatrix} \Delta_{iz} \\ \Theta_{ix} \\ \Theta_{iy} \end{bmatrix}$$

or

$$\delta_{ij} = T_{ij} \, \Delta_i$$

where

δ_{ij} is the element end displacement vector at end "*i*" of element "*i,j*" in the element axes system.

Δ_i is the nodal displacement vector at end "*i*" of element "*i,j*" in the global axes system.

T_{ij} is the transformation matrix for element "*i,j*".

The inverse of T_{ij} (shown below) effects the reverse transformation (i.e. from the element axes system to the global axes system)

$$T_{ij}^{-1} = \begin{bmatrix} 1 & 0 & 0 \\ 0 & cos\alpha & -sin\alpha \\ 0 & sin\alpha & cos\alpha \end{bmatrix}$$

9.4 The Solution Algorithm

By now it should be clear to the reader that one set of matrix equations governs the elastic behaviour of all types of skeletal structures. The relationship (in element axes) between element end forces and element end displacements is, in terms of submatrices

$$f_{ij} = k_{ii}^{j} \, \delta_{ij} + k_{ij} \, \delta_{ji}$$

and

$$f_{ji} = k_{ji} \, \delta_{ij} + k_{jj}^{i} \, \delta_{ji}$$

If the transformation matrix, T_{ij}, transforms vectors of nodal force and displacement at node "*i*" from the global axes system to the element axes system for element "*i,j*", then the equilibrium of node "*i*" is described by the following equation

$$P_i = K_{ii}^{a} \, \Delta_i + K_{ia} \, \Delta_a + K_{ii}^{b} \, \Delta_i + K_{ib} \, \Delta_b + \ldots\ldots + K_{ii}^{n} \, \Delta_i + K_{in} \, \Delta_n$$

or

$$P_i = K_{ii} \, \Delta_i + K_{ia} \, \Delta_a + K_{ib} \, \Delta_b + \ldots\ldots + K_{in} \, \Delta_n$$

where

$$K_{ii} = \sum_{j=a}^{n} K_{ii}^{j}$$

and

$$K_{ii}^{j} = T_{ij}^{-1}\, k_{ii}^{j}\, T_{ij}$$

$$= \begin{bmatrix} 1 & 0 & 0 \\ 0 & \cos\alpha & -\sin\alpha \\ 0 & \sin\alpha & \cos\alpha \end{bmatrix} \begin{bmatrix} \frac{12EI}{L^3} & 0 & -\frac{6EI}{L^2} \\ 0 & \frac{GJ}{L} & 0 \\ -\frac{6EI}{L^2} & 0 & \frac{4EI}{L} \end{bmatrix} \begin{bmatrix} 1 & 0 & 0 \\ 0 & \cos\alpha & \sin\alpha \\ 0 & -\sin\alpha & \cos\alpha \end{bmatrix}$$

$$= \begin{bmatrix} \frac{12EI}{L^3} & \frac{6EI}{L^2}\sin\alpha & -\frac{6EI}{L^2}\cos\alpha \\ \frac{6EI}{L^2}\sin\alpha & \left[\frac{GJ}{L}\cos^2\alpha + \frac{4EI}{L}\sin^2\alpha\right] & \left[\frac{GJ}{L} - \frac{4EI}{L}\right]\cos\alpha\sin\alpha \\ -\frac{6EI}{L^2}\cos\alpha & \left[\frac{GJ}{L} - \frac{4EI}{L}\right]\cos\alpha\sin\alpha & \left[\frac{GJ}{L}\sin^2\alpha + \frac{4EI}{L}\cos^2\alpha\right] \end{bmatrix}$$

Similarly

$$K_{ij} = T_{ij}^{-1}\, k_{ij}\, T_{ij}$$

$$= \begin{bmatrix} -\frac{12EI}{L^3} & \frac{6EI}{L^2}\sin\alpha & -\frac{6EI}{L^2}\cos\alpha \\ -\frac{6EI}{L^2}\sin\alpha & \left[-\frac{GJ}{L}\cos^2\alpha + \frac{2EI}{L}\sin^2\alpha\right] & \left[-\frac{GJ}{L} - \frac{2EI}{L}\right]\cos\alpha\sin\alpha \\ \frac{6EI}{L^2}\cos\alpha & \left[-\frac{GJ}{L} - \frac{2EI}{L}\right]\cos\alpha\sin\alpha & \left[-\frac{GJ}{L}\sin^2\alpha + \frac{2EI}{L}\cos^2\alpha\right] \end{bmatrix}$$

The K_{ji} and the K_{jj}^{i} submatrices can be found by similar expansions. In practice only the K_{ii}^{j} and the K_{ij} submatrices need be evaluated for each element as the other submatrices can be obtained by multiplying the corresponding elements of the K_{ii}^{j} and K_{ij} submatrices as follows

Find K_{jj}^{i} by multiplying corresponding elements of K_{ii}^{j} by

$$\begin{bmatrix} 1 & -1 & -1 \\ -1 & 1 & 1 \\ -1 & 1 & 1 \end{bmatrix}$$

Find K_{ji} by multiplying corresponding elements of K_{ij} by

$$\begin{bmatrix} 1 & -1 & -1 \\ -1 & 1 & 1 \\ -1 & 1 & 1 \end{bmatrix}$$

Note that this is not a matrix multiplication, but simply the multiplication of corresponding elements.

Writing the equation of nodal equilibrium for each node in turn produces simultaneous equations that are collectively known as the initial structure stiffness equation. Application of the boundary conditions eliminates a number of these equations and the solution of the reduced system of equations (known collectively as the final structure stiffness equation) yields the unknown nodal displacements. Once found, the nodal displacements are transformed into the element axes system and the element end forces are found using equations (9.2) and (9.3).

Example 9.1

Find the nodal displacements and check the equilibrium of node 2 for the grillage shown below given that $I = 80 \times 10^9$ mm^4, $J = 40 \times 10^9$ mm^4, $G = 6$ kN/mm^2, and $E = 15$ kN/mm^2.

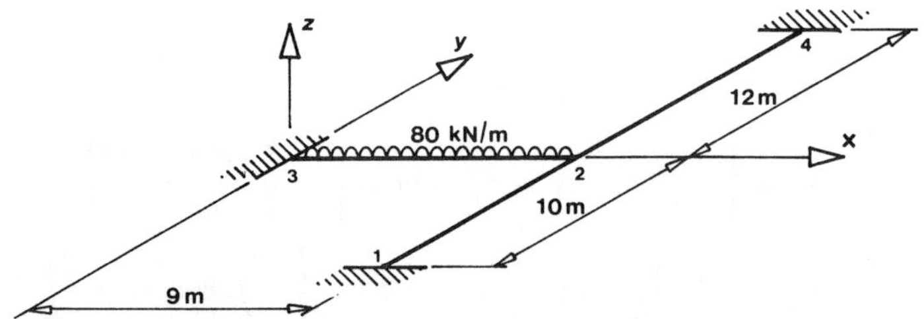

The initial stiffness equation, in terms of submatrices, is as follows

$$\begin{bmatrix} K_{11} & K_{12} & 0 & 0 \\ K_{21} & K_{22} & K_{23} & K_{24} \\ 0 & K_{32} & K_{33} & 0 \\ 0 & K_{42} & 0 & K_{44} \end{bmatrix} \begin{bmatrix} \Delta_1 \\ \Delta_2 \\ \Delta_3 \\ \Delta_4 \end{bmatrix} = \begin{bmatrix} P_1 \\ P_2 \\ P_3 \\ P_4 \end{bmatrix}$$

Applying the boundary conditions $\Delta_1 = \Delta_3 = \Delta_4 = 0$, as shown, yields the final structure stiffness equation.

$$K_{22}\,\Delta_2 = P_2 \qquad \text{where} \quad K_{22} = K_{22}^{1} + K_{22}^{3} + K_{22}^{4}$$

Loading

$$P_2 = P_2^n + P_{21}^e + P_{23}^e + P_{24}^e$$

$$= P_{23}^e = -T_{23}^{-1} f_{23}^f$$

$$= -\begin{bmatrix} 1 & 0 & 0 \\ 0 & -1 & 0 \\ 0 & 0 & -1 \end{bmatrix}\begin{bmatrix} 360 \\ 0 \\ -540000 \end{bmatrix} = \begin{bmatrix} -360 \\ 0 \\ -540000 \end{bmatrix}$$

Element 1,2

$I = 80 \times 10^9\ mm^4$, $J = 40 \times 10^9\ mm^4$, $L = 10000\ mm$,
$E = 15\ kN/mm^2$, $G = 6\ kN/mm^2$, $\alpha = 90°$

$$K_{22}^1 = T_{12}^{-1} k_{22}^1 T_{12}$$

$$= \begin{bmatrix} \frac{12EI}{L^3} & -\frac{6EI}{L^2}\sin\alpha & \frac{6EI}{L^2}\cos\alpha \\ -\frac{6EI}{L^2}\sin\alpha & \left(\frac{GJ}{L}\cos^2\alpha + \frac{4EI}{L}\sin^2\alpha\right) & \left(\frac{GJ}{L} - \frac{4EI}{L}\right)\cos\alpha\sin\alpha \\ \frac{6EI}{L^2}\cos\alpha & \left(\frac{GJ}{L} - \frac{4EI}{L}\right)\cos\alpha\sin\alpha & \left(\frac{GJ}{L}\sin^2\alpha + \frac{4EI}{L}\cos^2\alpha\right) \end{bmatrix}$$

$$= \begin{bmatrix} 14.4 & -72000 & 0 \\ -72000 & 480E6 & 0 \\ 0 & 0 & 24E6 \end{bmatrix}$$

Element 2,3

$I = 80 \times 10^9\ mm^4$, $J = 40 \times 10^9\ mm^4$, $L = 9000\ mm$,
$E = 15\ kN/mm^2$, $G = 6\ kN/mm^2$, $\alpha = 180°$

$$K_{22}^3 = T_{23}^{-1} k_{22}^3 T_{23}$$

$$= \begin{bmatrix} 19.8 & 0 & 88888 \\ 0 & 26.7E6 & 0 \\ 88888 & 0 & 533E6 \end{bmatrix}$$

Element 2,4

$I = 80 \times 10^9\ mm^4$, $J = 40 \times 10^9\ mm^4$, $L = 12000\ mm$,
$E = 15\ kN/mm^2$, $G = 6\ kN/mm^2$, $\alpha = 90°$

$$K_{22}^4 = T_{24}^{-1} k_{22}^4 T_{24}$$

$$= \begin{bmatrix} 8.33 & 50000 & 0 \\ 50000 & 400E6 & 0 \\ 0 & 0 & 200E6 \end{bmatrix}$$

Substituting numerical values into the final stiffness equation gives

$$\begin{bmatrix} 42.5 & -22000 & 88888 \\ -22000 & 906.7E6 & 0 \\ 88888 & 0 & 577E6 \end{bmatrix} \begin{bmatrix} \Delta_{2z} \\ \Theta_{2x} \\ \Theta_{2y} \end{bmatrix} = \begin{bmatrix} -360 \\ 0 \\ -540000 \end{bmatrix}$$

the solution to which is

$$\begin{bmatrix} \Delta_{2z} \\ \Theta_{2x} \\ \Theta_{2y} \end{bmatrix} = \begin{bmatrix} -9.791 \text{ mm} \\ -2.376E-4 \text{ rads} \\ 5.724E-4 \text{ rads} \end{bmatrix}$$

Element end forces (at node 2)

$$f_{21} = k_{21}\, T_{12}\, \Delta_1 + k_{22}^{1}\, T_{12}\, \Delta_2 + f_{21}^{f}$$

$$= k_{22}^{1}\, T_{12}\, \Delta_2 \qquad \text{as} \quad f_{21}^{f} = \mathbf{0} \quad \text{and} \quad \Delta_1 = \mathbf{0}$$

$$= \begin{bmatrix} 14.4 & 0 & 72000 \\ 0 & 24E6 & 0 \\ 72000 & 0 & 480E6 \end{bmatrix} \begin{bmatrix} 1 & 0 & 0 \\ 0 & 0 & 1 \\ 0 & -1 & 0 \end{bmatrix} \begin{bmatrix} -9.791 \\ -2.376E-4 \\ 5.724E-4 \end{bmatrix}$$

$$= \begin{bmatrix} -124 \text{ kN} \\ 13700 \text{ kN mm} \\ -591000 \text{ kN mm} \end{bmatrix}$$

$$f_{23} = k_{22}^{3}\, T_{23}\, \Delta_2 + k_{23}\, T_{23}\, \Delta_3 + f_{23}^{f}$$

$$= k_{22}^{1}\, T_{12}\, \Delta_2 + f_{23}^{f} \qquad \text{as} \quad \Delta_1 = \mathbf{0}$$

$$= \begin{bmatrix} 19.8 & 0 & -88888 \\ 0 & 26.7E6 & 0 \\ -88888 & 0 & 533E6 \end{bmatrix} \begin{bmatrix} 1 & 0 & 0 \\ 0 & -1 & 0 \\ 0 & 0 & -1 \end{bmatrix} \begin{bmatrix} -9.791 \\ -2.376E-4 \\ 5.724E-4 \end{bmatrix} +$$

$$\begin{bmatrix} 360 \\ 0 \\ -540000 \end{bmatrix}$$

$$= \begin{bmatrix} -142 \\ 6340 \\ 565000 \end{bmatrix} + \begin{bmatrix} 360 \\ 0 \\ -540000 \end{bmatrix} = \begin{bmatrix} 217 \text{ kN} \\ 6340 \text{ kN mm} \\ 25000 \text{ kN mm} \end{bmatrix}$$

$$f_{24} = k_{22}^{4}\, T_{24}\, \Delta_2 + k_{24}\, T_{24}\, \Delta_4 + f_{24}^{f}$$

$$= k_{22}^{4}\, T_{24}\, \Delta_2 \qquad \text{as} \quad f_{24}^{f} = \mathbf{0} \quad \text{and} \quad \Delta_4 = \mathbf{0}$$

$$= \begin{bmatrix} 8.33 & 0 & -50000 \\ 0 & 20E6 & 0 \\ -50000 & 0 & 400E6 \end{bmatrix} \begin{bmatrix} 1 & 0 & 0 \\ 0 & 0 & 1 \\ 0 & -1 & 0 \end{bmatrix} \begin{bmatrix} -9.791 \\ -2.376E-4 \\ 5.724E-4 \end{bmatrix}$$

$$= \begin{bmatrix} -93.4 \text{ kN} \\ -11400 \text{ kN mm} \\ 585000 \text{ kN mm} \end{bmatrix}$$

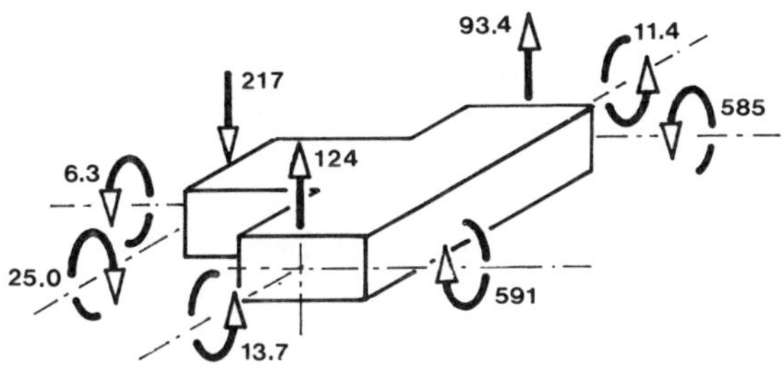

Noting that these forces are in the element axes systems, and that the forces experienced by the node are equal and opposite to the element end forces, allows the freebody diagram for node 2 to be drawn as shown (units are kN and m).

$$\sum \text{ Force in the global "}z\text{" direction} = 124 - 217 + 93.4 \approx 0$$

$$\sum \text{ Moments about the global "}x\text{" axis} = 6.3 - 591 + 585 \approx 0$$

$$\sum \text{ Moments about the global "}y\text{" axis} = 25.0 - 13.7 - 11.4 \approx 0$$

9.5 Element End Forces and Reactions

The following equation relates the element end forces and the nodal displacements and for all types of skeletal structures

$$f_{ij} = k_{ii}^{j}\, T_{ij}\, \Delta_i + k_{ij}\, T_{ij}\, \Delta_j$$

hence

$$\begin{bmatrix} f_{ijz} \\ m_{ijx} \\ m_{ijy} \end{bmatrix} = \begin{bmatrix} \frac{12EI}{L^3} & 0 & -\frac{6EI}{L^2} \\ 0 & \frac{GJ}{L} & 0 \\ -\frac{6EI}{L^2} & 0 & \frac{4EI}{L} \end{bmatrix} \begin{bmatrix} 1 & 0 & 0 \\ 0 & cos\alpha & sin\alpha \\ 0 & -sin\alpha & cos\alpha \end{bmatrix} \begin{bmatrix} \Delta_{iz} \\ \Theta_{ix} \\ \Theta_{iy} \end{bmatrix} +$$

$$\begin{bmatrix} -\frac{12EI}{L^3} & 0 & -\frac{6EI}{L^2} \\ 0 & -\frac{GJ}{L} & 0 \\ \frac{6EI}{L^2} & 0 & \frac{2EI}{L} \end{bmatrix} \begin{bmatrix} 1 & 0 & 0 \\ 0 & cos\alpha & sin\alpha \\ 0 & -sin\alpha & cos\alpha \end{bmatrix} \begin{bmatrix} \Delta_{jz} \\ \Theta_{jx} \\ \Theta_{jy} \end{bmatrix}$$

It is computationally advantageous to expand the above matrix equation, i.e.

$$f_{ijz} = \frac{12EI}{L^3}(\Delta_{iz} - \Delta_{jz}) + \frac{6EI}{L^2}[(\Theta_{ix} + \Theta_{jx})sin\alpha - (\Theta_{iy} + \Theta_{jy})cos\alpha]$$

$$m_{ijx} = \frac{GJ}{L}[(\Theta_{ix} - \Theta_{jx})cos\alpha + (\Theta_{iy} - \Theta_{jy})sin\alpha]$$

$$m_{ijy} = \frac{6EI}{L^2}[-\Delta_{iz} + \Delta_{jz}] - \frac{2EI}{L}[(2\Theta_{ix} + \Theta_{jx})sin\alpha - (2\Theta_{iy} + \Theta_{jy})cos\alpha]$$

By considering the equilibrium of the element it is a simple matter to show that the forces at end "*j*" of the element are as follows

$$f_{jiz} = -f_{ijz}$$
$$m_{jix} = -m_{ijx}$$
$$m_{jiy} = -m_{ijy} - Lf_{ijz}$$

Fixed end and equivalent joint forces can, of course, be used to deal element load, lack of fit, etc.

Reactions are calculated by using the appropriate matrices in the same equation as were used for plane trusses, space trusses and plane frames, i.e.

$$\boldsymbol{R}_i = \sum_{j=a}^{n} \boldsymbol{T}_{ij}^{-1} \boldsymbol{f}_{ij} - \boldsymbol{P}_i$$

where the summation is for all of the elements attached to node "i".

9.6 Grillage Program - GRID.BAS

In this book different programs are presented for different types of structural frameworks. This has been done for the sake of clarity. The reader should already have noticed the striking similarity between the programs presented thus far. Indeed, the same procedures have been used in the programs for the analysis of plane trusses, space trusses, and plane frames. The program for the automatic analysis of grillages, GRID.BAS, is a modification of the plane frame analysis program PFRAME.BAS. The facility to deal with element loads (which are very common on grillages) has been maintained, but the provision for local axes and for elements containing pins have been dropped.

Input

Unless GRID.BAS has been chained from PRE.BAS or DCHECK.BAS the user is prompted for the name of the data file to be read by the program. GRID.BAS then offers the user the following menu which drives the program.

Do you wish to:
(1) Analyse with output to the screen
(2) Analyse with output to the printer
(3) Analyse with output to a file
(4) Edit the data
(5) Check the data
(6) Read a new data file
(7) Stop

This data file should be constructed using the data preprocessor program PRE.BAS which is described in section 5.11. The program DCHECK.BAS can be used to check the data for obvious errors. DCHECK.BAS is presented in section 5.12.

Comments on the Algorithm

The length of each element and the cosine and sine of the angle it makes with the global "x" axis are calculated from the element node numbers and the nodal coordinates. Only the upper semi-band of the final stiffness matrix is assembled. Once the unknown nodal displacements have been found they are used to calculate the element end forces. The reactions are evaluated by summing the contributions from the elements connected to the restrained nodes.

Output

The data from the data file is echoed and messages are output to indicate the beginning of each solution phase. After solution of the final structure stiffness equation nodal displacements are output. The shear force, bending moment, and torsional moment at each end of each element are then calculated and output.

Finally the components of reaction (in the global axes system) at each restrained node are output. The user is given the option to send the program output to screen, printer or file.

Listing

```
1000 '=========================     GRID.BAS     =============================
1010 '
1020 COMMON DATAFILE$
1030 OPTION BASE 1 : KEY OFF
1040 TRUE% = -1 : FALSE% = 0
1050 DEVICE$ = "SCRN:"  : PAGELEN% = 23
1060 CLS
1070 PRINT
1080 PRINT "================================================================="
1090 PRINT
1100 PRINT " PROGRAM GRID                              Copyright (c) J.Balfour 1991"
1110 PRINT
1120 PRINT "         Program for the automatic analysis of grillages"
1130 PRINT "          Data generated with the data preprocessor PRE"
1140 PRINT
1150 PRINT "                For further information contact"
1160 PRINT " James A.D.Balfour, Heriot-Watt University, Riccarton, Edinburgh"
1170 PRINT "               Tel 031-449-5111, Fax 031-451-3170"
1180 PRINT
1190 PRINT "================================================================="
1200 '
1210 ' ... Note that variables are defined in Appendix A and the
1220 '     following statements set the maximum problem size
1230 '
1240 MAXNODE%  = 50                       '... Max no of nodes
1250 MAXELEM%  = 70                       '... Max no of elements
1260 MAXPROP%  = 20                       '... Max no of element properties
1270 MAXREST%  = 20                       '... Max no of restrained nodes
1280 MAXNLOAD% = 50                       '... Max no of nodal loads
1290 MAXELOAD% = 30                       '... Max no of element loads
1300 MAXBAND%  = 40                        '... Max semi-bandwidth
1310 '
1320 DIM NODE(MAXNODE%,3),  ELEM(MAXELEM%,7),       PROP(MAXPROP%,5)
1330 DIM REST(MAXREST%,5),  NLOAD(MAXNLOAD%,3),      ELOAD(MAXELOAD%,9)
1340 DIM FREE%(3*MAXNODE%), K(3*MAXNODE%, MAXBAND%), P(3*MAXNODE%)
1350 DIM ESTIFF(6,6),       EFREE%(6),               WKSP$(8)
1360 DIM TEMP(12)
1370 '
1380 ' ... Check if this program has been chained from PRE
1390 '
1400 IF DATAFILE$ <> "" THEN GOSUB 9860 : GOTO 1430
1410 PRINT
1420 INPUT "Name of the data file  =  ", DATAFILE$
1430 GOSUB 2150
1440 NOPTIONS% = 7
1450 WHILE (REPLY% <> 7)
1460   CLS : PRINT
1470   PRINT "Do you wish to:"
1480   PRINT "  (1)  Analyse with output to the screen"
1490   PRINT "  (2)  Analyse with output to the printer"
1500   PRINT "  (3)  Analyse with output to a file"
1510   PRINT "  (4)  Edit the data"
1520   PRINT "  (5)  Check the data"
1530   PRINT "  (6)  Read a new file"
1540   PRINT "  (7)  Stop"
1550   GOSUB 8960
1560   '
1570   ON REPLY% GOTO 1590, 1620, 1700, 1770, 1800, 1820, 1860
1580     '
1590     OPEN "SCRN:" FOR OUTPUT AS #2                      ' Reply = 1
1600     GOSUB 1910 : GOTO 1860
1610     '
1620     ON ERROR GOTO 1670                                 ' Reply = 2
1630     OPEN "LPT1:" FOR OUTPUT AS #2
1640     PRINT  : PRINT "Wait - looking for printer" : PRINT #2,
```

```
1650      ON ERROR GOTO 0 : DEVICE$ = "LPT1:" : GOSUB 1910
1660      PRINT #2, CHR$(12); : GOTO 1860
1670      CLS : PRINT : PRINT "** ERROR ** Failed to find the printer"
1680      GOSUB 9860 : RESUME 1860
1690      '
1700      ON ERROR GOTO 1740                              ' Reply = 3
1710      INPUT "Name of file for output = ", DEVICE$
1720      OPEN DEVICE$ FOR OUTPUT AS #2
1730      ON ERROR GOTO 0 : GOSUB 1910 : GOTO 1860
1740      CLS : PRINT : PRINT "** ERROR ** Failed to open output file"
1750      GOSUB 9860 : ON ERROR GOTO 0 : RESUME 1850
1760      '
1770      PRINT "Wait - chaining PRE"                       ' Reply = 4
1780      CHAIN "PRE"
1790      '
1800      PRINT "Wait - chaining DCHECK"                    ' Reply = 5
1810      CHAIN "DCHECK"
1820      PRINT                                             ' Reply = 6
1830      '
1840      INPUT "Name of the data file  =  ", DATAFILE$
1850      GOSUB 2150
1860      CLOSE #2 : DEVICE$="SCRN:" : ON ERROR GOTO 0
1870 WEND
1880 KEY ON
1890 END
1900 '
1910 ' *************************     ANALYSE     *******************************
1920 '
1930 IF DATAFILEOK% THEN GOTO 1980
1940 CLS : PRINT
1950 PRINT "** ERROR ** Analysis cannot proceed due to errors in data file";
1960 GOSUB 9860                           '... Pause
1970 GOTO 2130
1980 GOSUB 3650                           '... Print data
1990 PRINT : PRINT "Generating the freedom vector                 "
2000 GOSUB 5100
2010 PRINT : PRINT "Assembling the stiffness matrix, adding element :- ";
2020 ROW% = CSRLIN : COL% = POS(0)
2030 GOSUB 5430
2040 PRINT : PRINT "Generating the loading vector                 "
2050 GOSUB 6310
2060 PRINT
2070 PRINT "Solving equations - no. of equations = "; NDOF%
2080 PRINT "                  - semi-bandwidth   = "; BAND%
2090 GOSUB 6960
2100 GOSUB 9860                           '... Pause
2110 GOSUB 7560                           '... Output nodal displacements
2120 GOSUB 7790                           '... Output element forces & reactions
2130 RETURN
2140 '
2150 '***********************     READ DATA FILE     ***************************
2160 '
2170 ' ...  This subroutine reads the data from a data file
2180 '
2190 PRINT
2200 PRINT "Wait - Reading file "; DATAFILE$; " line";
2210 ROW% = CSRLIN : COL% = POS(0) : LCOUNT% = 0
2220 DATAFILEOK% = FALSE%
2230 DATAEND%    = FALSE%
2240 NNODE%      = 0 : NELEM%  = 0 : NPROP% = 0
2250 NREST%      = 0 : NNLOAD% = 0 : NELOAD%  = 0
2260 ON ERROR GOTO 2280 : OPEN DATAFILE$ FOR INPUT AS #1
2270 ON ERROR GOTO 0 : GOTO 2340
2280 CLS : PRINT
2290 PRINT "** ERROR ** Failed to open file: "; DATAFILE$; " for input"
2300 GOSUB 9860 : RESUME 3620
2310 '
2320 ' ... Read problem type, set up array sizes and no of nos in each row
2330 '
2340 TARGET$ = "GRID" : GOSUB 9610
2350 IF STRFOUND% THEN GOTO 2420
2360 CLS : PRINT
2370 PRINT "** ERROR ** Not a file for GRID"
2380 GOSUB 9860 : GOTO 3620
2390 '
2400 ' ... Read title and units strings
2410 '
```

```
2420 NLINES% = 2 : GOSUB 9770 : IF DATAEND% THEN GOTO 2490
2430 WHILE (MID$(WRKSTR$,I%,1) = " ")
2440   I% = I% + 1
2450 WEND
2460 TITLE$ = MID$(WRKSTR$, I%)
2470 TARGET$ = "Units" : GOSUB 9610
2480 IF STRFOUND% THEN GOTO 2520
2490 CLS : PRINT
2500 PRINT "** ERROR ** Error reading Title/Units strings"
2510 GOSUB 9860 : GOTO 2540
2520 UNITS$ = MID$(WRKSTR$, INSTR(WRKSTR$, ":- ") + 3, 24)
2530 DATAFILEOK% = TRUE%
2540 WHILE NOT DATAEND%
2550   '
2560   ' ... Get module string
2570   '
2580   IF LEFT$(WRKSTR$,4) = "+   " THEN GOTO 2600
2590   TARGET$ = "+   " : GOSUB 9610 : IF NOT STRFOUND% THEN GOTO 3610
2600   MODSTR$ = WRKSTR$
2610   '
2620   ' ...  Identify module
2630   '
2640   IF INSTR(MODSTR$, "NODAL COORDINATES") = 0 THEN GOTO 2790
2650   '
2660   ' ... Read nodal coordinates
2670   '
2680   TARGET$ = "NODE" : GOSUB 9610 : IF NOT STRFOUND% THEN GOTO 3540
2690   GOSUB 9240
2700   WHILE NNOS% > 0
2710     NNODE% = NNODE% + 1
2720     FOR J% = 1 TO NNOS%
2730       NODE(NNODE%,J%) = VAL(WKSP$(J%))
2740     NEXT J%
2750     GOSUB 9240
2760   WEND
2770   GOTO 3610
2780   '
2790   IF INSTR(MODSTR$,"ELEMENTS") = 0 THEN GOTO 2940
2800   '
2810   ' ... Read elements
2820   '
2830   TARGET$ = "ELEMENT" : GOSUB 9610 : IF NOT STRFOUND% THEN GOTO 3540
2840   GOSUB 9240
2850   WHILE NNOS% > 0
2860     NELEM% = NELEM% + 1
2870     FOR J% = 1 TO NNOS%
2880       ELEM(NELEM%,J%) = VAL(WKSP$(J%))
2890     NEXT J%
2900     GOSUB 9240
2910   WEND
2920   GOTO 3610
2930   '
2940   IF INSTR(MODSTR$,"PROPERTIES") = 0 THEN GOTO 3090
2950   '
2960   ' ... Read properties
2970   '
2980   TARGET$ = "PROPERTY" : GOSUB 9610 : IF NOT STRFOUND% THEN GOTO 3540
2990   GOSUB 9240
3000   WHILE NNOS% > 0
3010     NPROP% = NPROP% + 1
3020     FOR J% = 1 TO NNOS%
3030       PROP(NPROP%,J%) = VAL(WKSP$(J%))
3040     NEXT J%
3050     GOSUB 9240
3060   WEND
3070   GOTO 3610
3080   '
3090   IF INSTR(MODSTR$,"RESTRAINTS") = 0 THEN GOTO 3240
3100   '
3110   ' ... Read restraints
3120   '
3130   TARGET$ = "NODE" : GOSUB 9610 : IF NOT STRFOUND% THEN GOTO 3540
3140   GOSUB 9240
3150   WHILE NNOS% > 0
3160     NREST% = NREST% + 1
3170     FOR J% = 1 TO NNOS%
3180       REST(NREST%,J%) = VAL(WKSP$(J%))
```

```
3190      NEXT J%
3200      GOSUB 9240
3210    WEND
3220    GOTO 3610
3230    '
3240    IF INSTR(MODSTR$,"NODAL LOADS") = 0 THEN GOTO 3390
3250    '
3260    ' ... Read nodal loads
3270    '
3280    TARGET$ = "NODE" : GOSUB 9610 : IF NOT STRFOUND% THEN GOTO 3540
3290    GOSUB 9240
3300    WHILE NNOS% > 0
3310      NNLOAD% = NNLOAD% + 1
3320      FOR J% = 1 TO NNOS%
3330        NLOAD(NNLOAD%,J%) = VAL(WKSP$(J%))
3340      NEXT J%
3350      GOSUB 9240
3360    WEND
3370    GOTO 3610
3380    '
3390    IF INSTR(MODSTR$,"ELEMENT LOADS") = 0 THEN GOTO 3540
3400    '
3410    ' ... Read element loads
3420    '
3430    TARGET$ = "TYPE" : GOSUB 9610 : IF NOT STRFOUND% THEN GOTO 3540
3440    GOSUB 9240
3450    WHILE NNOS% > 0
3460      NELOAD% = NELOAD% + 1
3470      FOR J% = 1 TO NNOS%
3480        ELOAD(NELOAD%,J%) = VAL(WKSP$(J%))
3490      NEXT J%
3500      GOSUB 9240
3510    WEND
3520    GOTO 3610
3530    '
3540    CLS : PRINT
3550    PRINT "** ERROR ** in module "; MODSTR$; "run DCHECK for more info";
3560    GOSUB 9860
3570    PRINT
3580    PRINT "Wait - Reading file "; DATAFILE$; " line";
3590    ROW% = CSRLIN : COL% = POS(0)
3600    DATAFILEOK% = FALSE% : WRKSTR$ = ""
3610 WEND
3620 CLOSE #1 : ON ERROR GOTO 0
3630 RETURN
3640 '
3650 ' ***********************    PRINT DATA    ******************************
3660 '
3670 GOSUB 3760                          '... List header
3680 GOSUB 4090                          '... List nodal coordinates
3690 GOSUB 4240                          '... List element data
3700 GOSUB 4390                          '... List list element properties
3710 GOSUB 4580                          '... List restraint data
3720 GOSUB 4730                          '... List nodal loads
3730 GOSUB 4900                          '... List element loads
3740 RETURN
3750 '
3760 '************************    LIST HEADER    *****************************
3770 '
3780 CLS
3790 PRINT #2, : PRINT #2,
3800 PRINT #2, "======================================";
3810 PRINT #2, "======================================"
3820 PRINT #2,
3830 PRINT #2, "     PROGRAM GRID"; STRING$(26,32);
3840 PRINT #2, "Copyright(c) J.A.D.Balfour 1991"
3850 PRINT #2,
3860 FOR I% = 1 TO (76 - LEN(TITLE$)) / 2
3870   PRINT #2, " ";
3880 NEXT I%
3890 PRINT #2, TITLE$
3900 PRINT #2,
3910 PRINT #2, "     File Name :- "; DATAFILE$;
3920 FOR I% = 1 TO (40 - LEN(DATAFILE$))
3930   PRINT #2, " ";
3940 NEXT I%
3950 PRINT #2, "Date :- "; MID$(DATE$,4,3); MID$(DATE$,1,3);
```

```
3960 PRINT #2, MID$(DATE$,9,2)
3970 PRINT #2, "    Units     :- "; UNITS$;
3980 FOR I% = 1 TO (40 - LEN(UNITS$))
3990   PRINT #2, " ";
4000 NEXT I%
4010 PRINT #2, "Time :- "; TIME$
4020 PRINT #2,
4030 PRINT #2, "=======================================";
4040 PRINT #2, "======================================="
4050 PRINT #2,
4060 GOSUB 9860
4070 RETURN
4080 '
4090 '********************   LIST NODAL COORDINATES   **********************
4100 '
4110 PRINT #2,
4120 PRINT #2, "+ + + + + + + + + + + + + +"
4130 PRINT #2, "+   NODAL COORDINATES   +"
4140 PRINT #2, "+ + + + + + + + + + + + + +"
4150 PRINT #2,
4160 PRINT #2, "NODE           X              Y"
4170 FOR I% = 1 TO NNODE%
4180   GOSUB 9950
4190   PRINT #2, " "; NODE(I%,1), NODE(I%,2), NODE(I%,3)
4200 NEXT I%
4210 GOSUB 9860
4220 RETURN
4230 '
4240 '************************   LIST ELEMENTS   **************************
4250 '
4260 PRINT #2,
4270 PRINT #2, "+ + + + + + + + + +"
4280 PRINT #2, "+   ELEMENTS    +"
4290 PRINT #2, "+ + + + + + + + + +"
4300 PRINT #2,
4310 PRINT #2, "ELEMENT    PROPERTY"
4320 FOR I% = 1 TO NELEM%
4330   GOSUB 9950
4340   PRINT #2, USING "### ###      ###"; ELEM(I%,1); ELEM(I%,2); ELEM(I%,3)
4350 NEXT I%
4360 GOSUB 9860
4370 RETURN
4380 '
4390 '***********************   LIST PROPERTIES   *************************
4400 '
4410 PRINT #2,
4420 PRINT #2, "+ + + + + +  + + +"
4430 PRINT #2, "+   PROPERTIES   +"
4440 PRINT #2, "+ + + + + +  + + +"
4450 PRINT #2,
4460 PRINT #2, "PROPERTY        I             J             G             E"
4470 FOR I% = 1 TO NPROP%
4480   GOSUB 9950
4490   PRINT #2, USING "  ###  "; PROP(I%,1);
4500   FOR J% = 2 TO 5
4510     PRINT #2, USING "   #.####^^^^"; PROP(I%,J%);
4520   NEXT J%
4530   PRINT #2,
4540 NEXT I%
4550 GOSUB 9860
4560 RETURN
4570 '
4580 '*************************   LIST RESTRAINTS   ************************
4590 '
4600 PRINT #2,
4610 PRINT #2, "+ + + + + + + + + + + "
4620 PRINT #2, "+   RESTRAINTS    +"
4630 PRINT #2, "+ + + + + + + + + + + "
4640 PRINT #2,
4650 PRINT #2, "NODE    DIRECTION(S)"
4660 FOR I% = 1 TO NREST%
4670   GOSUB 9950
4680   PRINT #2, USING " ##     ######"; REST(I%,1); REST(I%,2)
4690 NEXT I%
4700 GOSUB 9860
4710 RETURN
4720 '
```

```
4730 '*********************** LIST NODAL LOADS ***************************
4740 '
4750 IF NNLOAD% = 0 THEN GOTO 4880
4760 PRINT #2,
4770 PRINT #2, "+ + + + + + + + + + +"
4780 PRINT #2, "+   NODAL LOADS   +"
4790 PRINT #2, "+ + + + + + + + + + +"
4800 PRINT #2,
4810 PRINT #2, "NODE    DIRECTION        VALUE"
4820 FOR I% = 1 TO NNLOAD%
4830   GOSUB 9950
4840   PRINT #2, USING "###        ###"; NLOAD(I%,1); NLOAD(I%,2);
4850   PRINT #2, USING "        #.####^^^^"; NLOAD(I%,3)
4860 NEXT I%
4870 GOSUB 9860
4880 RETURN
4890 '
4900 '*********************** LIST ELEMENT LOADS *************************
4910 '
4920 IF NELOAD% = 0 THEN GOTO 5080
4930 PRINT #2,
4940 PRINT #2, "+ + + + + + + + + + + + "
4950 PRINT #2, "+   ELEMENT LOADS   +    -    UDL = 1,    POINT LOAD = 2"
4960 PRINT #2, "+ + + + + + + + + + + + "
4970 PRINT #2,
4980 PRINT #2, "ELEMENT   LOAD     MAGNITUDE    DISTANCE FROM"
4990 PRINT #2, "          TYPE                  LOWER NODE"
5000 FOR I% = 1 TO NELOAD%
5010   GOSUB 9950
5020   PRINT #2, USING "### ###   ###"; ELOAD(I%,1); ELOAD(I%,2); ELOAD(I%,3);
5030   PRINT #2, USING "     #.####^^^^"; ELOAD(I%,4);
5040   IF ELOAD(I%,3) = 2 THEN PRINT #2, USING "    #.####^^^^"; ELOAD(I%,5)
5050   IF ELOAD(I%,3) <> 2 THEN PRINT #2, "        N/A"
5060 NEXT I%
5070 GOSUB 9860
5080 RETURN
5090 '
5100 ' *****************    GENERATE THE FREEDOM VECTOR    *******************
5110 '
5120 ' ... Zero the freedom vector
5130 '
5140 FOR I% = 1 TO 3 * NNODE%
5150   FREE%(I%) = 0
5160 NEXT I%
5170 '
5180 ' ... Loop for all restrained nodes
5190 '
5200 FOR I% = 1 TO NREST%
5210   K% = REST(I%,2)
5220   '
5230   ' ... Evaluate the freedom to be restrained from K%
5240   '
5250   J% = 3 * REST(I%,1) - 3 + (K% MOD 10)
5260   FREE%(J%) = 1
5270   K% = INT(K% / 10)
5280   IF K% > 0 THEN GOTO 5250
5290 NEXT I%
5300 '
5310 ' ... Number the freedoms (restraints set to zero)
5320 '
5330 NDOF% = 0
5340 FOR I% = 1 TO 3 * NNODE%
5350   IF FREE%(I%) = 1 THEN GOTO 5390
5360   NDOF% = NDOF% + 1
5370   FREE%(I%) = NDOF%
5380   GOTO 5400
5390   FREE%(I%) = 0
5400 NEXT I%
5410 RETURN
5420 '
5430 ' ***************    ASSEMBLE THE STIFFNESS MATRIX    ******************
5440 '
5450 ' ... Zero the augmented matrix
5460 '
5470 FOR I% = 1 TO NDOF%
5480   FOR J% = 1 TO MAXBAND%
5490     K(I%,J%) = 0
```

```
5500    NEXT J%
5510    P(I%) = 0
5520  NEXT I%
5530  '
5540  ' ... Loop for each element
5550  '
5560  BAND% = 0
5570  FOR K% = 1 TO NELEM%
5580    LOCATE ROW%, COL% : PRINT K%;
5590    IN% = ELEM(K%,1)
5600    JN% = ELEM(K%,2)
5610    PN% = ELEM(K%,3)
5620    '
5630    ' ... Calculate the element length (store in ELEM(K%,4))
5640    '
5650    ELEM(K%,4) = (NODE(JN%,2)-NODE(IN%,2))^2 + (NODE(JN%,3)-NODE(IN%,3))^2
5660    ELEM(K%,4) = SQR(ELEM(K%,4))
5670    '
5680    ' ... Calculate cosine and sine of the alpha angle store in ELEM
5690    '
5700    ELEM(K%,5) = (NODE(JN%,2)-NODE(IN%,2)) / ELEM(K%,4)
5710    ELEM(K%,6) = (NODE(JN%,3)-NODE(IN%,3)) / ELEM(K%,4)
5720    '
5730    ' ... Assemble the upper triangle of the element stiffness matrix
5740    '
5750    L   = ELEM(K%,4)
5760    C   = ELEM(K%,5)
5770    S   = ELEM(K%,6)
5780    EIL = PROP(PN%,5) * PROP(PN%,2) / ELEM(K%,4)
5790    GJL = PROP(PN%,4) * PROP(PN%,3) / ELEM(K%,4)
5800    ESTIFF(1,1) = 12 * EIL / L ^ 2
5810    ESTIFF(1,2) =  6 * EIL * S / L
5820    ESTIFF(1,3) = -6 * EIL * C / L
5830    ESTIFF(1,4) = -ESTIFF(1,1)
5840    ESTIFF(1,5) =  ESTIFF(1,2)
5850    ESTIFF(1,6) =  ESTIFF(1,3)
5860    ESTIFF(2,2) =  4 * EIL * S * S + GJL * C * C
5870    ESTIFF(2,3) = -(4 * EIL - GJL) * S * C
5880    ESTIFF(2,4) = -ESTIFF(1,2)
5890    ESTIFF(2,5) =  2 * EIL * S * S - GJL * C * C
5900    ESTIFF(2,6) = -(2 * EIL + GJL) * S * C
5910    ESTIFF(3,3) =  4 * EIL * C * C + GJL * S * S
5920    ESTIFF(3,4) = -ESTIFF(1,3)
5930    ESTIFF(3,5) =  ESTIFF(2,6)
5940    ESTIFF(3,6) =  2 * EIL * C * C - GJL * S * S
5950    ESTIFF(4,4) =  ESTIFF(1,1)
5960    ESTIFF(4,5) = -ESTIFF(1,2)
5970    ESTIFF(4,6) = -ESTIFF(1,3)
5980    ESTIFF(5,5) =  ESTIFF(2,2)
5990    ESTIFF(5,6) =  ESTIFF(2,3)
6000    ESTIFF(6,6) =  ESTIFF(3,3)
6010    '
6020    ' ... Set up the element code number
6030    '
6040    EFREE%(1) = FREE%(3*IN%-2)
6050    EFREE%(2) = FREE%(3*IN%-1)
6060    EFREE%(3) = FREE%(3*IN%)
6070    EFREE%(4) = FREE%(3*JN%-2)
6080    EFREE%(5) = FREE%(3*JN%-1)
6090    EFREE%(6) = FREE%(3*JN%)
6100    '
6110    ' ... Add the element stiffness to the structure stiffness matrix
6120    '
6130    FOR I% = 1 TO 6
6140      IF EFREE%(I%) = 0 THEN GOTO 6260
6150      FOR J% = I% TO 6
6160        IF EFREE%(J%) = 0 THEN GOTO 6250
6170        L% = EFREE%(I%)
6180        M% = EFREE%(J%) - L% + 1
6190        IF M% <= BAND% THEN GOTO 6240
6200        BAND% = M%
6210        IF BAND% <= MAXBAND% THEN GOTO 6240
6220        PRINT #2, "Max semi-bandwidth ("; STR$(MAXBAND%); ") exceeded"
6230        END
6240        K(L%,M%) = K(L%,M%) + ESTIFF(I%,J%)
6250      NEXT J%
6260    NEXT I%
```

```
6270 NEXT K%
6280 PRINT
6290 RETURN
6300 '
6310 '******************     SET UP THE LOADING VECTOR      *********************
6320 '
6330 ' ... Loop for all point loads
6340 '
6350 FOR I% = 1 TO NNLOAD%
6360   J% = 3 * NLOAD(I%,1) - 3 + NLOAD(I%,2)
6370   J% = FREE%(J%)
6380   IF J% > 0 THEN P(J%) = NLOAD(I%,3)
6390 NEXT I%
6400 '
6410 ' ... Add element loads to the loading vector
6420 '
6430 FOR I% = 1 TO NELOAD%
6440   IN% = ELOAD(I%,1)
6450   JN% = ELOAD(I%,2)
6460   '
6470   '  ... Search for element
6480   '
6490   FOR K% = 1 TO NELEM%
6500     IF ELEM(K%, 1) <> IN% THEN GOTO 6520
6510     IF ELEM(K%, 2) = JN% THEN J% = K%
6520   NEXT K%
6530   W = ELOAD(I%,4)
6540   L = ELEM(J%,4)
6550   X = ELOAD(I%,5)
6560   Y = L - X
6570   C = ELEM(J%,5)
6580   S = ELEM(J%,6)
6590   '
6600   ' ... Calc and store F.E.F.
6610   '
6620   ON ELOAD(I%,3) GOTO 6660, 6740
6630   '
6640   ' ... Uniformly distributed load
6650   '
6660   ELOAD(I%,6) = -W * L / 2
6670   ELOAD(I%,7) =  W * L^2 / 12
6680   ELOAD(I%,8) =  ELOAD(I%,6)
6690   ELOAD(I%,9) = -ELOAD(I%,7)
6700   GOTO 6810
6710   '
6720   ' ... Point load
6730   '
6740   ELOAD(I%,6) = -W * (Y/L)^2 * (1 + 2*X/L)
6750   ELOAD(I%,7) =  W * X * Y^2 / L^2
6760   ELOAD(I%,8) = -W * (X/L)^2 * (1 + 2*Y/L)
6770   ELOAD(I%,9) = -W * X^2 * Y / L^2
6780   '
6790   ' ... Add equivalent joint forces to the loading vector
6800   '
6810   J% = FREE%(3*IN%-2)
6820   IF J% <> 0 THEN P(J%) = P(J%) - ELOAD(I%,6)
6830   J% = FREE%(3*IN%-1)
6840   IF J% <> 0 THEN P(J%) = P(J%) + ELOAD(I%,7)*S
6850   J% = FREE%(3*IN%)
6860   IF J% <> 0 THEN P(J%) - P(J%)   ELOAD(I%,7)*C
6870   J% = FREE%(3*JN%-2)
6880   IF J% <> 0 THEN P(J%) = P(J%) - ELOAD(I%,8)
6890   J% = FREE%(3*JN%-1)
6900   IF J% <> 0 THEN P(J%) = P(J%) + ELOAD(I%,9)*S
6910   J% = FREE%(3*JN%)
6920   IF J% <> 0 THEN P(J%) = P(J%) - ELOAD(I%,9)*C
6930 NEXT I%
6940 RETURN
6950 '
6960 ' **********************    SOLVE EQUATIONS     **************************
6970 '
6980 ' ... This subroutine solves a system of symmetric, banded equations
6990 '     using Gaussian elimation with no row interchange.  Only the
7000 '     uppper semi-band of the coefficient matrix is stored.
7010 '
7020 ' ... Loop for all pivots
7030 '
```

```
7040 BB% = BAND%
7050 PRINT "                    - current pivot    = ";
7060 ROW% = CSRLIN: COL% = POS(0):
7070 FOR I% = 1 TO NDOF%
7080   LOCATE ROW%, COL%: PRINT I%;
7090   '
7100   ' ... Check if in the unused triangle
7110   '
7120   IF I% > NDOF%-BAND%+1 THEN BB% = NDOF%-I%+1
7130   PIVOT = K(I%,1)
7140   '
7150   ' ... Normalise
7160   '
7170   FOR J% = 1 TO BB%
7180     K(I%,J%) = K(I%,J%) / PIVOT
7190   NEXT J%
7200   P(I%) = P(I%) / PIVOT
7210   '
7220   ' ... Check if in the last row
7230   '
7240   IF BB% = 1 THEN GOTO 7430
7250   '
7260   ' ... Eliminate (within band) for all rows above the pivot
7270   '
7280   FOR K% = 2 TO BB%
7290     '
7300     ' ... Calculate the row number and evaluate the multiplier
7310     '
7320     L%   = I%+K%-1
7330     MULT = K(I%,K%) * PIVOT
7340     '
7350     ' ... Loop for all elements in the elimination row
7360     '
7370     FOR J% = K% TO BB%
7380       M% = J% - K% + 1
7390       K(L%,M%) = K(L%,M%) - MULT*K(I%,J%)
7400     NEXT J%
7410     P(L%) = P(L%) - MULT*P(I%)
7420   NEXT K%
7430 NEXT I%
7440 '
7450 ' ... Back substitute
7460 '
7470 FOR I% = 1 TO NDOF% - 1
7480   IF I% > BAND%-1 THEN BB% = BAND%-1 ELSE BB% = I%
7490   L% = NDOF% - I%
7500   FOR J% = 1 TO BB%
7510     P(L%) = P(L%) - K(L%,J%+1) * P(L%+J%)
7520   NEXT J%
7530 NEXT I%
7540 RETURN
7550 '
7560 '********************     OUTPUT THE DISPLACEMENTS    ********************
7570 '
7580 PRINT #2,
7590 PRINT #2, "* * * * * * * * * * * **"
7600 PRINT #2, "*   DISPLACEMENTS   *            * :-  RESTRAINT"
7610 PRINT #2, "* * * * * * * * * * * **"
7620 PRINT #2,
7630 PRINT #2, "NODE     Z-DISPLACEMENT      X-ROTATION       Y-ROTATION"
7640 FOR I% = 1 TO NNODE%
7650   GOSUB 9950
7660   PRINT #2, USING "###    "; I%;
7670   FOR J% = 1 TO 3
7680     K% = FREE%(3*I% + J% - 3)
7690     IF K% = 0 THEN GOTO 7720
7700     PRINT #2, USING "       #.####^^^^"; P(K%);
7710     GOTO 7730
7720     PRINT #2, "           *     ";
7730   NEXT J%
7740   PRINT #2,
7750 NEXT I%
7760 GOSUB 9860              '... Pause
7770 RETURN
7780 '
7790 '******   CALCULATE AND OUTPUT THE ELEMENT FORCES AND REACTIONS   ******
7800 '
```

```
7810 ' ... Zero the reaction components
7820 '
7830 FOR I% = 1 TO NREST%
7840   REST(I%,3) = 0
7850   REST(I%,4) = 0
7860   REST(I%,5) = 0
7870 NEXT I%
7880 PRINT #2,
7890 PRINT #2, "* * * * * * * * * * * *"
7900 PRINT #2, "*   ELEMENT FORCES   *"
7910 PRINT #2, "* * * * * * * * * * * *"
7920 PRINT #2,
7930 PRINT #2, "                .........  END I   .........";
7940 PRINT #2, "        .........  END J   ........."
7950 PRINT #2, "ELEMENT       SHEAR      TORSION      BENDING";
7960 PRINT #2, "        SHEAR      TORSION      BENDING"
7970 PRINT #2, "              FORCE       MOMENT       MOMENT";
7980 PRINT #2, "        FORCE      MOMENT       MOMENT"
7990 '
8000 ' ...  Loop for all elements
8010 '
8020 FOR K% = 1 TO NELEM%
8030   PN% = ELEM(K%,3)
8040   L   = ELEM(K%,4)
8050   C   = ELEM(K%,5)
8060   S   = ELEM(K%,6)
8070   EIL = PROP(PN%,5) * PROP(PN%,2) / ELEM(K%,4)
8080   GJL = PROP(PN%, 4) * PROP(PN%, 3) / ELEM(K%, 4)
8090   I%  = 3*ELEM(K%,1) - 3
8100   '
8110   ' ...  Extract the nodal displacement - store in TEMP(1)-(6)
8120   '
8130   FOR M% = 1 TO 6
8140     J% = FREE%(I%+M%)
8150     IF J% = 0 THEN TEMP(M%) = 0 ELSE TEMP(M%) = P(J%)
8160     IF M% = 3 THEN I% = 3 * ELEM(K%,2) - 6
8170   NEXT M%
8180   '
8190   ' ...  Calculate the element end forces - store in TEMP(7)-(12)
8200   '
8210   TEMP(7) = (TEMP(2)*S - TEMP(3)*C + TEMP(5)*S - TEMP(6)*C)
8220   TEMP(7) = 6 * EIL * (2*(TEMP(1)-TEMP(4))/L + TEMP(7)) / L
8230   TEMP(8) = GJL * (TEMP(2)*C + TEMP(3)*S - TEMP(5)*C - TEMP(6)*S)
8240   TEMP(9) = -2*TEMP(2)*S + 2*TEMP(3)*C - TEMP(5)*S + TEMP(6)*C
8250   TEMP(9) = 2 * EIL * (3*(TEMP(4)-TEMP(1))/L + TEMP(9))
8260   TEMP(10) = -TEMP(7)
8270   TEMP(11) = -TEMP(8)
8280   TEMP(12) = -L*TEMP(7) - TEMP(9)
8290   '
8300   ' ...  Add in fixed end forces
8310   '
8320   FOR I% = 1 TO NELOAD%
8330     IF ELEM(K%,1) <> ELOAD(I%,1) THEN GOTO 8390
8340     IF ELEM(K%,2) <> ELOAD(I%,2) THEN GOTO 8390
8350     TEMP(7)  = TEMP(7)  + ELOAD(I%,6)
8360     TEMP(9)  = TEMP(9)  + ELOAD(I%,7)
8370     TEMP(10) = TEMP(10) + ELOAD(I%,8)
8380     TEMP(12) = TEMP(12) + ELOAD(I%,9)
8390   NEXT I%
8400   GOSUB 9950
8410   PRINT #2, USING "## ##   "; ELEM(K%,1); ELEM(K%,2);
8420   PRINT #2, USING " ##.####^^^^"; TEMP(7); TEMP(8); TEMP(9);
8430   PRINT #2, USING " ##.####^^^^"; TEMP(10); TEMP(11); TEMP(12)
8440   '
8450   ' ...  Add any contribution from the element forces to the reactions
8460   '
8470   FOR I% = 1 TO NREST%
8480     IF (ELEM(K%,1) <> REST(I%,1)) THEN GOTO 8520
8490     REST(I%,3) = REST(I%,3) + TEMP(7)
8500     REST(I%, 4) = REST(I%,4) + C*TEMP(8) - S*TEMP(9)
8510     REST(I%, 5) = REST(I%,5) + S*TEMP(8) + C*TEMP(9)
8520     IF (ELEM(K%, 2) <> REST(I%, 1)) THEN GOTO 8560
8530     REST(I%, 3) = REST(I%,3) + TEMP(10)
8540     REST(I%, 4) = REST(I%,4) + C*TEMP(11) - S*TEMP(12)
8550     REST(I%, 5) = REST(I%,5) + S*TEMP(11) + C*TEMP(12)
8560   NEXT I%
8570 NEXT K%
```

```
8580 GOSUB 9860             '... Pause
8590 '
8600 ' ...   Add any applied loads
8610 '
8620 FOR I% = 1 TO NREST%
8630   FOR J% = 1 TO NNLOAD%
8640     IF REST(I%, 1) <> NLOAD(J%, 1) THEN GOTO 8670
8650     K% = NLOAD(J%,2)
8660     REST(I%,2+K%) = REST(I%,2+K%) - NLOAD(J%,3)
8670   NEXT J%
8680 NEXT I%
8690 '
8700 ' ...  Output the reactions
8710 '
8720 PRINT #2,
8730 PRINT #2, "* * * * * * * * * **"
8740 PRINT #2, "*   REACTIONS   *"
8750 PRINT #2, "* * * * * * * * * **"
8760 PRINT #2,
8770 PRINT #2, "                Z-COMPONENT       X-MOMENT        Y-MOMENT"
8780 TEMP(1) = 0 : TEMP(2) = 0 : TEMP(3) = 0
8790 '
8800 ' ... Extract reaction values calculated from element end forces
8810 '
8820 FOR I% = 1 TO NREST%
8830   GOSUB 9950
8840   PRINT #2, USING "NODE ##  "; REST(I%,1);
8850   PRINT #2, USING "    ##.####^^^^"; REST(I%,3); REST(I%,4); REST(I%,5)
8860   TEMP(1) = TEMP(1) + REST(I%,3)
8870   TEMP(2) = TEMP(2) + REST(I%,4)
8880   TEMP(3) = TEMP(3) + REST(I%,5)
8890 NEXT I%
8900 PRINT #2, "                ----------      ----------      ----------"
8910 PRINT #2, "TOTAL    ";
8920 PRINT #2, USING "     #.####^^^^"; TEMP(1); TEMP(2); TEMP(3)
8930 GOSUB 9860            '... Pause
8940 RETURN
8950 '
8960 ' *******************    GET KEYBOARD RESPONSE    *********************
8970 '
8980 ROW% = CSRLIN : COL% = POS(0)
8990 REPLY% = 0
9000 WHILE REPLY% = 0
9010   LOCATE ROW%, COL% : INPUT "", WRKSTR$ : REPLY% = VAL(WRKSTR$)
9020   IF (REPLY%<1)  OR  (REPLY%>NOPTIONS%) THEN GOTO 9050
9030   IF (REPLY%<10) AND (LEN(WRKSTR$)=1)   THEN GOTO 9060
9040   IF (REPLY%>9)  AND (LEN(WRKSTR$)=2)   THEN GOTO 9060
9050   SOUND 700, 4 : REPLY% = 0 : LOCATE ROW%, COL%
9060 WEND
9070 PRINT
9080 RETURN
9090 '
9100 '**********************    GET MENU RESPONSE    ***********************
9110 '
9120 ROW% = CSRLIN : COL% = POS(0)
9130 REPLY% = 0
9140 WHILE REPLY% = 0
9150   LOCATE ROW%, COL% : INPUT "", WRKSTR$ : REPLY% = VAL(WRKSTR$)
9160   IF (REPLY%<1)  OR  (REPLY%>NOPTIONS%) THEN GOTO 9190
9170   IF (REPLY%<10) AND (LEN(WRKSTR$)=1)   THEN GOTO 9200
9180   IF (REPLY%>9)  AND (LEN(WRKSTR$)=2)   THEN GOTO 9200
9190   SOUND 700, 4 : REPLY% = 0 : LOCATE ROW%, COL%
9200 WEND
9210 PRINT
9220 RETURN
9230 '
9240 '***********************    RECORD NUMBERS    ***************************
9250 '
9260 NNOS% = 0 : DONE% = TRUE%
9270 IF EOF(1) <> 0 THEN DATAEND% = TRUE% : GOTO 9590
9280 '
9290 ' ... Read next line and check if it is blank
9300 '
9310 LINE INPUT #1, WRKSTR$
9320 LCOUNT% = LCOUNT% + 1 : LOCATE ROW%, COL% : PRINT LCOUNT%;
9330 FOR I% = 1 TO LEN(WRKSTR$)
9340   IF MID$(WRKSTR$, I%, 1) <> " " THEN DONE%=FALSE% : I%=LEN(WRKSTR$)
```

```
9350 NEXT I%
9360 IF DONE% = TRUE% THEN GOTO 9590
9370 FOR I% = 1 TO 8
9380   WKSP$(I%) = ""
9390 NEXT I%
9400 '
9410 ' ... Search for numbers
9420 '
9430 FOR I% = 1 TO LEN(WRKSTR$)
9440   WHILE (MID$(WRKSTR$, I%, 1) = " ") AND I% <= LEN(WRKSTR$)
9450     I% = I% + 1
9460   WEND
9470   IF (I% > LEN(WRKSTR$)) THEN GOTO 9570
9480   IF NNOS% < 8 THEN GOTO 9530
9490   PRINT
9500   PRINT "*** ERROR *** Too many numbers on line - run DCHECK"
9510   '
9520   GOSUB 9860 : GOTO 9580
9530   NNOS% = NNOS% + 1 : WKSP$(NNOS%) = ""
9540   WHILE (MID$(WRKSTR$,I%,1) <> " ") AND I% <= LEN(WRKSTR$)
9550     WKSP$(NNOS%) = WKSP$(NNOS%) + MID$(WRKSTR$,I%,1) : I% = I% + 1
9560   WEND
9570 NEXT I%
9580 WRKSTR$ = ""
9590 RETURN
9600 '
9610 '***************************   FIND STRING    ***************************
9620 '
9630 STRFOUND% = FALSE% : DONE% = FALSE%
9640 WHILE DONE% = FALSE%
9650   IF EOF(1) <> 0 THEN DATAEND%=TRUE% : DONE%=TRUE% : GOTO 9740
9660   LINE INPUT #1, WRKSTR$
9670   LCOUNT% = LCOUNT% + 1 : LOCATE ROW%, COL% : PRINT LCOUNT%;
9680   IF INSTR(WRKSTR$,TARGET$) > 0 THEN STRFOUND% = TRUE% : DONE% = TRUE%
9690   IF(TARGET$="GRID") AND LCOUNT% > 4 THEN DONE% = TRUE%
9700   '
9710   ' ... Check for new module when looking for numbers
9720   '
9730   IF (LEFT$(WRKSTR$, 4) = "+   ") THEN DONE% = TRUE%
9740 WEND
9750 RETURN
9760 '
9770 '***************************   READ LINES    ****************************
9780 '
9790 FOR I% = 1 TO NLINES%
9800   IF EOF(1) <> 0 THEN DATAEND% = TRUE% : I% = NLINES% : GOTO 9830
9810   LINE INPUT #1, WRKSTR$
9820   LCOUNT% = LCOUNT% + 1 : LOCATE ROW%, COL% : PRINT LCOUNT%;
9830 NEXT I%
9840 RETURN
9850 '
9860 '*********************   ANY KEY TO CONTINUE    ***********************
9870 '
9880 IF DEVICE$<>"SCRN:" THEN GOTO 9930
9890 LOCATE PAGELEN% + 1, 1
9900 PRINT SPACE$(40); "Press any key to continue";
9910 WHILE INKEY$ = "" : WEND
9920 CLS
9930 RETURN
9940 '
9950 '**********************   CHECK FOR PAGE END    *********************
9960 '
9970 IF CSRLIN>=PAGELEN% THEN GOSUB 9880
9980 RETURN
9990 '
10000 '========================   GRID.BAS    ===========================
```

Sample Run

The following sample run shows the program GRID.BAS being used to analyse the structure from example 9.1

```
==============================================================================

    PROGRAM GRID                              Copyright(c) J.A.D.Balfour 1991

                                     Example 9.1

    File Name :- GRID1.DAT                                   Date :- 03-03-92
    Units     :- kN and mm                                   Time :- 21:23:02

==============================================================================


+ + + + + + + + + + + + +
+   NODAL COORDINATES   +
+ + + + + + + + + + + + +

NODE          X            Y
  1           9000         -10000
  2           9000         0
  3           0            0
  4           9000         12000

+ + + + + + + + +
+   ELEMENTS    +
+ + + + + + + + +

ELEMENT    PROPERTY
  1  2        1
  2  3        1
  2  4        1

+ + + + + +  + + +
+   PROPERTIES   +
+ + + + + +  + + +

PROPERTY         I            J            G             E
    1      0.8000E+11   0.4000E+11   0.6000E+01   0.1500E+02

+ + + + + + + + + +
+   RESTRAINTS    +
+ + + + + + + + + +

NODE    DIRECTION(S)
  1        123
  3        123
  4        123

+ + + + + + + + + + +
+   ELEMENT LOADS   +    -    UDL = 1,    POINT LOAD = 2
+ + + + + + + + + + +

ELEMENT   LOAD     MAGNITUDE    DISTANCE FROM
          TYPE                  LOWER NODE
  2   3    1      -.8000E-01       N/A


* * * * * * * * * * *
*   DISPLACEMENTS   *            * :-  RESTRAINT
* * * * * * * * * * *

NODE     Z-DISPLACEMENT      X-ROTATION      Y-ROTATION
  1             *                *               *
  2        -.9794E+01       -.2377E-03      0.5727E-03
  3             *                *               *
  4             *                *               *


* * * * * * * * * * *
*   ELEMENT FORCES  *
* * * * * * * * * * *

                ......... END I .........          ......... END J .........
ELEMENT          SHEAR     TORSION     BENDING        SHEAR     TORSION     BENDING
                 FORCE      MOMENT      MOMENT        FORCE      MOMENT      MOMENT
 1  2       1.2393E+02 -1.3744E+04 -6.4816E+05 -1.2393E+02  1.3744E+04 -5.9113E+05
 2  3       2.1743E+02  6.3376E+03  2.5197E+04  5.0257E+02 -6.3376E+03  1.2579E+06
 2  4      -9.3504E+01  1.1453E+04  5.8479E+05  9.3504E+01 -1.1453E+04  5.3726E+05
```

```
* * * * * * * * *
*   REACTIONS   *
* * * * * * * * *

             Z-COMPONENT       X-MOMENT        Y-MOMENT
NODE  1       1.2393E+02     6.4816E+05     -1.3744E+04
NODE  3       5.0257E+02     6.3376E+03     -1.2579E+06
NODE  4       9.3504E+01    -5.3726E+05     -1.1453E+04
              ----------     ----------      ----------
TOTAL         0.7200E+03     0.1192E+07      -.1283E+07
```

Chapter 10

Space Frames

10.1 Introduction

In practice structural frameworks are almost always three-dimensional. If, however, the structural action is predominantly two-dimensional then structural analysis using a two-dimensional mathematical model, such as a plane frame or a grillage, is usually sufficiently accurate. If, on the other hand, the behaviour of the structure is essentially three-dimensional then structural analysis should be conducted using a three-dimensional mathematical model. This chapter is concerned with the most general type of skeletal structure, the three-dimensional rigidly jointed framework.

A computer program capable of dealing with space frames can, of course, be used for the analysis of plane frames and grillages. Usually this will prove much less efficient than using a program specifically written for the type of structure under consideration.

10.2 The Element Stiffness Matrix in Element Axes

The displacement of each node of a space frame is described by three translational and three rotational components of displacement. Hence the degree of freedom of each unrestrained node is six.

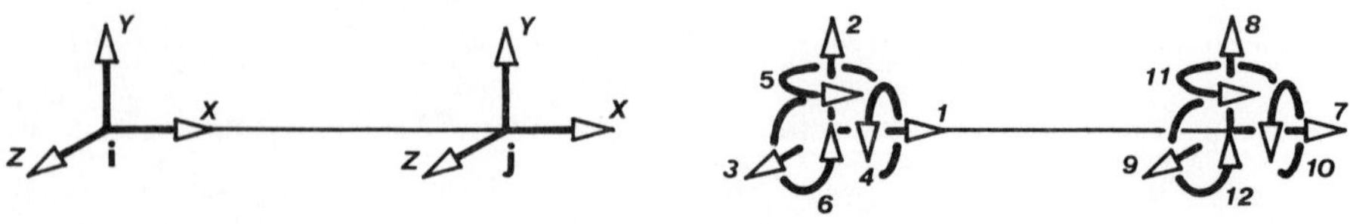

Figure 10.1 Element axes and freedoms for a space frame element

Figure 10.1 shows the element axes and the element freedoms adopted in this text. Previous chapters have shown that for any type of skeletal structure the element end forces are related to the element end displacements by the following matrix equations

$$\boldsymbol{f}_{ij} = \boldsymbol{k}_{ii}^{j}\,\boldsymbol{\delta}_{ij} + \boldsymbol{k}_{ij}\,\boldsymbol{\delta}_{ji}$$

and

$$\boldsymbol{f}_{ji} = \boldsymbol{k}_{ji}\,\boldsymbol{\delta}_{ij} + \boldsymbol{k}_{jj}^{i}\,\boldsymbol{\delta}_{ji}$$

At this point the reader should note that every stiffness coefficient associated with a space frame element has already been encountered when dealing with plane frames and grillages. Hence there is little difficulty in showing that the above equations take the following form when applied to a space frame element

$$\begin{bmatrix} f_{ijx} \\ f_{ijy} \\ f_{ijz} \\ m_{ijx} \\ m_{ijy} \\ m_{ijz} \end{bmatrix} = \begin{bmatrix} \frac{EA}{L} & 0 & 0 & 0 & 0 & 0 \\ 0 & \frac{12EI_z}{L^3} & 0 & 0 & 0 & \frac{6EI_z}{L^2} \\ 0 & 0 & \frac{12EI_y}{L^3} & 0 & -\frac{6EI_y}{L^2} & 0 \\ 0 & 0 & 0 & \frac{GJ}{L} & 0 & 0 \\ 0 & 0 & -\frac{6EI_y}{L^2} & 0 & \frac{4EI_y}{L} & 0 \\ 0 & \frac{6EI_z}{L^2} & 0 & 0 & 0 & \frac{4EI_z}{L} \end{bmatrix} \begin{bmatrix} \delta_{ijx} \\ \delta_{ijy} \\ \delta_{ijz} \\ \theta_{ijx} \\ \theta_{ijy} \\ \theta_{ijz} \end{bmatrix} +$$

$$\begin{bmatrix} -\frac{EA}{L} & 0 & 0 & 0 & 0 & 0 \\ 0 & -\frac{12EI_z}{L^3} & 0 & 0 & 0 & \frac{6EI_z}{L^2} \\ 0 & 0 & -\frac{12EI_y}{L^3} & 0 & -\frac{6EI_y}{L^2} & 0 \\ 0 & 0 & 0 & -\frac{GJ}{L} & 0 & 0 \\ 0 & 0 & \frac{6EI_y}{L^2} & 0 & \frac{2EI_y}{L} & 0 \\ 0 & -\frac{6EI_z}{L^2} & 0 & 0 & 0 & \frac{2EI_z}{L} \end{bmatrix} \begin{bmatrix} \delta_{jix} \\ \delta_{jiy} \\ \delta_{jiz} \\ \theta_{jix} \\ \theta_{jiy} \\ \theta_{jiz} \end{bmatrix}$$

$$\begin{bmatrix} f_{jix} \\ f_{jiy} \\ f_{jiz} \\ m_{jix} \\ m_{jiy} \\ m_{jiz} \end{bmatrix} = \begin{bmatrix} -\frac{EA}{L} & 0 & 0 & 0 & 0 & 0 \\ 0 & -\frac{12EI_z}{L^3} & 0 & 0 & 0 & -\frac{6EI_z}{L^2} \\ 0 & 0 & -\frac{12EI_y}{L^3} & 0 & \frac{6EI_y}{L^2} & 0 \\ 0 & 0 & 0 & -\frac{GJ}{L} & 0 & 0 \\ 0 & 0 & -\frac{6EI_y}{L^2} & 0 & \frac{2EI_y}{L} & 0 \\ 0 & \frac{6EI_z}{L^2} & 0 & 0 & 0 & \frac{2EI_z}{L} \end{bmatrix} \begin{bmatrix} \delta_{ijx} \\ \delta_{ijy} \\ \delta_{ijz} \\ \theta_{ijx} \\ \theta_{ijy} \\ \theta_{ijz} \end{bmatrix} +$$

$$\begin{bmatrix} \frac{EA}{L} & 0 & 0 & 0 & 0 & 0 \\ 0 & \frac{12EI_z}{L^3} & 0 & 0 & 0 & -\frac{6EI_z}{L^2} \\ 0 & 0 & \frac{12EI_y}{L^3} & 0 & \frac{6EI_y}{L^2} & 0 \\ 0 & 0 & 0 & \frac{GJ}{L} & 0 & 0 \\ 0 & 0 & \frac{6EI_y}{L^2} & 0 & \frac{4EI_y}{L} & 0 \\ 0 & -\frac{6EI_z}{L^2} & 0 & 0 & 0 & \frac{4EI_z}{L} \end{bmatrix} \begin{bmatrix} \delta_{jix} \\ \delta_{jiy} \\ \delta_{jiz} \\ \theta_{jix} \\ \theta_{jiy} \\ \theta_{jiz} \end{bmatrix}$$

10.3 Transformation of Force and Displacement

In Chapter 3 the matrix required to transform a vector of linear nodal displacements in three-dimensional space between two cartesian coordinate systems sharing a common origin was shown to be

$$\begin{bmatrix} \delta_{ijx} \\ \delta_{ijy} \\ \delta_{ijz} \end{bmatrix} = \begin{bmatrix} l_x & m_x & n_x \\ l_y & m_y & n_y \\ l_z & m_z & n_z \end{bmatrix} \begin{bmatrix} \Delta_{ix} \\ \Delta_{iy} \\ \Delta_{iz} \end{bmatrix}$$

In the previous chapter the transformation of rotational displacements was seen to be identical to the transformation of linear displacements, hence

$$\begin{bmatrix} \theta_{ijx} \\ \theta_{ijy} \\ \theta_{ijz} \end{bmatrix} = \begin{bmatrix} l_x & m_x & n_x \\ l_y & m_y & n_y \\ l_z & m_z & n_z \end{bmatrix} \begin{bmatrix} \Theta_{ix} \\ \Theta_{iy} \\ \Theta_{iz} \end{bmatrix}$$

As linear displacements are independent of rotational displacements, the transformation of a nodal displacement vector for a space frame, from the global to the element axes system is as follows

$$\begin{bmatrix} \delta_{ijx} \\ \delta_{ijy} \\ \delta_{ijz} \\ \theta_{ijx} \\ \theta_{ijy} \\ \theta_{ijz} \end{bmatrix} = \begin{bmatrix} l_x & m_x & n_x & 0 & 0 & 0 \\ l_y & m_y & n_y & 0 & 0 & 0 \\ l_z & m_z & n_z & 0 & 0 & 0 \\ 0 & 0 & 0 & l_x & m_x & n_x \\ 0 & 0 & 0 & l_y & m_y & n_y \\ 0 & 0 & 0 & l_z & m_z & n_z \end{bmatrix} \begin{bmatrix} \Delta_{ix} \\ \Delta_{iy} \\ \Delta_{iz} \\ \Theta_{ix} \\ \Theta_{iy} \\ \Theta_{iz} \end{bmatrix}$$

i.e.

$$\delta_{ij} = T_{ij}\,\Delta_i$$

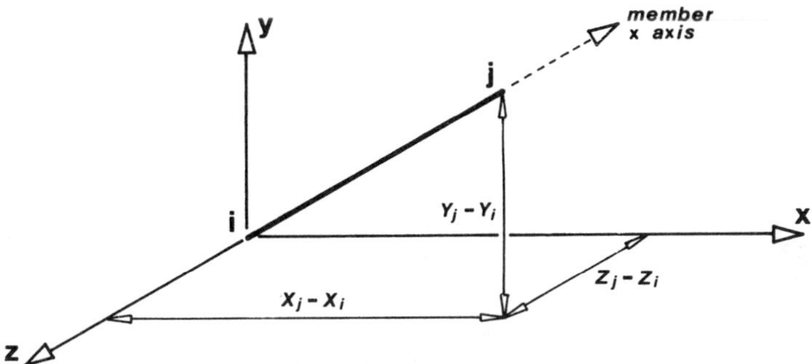

Figure 10.2 Calculation of the direction cosines for the element "x" axis

The same matrix will also transform vectors of nodal force, i.e.

$$f_{ij} = T_{ij} F_{ij}$$

l_x, m_x, and n_x are, as defined in Chapter 3, the cosines of the angles that the element "x" axis makes with the global "x", "y", and "z" axes respectively (i.e. the direction cosines of the element "x" axis), hence

$$l_x = \frac{x_j - x_i}{L_{ij}}$$

$$m_x = \frac{y_j - y_i}{L_{ij}}$$

$$n_x = \frac{z_j - z_i}{L_{ij}}$$

where (x_i, y_i, z_i) and (x_j, y_j, z_j) are the coordinates of nodes "i" and "j" (i.e. the coordinates of the ends of element "i,j"), and L_{ij} is the length of element "i,j" (see fig 10.2).

The problem here is to find the remaining elements of T_{ij}, as l_x, m_x, and n_x define only the orientation of the element "x" axis (i.e. the element "y" and "z" axes can be rotated about the element "x" axis to take up any orientation). To proceed the orientation of one of these axes must be defined. In this text the element "y" axis will be orientated to lie in the global "x,y" plane.

If, however, the element "x" axis lies along the global "z" axis then the condition that the element "y" axis must lie in the global "x,y" plane is automatically satisfied, and a further condition is necessary. In this text the element "y" axis of a element lying along the global "z" axis will be chosen to coincide with the global "y" axis. There are two possible cases, and these are illustrated in fig 10.3. The complete set of direction cosines can be found by inspection.

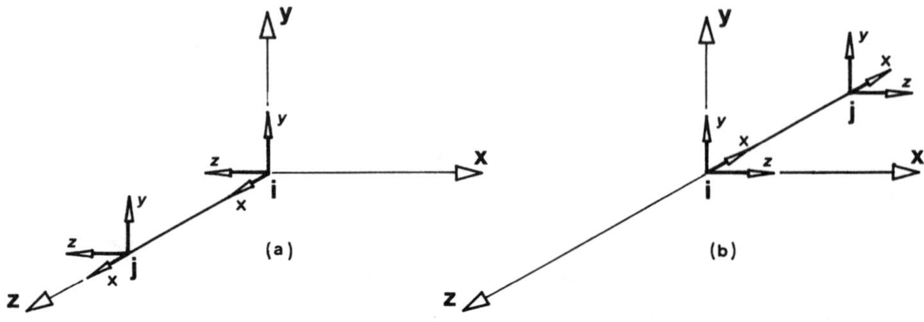

Figure 10.3 Element lying parallel to the global "z" axis

Case (a)

$$l_x = 0 \quad m_x = 0 \quad n_x = 1$$
$$l_y = 0 \quad m_y = 1 \quad n_y = 0$$
$$l_z = -1 \quad m_z = 0 \quad n_z = 0$$

Case (b)

$$l_x = 0 \quad m_x = 0 \quad n_x = -1$$
$$l_y = 0 \quad m_y = 1 \quad n_y = 0$$
$$l_z = 1 \quad m_z = 0 \quad n_z = 0$$

For elements not lying along the global "z" axis the evaluation of the direction cosines of the element "y" and "z" axes is less straightforward. The problem is best tackled using vector cross products. The cross product, $\vec{C}$, of two vectors, $\vec{A}$ and $\vec{A}$, is defined as follows

$$\vec{A} = \vec{i}\,x_a + \vec{j}\,y_a + \vec{k}\,z_a$$
$$\vec{B} = \vec{i}\,x_b + \vec{j}\,y_b + \vec{k}\,z_b$$

and

$$\vec{C} = \vec{A} \times \vec{B} = \begin{vmatrix} \vec{i} & \vec{j} & \vec{k} \\ x_a & y_a & z_a \\ x_b & y_b & z_b \end{vmatrix}$$

where

$\vec{C}$ is a third vector normal to the plane of $\vec{A}$ and $\vec{B}$ and is directed such that $\vec{A}$, $\vec{B}$, and $\vec{C}$ form a right-hand system. The length of $\vec{C}$ is $|\vec{A}|\,|\vec{B}|\,|sin(\vec{A},\vec{B})|$.

$sin(\vec{A},\vec{B})$ is the value of the sine of the angle between $\vec{A}$ and $\vec{B}$.

$\vec{i}, \vec{j}, \vec{k}$ are unit vectors in the direction of the "x", "y", and "z" axes respectively.

x_a, y_a, z_a are the magnitudes of the components of vector $\vec{A}$ in the directions of the "x", "y", and "z" axes respectively.

x_b, y_b, z_b are the magnitudes of the components of vector $\vec{B}$ in the directions of the "x", "y", and "z" axes respectively.

l_x, m_x, and n_x are, in effect, the components in the directions of the global "x", "y", and "z" axes of a unit vector lying along the element "x" axis. The element "y" axis is perpendicular to the element "x" axis and, if the element "y" axis lies in the global "x,y" plane, it must also be perpendicular to the global "z" axis. Hence the cross product of unit vectors lying along the global "z" and the element "x" axes must result in a vector, $\vec{B}$, in the direction of the element "y" axis (note that the unit vectors are expressed in terms of the global axes system, and that the components of the unit vector on the element "x" axis are equal to the direction cosines for that axis).

$$\vec{Y} = \begin{vmatrix} \vec{i} & \vec{j} & \vec{k} \\ 0 & 0 & 1 \\ l_x & m_x & n_x \end{vmatrix} = \vec{i}\, m_x + \vec{j}\, l_x + \vec{i}\, 0$$

The length of $\vec{Y} = |1||1||\sqrt{1 - n_x^2}|$ as $sin\theta = \sqrt{1 - cos^2\theta}$

l_y, m_y, and n_y are the components in the global axes system of a unit vector lying on the element "y" axis. Hence the direction cosines of the element "y" axis are found by scaling $\vec{Y}$ to make a unit vector

$$\frac{\vec{Y}}{\sqrt{1 - n_x^2}} = -\frac{m_x}{\sqrt{1 - n_x^2}}\,\vec{i} + \frac{l_x}{\sqrt{1 - n_x^2}}\,\vec{j} + \vec{k}\, 0$$

i.e.

$$l_y = -\frac{m_x}{\sqrt{1 - n_x^2}}$$

$$m_y = \frac{l_x}{\sqrt{1 - n_x^2}}$$

$$n_y = 0$$

A vector in the direction of the element "z" axis is generated by taking the cross product of unit vectors lying along the element "x" and "y" axes (note that the length of the resulting vector is unity).

$$\vec{Z} = \begin{vmatrix} \vec{i} & \vec{j} & \vec{k} \\ l_x & m_x & n_x \\ l_y & m_y & n_y \end{vmatrix}$$

$$= \vec{i}\,(m_x n_y - m_y n_x) - \vec{j}\,(l_x n_y - l_y n_x) + \vec{k}\,(l_x m_y - l_y m_x)$$

$$= -\frac{l_x n_x}{\sqrt{1-n_x^2}}\vec{i} - \frac{m_x n_x}{\sqrt{1-n_x^2}}\vec{j} + \frac{l_x^2 + n_x^2}{\sqrt{1-n_x^2}}\vec{k}$$

hence

$$l_z = -\frac{l_x n_x}{\sqrt{1-n_x^2}}$$

$$m_z = -\frac{m_x n_x}{\sqrt{1-n_x^2}}$$

$$n_z = \sqrt{1-n_x^2}$$

As all of the direction cosines have now been expressed in terms of l_x, m_x, and n_x the subscript "x" is no longer necessary and the transformation matrix becomes

$$T_{ij} = \begin{bmatrix} l & m & n & 0 & 0 & 0 \\ -m/D & l/D & 0 & 0 & 0 & 0 \\ -ln/D & -mn/D & D & 0 & 0 & 0 \\ 0 & 0 & 0 & l & m & n \\ 0 & 0 & 0 & -m/D & l/D & 0 \\ 0 & 0 & 0 & -ln/D & -mn/D & D \end{bmatrix} \tag{10.1}$$

where

$$D = \sqrt{1-n^2}$$

The transformation matrix shown in equation (10.1) tranforms vectors of nodal force and nodal displacement from the global axes system to the element axes system for space frame element "i,j". The inverse of that matrix, shown in equation (10.2), effects the reverse transformation (i.e. from the element to the global axes system).

$$T_{ij}^{-1} = \begin{bmatrix} l & -m/D & -ln/D & 0 & 0 & 0 \\ m & l/D & -mn/D & 0 & 0 & 0 \\ n & 0 & D & 0 & 0 & 0 \\ 0 & 0 & 0 & l & -m/D & -ln/D \\ 0 & 0 & 0 & m & l/D & -mn/D \\ 0 & 0 & 0 & n & 0 & D \end{bmatrix} \tag{10.2}$$

10.4 The Element Stiffness Matrix in Global Axes

As seen in previous chapters the element stiffness submatrices in the global axes system are obtained by the following triple matrix multiplications.

$$K_{ii}^{j} = T_{ij}^{-1}\, k_{ii}^{j}\, T_{ij}$$
$$K_{ij} = T_{ij}^{-1}\, k_{ij}\, T_{ij}$$
$$K_{ji} = T_{ij}^{-1}\, k_{ji}\, T_{ij}$$
$$K_{jj}^{i} = T_{ij}^{-1}\, k_{jj}^{i}\, T_{ij}$$

At this point it should be noted that the relationship between the element end forces and the element end displacements assumes that the element "y" and "z" axes coincide with the principal axes of the section. However, earlier in the chapter the element "y" axis was forced to lie in the global "x,y" plane, and this restriction means that the element "y" and "z" axes will not necessarily coincide with the principal axes (y^*, z^*) of the section. The general case is illustrated in fig 10.4. Note that fig 10.4 looks down element "i,j" from end "i" to end "j". Hence "β" is the anticlockwise rotation of the element (when viewed from end "i") that will make the principal axes coincide with the element axes. Note also that the principal "x" axis and the element "x" axis are always coincident.

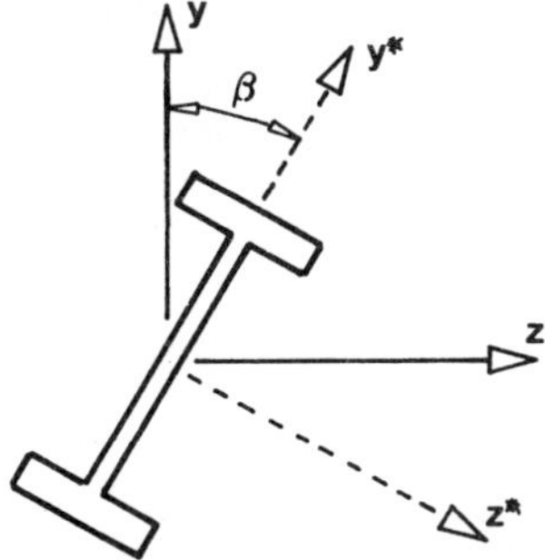

Figure 10.4 Principal axes

The reader should recognise that the required transformation is identical to the transformation from the global to the element axis system for a grillage element (see Chapter 9). Hence the transformation of element end displacements and rotations from the element axes system to the principal axes system is effected by the following matrix.

$$T_{ij}^{*} = \begin{bmatrix} 1 & 0 & 0 & 0 & 0 & 0 \\ 0 & \cos\beta & \sin\beta & 0 & 0 & 0 \\ 0 & -\sin\beta & \cos\beta & 0 & 0 & 0 \\ 0 & 0 & 0 & 1 & 0 & 0 \\ 0 & 0 & 0 & 0 & \cos\beta & \sin\beta \\ 0 & 0 & 0 & 0 & -\sin\beta & \cos\beta \end{bmatrix}$$

The reverse transformation is effected by

$$T_{ij}^{*-1} = \begin{bmatrix} 1 & 0 & 0 & 0 & 0 & 0 \\ 0 & cos\beta & -sin\beta & 0 & 0 & 0 \\ 0 & sin\beta & cos\beta & 0 & 0 & 0 \\ 0 & 0 & 0 & 1 & 0 & 0 \\ 0 & 0 & 0 & 0 & cos\beta & -sin\beta \\ 0 & 0 & 0 & 0 & sin\beta & cos\beta \end{bmatrix}$$

hence

$$\begin{bmatrix} \delta_{ijx}^* \\ \delta_{ijy}^* \\ \delta_{ijz}^* \\ \theta_{ijx}^* \\ \theta_{ijy}^* \\ \theta_{ijz}^* \end{bmatrix} = \begin{bmatrix} 1 & 0 & 0 & 0 & 0 & 0 \\ 0 & cos\beta & sin\beta & 0 & 0 & 0 \\ 0 & -sin\beta & cos\beta & 0 & 0 & 0 \\ 0 & 0 & 0 & 1 & 0 & 0 \\ 0 & 0 & 0 & 0 & cos\beta & sin\beta \\ 0 & 0 & 0 & 0 & -sin\beta & cos\beta \end{bmatrix} \begin{bmatrix} \delta_{ijx} \\ \delta_{ijy} \\ \delta_{ijz} \\ \theta_{ijx} \\ \theta_{ijy} \\ \theta_{ijz} \end{bmatrix}$$

As stated previously the element stiffness relationship is only valid in terms of the principal axes, i.e.

$$f_{ij}^* = k_{ii}^{j*} \delta_{ij}^* + k_{ij}^* \delta_{ij}^*$$

but

$$f_{ij} = T_{ij}^{*-1} f_{ij}^* \quad \text{and} \quad \delta_{ij}^* = T_{ij}^* \delta_{ij}$$

hence

$$f_{ij} = T_{ij}^{*-1} k_{ii}^{j*} T_{ij}^* \delta_{ij} + T_{ij}^{*-1} k_{ij}^* T_{ij}^* \delta_{ji}$$

The preceding equation gives the relationship between the element end forces and the element end displacements in terms of the element axes system, but using stiffness submatrices in the principal axes system (the stars denote matrices that are associated with the principal axes system). To express forces and displacements in the global axes system the transformation matrices shown in equations (10.1) and (10.2) are used

$$F_{ij} = T_{ij}^{-1} f_{ij}$$

and

$$\delta_{ij} = T_{ij} \Delta_i$$

hence

$$F_{ij} = T_{ij}^{-1} T_{ij}^{*-1} k_{ii}^{j*} T_{ij}^* T_{ij} \Delta_i + T_{ij}^{-1} T_{ij}^{*-1} k_{ij}^* T_{ij}^* T_{ij} \Delta_j \tag{10.3}$$

The equation of equilibrium for node "i" of a framed structure has been shown in previous chapters to be

$$P_i = K_{ii} \Delta_i + K_{ia} \Delta_a + K_{ib} \Delta_b + ... + K_{in} \Delta_n$$

where, for space frames,

$$K_{ii} = \sum_{j=a}^{n} K_{ii}^{j} = \sum_{j=a}^{n} T_{ij}^{-1} T_{ij}^{*-1} k_{ii}^{j*} T_{ij}^{*} T_{ij}$$

and

$$K_{ij} = T_{ij}^{-1} T_{ij}^{*-1} k_{ij}^{*} T_{ij}^{*} T_{ij}$$

From here the final structure stiffness equation is developed using the same procedures as were used for all other types of frameworks.

Third Node Method

The rotation that will make the principal axes coincide with the element axis cannot always be assessed by inspection. Take for example the element shown in fig 10.5(a). Figure 10.5(b) shows the same element viewed looking down the global *"y"* axis. The problem in hand is to calculate the angle through which the element must be rotated in order to make the principal axes coincide with the element axes (note that anticlockwise rotation of the element when viewed looking from end *"i"* to end *"j"* is positive). A convenient way forward is to locate a point *"p"* that lies in the principal *"x,y"* plane (but not on the principal *"x"* axis). This point is known as the *third node* and nodes *"i"*, *"j"* and *"p"* define the principal *"x,y"* plane. In the case of the element shown in fig 10.5 the global *"y"* axis lies in the principal *"x,y"* plane of the element. Thus any point, except the node *"i"*, on the global *"y"* axis will suffice for *"p"*.

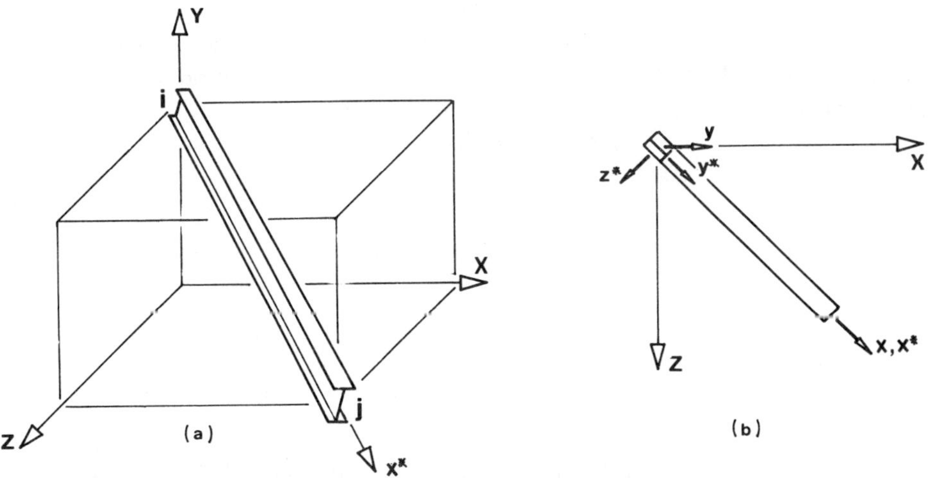

Figure 10.5 Principal and element axes

The cross product of two independent vectors lying on the principal *"x,y"* plane will result in a vector normal to that plane (i.e. in the direction of the principal "*z*" axis). The nodes "*i*" and "*j*" provide one vector, and the nodes "*i*" and "*p*" provide the other.

hence

$$\vec{Z}^* = \begin{vmatrix} \vec{i} & \vec{j} & \vec{k} \\ (x_j - x_i) & (y_j - y_i) & (z_j - z_i) \\ (x_p - x_i) & (y_p - y_i) & (z_p - z_i) \end{vmatrix}$$

$$= \vec{i}\,[(y_j - y_i)(z_p - z_i) - (y_p - y_i)(z_j - z_i)] - \vec{j}\,[(x_j - x_i)(z_p - z_i) - (x_p - x_i)(z_j - z_i)] + \vec{k}\,[(x_j - x_i)(y_p - y_i) - (x_p - x_i)(y_j - y_i)]$$

$$= \vec{i}\,X_z^* + \vec{j}\,Y_z^* + \vec{k}\,Z_z^*$$

The length of this vector is found by Pythagoras' Theorem, and is then used to find the direction cosines of the principal "*z*" axis.

$$L = \sqrt{X_z^{*2} + Y_z^{*2} + Z_z^{*2}}$$

hence

$$l_z^* = \frac{X_z^*}{L} \qquad m_z^* = \frac{Y_z^*}{L} \qquad n_z^* = \frac{Z_z^*}{L}$$

l_z^*, m_z^* and n_z^* are the components (in the global axes system) of a unit vector lying on the principal "*z*" axis. Similarly l_z, m_z and n_z are the components of a unit vector lying on the element "*z*" axis. The angle between these unit vectors can be found by the cosine rule

$$\beta = cos^{-1}\left(\frac{a^2 + b^2 - c^2}{2ab}\right)$$

but the lengths of "*a*" and "*b*" are unity, therefore

$$\beta = cos^{-1}\left(\frac{1 - c^2}{2}\right)$$

where

$$c^2 = (l_z^* - l_z)^2 + (m_z^* - m_z)^2 + (n_z^* - n_z)^2$$

There is, however, one further complication. If the angle ,"γ", between the element "*y*" axis and the principal "*z*" axis is less than 90° then the wrong angle will be found as illustrated in fig 10.6, and "β" must be set equal to (360 - β) degrees (note that the angle calculated in these cases is, in fact, the clockwise rotation necessary to make the principal axes coincide with the element axes). The direction cosines of the principal "*z*" axis are already known and the direction cosines of the element "y" axis are readily available, allowing the angle between these axes to be calculated using the cosine rule.

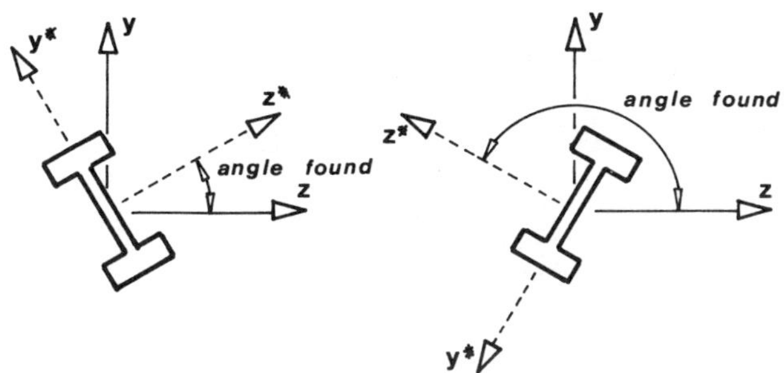

Figure 10.6 Cases where the wrong angle is evaluated

Note that this complication arises because the vector cross product always produces a right-hand axes system.

Example 10.1

Find "β" for the element shown in the following diagram if the coordinates of "i" and "j" are (0,10,0) and (10,0,10) respectively.

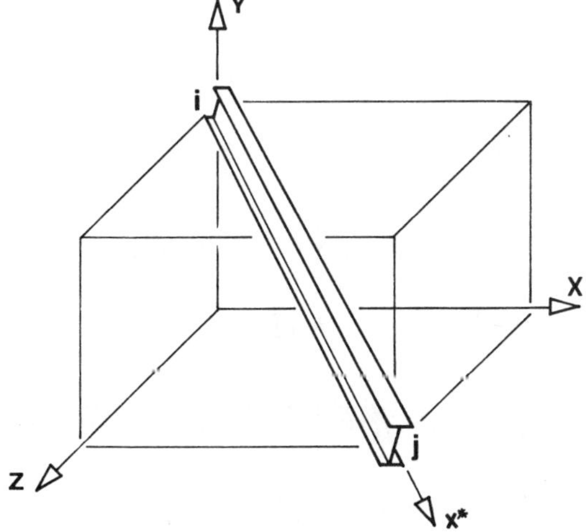

"p" can be located anywhere on the principal "x,y" plane (the principal "x" axis excepted). Here "p" will be located at (10,10,10) (i.e. directly above node "j"). Taking node "i" as the origin, the coordinates of nodes "j" and "p" become (10,-10,10) and (10,0,10) respectively. The following vector cross product generates a vector in the direction of the principal "z" axis.

$$\vec{Z}^* = \begin{vmatrix} \vec{i} & \vec{j} & \vec{k} \\ 10 & -10 & 10 \\ 10 & 0 & 10 \end{vmatrix} = -100\,\vec{i} - 0\,\vec{j} + 100\,\vec{k}$$

Calculate the length of the vector

$$L = \sqrt{100^2 + 100^2} = 141.4$$

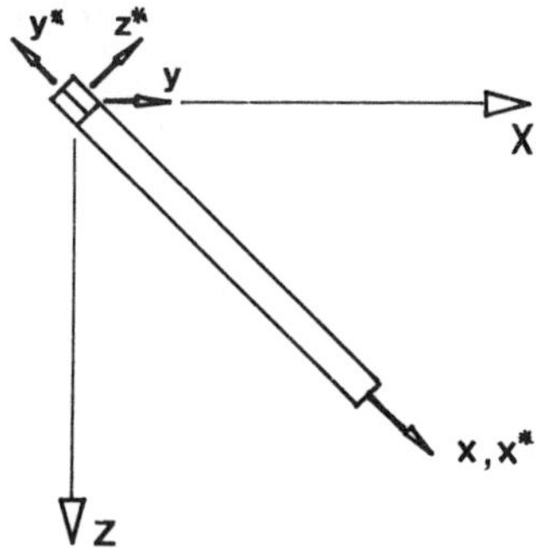

As the cross product of vectors produces a right-hand system, and the principal axes are themselves a right-hand system, the axes will point in the directions indicated in the preceding diagram which shows the element viewed looking down the global "y" axis towards the origin.

Evaluate the direction cosines of the principal "z" axis.

$$l_z^* = \frac{-100}{141.4} = -0.7072$$

$$m_z^* = \frac{-0}{141.4} = 0.0$$

$$n_z^* = \frac{100}{141.4} = 0.7072$$

The length of the element $= \sqrt{(10)^2 + (-10)^2 + (10)^2} = 17.32$

Hence the direction cosines of the element "x" axis are

$$l_x = \frac{10}{17.32} = 0.5774$$

$$m_x = \frac{-10}{17.32} = -0.5774$$

$$n_x = \frac{10}{17.32} = 0.5774$$

The direction cosines of the element "z" axis can now be calculated.

$$D = \sqrt{1 - n_x^2} = 0.8165$$

$$l_z = -\frac{l_x\, n_x}{D} = -0.4083$$

$$m_z = -\frac{m_x\, n_x}{D} = 0.4083$$

$$n_z = D = 0\,.8165$$

Now calculate β

$$\beta = cos^{-1}(1 - c^2/2)$$

where

$$c^2 = (l_z^* - l_z)^2 + (m_z^* - m_z)^2 + (n_z^* - n_z)^2$$

$$= (-0.7072+0.4083)^2 + (0\text{-}0.4083)^2 + (0.7072\text{-}0.8165)^2$$

$$= 0.268$$

hence

$$\beta = 30°$$

Check the angle between the element "y" axis and the principal "z" axis.

$$l_y = -\frac{m_x}{D} = 0.7072$$

$$m_y = \frac{l_x}{D} = 0.7072$$

$$n_y = 0$$

$$cos\gamma = (1 - c^2/2)$$

where

$$c^2 = (l_z^* - l_y)^2 + (m_z^* - m_y)^2 + (n_z^* - n_y)^2$$

$$= (-0.7072 - 0.7072)^2 + (0 - 0.7072)^2 + (0.7072 - 0.0)^2$$

$$= 3.000$$

hence

$$cos\gamma = -0.5 \text{ (negative, therefore } \gamma > 90°)$$

Therefore the true value of "β" is 30°. Note that if $cos\gamma$ is positive then the true value of "β" is (360 - β).

Example 10.2

Calculate "β" for the element considered in example 10.1, but choose "p" to be the point (10,-10,10) i.e. directly below node "j".

Make node "i" the origin then generate a vector in the direction of the principal "z" axis.

$$\vec{Z}^* = \begin{vmatrix} \vec{i} & \vec{j} & \vec{k} \\ 10 & -10 & 10 \\ 10 & -20 & 10 \end{vmatrix} = -100\,\vec{i} - 0\,\vec{j} - 100\,\vec{k}$$

The length of the vector $\vec{Z}^*$ is $L = \sqrt{100^2 + 100^2} = 141.4$

As the cross product of vectors produces a right-hand system, and the principal axes are themselves a right-hand system, the axes will point in the directions indicated in the following diagram which shows the element viewed looking down the global "y" axis towards the origin.

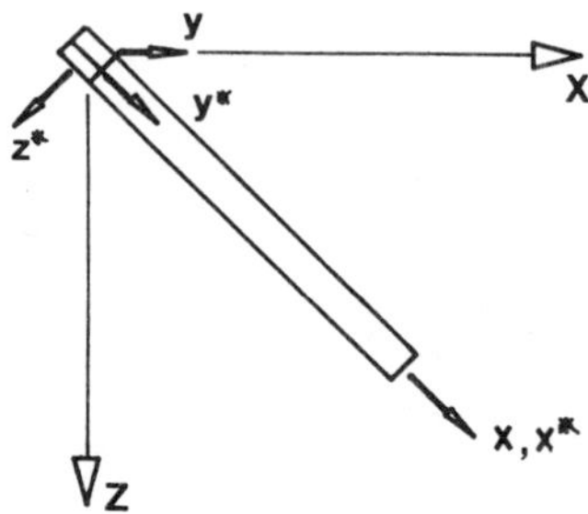

Evaluate the direction cosines of the principal "z" axis.

$$l_z^* = \frac{100}{141.4} = 0.7072$$

$$m_z^* = \frac{-0}{141.4} = 0.0$$

$$n_z^* = \frac{-100}{141.4} = -0.7072$$

The length of the element and the direction cosines of the element "x", "y" and "z" axes were found in the previous example and are copied here for convenience.

$$L = 17.32$$

$$l_x = 0.5774 \qquad l_y = 0.7072 \qquad l_z = -0.4083$$

$$m_x = -0.5774 \qquad m_y = 0.7072 \qquad m_z = 0.4083$$

$$n_x = 0.5774 \qquad l_y = 0.0 \qquad l_z = 0.8165$$

Calculate β

$$\beta = cos^{-1}(1 - c^2/2)$$

where

$$c^2 = (l_z^* - l_z)^2 + (m_z^* - m_z)^2 + (n_z^* - n_z)^2$$

$$= (0.7072+0.4083)^2 + (0 - 0.4083)^2 + (-0.7072 - 0.8165)^2$$

$$= 3.733$$

hence

$$\beta = 150°$$

Check the angle between the element "y" axis and the principal "z" axis (direction cosines of the element "y" axis are unchanged from the previous example).

$$cos\gamma = (1 - c^2/2)$$

where

$$c^2 = (l_z^* - l_y)^2 + (m_z^* - m_y)^2 + (n_z^* - n_y)^2$$

$$= (0.7072 - 0.7072)^2 + (0-0.7072)^2 + (-0.7072 - 0.0)^2$$

$$= 1.000$$

hence

$$cos\gamma = 0.5 \text{ (positive, therefore } \gamma < 90°)$$

Therefore the true value of "β" is (360 - 150) degrees (i.e. 210°). Note that the examples 10.1 and 10.2 have considered the same element and that the angles found from the two choices of "p" are 180 degrees different from each other.

Program TNODE.BAS

There follows the listing of a program that utilises the theory presented in this section to calculate "β" for a space frame element using the third node method.

```
1000 '=========================       TNODE.BAS       ===========================
1010 '
1020 OPTION BASE 1
1030 TRUE% = 1 : FALSE% = 0
1040 CLS
```

```
1050 PRINT
1060 PRINT "=================================================================="
1070 PRINT
1080 PRINT "PROGRAM TNODE.BAS                    Copyright (c) J.Balfour 1991"
1090 PRINT
1100 PRINT "        Program to calculate the angle between the principal"
1110 PRINT "      y-axis and the element y-axis for a space frame element"
1120 PRINT
1130 PRINT "                For further information contact"
1140 PRINT " James A.D.Balfour, Heriot-Watt University, Riccarton, Edinburgh"
1150 PRINT "             Tel 031-449-5111, Fax 031-451-3170"
1160 PRINT
1170 PRINT "=================================================================="
1180 '
1190 ' ... Note that variables are defined in Appendix A
1200 '
1210 DONE% = FALSE% : I% = 1
1220 WHILE DONE% = FALSE%
1230   PRINT
1240   PRINT "ELEMENT "; I%
1250   PRINT
1260   INPUT "X-Coordinate lower node  = ", XI
1270   INPUT "Y-Coordinate lower node  = ", YI
1280   INPUT "Z-Coordinate lower node  = ", ZI
1290   PRINT
1300   INPUT "X-Coordinate higher node = ", XJ
1310   INPUT "Y-Coordinate higer  node = ", YJ
1320   INPUT "Z-Coordinate higher node = ", ZJ
1330   PRINT
1340   INPUT "X-Coordinate third node  = ", XP
1350   INPUT "Y-Coordinate third node  = ", YP
1360   INPUT "Z-Coordinate third node  = ", ZP
1370   '
1380   ' ... Move origin to lower node
1390   '
1400   XJ = XJ - XI
1410   YJ = YJ - YI
1420   ZJ = ZJ - ZI
1430   XP = XP - XI
1440   YP = YP - YI
1450   ZP = ZP - ZI
1460   '
1470   ' ... Calculate the direction cosines of the principal z-axis
1480   '
1490   LZP = YJ*ZP - YP*ZJ
1500   MZP = XP*ZJ - XJ*ZP
1510   NZP = XJ*YP - XP*YJ
1520   L   = SQR(LZP^2 + MZP^2 + NZP^2)
1530   LZP = LZP / L
1540   MZP = MZP / L
1550   NZP = NZP / L
1560   '
1570   ' ... Calculate the direction cosines of the element x-axis
1580   '
1590   L  = SQR(XJ^2 + YJ^2 + ZJ^2)
1600   LX = XJ / L
1610   MX = YJ / L
1620   NX = ZJ / L
1630   '
1640   ' ... Calculate the direction cosines of the element z-axis
1650   '
1660   IF ABS(NX) > .9999 THEN GOTO 1710
1670   NZ =  SQR(1 - NX^2)
1680   LZ = -LX * NX / NZ
1690   MZ = -MX * NX / NZ
1700   GOTO 1770
1710   LZ = -NX
1720   MZ = 0
1730   NZ = 0
1740   '
1750   ' ... Calculate beta
1760   '
1770   X    = 1 - ((LZP-LZ)^2+(MZP-MZ)^2+(NZP-NZ)^2)/2
1780   IF ABS(X) > .9999 THEN BETA = 0
1790   IF ABS(X) > .9999 THEN GOTO 1890
1800   BETA = (180 * (1.5708-ATN(X/SQR(1-X*X)))) / (4 * ATN(1))
1810   '
```

```
1820    ' ...  Check angle between the element y-axis and the principal z-axis
1830    '
1840    LY  = -MX/NZ
1850    MY  =  LX/NZ
1860    CYZ = (1 - ((LZP-LY)^2 + (MZP-MY)^2 + (NZP)^2)/2)
1870    IF CYZ > 0 THEN BETA = 360 - BETA
1880    PRINT
1890    PRINT "Beta  =  "; BETA; " Degrees"
1900    PRINT
1910    INPUT "Another element (y)es or (n)o? ", TEMP$
1920    IF (INSTR("Y",TEMP$) = 0) AND (INSTR("y",TEMP$) = 0) THEN DONE% = TRUE%
1930    I% = I% + 1
1940 WEND
1950 END
1960 '
1970 '========================     TNODE.BAS     ===========================
```

Sample Run

There follows a sample run which shows program TNODE.BAS being used to solve examples 10.1 and 10.2. Input from the keyboard is shown underlined.

```
=================================================================

PROGRAM TNODE.BAS                        Copyright (c) J.Balfour 1991

         Program to calculate the angle between the principal
       y-axis and the element y-axis for a space frame element

                  For further information contact
  James A.D.Balfour, Heriot-Watt University, Riccarton, Edinburgh
                Tel 031-449-5111, Fax 031-451-3170

=================================================================


ELEMENT  1

X-Coordinate lower node  = 0
Y-Coordinate lower node  = 10
Z-Coordinate lower node  = 0

X-Coordinate higher node = 10
Y-Coordinate higer  node = 0
Z-Coordinate higher node = 10

X-Coordinate third node  = 10
Y-Coordinate third node  = 10
Z-Coordinate third node  = 10

Beta  =   30.00021  Degrees

Another element (y)es or (n)o? Y


ELEMENT  2

X-Coordinate lower node  = 0
Y-Coordinate lower node  = 10
Z-Coordinate lower node  = 0

X-Coordinate higher node = 10
Y-Coordinate higer  node = 0
Z-Coordinate higher node = 10

X-Coordinate third node  = 10
Y-Coordinate third node  = -10
Z-Coordinate third node  = 10

Beta  =   209.9998  Degrees

Another element (y)es or (n)o? N
```

10.5 Element End Forces and Reactions

The relationship between the forces at the ends of element "i,j" due to displacements at "i" and "j", in the principal axes system, has already been seen to be

$$f^*_{ij} = k^{j*}_{ii}\,\delta^*_{ij} + k^*_{ij}\,\delta^*_{ji}$$

but

$$\delta^*_{ij} = T^*_{ij}\,\delta_{ij} \quad \text{and} \quad \delta^*_{ji} = T^*_{ij}\,\delta_{ji}$$

$$\delta_{ij} = T_{ij}\,\Delta_i \quad \text{and} \quad \delta_{ji} = T_{ij}\,\Delta_j$$

hence

$$f^*_{ij} = k^{j*}_{ii}\,T^*_{ij}\,T_{ij}\,\Delta_i + k^*_{ij}\,T^*_{ij}\,T_{ij}\,\Delta_j \tag{10.4}$$

Once the final stiffness equation has been solved the nodal displacements are used to find element end forces at end "i" of each element using equation (10.4). An equation similar to equation (10.4) could be used to find the forces at end "j", but it is more efficient to consider the equilibrium of the element which shows that

$$f^*_{jix} = -f^*_{ijx}$$
$$f^*_{jiy} = -f^*_{ijy}$$
$$f^*_{jiz} = -f^*_{ijz}$$
$$m^*_{jix} = -m^*_{ijx}$$
$$m^*_{jiy} = -m^*_{ijy} - L_{ij}\,f^*_{ijz}$$
$$m^*_{jiz} = -m^*_{ijz} + L_{ij}\,f^*_{ijy}$$

If equivalent joint forces have been used to cater for settlement, lack of fit, etc., as described in section 8.2, then the final element end forces are found by superimposing the fixed end force system on the forces due to nodal displacements.

10.6 Space Frame Program - SFRAME.BAS

This section presents the program SFRAME.BAS for the automatic analysis of space frames. The program is similar in structure to the grillage analysis program GRID.BAS, except that the facility to deal with element loads has been omitted.

Menu System

Unless SFRAME.BAS has been chained from PRE.BAS or DCHECK.BAS the user is prompted for the name of the data file to be read by the program. SFRAME.BAS then offers the user the following menu which drives the program.

Do you wish to:

(1) Analyse with output to the screen
(2) Analyse with output to the printer
(3) Analyse with output to file
(4) Edit the data
(5) Check the data
(6) Read a new data file
(7) Stop

This data file should be constructed using the data preprocessor program PRE.BAS which is described in section 5.11. The program DCHECK.BAS can be used to check the data for obvious errors. DCHECK.BAS is presented in section 5.12.

Comments on the Algorithm

The length of each element and the direction cosines for the element "x" axis are calculated from the element node numbers and the nodal coordinates. The following equations were derived in section 10.3 and show how the element stiffness submatrices in the global axes system can be obtained from the element stiffness submatrices in the principal axes system using transformation matrices.

$$K_{ii}^{j} = T_{ij}^{-1} \, T_{ij}^{*-1} \, k_{ii}^{j*} \, T_{ij}^{*} \, T_{ij}$$

$$K_{ij} = T_{ij}^{-1} \, T_{ij}^{*-1} \, k_{ij}^{*} \, T_{ij}^{*} \, T_{ij}$$

From the above equations it can be seen that to evaluate one global element stiffness submatrix involves multiplying five 6 x 6 matrices together. Using standard matrix multiplication this requires 864 multiplications and 720 additions. This volume of arithmetic is unmanageable by hand. To set up the matrices and conduct the matrix multiplication on a computer is straightforward. However, setting up the matrices takes a fair amount of coding, and if a relatively slow computer is used then the time taken to evaluate the global element stiffness submatrices can become unacceptably long. Significant savings of computer time and memory can be made by expanding the matrix multiplication as follows.

$$T_{ij}^{-1} \, T_{ij}^{*-1} = \begin{bmatrix} T_1 & T_2 & T_3 & 0 & 0 & 0 \\ T_4 & T_5 & T_6 & 0 & 0 & 0 \\ T_7 & T_8 & T_9 & 0 & 0 & 0 \\ 0 & 0 & 0 & T_1 & T_2 & T_3 \\ 0 & 0 & 0 & T_4 & T_5 & T_6 \\ 0 & 0 & 0 & T_7 & T_8 & T_9 \end{bmatrix}$$

where

$$
\begin{array}{lcllcl}
T_1 & = & l & \text{or} & = & 0 \\
T_2 & = & -(m\, cos\beta \;+\; l\, n\, sin\beta)\;/\;D & \text{or} & = & -n\, sin\beta \\
T_3 & = & (m\, sin\beta \;-\; l\, n\, cos\beta)\;/\;D & \text{or} & = & -n\, cos\beta \\
T_4 & = & m & \text{or} & = & 0 \\
T_5 & = & (l\, cos\beta \;-\; m\, n\, sin\beta)\;/\;D & \text{or} & = & c \\
T_6 & = & -(l\, sin\beta \;+\; m\, n\, cos\beta)\;/\;D & \text{or} & = & -s \\
T_7 & = & n & \text{or} & = & n \\
T_8 & = & D\, sin\beta & \text{or} & = & 0 \\
T_3 & = & D\, cos\beta & \text{or} & = & 0
\end{array}
$$

(the right-hand expressions are those assumed for the special case where the element lies parallel to the coordinate "z" axis). Also

$$
T^*_{ij}\, T_{ij} \;=\; [\, T^{-1}_{ij}\, T^{*-1}_{ij} \,]^T
$$

$$
= \begin{bmatrix}
T_1 & T_4 & T_7 & 0 & 0 & 0 \\
T_2 & T_5 & T_8 & 0 & 0 & 0 \\
T_3 & T_6 & T_9 & 0 & 0 & 0 \\
0 & 0 & 0 & T_1 & T_4 & T_7 \\
0 & 0 & 0 & T_2 & T_5 & T_8 \\
0 & 0 & 0 & T_3 & T_6 & T_9
\end{bmatrix}
$$

Writing the k^{j*}_{ii}, k^*_{ij}, k^*_{ji} and k^{i*}_{jj} submatrices in terms of "K" factors allows the global stiffness submatrices to be evaluated in terms of the "T" and "K" factors as follows.

$$
k^{j*}_{ii} \;=\; \begin{bmatrix}
K_1 & 0 & 0 & 0 & 0 & 0 \\
0 & K_2 & 0 & 0 & 0 & K_7 \\
0 & 0 & K_3 & 0 & K_8 & 0 \\
0 & 0 & 0 & K_4 & 0 & 0 \\
0 & 0 & K_9 & 0 & K_5 & 0 \\
0 & K_{10} & 0 & 0 & 0 & K_6
\end{bmatrix}
$$

where

$$
\begin{array}{lcllcl}
K_1 & = & EA/L & K_2 & = & 12EI_z/L^3 \\
K_3 & = & 12EI_y/L^3 & K_4 & = & GJ/L \\
K_5 & = & 4EI_y/L & K_6 & = & 4EI_z/L \\
K_7 & = & 6EI_z/L^2 & K_8 & = & -6EI_y/L^2 \\
K_9 & = & -6EI_y/L^2 & K_{10} & = & 6EI_z/L^2
\end{array}
$$

$$
\boldsymbol{k}^*_{ij} = \begin{bmatrix} -K_1 & 0 & 0 & 0 & 0 & 0 \\ 0 & -K_2 & 0 & 0 & 0 & K_7 \\ 0 & 0 & -K_3 & 0 & K_8 & 0 \\ 0 & 0 & 0 & -K_4 & 0 & 0 \\ 0 & 0 & -K_9 & 0 & -K_5/2 & 0 \\ 0 & -K_{10} & 0 & 0 & 0 & -K_6/2 \end{bmatrix}
$$

$$
\boldsymbol{K}^j_{ii} = \boldsymbol{T}^{-1}_{ij}\, \boldsymbol{T}^{*-1}_{ij}\, \boldsymbol{k}^{j*}_{ii}\, \boldsymbol{T}^*_{ij}\, \boldsymbol{T}_{ij}
$$

$$
= \begin{bmatrix} k_{11} & k_{12} & k_{13} & k_{14} & k_{15} & k_{16} \\ k_{21} & k_{22} & k_{23} & k_{24} & k_{25} & k_{26} \\ k_{31} & k_{32} & k_{33} & k_{34} & k_{35} & k_{36} \\ k_{41} & k_{42} & k_{43} & k_{44} & k_{45} & k_{46} \\ k_{51} & k_{52} & k_{53} & k_{54} & k_{55} & k_{56} \\ k_{61} & k_{62} & k_{63} & k_{64} & k_{65} & k_{66} \end{bmatrix}
$$

where

$$
\begin{aligned}
k_{11} &= K_1 T_1 T_1 + K_2 T_2 T_2 + K_3 T_3 T_3 \\
k_{12} &= K_1 T_1 T_4 + K_2 T_2 T_5 + K_3 T_3 T_6 \\
k_{13} &= K_1 T_1 T_7 + K_2 T_2 T_8 + K_3 T_3 T_9 \\
k_{14} &= K_7 T_2 T_3 + K_8 T_3 T_2 \\
k_{15} &= K_7 T_2 T_6 + K_8 T_3 T_5 \\
k_{16} &= K_7 T_2 T_9 + K_8 T_3 T_8 \\
k_{21} &= K_1 T_4 T_1 + K_2 T_5 T_2 + K_3 T_6 T_3 \\
k_{22} &= K_1 T_4 T_4 + K_2 T_5 T_5 + K_3 T_6 T_6 \\
k_{23} &= K_1 T_4 T_7 + K_2 T_5 T_8 + K_3 T_6 T_9 \\
k_{24} &= K_7 T_5 T_3 + K_8 T_6 T_2 \\
k_{25} &= K_7 T_5 T_6 + K_8 T_6 T_5 \\
k_{26} &= K_7 T_5 T_9 + K_8 T_6 T_8 \\
k_{31} &= K_1 T_7 T_1 + K_2 T_8 T_2 + K_3 T_9 T_3 \\
k_{32} &= K_1 T_7 T_4 + K_2 T_8 T_5 + K_3 T_9 T_6 \\
k_{33} &= K_1 T_7 T_7 + K_2 T_8 T_8 + K_3 T_9 T_9 \\
k_{34} &= K_7 T_8 T_3 + K_8 T_9 T_2 \\
k_{35} &= K_7 T_8 T_6 + K_8 T_9 T_5 \\
k_{36} &= K_7 T_8 T_9 + K_8 T_9 T_8 \\
k_{41} &= K_9 T_2 T_3 + K_{10} T_3 T_2 \\
k_{42} &= K_9 T_2 T_6 + K_{10} T_3 T_5 \\
k_{43} &= K_9 T_2 T_9 + K_{10} T_3 T_8 \\
k_{44} &= K_4 T_1 T_1 + K_5 T_2 T_2 + K_6 T_3 T_3
\end{aligned}
$$

$$
\begin{aligned}
k_{45} &= K_4T_1T_4 + K_5T_2T_5 + K_6T_3T_6 \\
k_{46} &= K_4T_1T_7 + K_5T_2T_8 + K_6T_3T_9 \\
k_{51} &= K_9T_5T_3 + K_{10}T_6T_2 \\
k_{52} &= K_9T_5T_6 + K_{10}T_6T_5 \\
k_{53} &= K_9T_5T_9 + K_{10}T_6T_8 \\
k_{54} &= K_4T_4T_1 + K_5T_5T_2 + K_6T_6T_3 \\
k_{55} &= K_4T_4T_4 + K_5T_5T_5 + K_6T_6T_6 \\
k_{56} &= K_4T_4T_7 + K_5T_5T_8 + K_6T_6T_9 \\
k_{61} &= K_9T_8T_3 + K_{10}T_9T_2 \\
k_{62} &= K_9T_8T_6 + K_{10}T_9T_5 \\
k_{63} &= K_9T_8T_9 + K_{10}T_9T_8 \\
k_{64} &= K_4T_7T_1 + K_5T_8T_2 + K_6T_9T_3 \\
k_{65} &= K_4T_7T_4 + K_5T_8T_5 + K_6T_9T_6 \\
k_{66} &= K_4T_7T_7 + K_5T_8T_8 + K_6T_9T_9
\end{aligned}
$$

To evaluate one global element substiffness matrix using the above expansions reduces the arthmetic to 180 multiplications and 54 additions. Further savings can be made by taking advantage of the symmetry of the global element stiffness matrix and the similarity of the global element stiffness submatrices. In SFRAME.BAS to generate the upper triangle of the global element stiffness submatrix requires only 138 multiplications/divisions and 39 additions/ subtractions. Contrast this with the 3456 multiplications and 2880 additions required to evaluate the global element stiffness submatrices by matrix multiplication of five number 6 x 6 matrices.

Using code numbers the global element stiffness matrix for each element in turn is added to the upper triangle of the final structure stiffness matrix. The loading vector is then generated and the final stiffness equation is solved using the Gaussian elimination subroutine from Chapter 4. Once the unknown nodal displacements have been found they are used to calculate the element forces. Note that the "*T*" and "*K*" factors that were used to evaluate the element stiffness are also used to simplify the calculation of the element forces. The reactions are evaluated by summing the reaction contributions from the elements connected to the restrained nodes. If node "*i*" is restrained then the contribution, R_{ij}, from element "*i,j*" to the reactions at node "*i*" is (in the global axes system) is as follows

$$
\begin{bmatrix} R_{ijx} \\ R_{ijy} \\ R_{ijz} \\ RM_{ijx} \\ RM_{ijy} \\ RM_{ijz} \end{bmatrix}
=
\begin{bmatrix}
T_1 & T_2 & T_3 & 0 & 0 & 0 \\
T_4 & T_5 & T_6 & 0 & 0 & 0 \\
T_7 & T_8 & T_9 & 0 & 0 & 0 \\
0 & 0 & 0 & T_1 & T_2 & T_3 \\
0 & 0 & 0 & T_4 & T_5 & T_6 \\
0 & 0 & 0 & T_7 & T_8 & T_9
\end{bmatrix}
\begin{bmatrix} f^*_{ijx} \\ f^*_{ijy} \\ f^*_{ijz} \\ m^*_{ijx} \\ m^*_{ijy} \\ m^*_{ijz} \end{bmatrix}
$$

$$= \begin{vmatrix} T_1 f_{ijx}^* + T_2 f_{ijy}^* + T_3 f_{ijz}^* \\ T_4 f_{ijx}^* + T_5 f_{ijy}^* + T_6 f_{ijz}^* \\ T_7 f_{ijx}^* + T_8 f_{ijy}^* + T_9 f_{ijz}^* \\ T_1 m_{ijx}^* + T_2 m_{ijy}^* + T_3 m_{ijz}^* \\ T_4 m_{ijx}^* + T_5 m_{ijy}^* + T_6 m_{ijz}^* \\ T_7 m_{ijx}^* + T_8 m_{ijy}^* + T_9 m_{ijz}^* \end{vmatrix}$$

where

R_{ijx} is the reactive component of force in the global "x" direction.
R_{ijy} is the reactive component of force in the global "y" direction.
R_{ijz} is the reactive component of force in the global "z" direction.
RM_{ijx} is the reactive component of moment about the global "x" axis.
RM_{ijy} is the reactive component of moment about the global "y" axis.
RM_{ijz} is the reactive component of moment about the global "z" axis.

The final reaction at node "i" is given by the following equation

$$R_i = \sum_{j=a}^{n} F_{ij} - P_i$$

where the summation is for all of the elements attached to node "i".

Output

The data from the data file is echoed and messages are output to indicate the beginning of each solution phase. After solution of the stiffness equation nodal displacements are output. The axial force, shear forces (in two planes), bending moments (also in two planes), and torsional moment at each end of each element are then calculated and output. Finally the components of reaction (in the global axes system) at each restrained node are output.

Listing

```
1000 '=========================     SFRAME.BAS     ============================
1010 '
1020 COMMON DATAFILE$
1030 OPTION BASE 1 : KEY OFF
1040 TRUE% = -1 : FALSE% = 0 : DEVICE$ = "SCRN:" : PAGELEN% = 23
1050 CLS
1060 PRINT
1070 PRINT "================================================================"
1080 PRINT
1090 PRINT " PROGRAM SFRAME                        Copyright (c) J.Balfour 1991"
1100 PRINT
1110 PRINT "        Program for the automatic analysis of grillages"
1120 PRINT "       Data generated with the data preprocessor PRE"
1130 PRINT
1140 PRINT "                    For further information contact"
1150 PRINT " James A.D.Balfour, Heriot-Watt University, Riccarton, Edinburgh"
```

```
1160 PRINT "                    Tel 031-449-5111, Fax 031-451-3170"
1170 PRINT
1180 PRINT "======================================================================="
1190 '
1200 ' ... Note that variables are defined in Appendix A and the
1210 '     following statements set the maximum problem size
1220 '
1230 MAXNODE%  = 25                        '... Max no of nodes
1240 MAXELEM%  = 50                        '... Max no of elements
1250 MAXPROP%  = 20                        '... Max no of element properties
1260 MAXREST%  = 20                        '... Max no of restrained nodes
1270 MAXNLOAD% = 40                        '... Max no of nodal loads
1280 MAXBAND%  = 45                        '... Max semi-bandwidth
1290 '
1300 DIM NODE(MAXNODE%,4),        ELEM(MAXELEM%,8),    PROP(MAXPROP%,8)
1310 DIM REST(MAXREST%,8),        NLOAD(MAXNLOAD%,3),  FREE%(6*MAXNODE%)
1320 DIM K(6*MAXNODE%,MAXBAND%),  P(3*MAXNODE%),       ESTIFF(12,12)
1330 DIM EFREE%(12),              WKSP$(10),           U(24)
1340 '
1350 ' ... Check if this program has been chained from PRE
1360 '
1370 IF DATAFILE$ <> "" THEN GOSUB 9410 : GOTO 1400
1380 PRINT
1390 INPUT "Name of the data file  =  ", DATAFILE$
1400 GOSUB 2090
1410 NOPTIONS% = 7
1420 WHILE (REPLY% <> 7)
1430   CLS : PRINT
1440   PRINT "Do you wish to:"
1450   PRINT "  (1)  Analyse with output to the screen"
1460   PRINT "  (2)  Analyse with output to the printer"
1470   PRINT "  (3)  Analyse with output to a file"
1480   PRINT "  (4)  Edit the data"
1490   PRINT "  (5)  Check the data"
1500   PRINT "  (6)  Read a new data file
1510   PRINT "  (7)  Stop"
1520   GOSUB 8700
1530   ON REPLY% GOTO 1550, 1580, 1660, 1730, 1760, 1780, 1820
1540     '
1550     OPEN "SCRN:" FOR OUTPUT AS #2                      ' Reply = 1
1560     GOSUB 1860 : GOTO 1820
1570     '
1580     ON ERROR GOTO 1630                                 ' Reply = 2
1590     OPEN "LPT1:" FOR OUTPUT AS #2
1600     PRINT : PRINT "Wait - looking for printer" : PRINT #2,
1610     ON ERROR GOTO 0     DEVICE$ = "LPT1:" : GOSUB 1860
1620     PRINT #2, CHR$(12); : GOTO 1820
1630     CLS : PRINT : PRINT "** ERROR ** Failed to find the printer"
1640     GOSUB 9410 : RESUME 1820
1650     '
1660     ON ERROR GOTO 1700                                 ' Reply = 3
1670     INPUT "Name of file for output = ", DEVICE$
1680     OPEN DEVICE$ FOR OUTPUT AS #2
1690     ON ERROR GOTO 0 : GOSUB 1860 : GOTO 1820
1700     CLS : PRINT : PRINT "** ERROR ** Failed to open output file"
1710     GOSUB 9410 : RESUME 1820
1720     '
1730     PRINT "Wait - chaining PRE"                          ' Reply = 4
1740     CHAIN "PRE"
1750     '
1760     PRINT "Wait - chaining DCHECK"                       ' Reply = 5
1770     CHAIN "DCHECK"
1780     PRINT                                               ' Reply = 6
1790     '
1800     INPUT "Name of the data file  =  ", DATAFILE$
1810     GOSUB 2090
1820     CLOSE #2 : DEVICE$ = "SCRN:" : ON ERROR GOTO 0
1830 WEND
1840 END
1850 '
1860 ' ************************      ANALYSE     ******************************
1870 '
1880 IF DATAFILEOK% THEN GOTO 1930
1890 CLS : PRINT
1900 PRINT "** ERROR ** Analysis cannot proceed due to errors in data file";
1910 GOSUB 9410          'Pause
1920 GOTO  2070
```

```
1930 GOSUB 3410          'Print data
1940 PRINT : PRINT "Generating the freedom vector                              "
1950 GOSUB 4670
1960 PRINT : PRINT "Assembling the stiffness matrix, adding element :- ";
1970 GOSUB 5010
1980 PRINT : PRINT "Generating the loading vector                              "
1990 GOSUB 6150
2000 PRINT
2010 PRINT "Solving equations - no. of equations = "; NDOF%;
2020 PRINT "                   - semi-bandwidth    = "; BAND%
2030 GOSUB 6260
2040 GOSUB 9410
2050 GOSUB 6860    'Output the nodal displacements
2060 GOSUB 7090    'Output the element forces and reactions
2070 RETURN
2080 '
2090 '***********************   READ DATA FILE    ***************************
2100 '
2110 ' ...  This subroutine reads the data from a data file
2120 '
2130 PRINT
2140 PRINT "Wait - Reading file "; DATAFILE$; " line";
2150 ROW% = CSRLIN : COL% = POS(0) : LCOUNT% = 0
2160 DATAFILEOK% = FALSE : DATAEND% = FALSE%:
2170 NNODE%  = 0 : NELEM%  = 0 : NPROP% = 0 : NREST% = 0 : NELOAD% = 0
2180 ON ERROR GOTO 2200 : OPEN DATAFILE$ FOR INPUT AS #1
2190 ON ERROR GOTO 0 : GOTO 2260
2200 CLS : PRINT
2210 PRINT "** ERROR ** Failed to open file: "; DATAFILE$; " for input"
2220 GOSUB 9410 : RESUME 3380
2230 '
2240 ' ... Read problem type, set up array sizes and no of nos in each row
2250 '
2260 TARGET$ = "SFRAME" : GOSUB 9170
2270 IF STRFOUND% THEN GOTO 2340
2280 CLS : PRINT
2290 PRINT "** ERROR ** Not a file for SFRAME"
2300 GOSUB 9410 : GOTO 3380
2310 '
2320 ' ... Read title and units strings
2330 '
2340 NLINES% = 2 : GOSUB 9320 : IF DATAEND% THEN GOTO 2410
2350 WHILE (MID$(WRKSTR$,I%,1) = " ")
2360   I% = I% + 1
2370 WEND
2380 TITLE$  = MID$(WRKSTR$,I%)
2390 TARGET$ = "Units" : GOSUB 9170
2400 IF STRFOUND% THEN GOTO 2440
2410 CLS : PRINT
2420 PRINT "** ERROR ** Error reading Title/Units strings"
2430 GOSUB 9410 : GOTO 2460
2440 UNITS$ = MID$(WRKSTR$,INSTR(WRKSTR$,":- ")+3,24)
2450 DATAFILEOK% = TRUE%
2460 WHILE NOT DATAEND%
2470   '
2480   ' ... Get module string
2490   '
2500   IF LEFT$(WRKSTR$,4)="+   " THEN GOTO 2520
2510   TARGET$ = "+   " : GOSUB 9170 : IF DATAEND% THEN GOTO 3380
2520   MODSTR$ = WRKSTR$
2530   '
2540   ' ...  Identify module
2550   '
2560   IF INSTR(MODSTR$,"NODAL COORDINATES")=0 THEN GOTO 2710
2570   '
2580   ' ... Read nodal coordinates
2590   '
2600   TARGET$ = "NODE" : GOSUB 9170 : IF DATAEND% THEN GOTO 3310
2610   GOSUB 8840
2620   WHILE NNOS% > 0
2630     NNODE% = NNODE% + 1
2640     FOR J% = 1 TO NNOS%
2650       NODE(NNODE%,J%) = VAL(WKSP$(J%))
2660     NEXT J%
2670     GOSUB 8840
2680   WEND
2690   GOTO 3370
```

```
2700   '
2710   IF INSTR(MODSTR$,"ELEMENTS")=0 THEN GOTO 2860
2720   '
2730   ' ... Read elements
2740   '
2750   TARGET$ = "ELEMENT" : GOSUB 9170 : IF DATAEND% THEN GOTO 3310
2760   GOSUB 8840
2770   WHILE NNOS% > 0
2780     NELEM% = NELEM% + 1
2790     FOR J% = 1 TO NNOS%
2800       ELEM(NELEM%,J%) = VAL(WKSP$(J%))
2810     NEXT J%
2820     GOSUB 8840
2830   WEND
2840   GOTO 3370
2850   '
2860   IF INSTR(MODSTR$,"PROPERTIES")=0 THEN GOTO 3010
2870   '
2880   ' ... Read properties
2890   '
2900   TARGET$ = "PROPERTY" : GOSUB 9170 : IF DATAEND% THEN GOTO 3310
2910   GOSUB 8840
2920   WHILE NNOS% > 0
2930     NPROP% = NPROP% + 1
2940     FOR J% = 1 TO NNOS%
2950       PROP(NPROP%,J%) = VAL(WKSP$(J%))
2960     NEXT J%
2970     GOSUB 8840
2980   WEND
2990   GOTO 3370
3000   '
3010   IF INSTR(MODSTR$,"RESTRAINTS")=0 THEN GOTO 3160
3020   '
3030   ' ... Read restraints
3040   '
3050   TARGET$ = "NODE" : GOSUB 9170 : IF DATAEND% THEN GOTO 3310
3060   GOSUB 8840
3070   WHILE NNOS% > 0
3080     NREST% = NREST% + 1
3090     FOR J% = 1 TO NNOS%
3100       REST(NREST%,J%) = VAL(WKSP$(J%))
3110     NEXT J%
3120     GOSUB 8840
3130   WEND
3140   GOTO 3370
3150   '
3160   IF INSTR(MODSTR$,"NODAL LOADS")=0 THEN GOTO 3310
3170   '
3180   ' ... Read nodal loads
3190   '
3200   TARGET$ = "NODE" : GOSUB 9170 : IF DATAEND% THEN GOTO 3310
3210   GOSUB 8840
3220   WHILE NNOS% > 0
3230     NNLOAD% = NNLOAD% + 1
3240     FOR J% = 1 TO NNOS%
3250       NLOAD(NNLOAD%,J%) = VAL(WKSP$(J%))
3260     NEXT J%
3270     GOSUB 8840
3280   WEND
3290   GOTO 3370
3300   '
3310   CLS : PRINT
3320   PRINT "** ERROR ** in module "; MODSTR$ "run DCHECK for more info";
3330   GOSUB 9410
3340   PRINT
3350   PRINT "Wait - Reading file "; DATAFILE$; " line";
3360   ROW% = CSRLIN : COL% = POS(0) : WRKSTR$ = ""
3370 WEND
3380 CLOSE #1 : ON ERROR GOTO 0
3390 RETURN
3400 '
3410 ' ***********************    PRINT DATA    *****************************
3420 '
3430 GOSUB 3510                          '... List header
3440 GOSUB 3840                          '... List nodal coordinates
3450 GOSUB 3990                          '... List element data
3460 GOSUB 4150                          '... List list element properties
```

```
3470 GOSUB 4350                        '... List restraint data
3480 GOSUB 4500                        '... List nodal loads
3490 RETURN
3500 '
3510 ' ++++++++++++++++++++++++   LIST HEADER    ++++++++++++++++++++++++++++
3520 '
3530 CLS
3540 PRINT #2, : PRINT #2,
3550 PRINT #2, "======================================";
3560 PRINT #2, "======================================"
3570 PRINT #2,
3580 PRINT #2, "     PROGRAM SFRAME"; STRING$(24,32);
3590 PRINT #2, "Copyright(c) J.A.D.Balfour"
3600 PRINT #2,
3610 FOR I% = 1 TO (76-LEN(TITLE$))/2
3620   PRINT #2, " ";
3630 NEXT I%
3640 PRINT #2, TITLE$
3650 PRINT #2,
3660 PRINT #2, "     File Name :- "; DATAFILE$;
3670 FOR I% = 1 TO (40-LEN(DATAFILE$))
3680   PRINT #2, " ";
3690 NEXT I%
3700 PRINT #2, "Date :- "; MID$(DATE$,4,3); MID$(DATE$,1,3);
3710 PRINT #2, MID$(DATE$, 9, 2)
3720 PRINT #2, "     Units     :- "; UNITS$;
3730 FOR I% = 1 TO (40-LEN(UNITS$))
3740   PRINT #2, " ";
3750 NEXT I%
3760 PRINT #2, "Time :- "; TIME$
3770 PRINT #2,
3780 PRINT #2, "======================================";
3790 PRINT #2, "======================================"
3800 PRINT #2,
3810 GOSUB 9410
3820 RETURN
3830 '
3840 ' ++++++++++++++++++    LIST NODAL COORDINATES    ++++++++++++++++++++++
3850 '
3860 PRINT #2,
3870 PRINT #2, "+ + + + + + + + + + + + + +"
3880 PRINT #2, "+   NODAL COORDINATES   +"
3890 PRINT #2, "+ + + + + + + + + + + + + +"
3900 PRINT #2,
3910 PRINT #2, "NODE           X            Y            Z"
3920 FOR I% = 1 TO NNODE%
3930   GOSUB 9500
3940   PRINT #2, " "; NODE(I%,1), NODE(I%,2), NODE(I%,3), NODE(I%,4)
3950 NEXT I%
3960 GOSUB 9410
3970 RETURN
3980 '
3990 ' +++++++++++++++++++++++   LIST ELEMENTS    ++++++++++++++++++++++++++
4000 '
4010 PRINT #2,
4020 PRINT #2, "+ + + + + + + + + +"
4030 PRINT #2, "+   ELEMENTS    +"
4040 PRINT #2, "+ + + + + + + + + +"
4050 PRINT #2,
4060 PRINT #2, "ELEMENT    PROPERTY      BETA"
4070 FOR I% = 1 TO NELEM%
4080   GOSUB 9500
4090   PRINT #2, USING "### ###       ###"; ELEM(I%,1); ELEM(I%,2); ELEM(I%,3);
4100   PRINT #2, USING "          ###.#"; ELEM(I%,4)
4110 NEXT I%
4120 GOSUB 9410
4130 RETURN
4140 '
4150 ' ++++++++++++++++++++++   LIST PROPERTIES    +++++++++++++++++++++++++
4160 '
4170 PRINT #2,
4180 PRINT #2, "+ + + + + +  + + +"
4190 PRINT #2, "+   PROPERTIES   +"
4200 PRINT #2, "+ + + + + +  + + +"
4210 PRINT #2,
4220 PRINT #2, "PROPERTY       A          Iyy          Izz          J";
4230 PRINT #2, "           G          E"
```

```
4240 FOR I% = 1 TO NPROP%
4250   GOSUB 9500
4260   PRINT #2, USING " ###    "; PROP(I%,1);
4270   FOR J% = 2 TO 7
4280     PRINT #2, USING "  #.####^^^^"; PROP(I%,J%);
4290   NEXT J%
4300   PRINT #2,
4310 NEXT I%
4320 GOSUB 9410
4330 RETURN
4340 '
4350 ' +++++++++++++++++++++++    LIST RESTRAINTS     +++++++++++++++++++++++
4360 '
4370 PRINT #2,"
4380 PRINT #2, "+ + + + + + + + + + + "
4390 PRINT #2, "+   RESTRAINTS    +"
4400 PRINT #2, "+ + + + + + + + + + + "
4410 PRINT #2,
4420 PRINT #2, "NODE    DIRECTION(S)"
4430 FOR I% = 1 TO NREST%
4440   GOSUB 9500
4450   PRINT #2, USING "###      ######"; REST(I%,1); REST(I%,2)
4460 NEXT I%
4470 GOSUB 9410
4480 RETURN
4490 '
4500 ' +++++++++++++++++++    LIST NODAL LOADS    ++++++++++++++++++++++++
4510 '
4520 IF NNLOAD% = 0 THEN GOTO 4650
4530 PRINT #2,
4540 PRINT #2, "+ + + + + + + + + + +"
4550 PRINT #2, "+   NODAL LOADS   +"
4560 PRINT #2, "+ + + + + + + + + + +"
4570 PRINT #2,
4580 PRINT #2, "NODE    DIRECTION       VALUE"
4590 FOR I% = 1 TO NNLOAD%
4600   GOSUB 9500
4610   PRINT #2, USING "###          #"; NLOAD(I%,1); NLOAD(I%,2);
4620   PRINT #2, USING "         #.####^^^^"; NLOAD(I%,3)
4630 NEXT I%
4640 GOSUB 9410
4650 RETURN
4660 '
4670 ' *****************    GENERATE THE FREEDOM VECTOR    *******************
4680 '
4690 ' ... Zero the freedom vector
4700 '
4710 FOR I% = 1 TO 6*NNODE%
4720   FREE%(I%) = 0
4730 NEXT I%
4740 '
4750 ' ... Loop for all restrained nodes
4760 '
4770 FOR I% = 1 TO NREST%
4780   '
4790   ' ... Evaluate the freedom to be restrained
4800   '
4810   FOR J% = 1 TO LEN(STR$(REST(I%,2)))
4820     K% = VAL(MID$(STR$(REST(I%,2)),J%,1))
4830     IF K%<1 OR K%>6 THEN GOTO 4860
4840     K% = 6*(REST(I%,1)-1)  + VAL(MID$(STR$(REST(I%,2)),J%,1))
4850     FREE%(K%) = 1
4860   NEXT J%
4870 NEXT I%
4880 '
4890 ' ... Number the freedoms (restraints set to zero)
4900 '
4910 NDOF% = 0
4920 FOR I% = 1 TO 6*NNODE%
4930   IF FREE%(I%) = 1 THEN GOTO 4970
4940   NDOF% = NDOF% + 1
4950   FREE%(I%) = NDOF%
4960   GOTO 4980
4970   FREE%(I%) = 0
4980 NEXT I%
4990 RETURN
5000 '
```

```
5010 ' ****************    ASSEMBLE THE STIFFNESS MATRIX    *******************
5020 '
5030 ' ... Zero the augmented matrix
5040 '
5050 FOR I% = 1 TO NDOF%
5060   FOR J% = 1 TO MAXBAND%
5070     K(I%,J%) = 0
5080   NEXT J%
5090   P(I%) = 0
5100 NEXT I%
5110 ROW% = CSRLIN: COL% = POS(0)
5120 '
5130 ' ... Loop for each element
5140 '
5150 BAND% = 0
5160 FOR K% = 1 TO NELEM%
5170   LOCATE ROW%, COL%: PRINT K%;
5180   IN% = ELEM(K%,1)
5190   JN% = ELEM(K%,2)
5200   PN% = ELEM(K%,3)
5210   '
5220   ' ... Calculate the element length (store in ELEM(K%,5))
5230   '
5240   ELEM(K%,5) = (NODE(JN%,2)-NODE(IN%,2))^2 + (NODE(JN%,3)-NODE(IN%,3))^2
5250   ELEM(K%,5) = SQR((NODE(JN%,4)-NODE(IN%,4))^2 + ELEM(K%,5))
5260   '
5270   ' ... Calculate the direction cosines for the member x-axis
5280   '
5290   ELEM(K%,6) = (NODE(JN%,2) - NODE(IN%,2)) / ELEM(K%,5)
5300   ELEM(K%,7) = (NODE(JN%,3) - NODE(IN%,3)) / ELEM(K%,5)
5310   ELEM(K%,8) = (NODE(JN%,4) - NODE(IN%,4)) / ELEM(K%,5)
5320   '
5330   ' ... Assemble the upper triangle of the element stiffness matrix
5340   '
5350   GOSUB 8250
5360   ESTIFF(1,1)  =  K1*T1*T1 + K2*T2*T2 + K3*T3*T3
5370   ESTIFF(1,2)  =  K1*T1*T4 + K2*T2*T5 + K3*T3*T6
5380   ESTIFF(1,3)  =  K1*T1*T7 + K2*T2*T8 + K3*T3*T9
5390   ESTIFF(1,4)  =  K7*T2*T3 + K8*T3*T2
5400   ESTIFF(1,5)  =  K7*T2*T6 + K8*T3*T5
5410   ESTIFF(1,6)  =  K7*T2*T9 + K8*T3*T8
5420   ESTIFF(2,2)  =  K1*T4*T4 + K2*T5*T5 + K3*T6*T6
5430   ESTIFF(2,3)  =  K1*T4*T7 + K2*T5*T8 + K3*T6*T9
5440   ESTIFF(2,4)  =  K7*T5*T3 + K8*T6*T2
5450   ESTIFF(2,5)  =  K7*T5*T6 + K8*T6*T5
5460   ESTIFF(2,6)  =  K7*T5*T9 + K8*T6*T8
5470   ESTIFF(3,3)  =  K1*T7*T7 + K2*T8*T8 + K3*T9*T9
5480   ESTIFF(3,4)  =  K7*T8*T3 + K8*T9*T2
5490   ESTIFF(3,5)  =  K7*T8*T6 + K8*T9*T5
5500   ESTIFF(3,6)  =  K7*T8*T9 + K8*T9*T8
5510   ESTIFF(4,4)  =  K4*T1*T1 + K5*T2*T2 + K6*T3*T3
5520   ESTIFF(4,5)  =  K4*T1*T4 + K5*T2*T5 + K6*T3*T6
5530   ESTIFF(4,6)  =  K4*T1*T7 + K5*T2*T8 + K6*T3*T9
5540   ESTIFF(4,10) = (ESTIFF(4,4) - 3*K4*T1*T1) / 2
5550   ESTIFF(4,11) = (ESTIFF(4,5) - 3*K4*T1*T4) / 2
5560   ESTIFF(4,12) = (ESTIFF(4,6) - 3*K4*T1*T7) / 2
5570   ESTIFF(5,5)  =  K4*T4*T4 + K5*T5*T5 + K6*T6*T6
5580   ESTIFF(5,6)  =  K4*T4*T7 + K5*T5*T8 + K6*T6*T9
5590   ESTIFF(5,10) =  ESTIFF(4,11)
5600   ESTIFF(5,11) = (ESTIFF(5,5) - 3*K4*T4*T4) / 2
5610   ESTIFF(5,12) = (ESTIFF(5,6) - 3*K4*T4*T7) / 2
5620   ESTIFF(6,6)  =  K4*T7*T7 + K5*T8*T8 + K6*T9*T9
5630   ESTIFF(6,10) =  ESTIFF(4,12)
5640   ESTIFF(6,11) =  ESTIFF(5,12)
5650   ESTIFF(6,12) = (ESTIFF(6,6) - 3*K4*T7*T7) / 2
5660   FOR I% = 1 TO 3
5670     FOR J% = 1 TO 3
5680       ESTIFF(I%,J%+9)  =     ESTIFF(I%,J%+3)
5690       IF I% <= J% THEN       ESTIFF(I%,J%+6)  = - ESTIFF(I%,J%)
5700       IF J% <  I% THEN       ESTIFF(I%,J%+6)  = - ESTIFF(J%,I%)
5710       ESTIFF(I%+3,J%+6) = -  ESTIFF(J%,I%+3)
5720       ESTIFF(I%+6,J%+9) = -  ESTIFF(I%,J%+3)
5730       IF J% < I% THEN GOTO 5760
5740       ESTIFF(I%+6,J%+6) =    ESTIFF(I%,J%)
5750       ESTIFF(I%+9,J%+9) =    ESTIFF(I%+3,J%+3)
5760     NEXT J%
5770   NEXT I%
```

```
5780    '
5790    ' ... Set up the element code number
5800    '
5810    '
5820    EFREE%(1)  = FREE%(6*IN%-5)
5830    EFREE%(2)  = FREE%(6*IN%-4)
5840    EFREE%(3)  = FREE%(6*IN%-3)
5850    EFREE%(4)  = FREE%(6*IN%-2)
5860    EFREE%(5)  = FREE%(6*IN%-1)
5870    EFREE%(6)  = FREE%(6*IN%)
5880    EFREE%(7)  = FREE%(6*JN%-5)
5890    EFREE%(8)  = FREE%(6*JN%-4)
5900    EFREE%(9)  = FREE%(6*JN%-3)
5910    EFREE%(10) = FREE%(6*JN%-2)
5920    EFREE%(11) = FREE%(6*JN%-1)
5930    EFREE%(12) = FREE%(6*JN%)
5940    '
5950    ' ... Add the element stiffness to the structure stiffness matrix
5960    '
5970    FOR I% = 1 TO 12
5980      IF EFREE%(I%) = 0 THEN GOTO 6100
5990      FOR J% = I% TO 12
6000        IF EFREE%(J%) = 0 THEN GOTO 6090
6010        L% = EFREE%(I%)
6020        M% = EFREE%(J%) - L% + 1
6030        IF M%<=BAND% THEN GOTO 6080
6040        BAND% = M%
6050        IF BAND% <= MAXBAND% THEN GOTO 6080
6060        PRINT #2, "Max semi-bandwidth ("; STR$(MAXBAND%); ") exceeded"
6070        END
6080        K(L%,M%) = K(L%,M%) + ESTIFF(I%,J%)
6090      NEXT J%
6100    NEXT I%
6110 NEXT K%
6120 PRINT
6130 RETURN
6140 '
6150 '*****************     SET UP THE LOADING VECTOR     ********************
6160 '
6170 ' ... Loop for all point loads
6180 '
6190 FOR I% = 1 TO NNLOAD%
6200   J% = 6*NLOAD(I%,1) - 6 + NLOAD(I%,2)
6210   J% = FREE%(J%)
6220   IF J% > 0 THEN P(J%) = NLOAD(I%,3)
6230 NEXT I%
6240 RETURN
6250 '
6260 ' **********************    SOLVE EQUATIONS     **************************
6270 '
6280 ' ... This subroutine solves a system of symmetric, banded equations
6290 '     using Gaussian elimation with no row interchange.  Only the
6300 '     uppper semi-band of the coefficient matrix is stored.
6310 '
6320 ' ... Loop for all pivots
6330 '
6340 BB% = BAND%
6350 PRINT "                      - current pivot    = ";
6360 ROW% = CSRLIN: COL% = POS(0)
6370 FOR I% = 1 TO NDOF%
6380   LOCATE ROW%, COL% : PRINT I%;
6390   '
6400   ' ... Check if in the unused triangle
6410   '
6420   IF I% > NDOF%-BAND%+1 THEN BB% = NDOF%-I%+1
6430   PIVOT = K(I%,1)
6440   '
6450   ' ... Normalise
6460   '
6470   FOR J% = 1 TO BB%
6480     K(I%,J%) = K(I%,J%) / PIVOT
6490   NEXT J%
6500   P(I%) = P(I%) / PIVOT
6510   '
6520   ' ... Check if in the last row
6530   '
6540   IF BB% = 1 THEN GOTO 6730
```

```
6550    '
6560    ' ... Eliminate (within band) for all rows above the pivot
6570    '
6580    FOR K% = 2 TO BB%
6590      '
6600      ' ... Calculate the row number and evaluate the multiplier
6610      '
6620      L%   = I% + K% - 1
6630      MULT = K(I%,K%) * PIVOT
6640      '
6650      ' ... Loop for all elements in the elimination row
6660      '
6670      FOR J% = K% TO BB%
6680        M% = J% - K% + 1
6690        K(L%,M%) = K(L%,M%) - MULT * K(I%,J%)
6700      NEXT J%
6710      P(L%) = P(L%) - MULT*P(I%)
6720    NEXT K%
6730 NEXT I%
6740 '
6750 ' ... Back substitute
6760 '
6770 FOR I% = 1 TO NDOF%-1
6780   IF I% > BAND%-1 THEN BB% = BAND%-1 ELSE BB% = I%
6790   L% = NDOF%-I%
6800   FOR J%  = 1 TO BB%
6810     P(L%) = P(L%) - K(L%,J%+1)*P(L%+J%)
6820   NEXT J%
6830 NEXT I%
6840 RETURN
6850 '
6860 '*********************    OUTPUT THE DISPLACEMENTS   **********************
6870 '
6880 CLS
6890 PRINT #2,
6900 PRINT #2, "* * * * * * * * * * * * **"
6910 PRINT #2, "*   DISPLACEMENTS    *              * :-  RESTRAINT"
6920 PRINT #2, "* * * * * * * * * * * * **"
6930 PRINT #2,
6940 PRINT #2, "NODE         X-DISP.      Y-DISP.      Z-DISP.";
6950 PRINT #2, "      X-ROTN.      Y-ROTN.      Z-ROTN."
6960 PRINT #2,
6970 FOR I% = 1 TO NNODE%
6980   PRINT #2, USING "###     "; I%;
6990   FOR J% = 1 TO 6
7000     K% = FREE%(6*I%+J%-6)
7010     IF K% =  0 THEN PRINT #2, "        *    ";
7020     IF K% <> 0 THEN PRINT #2, USING "  #.####^^^^"; P(K%);
7030   NEXT J%
7040   PRINT #2,
7050 NEXT I%
7060 GOSUB 9410
7070 RETURN
7080 '
7090 ' ****** CALCULATE AND OUTPUT THE ELEMENT FORCES AND REACTIONS    ******
7100 '
7110 ' ... Zero the reaction components
7120 '
7130 FOR I% = 1 TO NREST%
7140   REST(I%,3) = 0
7150   REST(I%,4) = 0
7160   REST(I%,5) = 0
7170 NEXT I%
7180 CLS
7190 PRINT #2,
7200 PRINT #2, "* * * * * * * * * * * * **"
7210 PRINT #2, "*   ELEMENT FORCES  **"
7220 PRINT #2, "* * * * * * * * * * * * **"
7230 PRINT #2,
7240 PRINT #2, "ELEMENT         AXIAL        SHEAR         SHEAR";
7250 PRINT #2, "      TORSION      BENDING      BENDING"
7260 PRINT #2, "                FORCE       FORCE Y       FORCE Z";
7270 PRINT #2, "       MOMENT     MOMENT Y     MOMENT Z"
7280 '
7290 ' ...  Loop for all elements
7300 '
7310 FOR K% = 1 TO NELEM%
```

```
7320    I% = 6*ELEM(K%,1) - 6
7330    '
7340    ' ... Extract the nodal displacement store in U(1) to U(12)
7350    '
7360    FOR M% = 1 TO 12
7370      J% = FREE%(I%+M%)
7380      IF J% = 0 THEN U(M%) = 0 ELSE U(M%) = P(J%)
7390      IF M% = 6 THEN I% = 6*ELEM(K%,2) - 12
7400    NEXT M%
7410    '
7420    ' ... Calculate element end forces store in U(13) to U(24)
7430    '
7440    GOSUB 8250
7450    U(13) =         K1*((U(1)-U(7))*T1   +(U(2)-U(8))*T4  + (U(3)-U(9))*T7)
7460    U(14) =         K2*((U(1)-U(7))*T2   +(U(2)-U(8))*T5  + (U(3)-U(9))*T8)
7470    U(14) = U(14)+K7*((U(4)+U(10))*T3  +(U(5)+U(11))*T6 + (U(6)+U(12))*T9)
7480    U(15) =         K3*((U(1)-U(7))*T3   +(U(2)-U(8))*T6  + (U(3)-U(9))*T9)
7490    U(15) = U(15)+K8*((U(4)+U(10))*T2  +(U(5)+U(11))*T5 + (U(6)+U(12))*T8)
7500    U(16) =         K4*((U(4)-U(10))*T1  +(U(5)-U(11))*T4 + (U(6)-U(12))*T7)
7510    U(17) =         K5*((U(4)+U(10)/2)*T2+(U(5)+U(11)/2)*T5+(U(6)+U(12)/2)*T8)
7520    U(17) = U(17)+K8*((U(1)-U(7))*T3   +(U(2)-U(8))*T6  + (U(3)-U(9))*T9)
7530    U(18) =         K6*((U(4)+U(10)/2)*T3+(U(5)+U(11)/2)*T6+(U(6)+U(12)/2)*T9)
7540    U(18) = U(18)+K7*((U(1)-U(7))*T2   +(U(2)-U(8))*T5  + (U(3)-U(9))*T8)
7550    U(19) =-U(13)
7560    U(20) =-U(14)
7570    U(21) =-U(15)
7580    U(22) =-U(16)
7590    U(23) =-U(17) - U(15)*ELEM(K%,5)
7600    U(24) =-U(18) + U(14)*ELEM(K%,5)
7610    PRINT #2, USING "   ##  "; ELEM(K%,1);
7620    PRINT #2, USING "  #.####^^^^"; U(13),U(14),U(15),U(16),U(17),U(18)
7630    PRINT #2, USING "   ##  "; ELEM(K%,2);
7640    PRINT #2, USING "  #.####^^^^"; U(19),U(20),U(21),U(22),U(23),U(24)
7650    PRINT #2,
7660    '
7670    ' ... Add any contribution from the element forces to the reactions
7680    '
7690    FOR I% = 1 TO NREST%
7700      J% = 13
7710      IF ELEM(K%,1) = REST(I%,1) THEN GOTO 7740
7720      J% = 19
7730      IF ELEM(K%,2) <> REST(I%,1) THEN GOTO 7800
7740      REST(I%,3) = REST(I%,3) + T1*U(J%)   + T2*U(J%+1) + T3*U(J%+2)
7750      REST(I%,4) = REST(I%,4) + T4*U(J%)   + T5*U(J%+1) + T6*U(J%+2)
7760      REST(I%,5) = REST(I%,5) + T7*U(J%)   + T8*U(J%+1) + T9*U(J%+2)
7770      REST(I%,6) = REST(I%,6) + T1*U(J%+3) + T2*U(J%+4) + T3*U(J%+5)
7780      REST(I%,7) = REST(I%,7) + T4*U(J%+3) + T5*U(J%+4) + T6*U(J%+5)
7790      REST(I%,8) = REST(I%,8) + T7*U(J%+3) + T8*U(J%+4) + T9*U(J%+5)
7800    NEXT I%
7810 NEXT K%
7820 GOSUB 9410
7830 '
7840 ' ... Add any applied loads to the reactions
7850 '
7860 FOR I% = 1 TO NREST
7870   FOR J% = 1 TO NNLOD
7880   IF REST(I%,1) <> NLOD(J%,1) THEN GOTO 7910
7890   K%      = NLOD(J%,2)
7900   REST(I%,2+K%) = REST(I%,2+K%) - NLOD(J%,3)
7910   NEXT J%
7920 NEXT I%
7930 '
7940 ' ... OUTPUT THE REACTIONS
7950 '
7960 CLS
7970 PRINT #2,
7980 PRINT #2, "* * * * * * * * * * *"
7990 PRINT #2, "*   REACTIONS    *      -     GLOBAL AXES"
8000 PRINT #2, "* * * * * * * * * * *"
8010 PRINT #2,
8020 PRINT #2, "              X-FORCE     Y-FORCE      Z-FORCE";
8030 PRINT #2, "    X-MOMENT    Y-MOMENT     Z-MOMENT"
8040 TEMP(1) = 0 : TEMP(2) = 0 : TEMP(3) = 0
8050 TEMP(4) = 0 : TEMP(5) = 0 : TEMP(6) = 0
8060 FOR I%  = 1 TO NREST%
8070   PRINT #2, USING "NODE ##"; REST(I%,1);
```

```
8080    PRINT #2, USING "   #.####^^^^"; REST(I%,3), REST(I%,4), REST(I%,5),
8090    PRINT #2, USING "   #.####^^^^"; REST(I%,6), REST(I%,7), REST(I%,8)
8100    TEMP(1) = TEMP(1) + REST(I%,3)
8110    TEMP(2) = TEMP(2) + REST(I%,4)
8120    TEMP(3) = TEMP(3) + REST(I%,5)
8130    TEMP(4) = TEMP(4) + REST(I%,6)
8140    TEMP(5) = TEMP(5) + REST(I%,7)
8150    TEMP(6) = TEMP(6) + REST(I%,8)
8160 NEXT I%
8170 PRINT #2, "          ----------  ----------  ----------  ";
8180 PRINT #2, "----------  ----------  ----------"
8190 PRINT #2, "TOTAL   ";
8200 PRINT #2, USING "  #.####^^^^"; TEMP(1); TEMP(2); TEMP(3);
8210 PRINT #2, USING "  #.####^^^^"; TEMP(4); TEMP(5); TEMP(6)
8220 GOSUB 9410
8230 RETURN
8240 '
8250 '*****    EVALUATE THE STIFFNESS AND TRANSFORMATION COEFFICIENTS    *****
8260 '
8270 ' ...  Stiffness coefficients
8280 '
8290 PN% = ELEM(K%,3)
8300 K1 =      PROP(PN%,7) * PROP(PN%,2) / ELEM(K%,5)
8310 K2 = 12 * PROP(PN%,7) * PROP(PN%,4) / ELEM(K%,5)^3
8320 K3 = 12 * PROP(PN%,7) * PROP(PN%,3) / ELEM(K%,5)^3
8330 K4 =      PROP(PN%,6) * PROP(PN%,5) / ELEM(K%,5)
8340 K5 =  4 * PROP(PN%,7) * PROP(PN%,3) / ELEM(K%,5)
8350 K6 =  4 * PROP(PN%,7) * PROP(PN%,4) / ELEM(K%,5)
8360 K7 =  6 * PROP(PN%,7) * PROP(PN%,4) / ELEM(K%,5)^2
8370 K8 = -6 * PROP(PN%,7) * PROP(PN%,3) / ELEM(K%,5)^2
8380 '
8390 ' ... Transformation coefficients
8400 '
8410 AR = ATN(1)*ELEM(K%,4)/45
8420 C  = COS(AR)
8430 S  = SIN(AR)
8440 '
8450 ' ... Check if the element lies parallel to the coordinate z-axis
8460 '
8470 IF ABS(ELEM(K%,8)) > .999 THEN GOTO 8590
8480 D  = SQR(1 - ELEM(K%,8)^2)
8490 T1 =   ELEM(K%,6)
8500 T2 = (-ELEM(K%,7)*C - ELEM(K%,6)*ELEM(K%,8)*S) / D
8510 T3 = ( ELEM(K%,7)*S - ELEM(K%,6)*ELEM(K%,8)*C) / D
8520 T4 =   ELEM(K%,7)
8530 T5 = ( ELEM(K%,6)*C - ELEM(K%,7)*ELEM(K%,8)*S) / D
8540 T6 = (-ELEM(K%,6)*S - ELEM(K%,7)*ELEM(K%,8)*C) / D
8550 T7 =   ELEM(K%,8)
8560 T8 =   S * D
8570 T9 =   C * D
8580 GOTO 8680
8590 T1 =   0
8600 T2 = - ELEM(K%,8)*S
8610 T3 = - ELEM(K%,8)*C
8620 T4 =   0
8630 T5 =   C
8640 T6 = - S
8650 T7 =   ELEM(K%,8)
8660 T8 =   0
8670 T9 =   0
8680 RETURN
8690 '
8700 ' *****************    PROCESS KEYBOARD INPUT    **********************
8710 '
8720 ROW% = CSRLIN : COL% = POS(0)
8730 REPLY% = 0
8740 WHILE REPLY% = 0
8750   LOCATE ROW%, COL% : INPUT "", WRKSTR$ : REPLY% = VAL(WRKSTR$)
8760   IF (REPLY% < 1) OR (REPLY% > NOPTIONS%) THEN GOTO 8790
8770   IF (REPLY% < 10) AND (LEN(WRKSTR$) = 1) THEN GOTO 8800
8780   IF (REPLY% >  9) AND (LEN(WRKSTR$) = 2) THEN GOTO 8800
8790   SOUND 700, 4 : REPLY% = 0 : LOCATE ROW%, COL%
8800 WEND
8810 PRINT
```

```
8820 RETURN
8830 '
8840 '************************    RECORD NUMBERS    ******************************
8850 '
8860 NNOS% = 0 : DONE% = TRUE%
8870 IF EOF(1)<>0 THEN DATAEND% = TRUE% : GOTO 9150
8880 '
8890 ' ... Read next line and check if it is blank
8900 '
8910 LINE INPUT #1, WRKSTR$
8920 LCOUNT% = LCOUNT% + 1 : LOCATE ROW%, COL% : PRINT LCOUNT%;
8930 FOR I% = 1 TO LEN(WRKSTR$)
8940   IF MID$(WRKSTR$,I%,1)<> " " THEN DONE% = FALSE% : I% = LEN(WRKSTR$)
8950 NEXT I%
8960 IF DONE% = TRUE% THEN GOTO 9150
8970 FOR I% = 1 TO 10
8980   WKSP$(I%) = ""
8990 NEXT I%
9000 '
9010 ' ... Search for numbers
9020 '
9030 FOR I% = 1 TO LEN(WRKSTR$)
9040   WHILE (MID$(WRKSTR$, I%, 1) = " ") AND I% <= LEN(WRKSTR$)
9050     I% = I% + 1
9060   WEND
9070   IF (I% > LEN(WRKSTR$)) THEN GOTO 9130
9080   IF(TARGET$="SFRAME") AND LCOUNT% > 4 THEN DONE% = TRUE%
9090   NNOS% = NNOS% + 1: WKSP$(NNOS%) = ""
9100   WHILE (MID$(WRKSTR$, I%, 1) <> " ") AND I% <= LEN(WRKSTR$)
9110     WKSP$(NNOS%) = WKSP$(NNOS%) + MID$(WRKSTR$,I%,1): I% = I% + 1
9120   WEND
9130 NEXT I%
9140 WRKSTR$ = ""
9150 RETURN
9160 '
9170 '***************************    FIND STRING    ****************************
9180 '
9190 STRFOUND% = FALSE% : DONE% = FALSE%
9200 WHILE DONE% = FALSE%
9210   IF EOF(1)<>0 THEN DATAEND% = TRUE% : DONE% = TRUE% : GOTO 9290
9220   LINE INPUT #1, WRKSTR$
9230   LCOUNT% = LCOUNT% + 1 : LOCATE ROW%, COL% : PRINT LCOUNT%;
9240   IF INSTR(WRKSTR$,TARGET$)>0 THEN STRFOUND% = TRUE% : DONE% = TRUE%
9250   '
9260   ' ... Check for new module
9270   '
9280   IF (LEFT$(WRKSTR$,4) ="+   ") THEN DONE% = TRUE%
9290 WEND
9300 RETURN
9310 '
9320 '***************************    READ LINES    ******************************
9330 '
9340 FOR I% = 1 TO NLINES%
9350   IF EOF(1)<>0 THEN DATAEND% = TRUE% : I% = NLINES% : GOTO 9380
9360   LINE INPUT #1, WRKSTR$
9370   LCOUNT% = LCOUNT% + 1 : LOCATE ROW%, COL% : PRINT LCOUNT%;
9380 NEXT I%
9390 RETURN
9400 '
9410 '********************    ANY KEY TO CONTINUE    ************************
9420 '
9430 IF DEVICE$<>"SCRN:" THEN GOTO 9480
9440 LOCATE PAGELEN%+1, 1
9450 PRINT SPACE$(40); "Press any key to continue";
9460 WHILE INKEY$ = "" : WEND
9470 CLS
9480 RETURN
9490 '
9500 '**********************    CHECK FOR PAGE END    *********************
9510 '
9520 IF CSRLIN>=PAGELEN% THEN GOSUB 9410
9530 RETURN
9540 '
9550 '=======================    SFRAME.BAS    ===========================
```

Sample Run

The following sample run shows program SFRAME.BAS being used to analyse the cranked cantilever shown below.

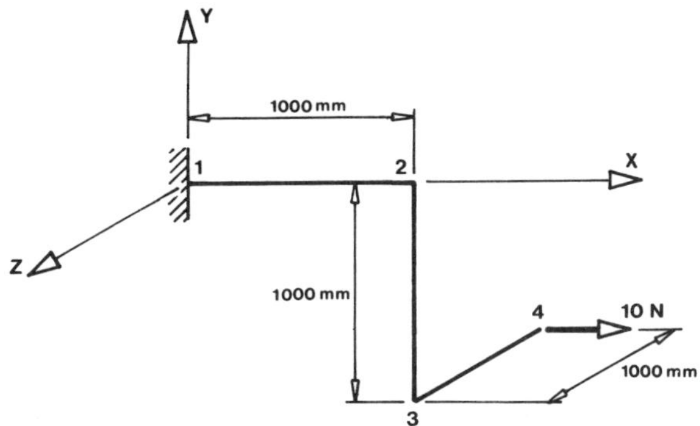

The structure is constructed from solid 10 mm x 20 mm steel sections. The following diagam shows how the elements are connected. Also shown are the principal axes and the element axes systems. Using these principal axes results in the following values of "β" ("β" is the angle through which the element must be rotated about its "*x*" axis to make the principal axes coincide with the element axes). Clockwise rotation is positive when the element is viewed looking from end "*i*" to end "*j*".

$$\beta_{12} = 0°$$

$$\beta_{23} = 0°$$

$$\beta_{34} = 90°$$

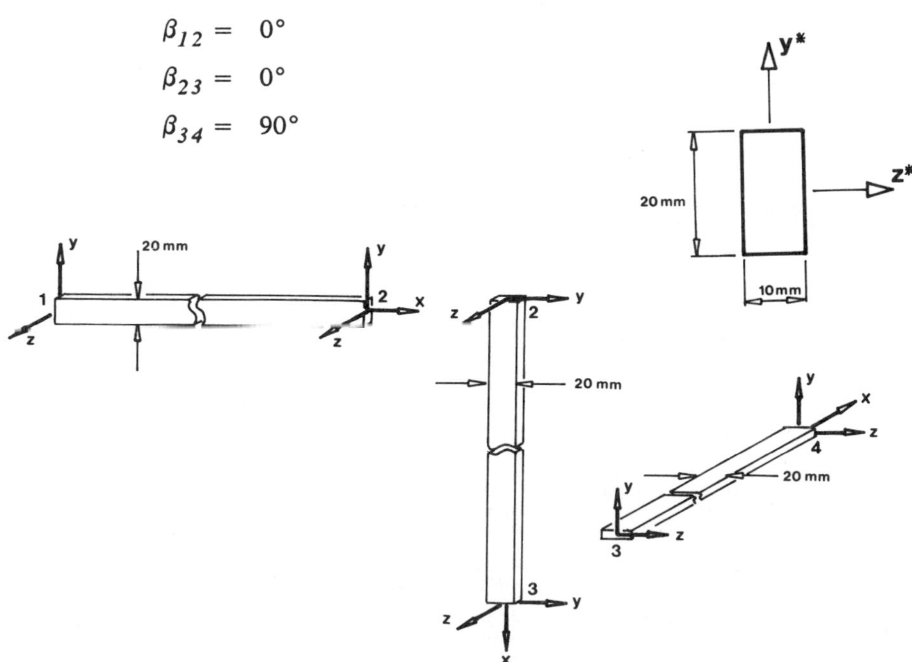

```
===============================================================================

    PROGRAM SFRAME                                 Copyright(c) J.A.D.Balfour

                    CASF 2nd Edition - Cranked Cantilever

    File Name :- SFRAME1.DAT                                   Date :- 04-03-92
    Units     :- N and mm                                      Time :- 10:09:21

===============================================================================


+ + + + + + + + + + + + + + +
+   NODAL COORDINATES   +
+ + + + + + + + + + + + + + +

NODE           X            Y            Z
  1            0            0            0
  2            1000         0            0
  3            1000         -1000        0
  4            1000         -1000        -1000

+ + + + + + + + +
+   ELEMENTS    +
+ + + + + + + + +

ELEMENT     PROPERTY      BETA
  1   2         1          0.0
  2   3         1          0.0
  3   4         1         90.0

+ + + + + +  + + +
+   PROPERTIES   +
+ + + + + +  + + +

PROPERTY         A          Iyy          Izz           J            G            E
   1      0.2000E+03  0.1670E+04  0.6670E+04  0.6670E+04  0.8000E+02  0.2090E+03

+ + + + + + + + + +
+   RESTRAINTS    +
+ + + + + + + + + +

NODE    DIRECTION(S)
  1     123456

+ + + + + + + + + +
+   NODAL LOADS   +
+ + + + + + + + + +

NODE    DIRECTION        VALUE
  4         1         0.1000E-01

* * * * * * * * * * *
*   DISPLACEMENTS   *                 * :-  RESTRAINT
* * * * * * * * * * *

NODE          X-DISP.      Y-DISP.      Z-DISP.      X-ROTN.      Y-ROTN.      Z-ROTN.

  1              *            *            *            *            *            *
  2       0.2392E-03  0.3588E+01  0.1433E+02  -.9166E-05  -.2866E-01  0.7176E-02
  3       0.9567E+01  0.3588E+01  0.1434E+02  -.1617E-04  -.4740E-01  0.1076E-01
  4       0.5936E+02  0.3572E+01  0.1434E+02  -.1617E-04  -.5098E-01  0.1076E-01

* * * * * * * * * * *
*   ELEMENT FORCES  *
* * * * * * * * * * *

ELEMENT        AXIAL        SHEAR        SHEAR      TORSION      BENDING      BENDING
               FORCE      FORCE Y      FORCE Z       MOMENT     MOMENT Y     MOMENT Z
    1     -.1000E-01  -.6001E-05  -.4888E-05  0.4891E-02  0.1000E+02  -.1001E+02
    2     0.1000E-01  0.6001E-05  0.4888E-05  -.4891E-02  -.1000E+02  0.1000E+02

    2     0.0000E+00  -.1000E-01  -.4899E-05  -.1000E+02  0.4895E-02  -.1000E+02
    3     0.0000E+00  0.1000E-01  0.4899E-05  0.1000E+02  0.3830E-05  -.2098E-04

    3     -.3986E-04  -.1000E-01  0.6898E-08  -.1491E-05  -.6236E-05  -.1000E+02
    4     0.3986E-04  0.1000E-01  -.6898E-08  0.1491E-05  -.6615E-06  -.1411E-03
```

```
* * * * * * * * *
*   REACTIONS   *      -     GLOBAL AXES
* * * * * * * * *

               X-FORCE      Y-FORCE      Z-FORCE     X-MOMENT     Y-MOMENT     Z-MOMENT
NODE  1     -.1000E-01   -.6001E-05   -.4888E-05  0.4891E-02  0.1000E+02   -.1001E+02
            ----------   ----------   ----------  ----------  ----------   ----------
TOTAL       -.1000E-01   -.6001E-05   -.4888E-05  0.4891E-02  0.1000E+02   -.1001E+02
```

Chapter 11

Structural Dynamics

11.1 Introduction

A great attraction of the stiffness method is that it is not limited to any particular structure or analysis type. It can easily be extended to deal with instability, dynamics, large deformations and non-linear materials. This chapter gives an introduction to the application of the stiffness method to the field of structural dynamics.

One consequence of recent improvements in structural materials and the more sophisticated analysis techniques offered by computers is that civil engineering structures have become increasingly light and flexible. This has led, in turn, to civil engineering structures that are more susceptible to dynamic effects from environmental loads such as wind. Large flexible structures such as long bridges and stadia roofs must be checked during design to ensure that dynamic oscillations due to wind gusting will never reach unacceptable levels. Earthquakes are another important source of dynamic loading. Earthquake disasters of recent years have shown that work still has to be done to make structures in earthquake zones more resistant to earthquake loading.

Structural dynamics is an extremely wide and sometimes complex subject. Consequently due to space limitations this chapter can only serve to give the reader an introduction to the subject. For more advanced topics and greater detail the reader is directed to other literature[6,10,25]. Particular recommendation is given to Clough and Penzien's text[6].

The starting point in many problems of structural dynamics is to find the mode shapes and frequencies of the structure. This chapter gives only brief coverage to the mathematical background to the subject and will concentrate on the practical aspects of extending the stiffness method to dynamic problems. Only plane trusses and plane frames will be covered, yet the reader should note how much of the material covered in previous chapters is reused here. It should also be obvious that the method developed for plane trusses and plane frames needs only minor modifications to cater for other framework types.

11.2 Single-Degree-of-Freedom Systems

Much can be learned about structural dynamics from the study of single-degree-of-freedom systems (SDOF systems). Consider the SDOF system shown in fig 11.1.

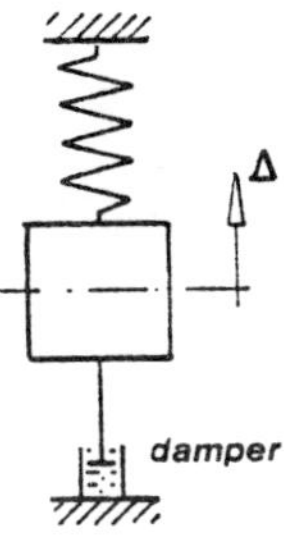

Figure 11.1 An SDOF system

Newton's second law of motion can be used to formulate the equation of motion.

$$Force \quad = \quad Rate\ of\ change\ of\ momentum$$

$$F(t) \quad = \quad \frac{d\left(m(t)\ d\Delta/dt\right)}{dt}$$

or

$$F(t)\ -\ \frac{d\left(m(t)\ d\Delta/dt\right)}{dt} \quad = \quad 0$$

where

$F(t)$ is the net force acting on the body .
Δ is the instantaneous displacement of the body.
$m(t)$ is the instantaneous mass of the body.

Mass is usually constant, therefore,

$$F(t)\ -\ m\,\ddot{\Delta} \ = \ 0$$

The term $m\,\ddot{\Delta}$ is called the *inertia* force resisting acceleration and the concept that a mass develops an inertia force proportional to acceleration and opposing it is known as d'Alembert's principle. This principle allows the equation of dynamic equilibrium of the mass to be written as

$$p(t)\ +\ f_i\ +\ f_d\ +\ f_s\ +\ f_g \quad = \quad 0 \tag{11.1}$$

and

$$f_i \quad = \quad -\,m\,\ddot{\Delta} \qquad \text{by d'Alembert's principle}$$

$$f_d = -c\dot{\Delta} \quad \text{if viscous damping, where "}c\text{" is a constant}$$

$$f_s = -k\Delta \quad \text{if the spring has linear stiffness "}k\text{"}$$

$$f_g = -mg \quad \text{gravitational force}$$

Hence equation (11.1) becomes

$$m\ddot{\Delta} + c\dot{\Delta} + k\Delta + mg = p(t) \tag{11.2}$$

Note that in this equation the displacement "Δ" of the system is measured from the zero spring force position.

The gravitational force is an unnecessary complication which can be eliminated if the displacement is measured from the static equilibrium position, i.e. let

$$\Delta' = \Delta + \Delta_s$$

where

Δ' is the instantaneous displacement measured from the static equilibrium position.

Δ is the instantaneous displacement measured from the zero spring force position.

Δ_s is the static displacement.

note

$$\dot{\Delta}' = \dot{\Delta} \quad \text{and} \quad \ddot{\Delta}' = \ddot{\Delta}$$

Hence equation (11.2) becomes

$$m\ddot{\Delta}' + c\dot{\Delta}' + k(\Delta' - \Delta_s) + mg = p(t)$$

but

$$k\Delta_s = mg$$

hence

$$m\ddot{\Delta}' + c\dot{\Delta}' + k\Delta' = p(t)$$

This equation will be simply written as

$$m\ddot{\Delta} + c\dot{\Delta} + k\Delta = p(t)$$

Where Δ is the displacement from the *static equilibrium position.*

Free Vibration Response

This chapter will consider only free vibration (i.e. $p(t) = 0$), hence

$$m\ddot{\Delta} + c\dot{\Delta} + k\Delta = 0$$

The solution to this a second order ordinary differential equation takes the form

$$\Delta(t) = G\, e^{st}$$

where s is a constant, hence

$$(m\, s^2 + c\, s + k)\, G\, e^{st} = 0$$

For a given system (known "m" and "k") the expression for "s" depends upon "c", the *damping coefficient*. Further investigation of this equation shows that there are two vibration regimens which are divided by a particular value of "c" known as the *critical* damping coefficient "c_c" where

$$c_c = 2\, m\, \omega$$

and

$$\omega = \sqrt{k/m}$$

which is the *undamped natural frequency* of the system.

11.3 Multi-Degree-of-Freedom Systems

In general the dynamic response of real structures cannot be described accurately by an SDOF model and multi-degree-of-freedom systems (MDOF systems) should be used. The complexity of such analyses is such that it is almost invariably conducted using a commercial finite element computer program. This section is intended to give only an insight into some of the techniques used and the issues involved.

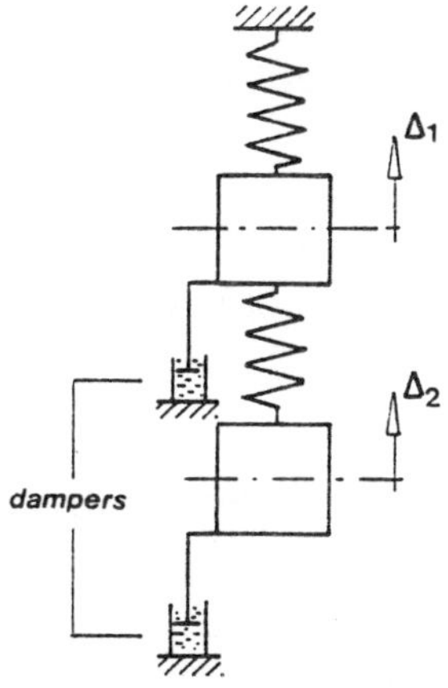

Figure 11.2 Two-degree-of-freedom system

Two-Degree-of-Freedom System

The simplest possible MDOF system is shown in fig 11.2 As with the SDOF model considered previously each mass will experience inertia, damping, elastic and applied forces. The equations of motion of that govern the behaviour of system can be formulated by considering the dynamic equilibrium of each mass in turn.

$$F_{I1} + F_{D1} + F_{E1} = P_1(t)$$
$$F_{I2} + F_{D2} + F_{E2} = P_2(t)$$

or in matrix form

$$\begin{bmatrix} F_{I1} \\ F_{I2} \end{bmatrix} + \begin{bmatrix} F_{D1} \\ F_{D2} \end{bmatrix} + \begin{bmatrix} F_{E1} \\ F_{E2} \end{bmatrix} = \begin{bmatrix} P_1(t) \\ P_2(t) \end{bmatrix}$$

or

$$\boldsymbol{F_I} + \boldsymbol{F_D} + \boldsymbol{F_E} = \boldsymbol{P(t)} \qquad (11.3)$$

where $\boldsymbol{F_I}$, $\boldsymbol{F_D}$ and $\boldsymbol{F_E}$ are the matrices of the instantaneous inertia, damping and elastic forces respectively. If the structural behaviour is linear then these forces can be expressed in terms of *influence coefficients*.

Elastic Forces

The elastic forces in terms of stiffness influence coefficients are

$$F_{S1} = K_{11}\Delta_1 + K_{12}\Delta_2$$
$$F_{S2} = K_{21}\Delta_1 + K_{22}\Delta_2$$

Where K_{ij} is the elastic force at freedom "i" corresponding to unit displacement at freedom "j" (i.e. it is a stiffness coefficient). Writing the above equations in matrix form gives

$$\boldsymbol{F_E} = \begin{bmatrix} F_{E1} \\ F_{E2} \end{bmatrix} = \begin{bmatrix} K_{11} & K_{12} \\ K_{21} & K_{22} \end{bmatrix} \begin{bmatrix} \Delta_1 \\ \Delta_2 \end{bmatrix} = \boldsymbol{K}\,\boldsymbol{\Delta} \qquad (11.4)$$

Damping Forces

If the damping is viscous (i.e. velocity dependent), then by analogy with the elastic force equation.

$$\boldsymbol{F_D} = \begin{bmatrix} F_{D1} \\ F_{D2} \end{bmatrix} = \begin{bmatrix} C_{11} & C_{12} \\ C_{21} & C_{22} \end{bmatrix} \begin{bmatrix} \dot{\Delta}_1 \\ \dot{\Delta}_2 \end{bmatrix} = \boldsymbol{C}\,\dot{\boldsymbol{\Delta}} \qquad (11.5)$$

Where C_{ij} is the damping force at freedom "i" corresponding to unit *velocity* at freedom "j" and is known as a *damping coefficient.*

Inertia Forces

The inertia forces are acceleration dependent and can be expressed in terms of influence coefficients known as *mass coefficients.*

$$\boldsymbol{F_I} = \begin{bmatrix} F_{I1} \\ F_{I2} \end{bmatrix} = \begin{bmatrix} M_{11} & M_{12} \\ M_{21} & M_{22} \end{bmatrix} \begin{bmatrix} \ddot{\Delta}_1 \\ \ddot{\Delta}_2 \end{bmatrix} = \boldsymbol{M} \, \ddot{\boldsymbol{\Delta}} \tag{11.6}$$

Where M_{ij} is the inertia force at freedom "i" corresponding to unit *acceleration* at freedom "j" and is known as an *inertia coefficient.*

Equations 11.4 to 11.6 allow equation 11.3 to be rewritten as follows

$$\boldsymbol{M} \, \ddot{\boldsymbol{\Delta}} + \boldsymbol{C} \, \dot{\boldsymbol{\Delta}} + \boldsymbol{K} \, \boldsymbol{\Delta} = \boldsymbol{P(t)} \tag{11.7}$$

Note that equation (11.7) is simply the matrix version of the SDOF equation of motion, and that if there is no motion then equation (11.7) reduces to the familiar stiffness equation $\boldsymbol{K} \, \boldsymbol{\Delta} = \boldsymbol{P}$.

Example 11.1

Find the equation of motion for the system shown in the following diagram if the motion of the masses is constrained to be vertical.

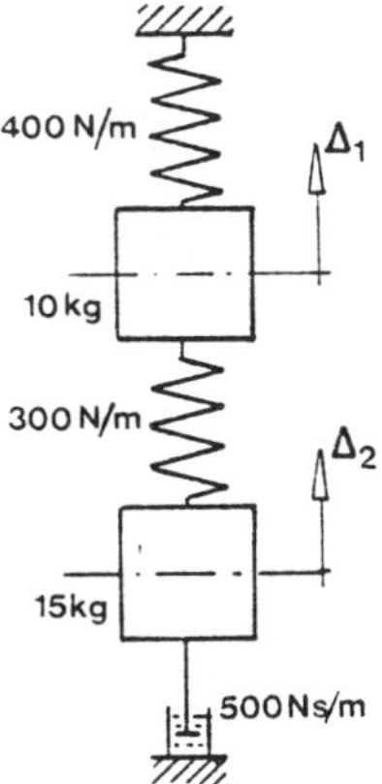

Stiffness Matrix

The system has the two degrees of freedom shown above, and the stiffness coefficients are found by considering the force system required to produce unit displacement at each freedom in turn.

$$K_{11} = 700 \text{ N/m} \qquad K_{12} = -300 \text{ N/m} \qquad K_{22} = 300 \text{ N/m}$$

hence

$$K = \begin{bmatrix} 700 & -300 \\ -300 & 300 \end{bmatrix}$$

Damping Matrix

There is only one damper and the damping coefficients are found by evalution of the force systems associated with unit velocity at each freedom in turn, hence, by inspection

$$C = \begin{bmatrix} 0 & 0 \\ 0 & 500 \end{bmatrix}$$

Mass Matrix

The mass coefficients are found by evalution of the force systems necessary to produce unit acceleration at each freedom in turn, hence, by inspection

$$M = \begin{bmatrix} 10 & 0 \\ 0 & 15 \end{bmatrix}$$

Equation of Motion

$$\begin{bmatrix} 10 & 0 \\ 0 & 15 \end{bmatrix}\begin{bmatrix} \ddot{\Delta}_1 \\ \ddot{\Delta}_2 \end{bmatrix} + \begin{bmatrix} 0 & 0 \\ 0 & 500 \end{bmatrix}\begin{bmatrix} \dot{\Delta}_1 \\ \dot{\Delta}_2 \end{bmatrix} + \begin{bmatrix} 700 & -300 \\ -300 & 300 \end{bmatrix}\begin{bmatrix} \Delta_1 \\ \Delta_2 \end{bmatrix} = \begin{bmatrix} P_1(t) \\ P_2(t) \end{bmatrix}$$

Undamped Free Vibrations of MDOF Systems

The matrix equation of motion for a freely vibration MDOF system can be obtained by omitting the damping matrix and the applied load vector from equation (11.7)

$$M\ddot{\Delta} + K\Delta = \mathbf{0} \tag{11.8}$$

It can be shown that if a linearly elastic structure is vibrating freely then every part of the structure will vibrate in phase with simple harmonic motion. Hence every freedom will vibrate with simple harmonic motion, i.e.

$$\Delta(t) = \Delta^i \sin(\omega_i t + \theta)$$

Where Δ^i represents the vibration mode shape for mode "i" of the system (which does not vary with time, only the amplitude varies), ω_i is the frequency, and θ is a phase angle. Differentiating twice yields the acceleration.

$$\ddot{\Delta}(t) = -\omega_i^2 \Delta^i \sin(\omega_i t + \theta) = -\omega^2 \Delta(t)$$

Hence equation (11.8) reduces to

$$- \omega_i^2 M \Delta^i \sin(\omega_i t + \theta) + K \Delta^i \sin(\omega_i t + \theta) = \mathbf{0}$$

i.e.

$$[K - \omega_i^2 M] \Delta^i = \mathbf{0} \tag{11.9}$$

This is a system of simultaneous equations (**A X** = **B**), and Cramer's rule can be used to find any unknown.

$$\Delta_k^i = \frac{|A_k^i|}{|A^i|}$$

Where $|A_k^i|$ is the determinant of the coefficient matrix A with column "k" replaced with the right-hand side vector B, and $|A^i|$ is the determinant of the coefficient matrix A. The "i" refers to the vibration frequency ω_i used to calculate A^i. The right hand-side vector in the system of equations shown in equation (11.9) comprises of only zeros which means that $|A_k|$ is equal to zero, therefore

$$\Delta_k^i = \frac{0}{|K - \omega_i^2 M|} \tag{11.10}$$

Hence for any non-trivial solution

$$|K - \omega_i^2 M| = 0 \tag{11.11}$$

The above equation is known as the *frequency equation* of the system, and is an *eigenvalue problem*. The determinant can be expanded to produce a polynomial known as the *characteristic equation*. The polynomial will be of degree "n", where "n" is the degree of freedom of the system, and its "n" roots are the vibration frequencies (squared) of the system.

Analysis of Vibration Mode Shapes

The free undamped equation of motion of an MDOF system vibrating with mode "i" is

$$[K - \omega_i^2 M] \Delta^i = \mathbf{0} \tag{11.12}$$

where ω_i is the ith vibration frequency and Δ^i is the ith vibration mode shape. Once the vibration frequencies have been found then the matrix

$$E^i = K - \omega_i^2 M$$

can be evaluated for ω_i and equation (11.12) becomes

$$E^i \Delta^i = \mathbf{0} \tag{11.13}$$

$\mathbf{E}^i$ is singular. This means that there is no unique, non-trivial, solution to equation (11.13) (i.e. the mode shape can be found, but not its amplitude). To find the mode shape one element of Δ^i is set to an arbitrary value - unity is convenient. This allows equation (11.13) to be rewritten as follows

$$\begin{bmatrix} e^i_{11} & e^i_{12} & e^i_{13} & \cdot & \cdot & e^i_{1n} \\ e^i_{21} & e^i_{22} & e^i_{23} & \cdot & \cdot & e^i_{2n} \\ \cdot & & & & & \\ \cdot & & & & & \\ e^i_{n1} & e^i_{n2} & e^i_{n3} & \cdot & \cdot & e^i_{nn} \end{bmatrix} \begin{bmatrix} 1 \\ \Delta^i_2 \\ \cdot \\ \cdot \\ \Delta^i_n \end{bmatrix} = \begin{bmatrix} 0 \\ 0 \\ \cdot \\ \cdot \\ 0 \end{bmatrix} \tag{11.14}$$

Rewriting the previous equation in terms of the submatrices indicated by the broken lines.

$$\begin{bmatrix} e^i_{11} & \mathbf{E}^i_{IO} \\ \mathbf{E}^i_{OI} & \mathbf{E}^i_{II} \end{bmatrix} \begin{bmatrix} 1 \\ \Delta^i_{OI} \end{bmatrix} = \begin{bmatrix} 0 \\ \mathbf{0} \end{bmatrix} \tag{11.15}$$

which gives

$$\mathbf{E}^i_{II}\, \Delta^i_{OI} = -\mathbf{E}^i_{OI} \tag{11.16}$$

This system of equations can be solved simultaneously to find the remaining displacement amplitudes Δ^i_{OI}. The displacement amplitudes are usually normalised by dividing each component of the vector by the largest component to produce the *mode shape vector* ϕ_i.

$$\phi_i = \begin{bmatrix} \phi_{1i} \\ \phi_{2i} \\ \cdot \\ \cdot \\ \phi_{ni} \end{bmatrix} = \frac{1}{\Delta^i_{max}} \begin{bmatrix} 1 \\ \Delta^i_2 \\ \cdot \\ \cdot \\ \Delta^i_n \end{bmatrix} \tag{11.17}$$

where Δ^i_{max} is the largest element of Δ^i.

Example 11.2

Find the undamped natural frequencies and mode shapes for the system analysed in example 11.1.

Natural Frequencies

The natural frequencies squared are the values of ω^2 for which

$$|K - \omega^2 M| = 0$$

i.e.

$$\left| \begin{bmatrix} 700 & -300 \\ -300 & 300 \end{bmatrix} - \omega^2 \begin{bmatrix} 10 & 0 \\ 0 & 15 \end{bmatrix} \right| = 0$$

hence

$$\begin{vmatrix} (700 - 10\omega^2) & -300 \\ -300 & (300 - 15\omega^2) \end{vmatrix} = 0$$

or

$$(700 - 10\omega^2)(300 - 15\omega^2) - (-300)(-300) = 0$$

$$\omega^4 - 90\omega^2 + 800 = 0$$

which yields

$$\omega_1^2 = 10 \qquad (\omega_1 = 3.16 \text{ rad/s})$$

or

$$\omega_2^2 = 80 \qquad (\omega_2 = 8.94 \text{ rad/s})$$

Mode Shapes

The free undamped equation of motion (for mode shape "i") is

$$[K - \omega_i^2 M] \Delta_i = \mathbf{0}$$

Substitute for ω_i^2 using the lowest natural frequency (the *fundamental frequency* ω_1)

$$\left[\begin{bmatrix} 700 & -300 \\ -300 & 300 \end{bmatrix} - 10 \begin{bmatrix} 10 & 0 \\ 0 & 15 \end{bmatrix} \right] \begin{bmatrix} \Delta_1^1 \\ \Delta_2^1 \end{bmatrix} = \begin{bmatrix} 0 \\ 0 \end{bmatrix}$$

$$\begin{bmatrix} 600 & -300 \\ -300 & 150 \end{bmatrix} \begin{bmatrix} \Delta_1^1 \\ \Delta_2^1 \end{bmatrix} = \begin{bmatrix} 0 \\ 0 \end{bmatrix} .$$

These equations are singular and only relative values can be found. As there are only two equations it is easiest to solve directly rather than use the general method described earlier. If Δ_1^1 is set to unity then Δ_2^1 is 2 and the normalised mode shape vector is

$$\phi_i = \begin{bmatrix} 0.5 \\ 1.0 \end{bmatrix}$$

The mode shape corresponding to the second natural frequency is found in exactly the same manner and can easily be shown to be

$$\phi_i = \begin{bmatrix} 1.0 \\ -0.333 \end{bmatrix}$$

Example 11.3

Find the undamped natural frequencies and mode shapes for the three-storey building shown below if the floors are are assumed to be rigid and weigh 20 tonnes each and the *EI* value for the columns is 25 x 10^{12} kN mm^2. This idealisation is commonly used in the analysis of tall buildings as it drastically reduces the degree of freedom without sacrificing the essential dynamic characteristics of the structure. The degree of freedom resulting from this idealisation is equal to the number of storeys as shown in the following diagram.

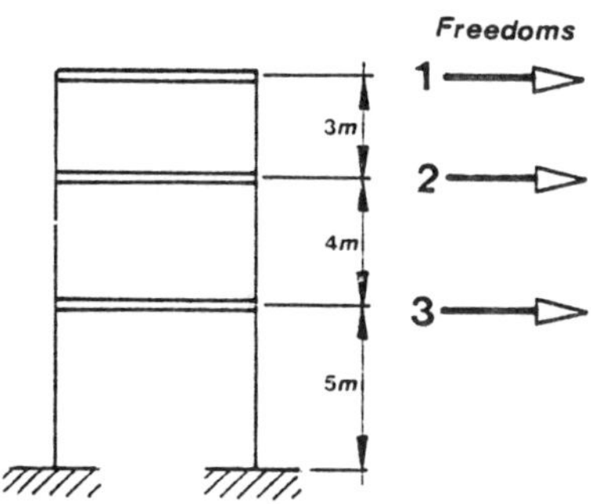

Convert Units

$$20 \text{ tonnes} = 20000 \text{ kg}$$

and

$$25 \times 10^{12} \text{ kN mm}^2 = 25 \times 10^9 \text{ N m}^2$$

Stiffness Matrix

Calculate the force system associated with unit displacement at freedom 1.

$$k_{11} = 2 \times 12\, EI / L^3 = 22.20 \times 10^9$$

$$k_{21} = -k_{11} = -22.20 \times 10^9$$

$$k_{31} = 0$$

Treating the other freedoms similarly yields

$$K = 10^9 \begin{bmatrix} 22.20 & -22.20 & 0 \\ -22.20 & 31.58 & -9.38 \\ 0 & -9.38 & 14.18 \end{bmatrix}$$

Mass Matrix

The mass coefficients are obtained by evaluating the force systems required to produce unit acceleration at each freedom in turn. This structure idealisation results in lumped masses, which allows the structure mass matrix to be written down by inspection.

$$M = \begin{bmatrix} 20000 & 0 & 0 \\ 0 & 20000 & 0 \\ 0 & 0 & 20000 \end{bmatrix}$$

Natural Frequencies

These are the values of ω^2 for which

$$\left| K - \omega^2 M \right| = 0$$

i.e.

$$\begin{vmatrix} (22.20E9 - 20000\omega^2) & -22.20E9 & 0 \\ -22.20E9 & (31.58E9 - 20000\omega^2) & -9.38E9 \\ 0 & -9.38E9 & (14.18E9 - 20000\omega^2) \end{vmatrix} = 0$$

which on expansion yields

$$999.6E27 - 17.66E24\,\omega^2 + 27.18E18\,\omega^4 - 8.0E12\,\omega^6 = 0$$

There are many algorithms that can be used to find roots of equations. For simplicity the equation will be solved here using trial and discard (with the assistance of a programmable calculator).

ω^2	$f(\omega^2)$
30000	494E27
50000	184E27
60000	36E27
70000	-106E27
65000	-35E27
62000	7.2E27
63000	-7.1E27

Take ω^2 as 62500 (i.e. ω = 250 rad/s) and then deflate the polynomial

$$(62500 - \omega^2)(16E24 - 26.68E18\,\omega^2 + 8E12\,\omega^4) = 0$$

Solution of the quadratic equation in the second set of brackets yields

$$\omega_1^2 = 784{,}000 \qquad (\omega_1 = 885 \text{ rad/s})$$

and

$$\omega_2^2 = 2{,}550{,}000 \qquad (\omega_2 = 1597 \text{ rad/s})$$

Mode Shapes

Now that the vibration frequencies have been found the mode shapes can be determined. Only the fundamental mode shape will be calcualted - the other two mode shapes can be found using exactly the same procedure.

$$[\boldsymbol{K} - \omega_i^2\,\boldsymbol{M}]\,\Delta^i = \mathbf{0}$$

Substitute for ω_i^2 using the lowest natural frequency (250 rad/s)

$$10^9 \begin{bmatrix} 20.95 & -22.20 & 0 \\ -22.20 & 30.33 & -9.38 \\ 0 & -9.38 & 12.93 \end{bmatrix} \begin{bmatrix} 1 \\ \Delta_2^1 \\ \Delta_3^1 \end{bmatrix} = \begin{bmatrix} 0 \\ 0 \\ 0 \end{bmatrix}$$

hence

$$\begin{bmatrix} -22.20 \\ 0 \end{bmatrix} + \begin{bmatrix} 30.33 & -9.38 \\ -9.38 & 12.93 \end{bmatrix} \begin{bmatrix} \Delta_2^1 \\ \Delta_3^1 \end{bmatrix} = \begin{bmatrix} 0 \\ 0 \end{bmatrix}$$

Solving the above equations simultaneously gives $\Delta_2^1 = 0.944$ and $\Delta_3^1 = 0.688$ and the normalised mode shape vector is

$$\phi_1 = \begin{bmatrix} 1.000 \\ 0.944 \\ 0.688 \end{bmatrix}$$

11.4 Element Mass Matrices

All of the problems thus far have involved *lumped masses*. Lumped masses are frequently employed in structural dynamics. In essence the mass of each element is distributed evenly to its nodes as point masses (note that point masses are usually assumed to have no rotational inertia). Lumped mass has the attraction of simplicity and results in diagonal mass matrix which has computational advantages.

The use of element mass matrices is generally recognised to be superior to the lumped mass approach. The mass matrix is analogous to the stiffness matrix - only, instead of relating nodal forces to nodal displacements, it relates nodal forces to nodal accelerations. The distribution of acceleration is assumed to be based up the same shape functions as the displacements. The resulting matrices are referred to as *consistant mass matrices*.

The reader should note that the structure mass matrix is assembled in exactly the same way as stiffness matrix, only element mass matrices are used rather than element stiffness matrices. Once the structure mass and stiffness matrices have been assembled the frequency equation can be solved to find the undamped mode shapes and frequencies for the structure.

In this section element mass matrices for plane trusses and plane frames will be developed using shape functions and the principle of virtual work (see Chapter 2).

Plane Truss

To illustrate how the principle of virtual work is used to find mass coefficients the mass coefficient k_{31} for a plane truss element will be found. Figure 11.3 shows the freedoms for a plane truss element and the reader should note that these are the same as the freedoms used for the static analysis of plane trusses (see Chapter 5). The mass coefficient m_{31} is the force in the direction of freedom 3 associated with unit acceleration at freedom 1 (the acceleration at the other freedoms is zero). Note that the acceleration throughout the element is assumed to be governed by the same shape function as governs displacement. Hence the longitudinal acceleration (and inertia force) will vary linearly from unit acceleration at "*i*" to zero acceleration at "*j*" as illustrated in fig 11.3.

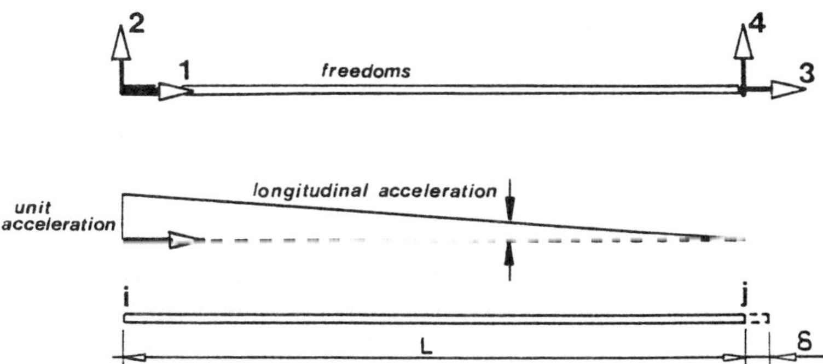

Figure 11.3 Unit acceleration at freedom 1 for a plane truss element

When a virtual displacement "δ" is applied the direction of freedom 3 as shown there is no change in strain energy as there is no stress in the element during the virtual displacement. Work will be done, however, by the external forces and internal inertia forces, and the resulting virtual work equation is *work done by the external forces + work done by the inertia forces = 0*

hence

$$m_{31}\,\delta + \int_0^L \textit{inertia force x distance moved} = 0$$

Consider an element of length dx long located "x" from node "i".

$$\text{inertia force} = \text{mass x acceleration} = m\,dx\,\frac{L - x}{L}$$

and

$$\textit{distance moved} = \frac{x}{L}\,\delta$$

hence

$$m_{31}\,\delta + \int_0^L m\,\frac{L - x}{L}\,\frac{x}{L}\,\delta\,dx = 0$$

which yields

$$m_{31} = \frac{mL}{6}$$

Where "m" is the mass per unit length of element. Treating other freedoms similarly yields the element mass matrix which relates nodal forces and accelerations as shown in the following equation.

$$\begin{bmatrix} f_{ijx} \\ f_{ijy} \\ f_{jix} \\ f_{jiy} \end{bmatrix} = \begin{bmatrix} \frac{mL}{3} & 0 & \frac{mL}{6} & 0 \\ 0 & \frac{mL}{3} & 0 & \frac{mL}{6} \\ \frac{mL}{6} & 0 & \frac{mL}{3} & 0 \\ 0 & \frac{mL}{6} & 0 & \frac{mL}{3} \end{bmatrix} \begin{bmatrix} \ddot{\delta}_{ijx} \\ \ddot{\delta}_{ijy} \\ \ddot{\delta}_{jix} \\ \ddot{\delta}_{jiy} \end{bmatrix}$$

Plane Frame

Plane frames are more complex than plane trusses as their elements deform in flexure as well axially. In keeping with other aspects of the stiffness method the same techniques that were used for plane trusses can be used for plane frames. This will be illustrated by using the principle of virtual work to find the mass coefficient m_{23}. Figure 11.4 shows the freedoms for a plane frame element and the force system associated with unit acceleration at freedom 3 (no acceleration at any other freedom). The transverse acceleration at any point on the element is assumed to be described by the same shape function as was used to describe transverse displacements when unit displacement was imposed in the direction of freedom 3 - i.e. this is a consistent mass formulation.

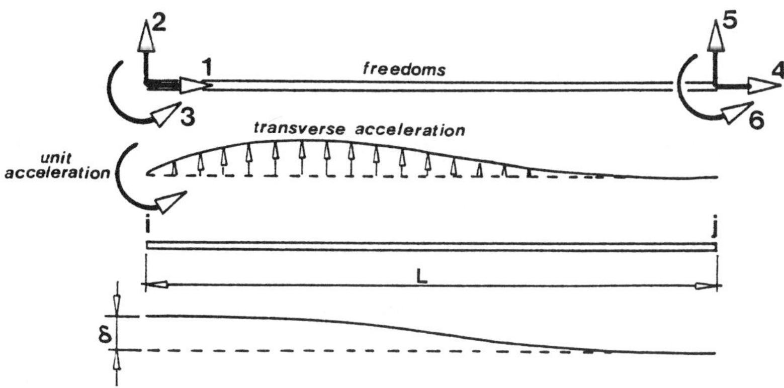

Figure 11.4 Unit acceleration at freedom 3 for a plane frame element

Apply the virtual displacement "δ" to the equilibrium force system shown in fig 11.4 Note that the virtual displacement is made in the direction of freedom 2 which means that the mass coefficient m_{23} is the only mass coefficient to do work during the virtual displacement. The transverse virtual displacement at any point on the element is described by the same shape function as was used to describe transverse displacements when unit displacement was imposed in the direction of freedom 2.

When a virtual displacement "δ" is applied in the direction of freedom 2 there is no change in the strain energy of the system as there is no stress in the element during the virtual displacement. Work will be done, however, by the external forces and internal inertia forces and the virtual work equation is

work done by the external forces + work done by the inertia forces = 0

hence

$$m_{23}\,\delta + \int_0^L \textit{inertia force x translation due to } \delta = 0$$

As a consistent mass formulation is being employed the inertia force acting on a differential element dx is $-m\,\phi_3(x)$, where m is the mass per unit length and $\phi_3(x)$ is the shape function associated with unit displacement in the direction of freedom 3. The negative sign arises because the inertial forces oppose acceleration.

The distance moved by the differential element during the virtual displacement is $\delta\,\phi_2(x)$, where "δ" is the virtual displacement and $\phi_2(x)$ is the shape function associated with unit displacement in the direction of freedom 2 (see fig 11.4). The equation of virtual work can, therefore, be rewritten as

$$m_{23}\,\delta = \int_0^L m\,\phi_2(x)\,\phi_3(x)\,\delta\,dx \qquad (11.18)$$

i.e.

$$m_{23} = m \int_0^L \phi_2(x)\, \phi_3(x)\, dx$$

In Chapter 2 the shape function for unit rotation at the left-hand end of line element was shown to be

$$\phi_3(x) = \frac{x^3}{L^2} - \frac{2x^2}{L} + x \qquad (11.19)$$

Chapter 2 also showed that the shape function for unit sway at the left-hand end of line element is

$$\phi_2(x) = \frac{2x^3}{L^3} - \frac{3x^2}{L^2} + 1 \qquad (11.20)$$

$$m_{23}\, \delta = m \int_0^L \phi_1(x)\, \phi_3(x)\, \delta\, dx$$

Substituting equation (11.19) and (11.20) into equation (11.18) gives

$$= m \int_0^L \left[\frac{2x^3}{L^3} - \frac{3x^2}{L^2} + 1\right] \left[\frac{x^3}{L^2} - \frac{2x^2}{L} + x\right] dx$$

and evaluation of this inetegral yields

$$m_{23} = \frac{11}{210}\, m\, L^2$$

Where "m" is the mass per unit length of element. Treating other freedoms similarly yields the element mass matrix which relates nodal forces and accelerations as shown in the following equation.

$$\begin{bmatrix} f_{ijx} \\ f_{ijy} \\ m_{ij} \\ f_{jix} \\ f_{jiy} \\ m_{ji} \end{bmatrix} = \frac{mL}{420} \begin{bmatrix} 140 & 0 & 0 & 70 & 0 & 0 \\ 0 & 156 & 22L & 0 & 54 & -13L \\ 0 & 22L & 4L^2 & 0 & 13L & -3L^2 \\ 70 & 0 & 0 & 140 & 0 & 0 \\ 0 & 54 & 13L & 0 & 156 & -22L \\ 0 & -13L & -3L^2 & 0 & -22L & 4L^2 \end{bmatrix} \begin{bmatrix} \ddot{\delta}_{ijx} \\ \ddot{\delta}_{ijy} \\ \ddot{\theta}_{ij} \\ \ddot{\delta}_{jix} \\ \ddot{\delta}_{jiy} \\ \ddot{\theta}_{ji} \end{bmatrix}$$

Example 11.4

Calculate the natural frequencies and mode shapes for the beam shown in the following diagram using a consistent mass formulation. To reduce the problem size axial rigidity may be assumed.

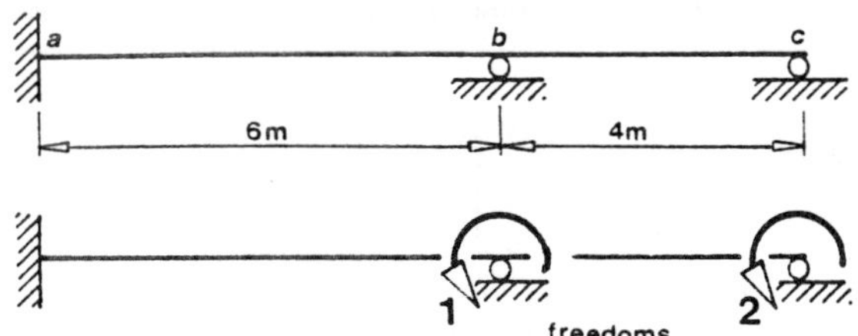

Stiffness and Mass Coefficients

Freedom 1

$$k_{11} = 4EI/L_{ab} + 4EI/L_{bc} = 416700 \text{ Nm/rad}$$

$$k_{21} = 2EI/L_{bc} = 125000 \text{ Nm/rad}$$

$$m_{11} = 4mL_{ab}^3/420 + 4mL_{bc}^3/420 = 1333 \text{ Nms}^2\text{/rad}$$

$$m_{21} = -3mL_{bc}^3/420 = -228.6 \text{ Nms}^2\text{/rad}$$

Freedom 2

$$k_{22} = 4EI/L_{bc} = 250000 \text{ Nm/rad}$$

$$m_{22} = 4mL_{bc}^3/420 = 304.7 \text{ Nms}^2\text{/rad}$$

Natural Frequencies

Find the values of ω^2 for which

$$|K - \omega^2 M| = 0$$

i.e.

$$\left| \begin{bmatrix} 416700 & 125000 \\ 125000 & 250000 \end{bmatrix} - \omega^2 \begin{bmatrix} 1333 & -228.6 \\ -228.6 & 304.7 \end{bmatrix} \right| = 0$$

hence

$$(416700-1333\omega^2)(250000-304.7\omega^2) - (125000+228.6\omega^2)^2 = 0$$

$$\omega^4 - 1462\omega^2 + 250400 = 0$$

which yields

$$\omega_1 = 14.1 \text{ rad/s} \qquad \text{and} \qquad \omega_2 = 35.5 \text{ rad/s}$$

Mode Shapes

The free undamped equation of motion is (for mode shape "*i*") is

$$[K - \omega_i^2 M]\,\Delta^i = \mathbf{0}$$

Substitute for ω_i^2 using the lowest natural frequency (the *fundamental frequency*, ω_1)

$$\begin{bmatrix} 152800 & 170300 \\ 170300 & 189700 \end{bmatrix} \begin{bmatrix} \Delta_1^1 \\ \Delta_2^1 \end{bmatrix} = \begin{bmatrix} 0 \\ 0 \end{bmatrix}$$

These equations are singular and only relative values can be found. As there are only two equations it is easiest to solve directly rather than use the general method described earlier. If Δ_1^1 is set to unity then Δ_2^1 is -0.9 and the normalised mode shape vector is

$$\phi_i = \begin{bmatrix} 1.0 \\ -0.9 \end{bmatrix}$$

The mode shape corresponding to the second natural frequency is found in exactly the same manner and can easily be shown to be

$$\phi_i = \begin{bmatrix} 0.327 \\ 1.0 \end{bmatrix}$$

The reader should sketch these mode shapes to see if they make physical sense.

11.5 The Eigenvalue Problem

From the preceding sections it is evident that finding mode shapes and natural frequencies for structural frameworks is a very computationally intensive procedure. Thus far the eigenvalues have been found by expanding the frequency equation to obtain the characteristic equation and then finding the roots of that equation. Expansion of the determinant in the frequency equation is fine when there is only a very small number of freedoms. More efficient methods are required for the solution of practical problems.

In the field of structural dynamics eigenvalues are routinely found by a number of commonly used algorithms. The optimum algorithm for any particular problem depends upon the problem size, whether an "in memory" solution is possible, the bandwidth of the matrices and the number of eigenpairs required. Eigenvalues are important in many fields of science and engineering and the numerical solution of eigenvalue problems is a wide field with a great deal of published literature. For further reading the reader is referred to Wilkinson's definitive work[28]. A description of the most efficient methods currently used for the solution of the frequency equation was given in a paper by Bathe and Wilson[2], and the main findings of that paper are summarised here.

Householder-QR-Inverse Iteration

This is a very efficient procedure that is generally regarded as being the best method for finding the complete eigensystems. Before this method can be applied, the frequency equation must be transformed into the standard form shown in equation (11.21)

$$\tilde{K}\,\phi = \omega^2\,\phi \tag{11.21}$$

This transformation can only be carried out if the mass matrix M is positive definite. In a lumped mass formulation where zero elements appear on the diagonal the mass matrix is made positive definite by using static condensation to eliminate the massless degrees of freedom.

Generalised Jacobi Iteration

The generalised Jacobi iteration method finds directly all of the eigenvectors without having to transform the problem to the standard form or having to condense out any massless degrees of freedom. It is well suited to relatively small problems that can be solved in memory. When the off-diagonal elements are small or sparse this method can be very efficient. The algorithm is very elegant and requires relatively little coding.

Determinant Search

When solving medium to large eigenvalue problems the number of required eigenpairs is usually much smaller than the order of the matrices. In this case it is more economical to find only the required eigensystem. If the bandwidth of the matrices is small a determinant search based upon a combination of $U\,D\,U^T$ factorisation and vector inverse iteration provides a very efficient solution. There is no need to reduce the problem to the standard form shown in equation (11.21), nor are special measures needed to deal with massless degrees of freedom. The required eigenpairs are found in succession.

Subspace Iteration

This method calculates very efficiently the lowest eigenpairs of systems too large to be stored in fast access memory. If "k" eigenpairs are required the alogorithm iterates simultaneously with "l" linearly independent vectors, where "l" is greater than "k". The convergence rate is very much governed by the initial choice of the iteration vectors. This method is commonly used in general finite element programs as it allows large systems to be solved on relatively small computers.

Program JACOBI.BAS

This program is based upon the generalised Jacobi method described earlier in this section. Only an overview of the method is given here. More detail and further references can be found in reference[2].

Comments on the Algorithm

Starting with $K_1 = K$ and $M_1 = M$ iterate using a rotation matrix P_k as follows

$$K_{k+1} = P_k^T K_k P_k, \qquad M_{k+1} = P_k^T M_k P_k$$

where

$$P_k = \begin{bmatrix} 1 & & & & \\ & \cdot & & & \\ & & 1 & \alpha & \\ & & \gamma & 1 & \\ & & & & \cdot \\ & & & & & 1 \end{bmatrix} \begin{matrix} \\ \\ i \\ j \\ \\ \\ \end{matrix}$$

(columns i, j; rows i, j)

The variables "α" and "γ" are selected to zero simultaneously the (i,j) and (j,i) elements in K_k and M_k. If M is positive definite then

$$K_k \rightarrow \text{diag}(K_r) \quad \text{and} \quad M_k \rightarrow \text{diag}(M_r) \quad \text{as } k \rightarrow \infty$$

The required eigenvalues are

$$\Omega^2 = \text{diag}\left\{\frac{K_r}{M_r}\right\}$$

and if convergence was achieved after "n" iterations then the eigenvectors are

$$\Pi = P_1 P_2 P_3 \dots P_n \,\text{diag}(M_r^{-½})$$

If M contains zero elements on the diagonal then so too will diag(M_r) and the algorithm should ignore these values when calculating eigenvalues and eigenvectors.

The efficiency of the method is improved if the algorithm is implemented as a threshold iteration in which rotation is applied only either of coupling factors

$$\frac{k_{ij}^{(k)2}}{k_{ii}^{(k)2}\, k_{jj}^{(k)2}} \quad \text{and} \quad \frac{m_{ij}^{(k)2}}{m_{ii}^{(k)2}\, m_{jj}^{(k)2}}$$

are bigger than the current threshold. This refinement has be incorporated in the program JACOBI.BAS.

Menu System

JACOBI.BAS has a flexible user interface which is driven by a menu system that offers the following choices

(1) Input new data from the keyboard
(2) Read new data from a file
(3) Edit banner
(4) Edit the stiffness matrix
(5) Edit the mass matrix
(6) List the matrices
(7) Solve the current problem
(8) Write the data to disk
(9) Stop

Listing

```
1000 '========================    JACOBI.BAS    ============================
1010 '
1020 OPTION BASE 1
1030 COMMON PTYPE$, INPUTFILE$, PROGNAME$
1040 TRUE% = 1 : FALSE% = 0 : OK% = 1 : FAIL% = 0 : PROGNAME$ = "SFRAME.BAS"
1050 CLS
1060 PRINT
1070 PRINT "==================================================================="
1080 PRINT
1090 PRINT "PROGRAM JACOBI.BAS                     Copyright (c) J.Balfour 1991"
1100 PRINT
1110 PRINT "      Program for the calculation of the all of the natural"
1120 PRINT "      frequecies and mode shapes for a plane frame structure"
1130 PRINT "                  using Jacobi iteration"
1140 PRINT
1150 PRINT "                For further information contact"
1160 PRINT " James A.D.Balfour, Heriot-Watt University, Riccarton, Edinburgh"
1170 PRINT "               Tel 031-449-5111, Fax 031-451-3170"
1180 PRINT
1190 PRINT "==================================================================="
1200 '
1210 MAXNDOF%  = 40                      '... Max number of degrees of freedoms
1220 PRINT : PRINT
1230 PRINT "Maximum number of freedoms          =  "; MAXNDOF%
1240 '
1250 DIM K1(MAXNDOF%,MAXNDOF%),  M1(MAXNDOF%,MAXNDOF%)
1260 DIM K2(MAXNDOF%,MAXNDOF%),  M2(MAXNDOF%,MAXNDOF%)
1270 DIM PHI(MAXNDOF%,MAXNDOF%), THETA(MAXNDOF%,2)
1280 GOSUB 4730                          '... Any key to continue
1290 WHILE REPLY% <> 9
1300   CLS : PRINT :  PRINT
1310   PRINT "Do you wish to:"
1320   PRINT "   (1)  Input data from keyboard"
1330   PRINT "   (2)  Read data from disk"
1340   PRINT "   (3)  Edit banner
1350   PRINT "   (4)  Edit the stiffness matrix"
1360   PRINT "   (5)  Edit the mass matrix
1370   PRINT "   (6)  List the matrices
1380   PRINT "   (7)  Solve the current problem"
1390   PRINT "   (8)  Write data to disk"
1400   PRINT "   (9)  Stop"
1410   NOPTIONS% = 9 : GOSUB 4590
1420   ON REPLY% GOTO 1430, 1470, 1480, 1490, 1500, 1510, 1540, 1550, 1560
1430   PRINT
1440   INPUT "Number of freedoms = ", NDOF%
1450   STIFFMAT% = TRUE%  : GOSUB 1970
1460   STIFFMAT% = FALSE% : GOSUB 1970 : GOTO 1560 '... Input keyboard data
1470   GOSUB 1660 : GOTO 1560                      '... Read data file
```

```
1480   GOSUB 2250 : GOTO 1560                        '... Edit the banner
1490   STIFFMAT% = TRUE%  : GOSUB 2340 : GOTO 1560 '... Edit stiffness matrix
1500   STIFFMAT% = FALSE% : GOSUB 2340 : GOTO 1560 '... Edit mass matrix
1510   STIFFMAT% = TRUE%  : GOSUB 2570 : GOSUB 4730'... List the matrices
1520   STIFFMAT% = FALSE% : GOSUB 2570 : GOSUB 4730
1530   GOTO 1560
1540   GOSUB 3010 : GOSUB 4180 : GOTO 1560        ' Solve problem
1550   GOSUB 2750 : GOTO 1560                      ' Write data to disk
1560 WEND
1570 END
1580 '
1590 '***********************    GENERAL ERROR TRAP    ***********************
1600 '
1610 PRINT
1620 PRINT "** FATAL ERROR **"
1630 PRINT
1640 ON ERROR GOTO 0
1650 '
1660 '*******************    READ DATA FROM DISK FILE    *********************
1670 '
1680 PRINT
1690 INPUT "Name of data file  =  ", INPUTFILE$
1700 ON ERROR GOTO 1720 : OPEN INPUTFILE$ FOR INPUT AS #1
1710 ON ERROR GOTO 1590 : GOTO 1740
1720 CLS : PRINT : PRINT "** ERROR ** Failed to open file" : INPUTFILE$ = ""
1730 ON ERROR GOTO 1890 : GOSUB 4730 : RESUME 1940
1740 LINE INPUT#1, TEMP$
1750 IF TEMP$ = "CASF eigenproblem data" THEN GOTO 1780
1760 PRINT
1770 PRINT "Not a data file for this program" : GOSUB 4730 : GOTO 1940
1780 PRINT
1790 PRINT "Reading data from file "; INPUTFILE$
1800 LINE INPUT#1, BANNER$
1810 INPUT#1, NDOF%
1820 LINE INPUT#1, TEMP$
1830 FOR I% = 1 TO NDOF%
1840   FOR J% = 1 TO NDOF%
1850     INPUT#1, K1(I%,J%)
1860   NEXT J%
1870 NEXT I%
1880 LINE INPUT#1, TEMP$
1890 FOR I% = 1 TO NDOF%
1900   FOR J% = 1 TO NDOF%
1910     INPUT#1, M1(I%,J%)
1920   NEXT J%
1930 NEXT I%
1940 CLOSE#1
1950 RETURN
1960 '
1970 '*************    INPUT THE MASS AND STIFFNESS MATRIX    ****************
1980 '
1990 CLS : PRINT
2000 IF STIFFMAT% = FALSE% THEN GOTO 2090
2010 PRINT "Enter an identification banner"
2020 PRINT
2030 INPUT "Banner = ", BANNER$
2040 CLS: PRINT
2050 PRINT "Enter the stiffness matrix"
2060 PRINT
2070 INPUT "Semi-bandwidth - stiffness matrix = ", BAND%
2080 GOTO 2120
2090 PRINT "Enter the mass matrix"
2100 PRINT
2110 INPUT "Semi-bandwidth - mass matrix = ", BAND%
2120 PRINT
2130 FOR I% = 1 TO NDOF%
2140   FOR J% = I% TO NDOF%
2150     PRINT USING "(##_,##) = "; I%, J%; :
2160     IF STIFFMAT% = FALSE% THEN GOTO 2190
2170     INPUT "", K1(I%, J%): K1(J%, I%) = K1(I%, J%)
2180     GOTO 2200
2190     INPUT "", M1(I%,J%): M1(J%,I%) = M1(I%,J%)
2200     IF J% >= I%+BAND%-1 THEN J% = NDOF%
2210   NEXT J%
2220 NEXT I%
2230 RETURN
2240 '
```

```
2250 '**********************    EDIT THE BANNER    ***************************
2260 '
2270 PRINT "Edit banner - <return> to retain"
2280 PRINT
2290 PRINT "Current banner = "; BANNER$
2300 INPUT "New banner     = ", TEXT$
2310 IF LEN(TEXT$) > 0 THEN BANNER$ = TEXT$
2320 RETURN
2330 '
2340 '******************    EDIT STIFFNESS/MASS MATRIX    *******************
2350 '
2360 DONE% = FALSE%
2370 WHILE DONE% = FALSE%
2380   GOSUB 2570
2390   PRINT
2400   INPUT "Row (0 to stop) = ", I%
2410   IF (I%>=1) AND (I%<=NDOF%) AND (J%>=1) AND (J%<=NDOF%) THEN GOTO 2520
2420   IF I% = 0 THEN DONE% = TRUE% : GOTO 2540
2430   IF (I%<1) OR (I%>NDOF%) THEN GOTO 2480
2440   INPUT "Column          = ", J%
2450   IF (J%<1) OR (J%>NDOF%) THEN GOTO 2480
2460   INPUT "Revised value   = ", TEMP
2470   GOTO 2520
2480   PRINT "Invalid location"
2490   IF CSRLIN<=20 THEN LOCATE 20, 1
2500   PRINT : PRINT SPACE$(40); "Press any key to continue"
2510   WHILE INKEY$ = "" : WEND : GOTO 2540
2520   IF STIFFMAT% =  TRUE% THEN K1(I%,J%) = TEMP : K1(J%,I%) = TEMP
2530   IF STIFFMAT% <> TRUE% THEN M1(I%,J%) = TEMP : M1(J%,I%) = TEMP
2540 WEND
2550 RETURN
2560 '
2570 '******************    PRINT STIFFNESS/MASS MATRIX    ********************
2580 '
2590 CLS : PRINT
2600 IF STIFFMAT% = TRUE%  THEN PRINT "STIFFNESS MATRIX"
2610 IF STIFFMAT% = FALSE% THEN PRINT "MASS MATRIX"
2620 PRINT
2630 FOR I% = 1 TO NDOF%
2640   PRINT USING "Row ###"; I%;
2650   FOR J% = 1 TO NDOF%
2660     IF STIFFMAT% = TRUE%  THEN PRINT USING "    #.####^^^^"; K1(I%,J%);
2670     IF STIFFMAT% = FALSE% THEN PRINT USING "    #.####^^^^"; M1(I%,J%);
2680     IF (J% = NDOF%) AND (J% MOD 5 =  0) THEN PRINT
2690     IF (J% = NDOF%) AND (J% MOD 5 <> 0) THEN PRINT : PRINT
2700     IF (J% MOD 5 = 0) THEN PRINT
2710   NEXT J%
2720 NEXT I%
2730 RETURN
2740 '
2750 '*********************    WRITE DATA TO DISK    ************************
2760 '
2770 PRINT : INPUT "Name of data file  =  ", OUTPUTFILE$
2780 ON ERROR GOTO 1720 : OPEN OUTPUTFILE$ FOR OUTPUT AS #1
2790 ON ERROR GOTO 1590 : GOTO 2820
2800 CLS : PRINT : PRINT "** ERROR ** Failed to open output file"
2810 ON ERROR GOTO 1890 : GOSUB 4730 : RESUME 2970
2820 PRINT#1, "CASF eigenproblem data"
2830 PRINT#1, BANNER$
2840 PRINT#1, NDOF%
2850 PRINT#1,
2860 FOR I% = 1 TO NDOF%
2870   FOR J% = 1 TO NDOF%
2880     PRINT#1, K1(I%,J%)
2890   NEXT J%
2900 NEXT I%
2910 PRINT#1,
2920 FOR I% = 1 TO NDOF%
2930   FOR J% = 1 TO NDOF%
2940     PRINT#1, M1(I%,J%)
2950   NEXT J%
2960 NEXT I%
2970 CLOSE#1
2980 PRINT "Data written to file "; OUTPUTFILE$
2990 RETURN
3000 '
3010 '*****************************    SOLVE    ********************************
```

```
3020 '
3030 NN% = 0 : TOL=1000000! : ABSTOL = 1E-08 : DONE% = FALSE% : PRINT
3040 PRINT : PRINT USING "Tolerance #.####^^^^      Sweep "; TOL;
3050 ROW%  = CSRLIN : COL%   = POS(0) : PRINT NN%;
3060 '
3070 ' ... Copy [K1] to [K2], [M1] to [M2] and set [PHI] to the idenity matrix
3080 '
3090 FOR I% = 1 TO NDOF%
3100   FOR J% = 1 TO NDOF%
3110     K2(I%,J%) = K1(I%,J%) : M2(I%,J%) = M1(I%,J%)
3120     IF I% = J% THEN PHI(I%,J%) = 1 ELSE PHI(I%,J%) = 0
3130   NEXT J%
3140 NEXT I%
3150 WHILE DONE% = FALSE%
3160   NN% = NN%+1 : LOCATE ROW%, COL% : PRINT NN%;
3170   DONE% = TRUE%
3180   FOR I% = 1 TO NDOF%
3190     FOR J% = I%+1 TO NDOF%
3200     IF (ABS(K2(I%,J%)) < TOL) AND (ABS(M2(I%,J%)) < TOL) THEN GOTO 3240
3210       DONE = FALSE%
3220       GOSUB 3340                              '...  Calc alpha and beta
3230       GOSUB 3450                              '...  Transform matrices
3240     NEXT J%
3250   NEXT I%
3260   IF DONE% = FALSE% THEN GOTO 3300
3270   IF TOL   < ABSTOL THEN GOTO 3300
3280   TOL = TOL*.001 : DONE% = FALSE%
3290   LOCATE ROW%, 11 : PRINT USING "#.####^^^^"; TOL;
3300 WEND
3310 GOSUB 3840                                    '...  Calc eigenvalues
3320 RETURN
3330 '
3340 '*************************    ALPHA & BETA    ***************************
3350 '
3360 KKII  = K2(I%,I%)*M2(I%,J%) - M2(I%,I%)*K2(I%,J%)
3370 KKJJ  = K2(J%,J%)*M2(I%,J%) - M2(J%,J%)*K2(I%,J%)
3380 KK    = K2(I%,I%)*M2(J%,J%) - K2(J%,J%)*M2(I%,I%)
3390 IF(KK > 0) THEN MULT=1 ELSE MULT=-1
3400 X     = KK/2 + MULT*SQR(KK*KK/4 + KKII*KKJJ)
3410 ALPHA =  KKJJ/X
3420 BETA  = -KKII/X
3430 RETURN
3440 '
3450 '***********************    TRANSFORM MATRICES    ***********************
3460 '
3470 ' ...  Transform the stiffness matrix
3480 '
3490 FOR K% = 1 TO NDOF%
3500   TEMP      = K2(I%,K%)
3510   K2(I%,K%) = TEMP + BETA*K2(J%,K%)
3520   K2(J%,K%) = ALPHA*TEMP + K2(J%,K%)
3530 NEXT K%
3540 '
3550 FOR K% = 1 TO NDOF%
3560   TEMP      = K2(K%,I%)
3570   K2(K%,I%) = TEMP + BETA*K2(K%,J%)
3580   K2(K%,J%) = ALPHA*TEMP + K2(K%,J%)
3590 NEXT K%
3600 '
3610 ' ...  Transform the mass matrix
3620 '
3630 FOR K% = 1 TO NDOF%
3640   TEMP      = M2(I%,K%)
3650   M2(I%,K%) = TEMP + BETA*M2(J%,K%)
3660   M2(J%,K%) = ALPHA*TEMP + M2(J%,K%)
3670 NEXT K%
3680 '
3690 FOR K% = 1 TO NDOF%
3700   TEMP      = M2(K%,I%)
3710   M2(K%,I%) = TEMP + BETA*M2(K%,J%)
3720   M2(K%,J%) = ALPHA*TEMP + M2(K%,J%)
3730 NEXT K%
3740 '
3750 ' ...  Transform PHI
3760 '
3770 FOR K% = 1 TO NDOF%
3780   TEMP      =  PHI(K%,I%)
```

```
3790   PHI(K%,I%)=  TEMP+BETA*PHI(K%,J%)
3800   PHI(K%,J%)= ALPHA*TEMP+PHI(K%,J%)
3810 NEXT K%
3820 RETURN
3830 '
3840 '**********************    CALCULATE EIGENVALUES    *********************
3850 '
3860 CLS : PRINT
3870 FOR K% = 1 TO NDOF%
3880   THETA(K%,1) = SQR(ABS(K2(K%,K%)/M2(K%,K%)))
3890   THETA(K%,2) = K%
3900 NEXT K%
3910 '
3920 ' ...  Normalise eigenvectors
3930 '
3940 FOR K% = 1 TO NDOF%
3950   TEMP = 0
3960   FOR L% = 1 TO NDOF%
3970     IF ABS(PHI(L%,K%))>ABS(TEMP) THEN TEMP=PHI(L%,K%)
3980   NEXT L%
3990   FOR L% = 1 TO NDOF%
4000     PHI(L%,K%) = PHI(L%,K%)/TEMP
4010   NEXT L%
4020 NEXT K%
4030 '
4040 ' ... Sort eigenvalues into order
4050 '
4060 DONE% = FALSE%
4070 WHILE DONE% = FALSE%
4080   DONE% = TRUE%
4090   FOR K% = 1 TO NDOF%-1
4100     IF THETA(K%,1)<=THETA(K%+1,1) THEN GOTO 4140
4110     DONE% = FALSE%
4120     TEMP  =THETA(K%,1) : THETA(K%,1)=THETA(K%+1,1) : THETA(K%+1,1)=TEMP
4130     TEMP  =THETA(K%,2) : THETA(K%,2)=THETA(K%+1,2) : THETA(K%+1,2)=TEMP
4140   NEXT K%
4150 WEND
4160 RETURN
4170 '
4180 '********************    OUTPUT THE EIGENPAIRS    **********************
4190 '
4200 IF NDOF% > 1 THEN GOTO 4230
4210 CLS : PRINT
4220 PRINT "No problem loaded" : GOSUB 4730 : GOTO 4570
4230 CLS : PRINT
4240 PRINT "* * * * * * * * * * * * *"
4250 PRINT "*   EIGENVALUES      *    "; BANNER$
4260 PRINT "* * * * * * * * * * * * *"
4270 L% = 0 : K% = 1
4280 WHILE L% < NDOF%
4290   IF K%+4 > NDOF% THEN L% = NDOF% ELSE L% = K%+4
4300   PRINT
4310   PRINT "NUMBER              ";
4320   FOR I% = K% TO L%
4330     PRINT USING "     (##)    "; I%;
4340   NEXT I%
4350   PRINT
4360   PRINT "FREQUENCY (rad/s)";
4370   FOR I% = K% TO L%
4380     PRINT USING "  #.####^^^^"; THETA(I%,1);
4390   NEXT I%
4400   PRINT
4410   PRINT "FREQUENCY^2       ";
4420   FOR I% = K% TO L%
4430     PRINT USING "  #.####^^^^"; THETA(I%,1)^2;
4440   NEXT I%
4450   PRINT : PRINT
4460   PRINT "FREEDOM           "
4470   FOR I% = 1 TO NDOF%
4480   PRINT USING "  ##              ";  I%;
4490     FOR J% = K% TO L%
4500       PRINT USING "  #.####^^^^"; PHI(I%,THETA(J%,2));
4510     NEXT J%
4520     PRINT
```

```
4530   NEXT I%
4540   GOSUB 4730
4550   K% = L%+1
4560 WEND
4570 RETURN
4580 '
4590 ' *******************    GET KEYBOARD RESPONSE    *********************
4600 '
4610 ROW% = CSRLIN: COL% = POS(0)
4620 REPLY% = 0
4630 WHILE REPLY% = 0
4640   LOCATE ROW%, COL%: INPUT "", WRKSTR$: REPLY% = VAL(WRKSTR$)
4650   IF (REPLY% < 1) OR (REPLY% > NOPTIONS%) THEN GOTO 4680
4660   IF (REPLY% < 10) AND (LEN(WRKSTR$) = 1) THEN GOTO 4690
4670   IF (REPLY% > 9)  AND (LEN(WRKSTR$) = 2) THEN GOTO 4690
4680   SOUND 700, 4: REPLY% = 0: LOCATE ROW%, COL%
4690 WEND
4700 PRINT
4710 RETURN
4720 '
4730 '********************    ANY KEY TO CONTINUE    **********************
4740 '
4750 IF CSRLIN<=23 THEN LOCATE 23, 42
4760 PRINT "Press any key to continue"
4770 WHILE INKEY$ = "" : WEND
4780 RETURN
4790 '
4800 '
4810 '=========================    JACOBI.BAS    ============================
```

Sample Run

The following sample run shows the program JACOBI.BAS being used to solve the eigenvalue problem from example 11.3. Input from the keyboard is underlined.

```
=================================================================

PROGRAM JACOBI.BAS                         Copyright (c) J.Balfour 1991

      Program for the calculation of the all of the natural
      frequecies and mode shapes for a plane frame structure
                    using Jacobi iteration

                For further information contact
 James A.D.Balfour, Heriot-Watt University, Riccarton, Edinburgh
           Tel 031-449-5111, Fax 031-451-3170

=================================================================


Maximum number of freedoms          =    40

Do you wish to:
  (1)  Input data from keyboard
  (2)  Read data from disk
  (3)  Edit banner
  (4)  Edit the stiffness matrix
  (5)  Edit the mass matrix
  (6)  List the matrices
  (7)  Solve the current problem
  (8)  Write data to disk
  (9)  Stop
1

Number of freedoms = 3

Enter an identification banner

Banner = CASF 2nd Edition - Example 11.3
```

```
Enter the stiffness matrix

Semi-bandwidth - stiffness matrix = 2

( 1, 1) = 22.2
( 1, 2) = -22.2E9
( 2, 2) = 31.58E9
( 2, 3) = -9.38E9
( 3, 3) = 14.18E9

Enter the mass matrix

Semi-bandwidth - mass matrix = 1

( 1, 1) = 20000
( 2, 2) = 20000
( 3, 3) = 20000

Do you wish to:
  (1)  Input data from keyboard
  (2)  Read data from disk
  (3)  Edit banner
  (4)  Edit the stiffness matrix
  (5)  Edit the mass matrix
  (6)  List the matrices
  (7)  Solve the current problem
  (8)  Write data to disk
  (9)  Stop
4

STIFFNESS MATRIX

Row   1     0.2220E+02    -.2220E+11    0.0000E+00

Row   2     -.2220E+11    0.3158E+11    -.9380E+10

Row   3     0.0000E+00    -.9380E+10    0.1418E+11

Row (0 to stop) = 1
Column          = 1
Revised value   = 22.2E9

STIFFNESS MATRIX

Row   1     0.2220E+11    -.2220E+11    0.0000E+00

Row   2     -.2220E+11    0.3158E+11    -.9380E+10

Row   3     0.0000E+00    -.9380E+10    0.1418E+11

Row (0 to stop) = 0

Do you wish to:
  (1)  Input data from keyboard
  (2)  Read data from disk
  (3)  Edit banner
  (4)  Edit the stiffness matrix
  (5)  Edit the mass matrix
  (6)  List the matrices
  (7)  Solve the current problem
  (8)  Write data to disk
  (9)  Stop
7

* * * * * * * * * * * * *
*   EIGENVALUES      *     CASF 2nd Edition - Example 11.3
* * * * * * * * * * * * *

NUMBER                   ( 1)          ( 2)          ( 3)
FREQUENCY (rad/s)   0.2500E+03    0.8849E+03    0.1598E+04
FREQUENCY^2        (0.6251E+05)  (0.7830E+06)  (0.2552E+07)

FREEDOM
   1                0.1000E+01    -.5357E+00    -.7695E+00
   2                0.9437E+00    -.1578E+00    0.1000E+01
   3                0.6846E+00    0.1000E+01    -.2544E+00
```

```
                                     3 eigenvalues output

Do you wish to:
   (1)  Input data from keyboard
   (2)  Read data from disk
   (3)  Edit banner
   (4)  Edit the stiffness matrix
   (5)  Edit the mass matrix
   (6)  List the matrices
   (7)  Solve the current problem
   (8)  Write data to disk
   (9)  Stop
9
```

Program DSEARCH.BAS

This program uses determinant search to find the first *"n"* eigenpairs of the frequency equation where *"n"* is defined by the user.

The basic concept of determinant search is extremely simple; the complexity of the final algorithm arises from the need to satisfy the conflicting requirements that the algorithm should be both robust and efficient. The eigenvalues of the frequency equation

$$|K - \omega^2 M| = 0$$

are the roots λ_1, λ_2, λ_3 .... λ_n of the characteristic polynomial obtained by expanding the above determinant.

$$p(\mu) = |K - \mu M|$$

where

$$\lambda_1 = \omega_1^2 \qquad \text{(i.e. } |K - \lambda_1 M| = 0$$

$$\lambda_2 = \omega_2^2 \qquad \text{(i.e. } |K - \lambda_2 M| = 0$$

$$\vdots$$

$$\lambda_n = \omega_n^2 \qquad \text{(i.e. } |K - \lambda_n M| = 0$$

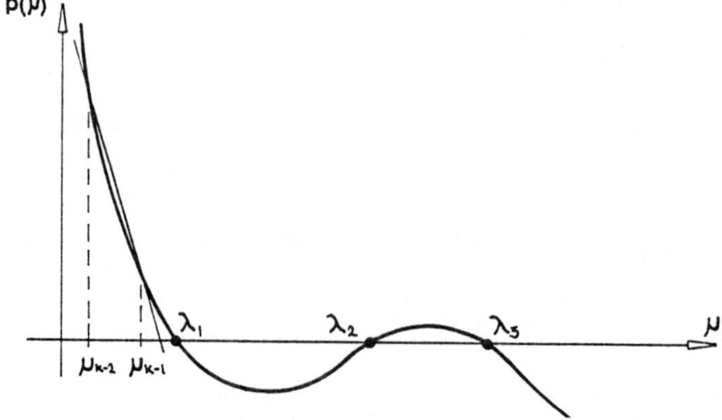

Figure 11.5 Characteristic polynomial

If sketched, the characteristic polynomial might look similar to the function shown in fig 11.5. The polynomial can be evalutated at μ using $\boldsymbol{U D U^T}$ factorisation of $\boldsymbol{K} - \mu \boldsymbol{M}$.

$$\boldsymbol{K} - \mu \boldsymbol{M} = \boldsymbol{U D U^T} \tag{11.22}$$

$p(\mu)$ is simply the product of the diagonal elements of $\boldsymbol{D}$. There are many numerical methods to calculate the roots of a polynomial from a knowledge of $p(\mu)$. The algorithm used in DSEARCH.BAS is the secant method. In essence, if $p(\mu)$ is known for two different values of "μ", a better estimate of the root can be obtained by projecting a straight line through the known points onto the "μ" axis as shown in fig 11.5. Fortunately the characteristic polynomials obtained from the leading principal minors of $\boldsymbol{K} - \mu \boldsymbol{M}$ form a Sturm sequence[3,28] from which it follows that in equation 11.22 the number of negative diagonal elements in $\boldsymbol{D}$ is equal to the number of eigenvalues smaller than "μ". This is extremely important, as it is the ability to check, at each factorisation, how many eigenvalues are to the left of "μ" that makes the algorithm viable.

Unfortunately the rate of convergence of the standard secant method can prove slow in the case of multiple roots. This, coupled with the fact that triangular factorisation is a computationally intensive procedure, suggests that the secant method may not be the most appropriate method once a root is approached.

Vector inverse iteration is generally recognised to be an efficient way to obtain an eigenvector if "μ" is close to a root. The paper by Bathe and Wilson[3] indicates that for a semi-bandwidth of 30 the labour associated with one triangular factorisation is equivalent to six cycles of vector inverse iteration. In view of this, the determinant search algorithm employs accelerated secant iteration to jump over a root (or a cluster of roots) followed by vector inverse iteration to find the eigenpairs jumped. The number of roots jumped is detected by the Sturm sequence count, and the algorithm ensures that the jump does not take "μ" very far to the right of the root(s). The accelerated secant method is based upon the algorithm shown in the following equation.

$$\mu_k = \mu_{k-1} - \eta \frac{p(\mu_{k-1})(\mu_{k-1} - \mu_{k-2})}{p(\mu_{k-1}) - p(\mu_{k-2})}$$

Where μ_{k-1} and μ_{k-2} are approximations to the eigenvalue λ_l and $\mu_{k-2} < \mu_{k-1} < \lambda_l$. "$\eta$" is initally set equal to 2 and is doubled each time successive iterates do not change in their first two most significant figures.

Once one or more roots have been jumped, vector inverse iteration is used until all of the roots to the left of "μ" have been found. Sometimes the iterative procedure will converge to a root to the right of "μ". This causes no problems and will generally hasten the solution of the problem.

To ensure that all roots are found in the case of multiple roots the iteration vector is orthogonalised with respect to the last six calculated eigenvectors using Gram-Schmidt orthogonalisation. Once all roots to the left of the current value of "μ" have been found the algorithm returns to accelerated secant method. The procedure to find the remaining roots is exactly the same as was used to find the first root (or roots), except that now the value of the polynomial must be deflated using all known roots according to the following formula.

$$p_i(\mu) = \frac{p(\mu)}{(\mu - \lambda_1)(\mu - \lambda_2) \; .... \; (\mu - \lambda_i)}$$

where λ_1, λ_2 .... λ_i have been found previously.

To start each secant iteration two bounds on λ_l are needed. The first bound is $\mu_1 = 0$ and the second bound is found by inverse iteration with "μ" equal to zero. If after "k" cycles of inverse iteration, the iteration vector is "$\boldsymbol{x}_k$", then the other bound is found as follows

$$\mu_2 = (1 - 0.01) \frac{\boldsymbol{x}_k^T \boldsymbol{K} \boldsymbol{x}_k}{\boldsymbol{x}_k^T \boldsymbol{M} \boldsymbol{x}_k}$$

This equation may produce a value of μ_2 that is greater λ_l. This situation is detected by the Sturm sequence count. Bathe and Wilson[3] suggest should this situation occur then μ_2 should be divided by $(\gamma + 1)$ until γ equals zero, where γ is the number of negative terms on the diagonal of $\boldsymbol{D}$. They also suggest that all terms in the initial trial vector for vector inverse iteration should be set to unity. It is the author's experience that this can, on rare occasions, cause problems and in the program listed below the initial trial vector is set up using random numbers between zero and unity.

In practice it is likely that $p(\mu)$ will be large enough to cause exponent overflow in the computer. This problem is avoided by storing $p(\mu)$ in the following form

$$p(\mu) = d \times 2^n$$

Where the absolute value of "d" is in the range 0.0625 to 1.0, and "n" is an integer.

Menu System

DSEARCH.BAS has the same user interface as JACOBI.BAS. The main menu offers the following choices

(1) Input new data from the keyboard
(2) Read new data from a file

(3) Edit the banner
(4) Edit the stiffness matrix
(5) Edit the mass matrix
(6) List the matrices
(7) Solve the current problem
(8) Write the data to disk
(9) Stop

Listing

```
1000 '========================    DSEARCH.BAS    ===========================
1010 '
1020 OPTION BASE 1
1030 COMMON PTYPE$, INPUTFILE$, PROGNAME$
1040 TRUE% = -1 : FALSE% = 0 : OK% = 1 : FAIL% = 0 : PROGNAME$ = "SFRAME.BAS"
1050 CLS
1060 PRINT
1070 PRINT "================================================================="
1080 PRINT
1090 PRINT "PROGRAM DSEARCH.BAS                    Copyright (c) J.Balfour 1991"
1100 PRINT
1110 PRINT "      Program for the calculation of the natural frequencies"
1120 PRINT "  and mode shapes from the structure stiffness and mass using the"
1130 PRINT "     method of determinate search and vector inverse iteration"
1140 PRINT
1150 PRINT "                For further information contact"
1160 PRINT " James A.D.Balfour, Heriot-Watt University, Riccarton, Edinburgh"
1170 PRINT "               Tel 031-449-5111, Fax 031-451-3170"
1180 PRINT
1190 PRINT "================================================================="
1200 PRINT
1210 '
1220 ' ... Max problem size set by the following dimension statements
1230 '
1240 MAXNONODES%   = 20                 ' Maximum number of nodes
1250 MAXBAND%      = 20                 ' Maximum semi-bandwidth
1260 MAXNDOF%      = 3*MAXNONODES% - 3  ' Maximum number of degrees of freedom
1270 MAXEVALUES%   = 20                 ' Maximum no of eigenvalues
1280 XXXX          = .00001             ' Set eigenvalue tolerance
1290 '
1300 ' ... Set up arrays
1310 '
1320 DIM K(MAXNDOF%,MAXBAND%)           ' Final Stiffness Matrix
1330 DIM M(MAXNDOF%,MAXBAND%)           ' Mass matrix
1340 DIM A(MAXNDOF%,MAXBAND%)           ' Stores K - Lamba*M
1350 DIM X(MAXNDOF%), Y(MAXNDOF%)       ' Workspace array
1360 DIM Y1(MAXNDOF%), Y2(MAXNDOF%)     ' Workspace arrays
1370 DIM EVALUE(MAXEVALUES%,2)          ' Array of eigenvalues
1380 DIM EVECTOR(MAXNDOF%, MAXEVALUES%)' Array of eigenvectors
1390 DIM W(MAXNDOF%, 6)                 ' Workspace array
1400 PRINT
1410 PRINT "Maximum number of equations              =  "; MAXNDOF%
1420 PRINT "Maximum semi-bandwidth - stiffness matrix =  "; MAXBAND%
1430 PRINT "Maximum semi-bandwidth - mass matrix      =  "; MAXBAND%
1440 PRINT
1450 GOSUB 7570
1460 WHILE REPLY% <> 9
1470   CLS : PRINT :  PRINT
1480   PRINT "Do you wish to:"
1490   PRINT "  (1)  Input new data from keyboard"
1500   PRINT "  (2)  Read new data from file"
1510   PRINT "  (3)  Edit the banner"
1520   PRINT "  (4)  Edit the stiffness matrix"
1530   PRINT "  (5)  Edit the mass matrix
1540   PRINT "  (6)  List the matrices
1550   PRINT "  (7)  Solve the current problem"
1560   PRINT "  (8)  Write data to disk"
1570   PRINT "  (9)  Stop"
1580   NOPTIONS% = 9 : GOSUB 7430
1590   ON REPLY% GOTO 1600, 1640, 1650, 1660, 1670, 1680, 1710, 1790, 1800
1600   PRINT
```

```
1610   INPUT "Number of freedoms = ", NDOF%
1620   STIFFMAT% = TRUE%  : GOSUB 2360
1630   STIFFMAT% = FALSE% : GOSUB 2360 : GOTO  1800 '... Input keyboard data
1640   GOSUB 1900 : GOTO 1800                      '... Read data file
1650   GOSUB 2700 : GOTO 1800                      '... Edit the banner
1660   STIFFMAT% = TRUE%  : GOSUB 2790 : GOTO  1800 '... Edit stiffness matrix
1670   STIFFMAT% = FALSE% : GOSUB 2790 : GOTO  1800 '... Edit mass matrix
1680   STIFFMAT% = TRUE%  : GOSUB 3010 : GOSUB 7570 '... List stiffness matrix
1690   STIFFMAT% = FALSE% : GOSUB 3010 : GOSUB 7570 '... List mass matrix
1700   GOTO 1800
1710   IF NDOF% > 1 THEN GOTO 1740
1720   PRINT
1730   PRINT "No problem loaded" : GOSUB 7570 : GOTO 1800
1740   PRINT
1750   IF NDOF% < MAXEVALUES% THEN I%=NDOF% ELSE I%=MAXEVALUES%
1760   PRINT USING "Number of eigenvalues - max ### = "; I%;
1770   INPUT "", NEIGEN%
1780   GOSUB 3530 : GOSUB 7050  : GOTO 1800         '... Solve problem
1790   GOSUB 3200 : GOTO 1800                      '... Write data to disk
1800 WEND
1810 END
1820 '
1830 '**********************    GENERAL ERROR TRAP   ************************
1840 '
1850 PRINT
1860 PRINT "** FATAL ERROR **"
1870 PRINT
1880 ON ERROR GOTO 0
1890 '
1900 '******************    READ DATA FROM DISK FILE    *********************
1910 '
1920 PRINT
1930 PRINT
1940 INPUT "Name of data file  =  ", INPUTFILE$
1950 ON ERROR GOTO 1990 : OPEN INPUTFILE$ FOR INPUT AS #1
1960 ON ERROR GOTO 1830 : GOTO 1990
1970 CLS : PRINT : PRINT "** ERROR ** Failed to open file" : INPUTFILE$ = ""
1980 ON ERROR GOTO 1830 : GOSUB 7570 : RESUME 2330
1990 LINE INPUT#1, TEXT$
2000 IF TEXT$ = "CASF eigenproblem data" THEN GOTO 2030
2010 PRINT
2020 PRINT "Not a data file for this program" : GOSUB 7570 : GOTO 2330
2030 PRINT
2040 PRINT "Reading data from file "; INPUTFILE$
2050 LINE INPUT#1, BANNER$
2060 INPUT#1, NDOF%
2070 LINE INPUT#1, TEXT$
2080 '
2090 ' ... Get the bandwidths
2100 '
2110 STIFFBAND% = 0
2120 FOR I% = 1 TO NDOF%
2130   FOR J% = 1 TO NDOF%
2140     INPUT#1, TEMP : IF J% <= MAXBAND% THEN K(I%,J%) = 0
2150     IF (J% < I%) OR  (ABS(TEMP)<XXXX) THEN GOTO 2180
2160     IF J%-I%+1 > STIFFBAND% THEN STIFFBAND% = J%-I%+1
2170     K(I%,J%-I%+1)= TEMP
2180   NEXT J%
2190 NEXT I%
2200 '
2210 LINE INPUT#1, TEXT$
2220 MASSBAND% = 0
2230 FOR I% = 1 TO NDOF%
2240   FOR J% = 1 TO NDOF%
2250     INPUT#1, TEMP : IF J% <= MAXBAND% THEN M(I%,J%) = 0
2260     IF (J% < I%) OR  (ABS(TEMP)<XXXX) THEN GOTO 2290
2270     IF J%-I%+1 > MASSBAND% THEN MASSBAND% = J%-I%+1
2280     M(I%,J%-I%+1)= TEMP
2290   NEXT J%
2300 NEXT I%
2310 '
2320 IF MASSBAND%>STIFFBAND% THEN BAND% = BAND% ELSE BAND%=STIFFBAND%
2330 CLOSE#1
2340 RETURN
2350 '
2360 '*************    INPUT THE MASS AND STIFFNESS MATRIX    ****************
2370 '
```

```
2380 CLS : PRINT
2390 IF STIFFMAT% = FALSE% THEN GOTO 2490
2400 PRINT "Enter an identification banner"
2410 PRINT
2420 INPUT "Banner = ", BANNER$
2430 CLS: PRINT
2440 PRINT "Enter the stiffness matrix"
2450 PRINT
2460 INPUT "Semi-bandwidth - stiffness matrix = ", STIFFBAND%
2470 BAND% = STIFFBAND%
2480 GOTO 2530
2490 PRINT "Enter the mass matrix"
2500 PRINT
2510 INPUT "Semi-bandwidth - mass matrix = ", MASSBAND%
2520 BAND% = MASSBAND%
2530 PRINT
2540 FOR I% = 1 TO NDOF%
2550   FOR J% = 1 TO MAXBAND%
2560     IF STIFFMAT% THEN K(I%,J%) = 0 ELSE M(I%,J%) = 0
2570   NEXT J%
2580   FOR J% = I% TO NDOF%
2590     PRINT USING "(##_,##) = "; I%, J%; :
2600     IF STIFFMAT% = FALSE% THEN GOTO 2630
2610     INPUT "", K(I%, J%-I%+1)
2620     GOTO 2640
2630     INPUT "", M(I%,J%-I%+1)
2640     IF J% >= I%+BAND%-1 THEN J% = NDOF%
2650   NEXT J%
2660 NEXT I%
2670 IF MASSBAND%>STIFFBAND% THEN BAND% = BAND% ELSE BAND%=STIFFBAND%
2680 RETURN
2690 '
2700 '**********************    EDIT THE BANNER    **************************
2710 '
2720 PRINT "Edit banner - <return> to retain"
2730 PRINT
2740 PRINT "Current banner = "; BANNER$
2750 INPUT "New banner     = ", TEXT$
2760 IF LEN(TEXT$) > 0 THEN BANNER$ = TEXT$
2770 RETURN
2780 '
2790 '******************    EDIT STIFFNESS/MASS MATRIX    *******************
2800 '
2810 DONE% = FALSE%
2820 WHILE DONE% = FALSE%
2830   GOSUB 3010
2840   PRINT
2850   PRINT "Enter row and column numbers as they appear on the screen"
2860   PRINT
2870   INPUT "Row (0 to stop) = ", I%
2880   IF (I%>=1) AND (I%<=NDOF%) THEN GOTO 2910
2890   IF I% = 0 THEN DONE% = TRUE% : GOTO 2980
2900   IF (I%<1) OR (I%>NDOF%) THEN GOTO 2950
2910   INPUT "Column          = ", J%
2920   IF (J%<1) OR (J%>TBAND%) THEN GOTO 2950
2930   INPUT "Revised value   = ", TEMP
2940   GOTO 2960
2950   PRINT : PRINT "Invalid location" : GOSUB 7570 : GOTO 2980
2960   IF STIFFMAT% =  TRUE% THEN K(I%,J%) = TEMP
2970   IF STIFFMAT% <> TRUE% THEN M(I%,J%) = TEMP
2980 WEND
2990 RETURN
3000 '
3010 '******************    PRINT STIFFNESS/MASS MATRIX    ******************
3020 '
3030 CLS : PRINT
3040 IF STIFFMAT% = TRUE%   THEN PRINT "UPPER SEMI-BAND OF STIFFNESS MATRIX"
3050 IF STIFFMAT% = FALSE%  THEN PRINT "UPPER SEMI-BAND OF MASS MATRIX"
3060 IF STIFFMAT% = TRUE%   THEN TBAND% = STIFFBAND% ELSE TBAND% = MASSBAND%
3070 PRINT
3080 FOR I% = 1 TO NDOF%
3090   PRINT USING "Row ###"; I%;
3100   FOR J% = 1 TO TBAND%
3110     IF STIFFMAT%     THEN PRINT USING "    #.####^^^^"; K(I%,J%);
3120     IF NOT STIFFMAT% THEN PRINT USING "    #.####^^^^"; M(I%,J%);
3130     IF (J% = TBAND%) AND (J% MOD 5 =  0) THEN PRINT
3140     IF (J% = TBAND%) AND (J% MOD 5 <> 0) THEN PRINT : PRINT
```

```
3150      IF (J% MOD 5 = 0) THEN PRINT
3160    NEXT J%
3170 NEXT I%
3180 RETURN
3190 '
3200 '*********************     WRITE DATA TO DISK     ************************
3210 '
3220 PRINT : INPUT "Name of data file  =  ", OUTPUTFILE$
3230 ON ERROR GOTO 3250 : OPEN OUTPUTFILE$ FOR OUTPUT AS #1
3240 ON ERROR GOTO 1830 : GOTO 3270
3250 CLS : PRINT : PRINT "** ERROR ** Failed to open output file"
3260 ON ERROR GOTO 1830 : GOSUB 7570 : RESUME 3500
3270 PRINT#1, "CASF eigenproblem data"
3280 PRINT#1, BANNER$
3290 PRINT#1, NDOF%
3300 PRINT#1,
3310 FOR I% = 1 TO NDOF%
3320   FOR J% = 1 TO NDOF%
3330     IF J% <= I%-STIFFBAND%   THEN PRINT#1, STR$(0) : GOTO 3370
3340     IF J% >  I%+STIFFBAND%-1 THEN PRINT#1, STR$(0) : GOTO 3370
3350     IF J% <  I% THEN PRINT#1, K(J%,I%-J%+1)
3360     IF J% >= I% THEN PRINT#1, K(I%,J%-I%+1)
3370   NEXT J%
3380 NEXT I%
3390 '
3400 PRINT#1,
3410 FOR I% = 1 TO NDOF%
3420   FOR J% = 1 TO NDOF%
3430     IF J% <= I%-MASSBAND%   THEN PRINT#1, STR$(0) : GOTO 3470
3440     IF J% >  I%+MASSBAND%-1 THEN PRINT#1, STR$(0) : GOTO 3470
3450     IF J% <  I% THEN PRINT#1, M(J%,I%-J%+1)
3460     IF J% >= I% THEN PRINT#1, M(I%,J%-I%+1)
3470   NEXT J%
3480 NEXT I%
3490 PRINT "Data written to file "; OUTPUTFILE$
3500 CLOSE#1
3510 RETURN
3520 '
3530 '*****************************     SOLVE     ******************************
3540 '
3550 '   This subroutine solves the eigenvalue problem [A][X] = LAMBA[B][X]
3560 '   for the smallest NEIGEN% eigenpairs using a combination of
3570 '   determinant search and inverse iteration
3580 '
3590 ' ... Zero start values
3600 '
3610 NFOUND% = 0 : NCALC% = 0 : U1 = 0 : U2 = 0:
3620 NFACT%  = 0 : NITER% = 0 : ROW% = CSRLIN
3630 '
3640 WHILE (NFOUND% < NEIGEN%)
3650   PRINT : PRINT USING "Eigenvector ##    Factorisations "; NCALC%+1;
3660   ROW% = CSRLIN  : COL1% = POS(0) : PRINT " 0     Iterations ";
3670   COL2% = POS(0) : PRINT " 0";
3680   GOSUB 4100                       ' Get start values
3690   GOSUB 4620                       ' Accelerated secant method
3700   NCALC% = NFOUND%                 ' No of eigenvalues less than U1
3710   WHILE (NCALC% < NLESS%)          ' Inverse iteration loop
3720     '
3730     ' ... Set up initial trial vector
3740     '
3750     RANDOMIZE TIMER
3760     FOR I% = 1 TO NDOF%
3770       Y1(I%) = RND
3780       WHILE Y1(I%)=0 : Y1(I%) = RND : WEND
3790     NEXT I%
3800     DONE = FALSE%
3810     WHILE DONE = FALSE%
3820       GOSUB 4970                   ' Inverse iteration
3830       IF (ABS((U3-U2)/U2)) < XXXX THEN DONE = TRUE%
3840       U2 = U3
3850     WEND
3860     GOSUB 5690                     ' Find eigenvector
3870     IF U2 <= U1 THEN NCALC% = NCALC% + 1
3880     NFACT% = 0: NITER% = 0 : IF NCALC%=NLESS% THEN GOTO 3920
3890     PRINT : PRINT USING "Eigenvector ##*   Factorisations "; NCALC%+1;
3900     ROW%  = CSRLIN : COL1% = POS(0) : PRINT " 0     Iterations ";
3910     COL2% = POS(0) : PRINT " 0";
```

```
3920   WEND
3930 WEND
3940 '
3950 ' ... Sort eigenvalues into order
3960 '
3970 PRINT : DONE% = FALSE%
3980 WHILE DONE% = FALSE%
3990   DONE% = TRUE%
4000   FOR K% = 1 TO NFOUND%-1
4010     IF EVALUE(K%,1)<=EVALUE(K%+1,1) THEN GOTO 4050
4020     DONE% = FALSE%
4030     TEMP=EVALUE(K%,1) : EVALUE(K%,1)=EVALUE(K%+1,1) : EVALUE(K%+1,1)=TEMP
4040     TEMP=EVALUE(K%,2) : EVALUE(K%,2)=EVALUE(K%+1,2) : EVALUE(K%+1,2)=TEMP
4050   NEXT K%
4060 WEND
4070 GOSUB 7570
4080 RETURN
4090 '
4100 '*********************     GET START VALUES     ***************************
4110 '
4120 '   This subroutine sets up the start parameters for the accelerated
4130 '   secant method
4140 '
4150 U1 = 0 : U2 = 0
4160 FOR I% = 1 TO NDOF%                   ' Set up initial coefficient matrix
4170   FOR J% = 1 TO BAND%
4180     A(I%,J%) = K(I%,J%)
4190   NEXT J%
4200 NEXT I%
4210 '
4220 ' ... Factor initial coefficient matrix then calc and deflate determinant
4230 '
4240 GOSUB 6190
4250 GOSUB 6760
4260 DET1 = DET: POWER1 = POWER
4270 '
4280 ' ... Set up initial trial vector using random nos (not all zeros)
4290 '
4300 RANDOMIZE TIMER
4310 FOR I% = 1 TO NDOF%
4320   Y1(I%) = RND
4330   WHILE Y1(I%)=0 : Y1(I%) = RND : WEND
4340 NEXT I%
4350 '
4360 ' ... Use inverse iteration to calculate U2
4370 '
4380 WHILE NITER% < 5
4390   GOSUB 4970
4400   U2 = U3
4410 WEND
4420 '
4430 ' ... Check that U2 has not jumped any eigenvalues. If so then divide U2
4440 '     by no. jumped plus 1 and redo until U2 is to the left of the next
4450 '     eigenvalue
4460 '
4470 DONE% = FALSE%
4480 WHILE DONE% = FALSE%
4490   FOR I% = 1 TO NDOF%                 ' Set up initial coefficient matrix
4500     FOR J% = 1 TO BAND%
4510       A(I%,J%) = K(I%,J%) - U2*M(I%,J%)
4520     NEXT J%
4530   NEXT I%
4540   GOSUB 6190                          ' Factorise
4550   IF SINGULAR% = TRUE% THEN U2 = .9999 * U2: GOTO 4580
4560   GOSUB 6760                          ' Calc determinant and count values
4570   IF NLESS%=0 THEN DONE% = TRUE% ELSE U2 = U2/(NLESS%+1)
4580 WEND
4590 DET2 = DET: POWER2 = POWER
4600 RETURN
4610 '
4620 '*****************     ACCELERATED SECANT METHOD     *********************
4630 '
4640 '   This subroutine uses the accelerated secant method to jump over
4650 '   one or more eigenvalues
4660 '
4670 ACC = 2                               ' Set initial acceleration value
4680 WHILE (NLESS% <= NFOUND%)
```

```
4690    FOR I% = 1 TO NDOF%              ' Calculate coefficient matrix
4700      FOR J% = 1 TO BAND%
4710        A(I%,J%) = K(I%,J%) - U2 * M(I%,J%)
4720      NEXT J%
4730    NEXT I%
4740    GOSUB 6190                       ' Factor the coefficient matrix
4750    IF SINGULAR% = TRUE% THEN U2 = .9999 * U2: GOTO 4690
4760    GOSUB 6760                       ' Calc determinant count eigenvalues
4770    DET2 = DET
4780    POWER2 = POWER
4790    '
4800    ' ... If method has converged then to avoid the risk of division by
4810    '     use simple scaling
4820    '
4830    IF (ABS((U2-U1)/U2)>XXXX) OR (NLESS%>NFOUND%) GOTO 4850
4840    U3 = 1.0001*U2 : GOTO 4900
4850    U3 = U2 - ACC*DET2*(U2 - U1)/(DET2 - DET1*2^(POWER1 - POWER2))
4860    '
4870    ' ... If change less than 1% then accelerate the method
4880    '
4890    IF ABS((U3-U2)/U2) < .01 THEN ACC = 2*ACC
4900    U1 = U2
4910    DET1 = DET2
4920    POWER1 = POWER2
4930    U2 = U3
4940 WEND
4950 RETURN
4960 '
4970 ' **********************     INVERSE ITERATION     ************************
4980 '
4990 '  This subroutine performs one cycle of vector inverse iteration and
5000 '  vector orthogonalises the trial vector to the last six eigenvalues
5010 '  to allow for multiple roots
5020 '
5030 '
5040 ' ... Solve the equations for [X]
5050 '
5060 FOR I% = 1 TO NDOF%
5070   Y(I%) = Y1(I%)
5080 NEXT I%
5090 GOSUB 6490                          ' Solution subroutine
5100 '
5110 ' ... Calculate the weighted vector
5120 '
5130 FOR I% = 1 TO NDOF%
5140   Y(I%) = M(I%,1) * X(I%)
5150   FOR J% = 1 TO MASSBAND% - 1
5160     IF I% - J% > 0       THEN Y(I%) = Y(I%) + M(I%-J%,J%+1) * X(I%-J%)
5170     IF I% + J% <= NDOF% THEN Y(I%) = Y(I%) + M(I%,J%+1)     * X(I%+J%)
5180   NEXT J%
5190 NEXT I%
5200 '
5210 ' ... Calculate new trial vector based upon the weighted vector
5220 '
5230 FOR I% = 1 TO NDOF%
5240   Y2(I%) = Y(I%)
5250 NEXT I%
5260 IF (NFOUND% < 6) THEN FINISH% = NFOUND% ELSE FINISH% = 6
5270 '
5280 ' ... Orthogonalise for last six vectors
5290 '
5300 FOR J% = 1 TO FINISH%
5310   ALPHA = 0
5320   '
5330   ' ... Calculate the weighting factor
5340   '
5350   FOR I% = 1 TO NDOF%
5360     ALPHA = ALPHA + X(I%) * W(I%,J%)
5370   NEXT I%
5380   FOR I% = 1 TO NDOF%
5390     Y2(I%) = Y2(I%) - ALPHA * W(I%,J%)
5400   NEXT I%
5410 NEXT J%
5420 '
5430 ' ... Calculate the denominator
5440 '
5450 DEN = 0
```

```
5460 FOR I% = 1 TO NDOF%
5470   DEN = DEN + X(I%) * Y(I%)
5480 NEXT I%
5490 '
5500 ' ... Calculate the new trial vector
5510 '
5520 DEN = SQR(ABS(DEN))
5530 FOR I% = 1 TO NDOF%
5540   Y2(I%) = Y2(I%) / DEN
5550 NEXT I%
5560 NUM = 0
5570 '
5580 ' ... Calculate new estimate of LAMDA (U3) and prepare for the next iter.
5590 '
5600 DEN = DEN * DEN
5610 FOR I% = 1 TO NDOF%
5620   NUM = NUM + X(I%) * Y1(I%)
5630   Y1(I%) = Y2(I%)
5640 NEXT I%
5650 U3 = U1 + NUM/DEN
5660 NITER% = NITER%+1 : LOCATE ROW%, COL2% : PRINT NITER%;
5670 RETURN
5680 '
5690 ' *********************     FIND EIGENVECTOR     **************************
5700 '
5710 '    This subroutine calculates the eigenvector once the eigenvalue has
5720 '    been found by inverse iteration. This subroutine also calculates
5730 '    weighted vector (note that weighted vectors for the last six
5740 '    calculated eigenvalues are stored in [W])
5750 '
5760 NFOUND% = NFOUND%+1
5770 EVALUE(NFOUND%,1) = U2 : EVALUE(NFOUND%,2) = NFOUND%
5780 '
5790 ' ... Calculate the denominator
5800 '
5810 DEN = 0
5820 FOR I% = 1 TO NDOF%
5830   DEN = DEN + X(I%)*Y(I%)
5840 NEXT I%
5850 DEN = SQR(ABS(DEN)): MAX = 0
5860 '
5870 ' ... Calculate the vector and the max element
5880 '
5890 FOR I% = 1 TO NDOF%
5900  EVECTOR(I%, NFOUND%) = X(I%)/DEN
5910  IF ABS(EVECTOR(I%,NFOUND%)) > ABS(MAX) THEN MAX = EVECTOR(I%,NFOUND%)
5920 NEXT I%
5930 '
5940 ' ... Schmidt orthogonalisation
5950 '
5960 K% = NFOUND% - 6 * ((NFOUND% - 1) \ 6)
5970   '
5980 FOR I% = 1 TO NDOF%
5990   '
6000   ' ... Set up the weighted vector
6010   '
6020   W(I%, K%) = M(I%, 1) * EVECTOR(I%, NFOUND%)
6030   FOR J% = 1 TO MASSBAND% - 1
6040     IF I%-J%>0   THEN W(I%,K%)=W(I%,K%)+M(I%-J%,J%+1)*EVECTOR(I%-J%,NFOUND%)
6050     IF I%+J%<=NDOF% THEN W(I%,K%)=W(I%,K%)+M(I%,J%+1)*EVECTOR(I%+J%,NFOUND%)
6060   NEXT J%
6070 NEXT I%
6080 '
6090 ' ... Normalise the eigenvector
6100 '
6110 FOR I% = 1 TO NDOF%                ' Normalise the eigenvector
6120   EVECTOR(I%,NFOUND%) = EVECTOR(I%,NFOUND%)/MAX
6130 NEXT I%
6140 RETURN
6150 '
6160 ' ****************************     FACTOR     *****************************
6170 '
6180 '     This subroutine decomposes a symmetric, positive definite, banded
6190 '     matrix to the form [Ut][D][U].  The upper semi-band of the coeff.
6200 '     matrix is stored in A.  On return A contains [D] in the
6210 '     leftmost column and [U] is stored in the remaining columns
6220 '
```

```
6230 ' ... Loop for all pivots
6240 '
6250 SINGULAR% = FALSE%
6260 FOR I% = 1 TO NDOF%
6270   IF I%<BAND% THEN FINISH% = I%-1 ELSE FINISH% = BAND%-1
6280   '
6290   ' ... Evaluate the the diagonal element
6300   '
6310   FOR K% = 1 TO FINISH%
6320     A(I%, 1) = A(I%,1) - A(I%-K%,K%+1)*A(I%-K%,K%+1)*A(I%-K%,1)
6330     IF ABS(A(I%,1)) < XXXX THEN SINGULAR%=TRUE% : GOTO 6470
6340   NEXT K%
6350   '
6360   ' ... Evaluate the off-diagonal elements
6370   '
6380   FOR J% = 2 TO BAND%
6390     IF I%<BAND% THEN FINISH% = I%-1 ELSE FINISH% = BAND%-J%
6400     FOR K% = 1 TO FINISH%
6410       A(I%,J%) = A(I%,J%) - A(I%-K%,K%+1)*A(I%-K%,J%+K%)*A(I%-K%,1)
6420     NEXT K%
6430     A(I%,J%) = A(I%,J%)/A(I%,1)
6440   NEXT J%
6450 NEXT I%
6460 NFACT% = NFACT%+1 : LOCATE ROW%, COL1% : PRINT NFACT%;
6470 RETURN
6480 '
6490 ' ****************************     SOLVE     ********************************
6500 '
6510 '    This subroutine solves a system of linear equations where the
6520 '    coeff. matrix has been decomposed using [L][D][Lt] factorisation
6530 '
6540 ' ... Forward substitution
6550 '
6560 FOR I% = 1 TO NDOF%
6570   IF I% > BAND%-1 THEN FINISH% = BAND% - 1 ELSE FINISH% = I% - 1
6580   SUM = 0
6590   IF I% = 1 THEN GOTO 6630
6600   FOR J% = 1 TO FINISH%
6610     SUM = SUM + A(I%-J%,J%+1) * A(I%-J%,1) * X(I%-J%)
6620   NEXT J%
6630   X(I%) = (Y(I%) - SUM)/A(I%,1)
6640 NEXT I%
6650 '
6660 ' ... Backward substitution
6670 '
6680 FOR I% = 1 TO (NDOF%-1)
6690   IF I% > BAND%-1 THEN FINISH% = BAND% - 1 ELSE FINISH% = I%
6700   FOR J% = 1 TO FINISH%
6710     X(NDOF%-I%) = X(NDOF%-I%) - A(NDOF%-I%,J%+1)*X(NDOF%-I%+J%)
6720   NEXT J%
6730 NEXT I%
6740 RETURN
6750 '
6760 ' *******************   CALCULATE DETERMINANT    ************************
6770 '
6780 '    This subroutine calculates the (scaled) value of the determinant,
6790 '    counts the number of eigenvalues smaller than U2, and then
6800 '    deflates the polynomial w.r.t. the eigenvalues already found
6810 '
6820 NLESS% = 0: DET = 1: POWER = 0
6830 FOR I% = 1 TO NDOF%
6840   IF (A(I%, 1) < 0) THEN NLESS% = NLESS% + 1
6850   DET = DET * A(I%,1)
6860   '
6870   ' ... Deflate the determinant
6880   '
6890   IF (I% <= NFOUND%) THEN DET = DET / (U2 - EVALUE(I%,1))
6900   '
6910   ' ... Scale determinant    Note: Determinant = DET*2^POWER
6920   '                              where         0.0625 >= ABS(DET) <= 1
6930   '
6940   WHILE (ABS(DET) > 1)            ' Reduce mantissa if greater than 1
6950     DET   = DET * .0625
6960     POWER = POWER + 4
6970   WEND
6980   WHILE (ABS(DET) < .0625)        ' Increase mantissa if less than 0.0625
6990     DET   = DET * 16
```

```
7000       POWER = POWER - 4
7010     WEND
7020 NEXT I%
7030 RETURN
7040 '
7050 '********************     OUTPUT THE EIGENPAIRS     **********************
7060 '
7070 CLS : PRINT
7080 PRINT "* * * * * * * * * * * * **"
7090 PRINT "*    EIGENVALUES      *     "; BANNER$
7100 PRINT "* * * * * * * * * * * * **"
7110 L% = 0 : K% = 1
7120 WHILE L% < NEIGEN%
7130   IF K%+4 > NEIGEN% THEN L% = NEIGEN% ELSE L% = K%+4
7140   PRINT
7150   PRINT "NUMBER                ";
7160   FOR I% = K% TO L%
7170     PRINT USING "      (##)    "; I%;
7180   NEXT I%
7190   PRINT
7200   PRINT "FREQUENCY (rad/s)";
7210   FOR I% = K% TO L%
7220     PRINT USING " #.####^^^^ "; SQR(ABS(EVALUE(I%,1)));
7230   NEXT I%
7240   PRINT
7250   PRINT "FREQUENCY^2        ";
7260   FOR I% = K% TO L%
7270     PRINT USING " #.####^^^^ "; EVALUE(I%,1);
7280   NEXT I%
7290   PRINT : PRINT
7300   PRINT "FREEDOM                 "
7310   FOR I% = 1 TO NDOF%
7320   PRINT USING "  ##              ";  I%;
7330     FOR J% = K% TO L%
7340       PRINT USING "   #.####^^^^"; EVECTOR(I%,EVALUE(J%,2));
7350     NEXT J%
7360     PRINT
7370   NEXT I%
7380   GOSUB 7570
7390   K% = L%+1
7400 WEND
7410 RETURN
7420 '
7430 '**********************     GET MENU RESPONSE     ***********************
7440 '
7450 ROW% = CSRLIN: COL% = POS(0)
7460 REPLY% = 0
7470 WHILE REPLY% = 0
7480   LOCATE ROW%, COL%: INPUT "", WRKSTR$: REPLY% = VAL(WRKSTR$)
7490   IF (REPLY% < 1) OR (REPLY% > NOPTIONS%) THEN GOTO 7520
7500   IF (REPLY% < 10) AND (LEN(WRKSTR$) = 1) THEN GOTO 7530
7510   IF (REPLY% > 9)  AND (LEN(WRKSTR$) = 2) THEN GOTO 7530
7520   SOUND 700, 4: REPLY% = 0: LOCATE ROW%, COL%
7530 WEND
7540 PRINT
7550 RETURN
7560 '
7570 '**********************     ANY KEY TO CONTINUE     *********************
7580 '
7590 IF CSRLIN<=22 THEN LOCATE 22, 1
7600 PRINT
7610 PRINT SPACE$(40); "Press any key to continue"
7620 WHILE INKEY$ = "" : WEND
7630 RETURN
7640 '
7650 '========================     DSEARCH.BAS     ===========================
```

Sample Run

The following sample run shows the program DSEARCH.BAS being used to find the eigenvalues and eigenvectors from example 11.4. Input from the keyboard is underlined.

```
======================================================================

PROGRAM DSEARCH.BAS                        Copyright (c) J.Balfour 1991

      Program for the calculation of the natural frequencies
  and mode shapes from the structure stiffness and mass using the
     method of determinate search and vector inverse iteration

                 For further information contact
 James A.D.Balfour, Heriot-Watt University, Riccarton, Edinburgh
             Tel 031-449-5111, Fax 031-451-3170

======================================================================


Maximum number of equations                  =   27
Maximum semi-bandwidth - stiffness matrix =   10
Maximum semi-bandwidth - mass matrix         =   10


Do you wish to:
  (1)  Input new data from keyboard
  (2)  Read new data from file
  (3)  Edit the banner
  (4)  Edit the stiffness matrix
  (5)  Edit the mass matrix
  (6)  List the matrices
  (7)  Solve the current problem
  (8)  Write data to disk
  (9)  Stop
1

Number of freedoms = 2


Enter an identification banner

Banner = CASF 2nd Edition Example 11.4


Enter the stiffness matrix

Semi-bandwidth - stiffness matrix = 2

( 1, 1) = 416700
( 1, 2) = 125000
( 2, 2) = 250000


Enter the mass matrix

Semi-bandwidth - mass matrix = 2

( 1, 1) = 1333
( 1, 2) = -228.6
( 2, 2) = 304.7



Do you wish to:
  (1)  Input new data from keyboard
  (2)  Read new data from file
  (3)  Edit the banner
  (4)  Edit the stiffness matrix
  (5)  Edit the mass matrix
  (6)  List the matrices
  (7)  Solve the current problem
  (8)  Write data to disk
  (9)  Stop
7

Number of eigenvalues - max   2 = 2

* * * * * * * * * * * *
*   EIGENVALUES      *    CASF 2nd Edition Example 11.4
* * * * * * * * * * * *
```

```
NUMBER                    ( 1)          ( 2)
FREQUENCY (rad/s)   0.1407E+02    0.3555E+02
FREQUENCY^2        (0.1980E+03)  (0.1264E+04)

FREEDOM
   1                0.1000E+01    0.3264E+00
   2                -.8979E+00    0.1000E+01


Do you wish to:
   (1)  Input new data from keyboard
   (2)  Read new data from file
   (3)  Edit the banner
   (4)  Edit the stiffness matrix
   (5)  Edit the mass matrix
   (6)  List the matrices
   (7)  Solve the current problem
   (8)  Write data to disk
   (9)  Stop
9
```

11.6 Dynamic Analysis Program PFDYNAM.BAS

This section puts together much of the material presented in this chapter to develop a program for the dynamic analysis of plane frame structures.

Comments on the Algorithm

PFDYNAM.BAS is based upon the program PFRAME.BAS presented in Chapter 8 and it has essentially the same functionality (including elements with pins and local axes). The program uses code from PFRAME.BAS to assemble the stiffness matrix and additional code has been added to assemble the mass matrix from the element masses and any added masses. It uses the solution subroutine from the program JACOBI.BAS (presented earlier in this chapter) to solve the eigenvalue problem and thus find the natural frequencies and mode shapes.

PFDYNAM.BAS caters for local axes and elements with pins. The mass coefficients for elements containing pins are developed in exactly the same way as the mass coefficients for rigidly connected elements, only different shape functions are used (see the section on elements with pins in Chapter 8). In the program the variables M_1 - M_8 are used to deal with mass coefficients for elements containing pins as shown in the following equation.

$$\begin{bmatrix} f_{ijx} \\ f_{ijy} \\ m_{ij} \\ f_{jix} \\ f_{jiy} \\ m_{ij} \end{bmatrix} = \frac{mL}{840} \begin{bmatrix} 280 & 0 & 0 & 140 & 0 & 0 \\ & M_1 & M_2L & 0 & M_3 & M_4L \\ & & M_5L^2 & 0 & M_6L & M_7L^2 \\ & \textit{symmetric} & & 280 & 0 & 0 \\ & & & & -M_8 & M_9L \\ & & & & & M_{10}L^2 \end{bmatrix} \begin{bmatrix} \ddot{\delta}_{ijx} \\ \ddot{\delta}_{ijy} \\ \ddot{\theta}_{ij} \\ \ddot{\delta}_{jix} \\ \ddot{\delta}_{jiy} \\ \ddot{\theta}_{ji} \end{bmatrix}$$

Table 11.1 Mass coefficients

Element Type	Node "i"	Node "j"	M_1	M_2	M_3	M_4	M_5	M_6	M_7	M_8	M_9	M_{10}
1	Rigid	Rigid	312	44	108	-26	8	26	-6	312	-44	8
2	Rigid	Pinned	408	72	117	0	16	33	0	198	0	0
3	Pinned	Rigid	198	0	117	-33	0	0	0	408	-72	16
4	Pinned	Pinned	0	0	0	0	0	0	0	0	0	0

Menu System

Unless PFDYNAM.BAS has been chained from PRE.BAS or DCHECK.BAS the user is prompted for the name of the data file to be read by the program. PFRAME.BAS then offers the user the following menu which drives the program.

Do you wish to:
(1) Analyse with output to the screen
(2) Analyse with output to the printer
(3) Analyse with output to a file
(4) Edit the data
(5) Check the data
(6) Read new data file
(7) Stop

This data file should be constructed using the data preprocessor program PRE.BAS which is described in section 5.11. The program DCHECK.BAS can be used to check the data for obvious errors. DCHECK.BAS is presented in section 5.12.

Output

The natural frequencies and mode shapes of the structure are output. The eigenvalues of course give the relative values of the vibration magnitude in the directions of the freedoms.

Listing

```
1000 '=======================        PFDYNAM.BAS      ==============================
1010 '
1020 COMMON DATAFILE$
1030 OPTION BASE 1 : KEY OFF
1040 TRUE% = -1 : FALSE% = 0 : DEVICE$ = "SCRN:" : PAGELEN% = 23
1050 CLS
1060 PRINT
1070 PRINT "=============================================================="
1080 PRINT
1090 PRINT " PROGRAM PFDYNAM                          Copyright (c) J.Balfour 1991"
1100 PRINT
1110 PRINT "   Program for the automatic dynamic analysis of plane frames"
1120 PRINT "       Data generated with the data preprocessor PRE"
```

```
1130 PRINT
1140 PRINT "                For further information contact"
1150 PRINT " James A.D.Balfour, Heriot-Watt University, Riccarton, Edinburgh"
1160 PRINT "               Tel 031-440 5111, Fax 031-451-3170"
1170 PRINT
1180 PRINT "================================================================="
1190 '
1200 ' ... Note that variables are defined in Appendix A and the
1210 '     following statements set the maximum problem size
1220 '
1230 MAXNODE%  = 16                   '... Max no of nodes
1240 MAXELEM%  = 25                   '... Max no of elements
1250 MAXPROP%  = 10                   '... Max no of element properties
1260 MAXREST%  = 10                   '... Max no of restrained nodes
1270 MAXLAXES% = 10                   '... Max no of local axes
1280 MAXAMASS% = 30                   '... Max no of added masses
1290 '
1300 DIM NODE(MAXNODE%,4),          ELEM(MAXELEM%,9),           PROP(MAXPROP%,5)
1310 DIM LAXES(MAXLAXES%,2),        REST(MAXREST%,5),           AMASS(MAXAMASS%,3)
1320 DIM FREE%(3*MAXNODE%),         ESTIFF(6,6),                EMASS(6,6)
1330 DIM K(3*MAXNODE%,3*MAXNODE%),M(3*MAXNODE%,3*MAXNODE%),  EFREE%(6)
1340 DIM THETA(3*MAXNODE%,2),       PHI(3*MAXNODE%,3*MAXNODE%),WKSP$(8)
1350 '
1360 ' ... Check if this program has been chained from PRE
1370 '
1380 IF DATAFILE$ <> "" THEN GOSUB 9360 : GOTO 1400
1390 PRINT : INPUT "Name of the data file  =  ", DATAFILE$
1400 GOSUB 2050
1410 NOPTIONS% = 7
1420 WHILE (REPLY% <> 7)
1430   CLS : PRINT
1440   PRINT "Do you wish to:"
1450   PRINT "  (1)  Analyse with output to the screen"
1460   PRINT "  (2)  Analyse with output to the printer"
1470   PRINT "  (3)  Analyse with output to a file"
1480   PRINT "  (4)  Edit the data"
1490   PRINT "  (5)  Check the data"
1500   PRINT "  (6)  Read new data file"
1510   PRINT "  (7)  Stop"
1520   GOSUB 8610
1530   ON REPLY% GOTO 1550, 1580, 1660, 1730, 1760, 1790, 1820
1540     '
1550     OPEN "SCRN:" FOR OUTPUT AS #2                    ' Reply = 1
1560     GOSUB 1860 : GOTO 1810
1570     '
1580     ON ERROR GOTO 1630                               ' Reply = 2
1590     OPEN "LPT1:" FOR OUTPUT AS #2
1600     PRINT : PRINT "Wait - looking for printer" : PRINT #2,
1610     ON ERROR GOTO 0     : DEVICE$ = "LPT1:" : GOSUB 1860
1620     PRINT #2, CHR$(12); : GOTO 1810
1630     CLS : PRINT : PRINT "** ERROR ** Failed to find the printer"
1640     GOSUB 9360 : RESUME 1810
1650     '
1660     ON ERROR GOTO 1700                               ' Reply = 3
1670     INPUT "Name of file for output = ", DEVICE$
1680     OPEN DEVICE$ FOR OUTPUT AS #2
1690     ON ERROR GOTO 0 : GOSUB 1860 : GOTO 1810
1700     CLS : PRINT : PRINT "** ERROR ** Failed to open output file"
1710     GOSUB 9360 : RESUME 1810
1720     '
1730     PRINT "Wait - chaining PRE"                          ' Reply = 4
1740     CHAIN "PRE"
1750     '
1760     PRINT "Wait - chaining DCHECK"                       ' Reply = 5
1770     CHAIN "DCHECK"
1780     '
1790     INPUT "Name of the data file  =  ", DATAFILE$     ' Relpy = 6
1800     GOSUB 2050
1810     CLOSE #2 : DEVICE$ = "SCRN:" : ON ERROR GOTO 0
1820 WEND
1830 KEY ON
1840 END
1850 '
1860 '****************************   ANALYSE   ******************************
1870 '
1880 IF DATAFILEOK% THEN GOTO 1930
1890 CLS : PRINT
```

```
1900 PRINT "** ERROR ** Analysis cannot proceed due to errors in data file";
1910 GOSUB 9360          'Pause
1920 GOTO  2030
1930 GOSUB 3520          '... Print data
1940 PRINT : PRINT "Generating the freedom vector                    "
1950 GOSUB 4980
1960 PRINT
1970 PRINT "Assembling the stiffness and mass matrices, adding element :- ";
1980 ROW% = CSRLIN : COL% = POS(0) : GOSUB 5310
1990 PRINT
2000 PRINT "Finding frequencies and mode shapes"
2010 GOSUB 6740          '... Analyse
2020 GOSUB 7820          '... Output the results
2030 RETURN
2040 '
2050 '************************   READ DATA FILE    ****************************
2060 '
2070 ' ...  This subroutine reads the data from a data file
2080 '
2090 PRINT
2100 PRINT "Wait - Reading file "; DATAFILE$; " line";
2110 ROW% = CSRLIN : COL% = POS(0) : LCOUNT% = 0
2120 DATAFILEOK% = FALSE% : DATAEND% = FALSE%
2130 NNODE%  = 0 : NELEM%  = 0 : NPROP% = 0 : NREST% = 0 : NLAXES% = 0
2140 NAMASS% = 0
2150 ON ERROR GOTO 2170 : OPEN DATAFILE$ FOR INPUT AS #1
2160 ON ERROR GOTO 0 : GOTO 2230
2170 CLS : PRINT
2180 PRINT "** ERROR ** Failed to open file: "; DATAFILE$; " for input"
2190 GOSUB 9360 : RESUME 3490
2200 '
2210 ' ... Check that this is data file for PFDYNAM
2220 '
2230 TARGET$ = "PFDYNAM" : GOSUB 9110
2240 IF STRFOUND% THEN GOTO 2310
2250 CLS : PRINT
2260 PRINT "** ERROR ** Not a file for PFDYNAM"
2270 GOSUB 9360 : GOTO 3490
2280 '
2290 ' ... Read title and units strings
2300 '
2310 NLINES% = 2 : GOSUB 9270 : IF DATAEND% THEN GOTO 2380
2320 WHILE (MID$(WRKSTR$,I%,1) = " ")
2330   I% = I% + 1
2340 WEND
2350 TITLE$  = MID$(WRKSTR$,I%)
2360 TARGET$ = "Units" : GOSUB 9110
2370 IF STRFOUND% THEN GOTO 2410
2380 CLS : PRINT
2390 PRINT "** ERROR ** Error reading Title/Units strings"
2400 GOSUB 9360 : GOTO 2430
2410 UNITS$ = MID$(WRKSTR$,INSTR(WRKSTR$,":- ")+3,24)
2420 DATAFILEOK% = TRUE%
2430 WHILE NOT DATAEND%
2440   '
2450   ' ... Get module string
2460   '
2470   IF LEFT$(WRKSTR$,4)="+   " THEN GOTO 2490
2480   TARGET$ = "+   " : GOSUB 9110 : IF NOT STRFOUND% THEN GOTO 3480
2490   MODSTR$ = WRKSTR$
2500   '
2510   ' ...  Identify module
2520   '
2530   IF INSTR(MODSTR$,"NODAL COORDINATES")=0 THEN GOTO 2680
2540   '
2550   ' ... Read nodal coordinates
2560   '
2570   TARGET$ = "NODE" : GOSUB 9110 : IF NOT STRFOUND% THEN GOTO 3420
2580   GOSUB 8750
2590   WHILE NNOS% > 0
2600     NNODE% = NNODE% + 1
2610     FOR J% = 1 TO NNOS%
2620       NODE(NNODE%,J%) = VAL(WKSP$(J%))
2630     NEXT J%
2640     GOSUB 8750
2650   WEND
2660   GOTO 3480
```

```
2670    '
2680    IF INSTR(MODSTR$,"ELEMENTS")=0 THEN GOTO 2830
2690    '
2700    ' ... Read elements
2710    '
2720    TARGET$ = "ELEMENT" : GOSUB 9110 : IF NOT STRFOUND% THEN GOTO 3420
2730    GOSUB 8750
2740    WHILE NNOS% > 0
2750      NELEM% = NELEM% + 1
2760      FOR J% = 1 TO NNOS%
2770        ELEM(NELEM%,J%) = VAL(WKSP$(J%))
2780      NEXT J%
2790      GOSUB 8750
2800    WEND
2810    GOTO 3480
2820    '
2830    IF INSTR(MODSTR$,"PROPERTIES")=0 THEN GOTO 2980
2840    '
2850    ' ... Read properties
2860    '
2870    TARGET$ = "PROPERTY" : GOSUB 9110 : IF NOT STRFOUND% THEN GOTO 3420
2880    GOSUB 8750
2890    WHILE NNOS% > 0
2900      NPROP% = NPROP% + 1
2910      FOR J% = 1 TO NNOS%
2920        PROP(NPROP%,J%) = VAL(WKSP$(J%))
2930      NEXT J%
2940      GOSUB 8750
2950    WEND
2960    GOTO 3480
2970    '
2980    IF INSTR(MODSTR$,"RESTRAINTS")=0 THEN GOTO 3130
2990    '
3000    ' ... Read restraints
3010    '
3020    TARGET$ = "NODE" : GOSUB 9110 : IF NOT STRFOUND% THEN GOTO 3420
3030    GOSUB 8750
3040    WHILE NNOS% > 0
3050      NREST% = NREST% + 1
3060      FOR J% = 1 TO NNOS%
3070        REST(NREST%,J%) = VAL(WKSP$(J%))
3080      NEXT J%
3090      GOSUB 8750
3100    WEND
3110    GOTO 3480
3120    '
3130    IF INSTR(MODSTR$,"LOCAL AXES")=0 THEN GOTO 4250
3140    '
3150    ' ... Read local axes
3160    '
3170    TARGET$ = "NODE" : GOSUB 9110 : IF NOT STRFOUND% THEN GOTO 3420
3180    GOSUB 8750
3190    WHILE NNOS% > 0
3200      NLAXES% = NLAXES% + 1
3210      FOR J% = 1 TO NNOS%
3220        LAXES(NLAXES%,J%) = VAL(WKSP$(J%))
3230      NEXT J%
3240      GOSUB 8750
3250    WEND
3260    GOTO 3480
0270    '
3280    IF INSTR(MODSTR$,"ADDED MASS")=0 THEN GOTO 3420
3290    '
3300    ' ... Read added mass
3310    '
3320    TARGET$ = "NODE" : GOSUB 9110 : IF NOT STRFOUND% THEN GOTO 3420
3330    GOSUB 8750
3340    WHILE NNOS% > 0
3350      NAMASS% = NAMASS% + 1
3360      FOR J% = 1 TO NNOS%
3370        AMASS(NAMASS%,J%) = VAL(WKSP$(J%))
3380      NEXT J%
3390      GOSUB 8750
3400    WEND
3410    GOTO 3480
3420    CLS : PRINT
3430    PRINT "** ERROR ** in module "; MODSTR$ "run DCHECK for more info";
```

```
3440   GOSUB 9360
3450   PRINT
3460   PRINT "Wait - Reading file "; DATAFILE$; " line";
3470   ROW% = CSRLIN : COL% = POS(0)
3480 WEND
3490 CLOSE #1 : ON ERROR GOTO 0
3500 RETURN
3510 '
3520 '*************************     PRINT DATA     ******************************
3530 '
3540 GOSUB 3620                         '... List header
3550 GOSUB 3950                         '... List nodal coordinates
3560 GOSUB 4120                         '... List element data
3570 GOSUB 4290                         '... List list element properties
3580 GOSUB 4650                         '... List restraint data
3590 GOSUB 4810                         '... List added masses
3600 RETURN
3610 '
3620 ' +++++++++++++++++++++++   LIST HEADER    ++++++++++++++++++++++++++
3630 '
3640 CLS
3650 PRINT #2, : PRINT #2,
3660 PRINT #2, "======================================";
3670 PRINT #2, "======================================"
3680 PRINT #2,
3690 PRINT #2, "    PROGRAM PFDYNAM"; STRING$(23,32);
3700 PRINT #2, "Copyright(c) J.A.D.Balfour 1991"
3710 PRINT #2,
3720 FOR I% = 1 TO (76-LEN(TITLE$))/2
3730   PRINT #2, " ";
3740 NEXT I%
3750 PRINT #2, TITLE$
3760 PRINT #2,
3770 PRINT #2, "    File Name :- "; DATAFILE$;
3780 FOR I% = 1 TO (40-LEN(DATAFILE$))
3790   PRINT #2, " ";
3800 NEXT I%
3810 PRINT #2, "Date :- "; MID$(DATE$,4,3); MID$(DATE$,1,3);
3820 PRINT #2, MID$(DATE$, 9, 2)
3830 PRINT #2, "    Units     :- "; UNITS$;
3840 FOR I% = 1 TO (40-LEN(UNITS$))
3850   PRINT #2, " ";
3860 NEXT I%
3870 PRINT #2, "Time :- "; TIME$
3880 PRINT #2,
3890 PRINT #2, "======================================";
3900 PRINT #2, "======================================"
3910 PRINT #2,
3920 GOSUB 9360
3930 RETURN
3940 '
3950 ' +++++++++++++++++    LIST NODAL COORDINATES    +++++++++++++++++++++
3960 '
3970 PRINT #2,
3980 PRINT #2, "+ + + + + + + + + + + + + +"
3990 PRINT #2, "+   NODAL COORDINATES    +"
4000 PRINT #2, "+ + + + + + + + + + + + + +"
4010 PRINT #2,
4020 PRINT #2, "NODE           X              Y"
4030 LCOUNT% = 6
4040 FOR I% = 1 TO NNODE%
4050   IF (LCOUNT%>=PAGELEN%) THEN GOSUB 9450
4060   LCOUNT%=LCOUNT%+1
4070   PRINT #2, " "; NODE(I%,1), NODE(I%,2), NODE(I%,3)
4080 NEXT I%
4090 GOSUB 9360
4100 RETURN
4110 '
4120 ' ++++++++++++++++++++++   LIST ELEMENTS    +++++++++++++++++++++++++
4130 '
4140 PRINT #2,
4150 PRINT #2, "+ + + + + + + + + +"
4160 PRINT #2, "+   ELEMENTS     +"
4170 PRINT #2, "+ + + + + + + + + +"
4180 PRINT #2,
4190 PRINT #2,         "ELEMENT    PROPERTY      TYPE"
4200 LCOUNT% = 6
```

```
4210 FOR I% = 1 TO NELEM%
4220   GOSUB 9450
4230   PRINT #2, USING "### ###       ##"; ELEM(I%,1); ELEM(I%,2); ELEM(I%,3);
4240   PRINT #2, USING "            ##"; ELEM(I%,4)
4250 NEXT I%
4260 GOSUB 9360
4270 RETURN
4280 '
4290 ' ++++++++++++++++++++++   LIST PROPERTIES    ++++++++++++++++++++++++
4300 '
4310 PRINT #2,
4320 PRINT #2, "+ + + + + + +  + + +"
4330 PRINT #2, "+   PROPERTIES   +"
4340 PRINT #2, "+ + + + + + +  + + +"
4350 PRINT #2,
4360 PRINT #2, "PROPERTY         A              I            E            m"
4370 LCOUNT% = 6
4380 FOR I% = 1 TO NPROP%
4390   GOSUB 9450
4400   PRINT #2, USING "  ###     #.####^^^^  ";     PROP(I%,1); PROP(I%,2);
4410   PRINT #2, USING "   #.####^^^^   #.####^^^^"; PROP(I%,3); PROP(I%,4);
4420   PRINT #2, USING "   #.####^^^^"; PROP(I%,5)
4430 NEXT I%
4440 GOSUB 9360
4450 RETURN
4460 '
4470 ' ++++++++++++++++++++++   LIST LOCAL AXES    ++++++++++++++++++++++++
4480 '
4490 IF NLAXES% = 0 THEN GOTO 4630
4500 LCOUNT% = 6
4510 PRINT #2,
4520 PRINT #2, "+ + + + + + + + + + +"
4530 PRINT #2, "+   LOCAL AXES    +"
4540 PRINT #2, "+ + + + + + + + + + +"
4550 PRINT #2,
4560 PRINT #2, "NODE      ANGLE"
4570 LCOUNT% = 6
4580 FOR I% = 1 TO NLAXES%
4590   GOSUB 9450
4600   PRINT #2, USING " ###      ####.#"; LAXES(I%,1); LAXES(I%,2)
4610 NEXT I%
4620 GOSUB 9360
4630 RETURN
4640 '
4650 ' ++++++++++++++++++++++++   LIST RESTRAINTS    ++++++++++++++++++++++++
4660 '
4670 PRINT #2,
4680 PRINT #2, "+ + + + + + + + + + +"
4690 PRINT #2, "+   RESTRAINTS    +"
4700 PRINT #2, "+ + + + + + + + + + +"
4710 PRINT #2,
4720 PRINT #2, "NODE    DIRECTION(S)"
4730 LCOUNT% = 6
4740 FOR I% = 1 TO NREST%
4750   GOSUB 9450
4760   PRINT #2, USING "###      ######"; REST(I%,1); REST(I%,2)
4770 NEXT I%
4780 GOSUB 9360
4790 RETURN
4800 '
4810 ' ++++++++++++++++++++++   LIST ADDED MASSES    ++++++++++++++++++++++
4820 '
4830 IF NAMASS% = 0 THEN GOTO 4960
4840 PRINT #2,
4850 PRINT #2, "+ + + + + + + + + + + +"
4860 PRINT #2, "+   ADDED MASSES    +"
4870 PRINT #2, "+ + + + + + + + + + + +"
4880 PRINT #2,
4890 PRINT #2, "NODE       MASS"
4900 LCOUNT% = 6
4910 FOR I% = 1 TO NAMASS%
4920   GOSUB 9450
4930   PRINT #2, USING "###   #.###^^^^"; AMASS(I%,1); AMASS(I%,2)
4940 NEXT I%
4950 GOSUB 9360
4960 RETURN
4970 '
```

```
4980 '******************     GENERATE THE FREEDOM VECTOR     *******************
4990 '
5000 ' ... Zero the freedom vector
5010 '
5020 FOR I% = 1 TO 3*NNODE%
5030   FREE%(I%) = 0
5040 NEXT I%
5050 '
5060 ' ... Loop for all restrained nodes
5070 '
5080 FOR I% = 1 TO NREST%
5090   K% = REST(I%,2)
5100   '
5110   ' ... Evaluate the freedom to be restrained from K%
5120   '
5130   J% = 3 * REST(I%,1) - 3 + (K% MOD 10)
5140   FREE%(J%) = 1
5150   K% = INT(K%/10)
5160   IF K% > 0 THEN GOTO 5130
5170 NEXT I%
5180 '
5190 ' ... Number the freedoms (restraints set to zero)
5200 '
5210 NDOF% = 0
5220 FOR I% = 1 TO 3*NNODE%
5230   IF FREE%(I%) = 1 THEN GOTO 5270
5240   NDOF% = NDOF% + 1
5250   FREE%(I%) = NDOF%
5260   GOTO 5280
5270   FREE%(I%) = 0
5280 NEXT I%
5290 RETURN
5300 '
5310 '***********     ASSEMBLE THE STIFFNESS AND MASS MATRICES     ************
5320 '
5330 ' ... Zero the stiffness and mass matrices
5340 '
5350 FOR I% = 1 TO NDOF%
5360   FOR J% = 1 TO NDOF%
5370     K(I%,J%) = 0 : M(I%,J%) = 0
5380   NEXT J%
5390 NEXT I%
5400 '
5410 ' ... Loop for each element
5420 '
5430 FOR K% = 1 TO NELEM%
5440   LOCATE ROW%, COL% : PRINT K%;
5450   IN% = ELEM(K%,1)
5460   JN% = ELEM(K%,2)
5470   PN% = ELEM(K%,3)
5480   '
5490   ' ... Calculate the element length (store in ELEM(K%,5))
5500   '
5510   ELEM(K%,5) = (NODE(JN%,2)-NODE(IN%,2))^2 + (NODE(JN%,3)-NODE(IN%,3))^2
5520   ELEM(K%,5) = SQR(ELEM(K%,5))
5530   '
5540   ' ... Calculate the the direction cosines of the member x-axes
5550   '
5560   C = (NODE(JN%,2) - NODE(IN%,2)) / ELEM(K%,5)
5570   S = (NODE(JN%,3) - NODE(IN%,3)) / ELEM(K%,5)
5580   '
5590   ' ... CHECK FOR LOCAL AXES
5600   '
5610   BETAI = 0 : BETAJ = 0
5620   FOR I% = 1 TO NLAXES%
5630     IF LAXES(I%,1) = IN% THEN BETAI = LAXES(I%,2) * 3.142 / 180
5640     IF LAXES(I%,1) = JN% THEN BETAJ = LAXES(I%,2) * 3.142 / 180
5650   NEXT I%
5660   '
5670   '  ...  Store cosine and sine of alpha I and alpha J in ELEM(K%,6-9)
5680   '
5690   ELEM(K%,6) = C * COS(BETAI) + S * SIN(BETAI)
5700   ELEM(K%,7) = S * COS(BETAI) - C * SIN(BETAI)
5710   ELEM(K%,8) = C * COS(BETAJ) + S * SIN(BETAJ)
5720   ELEM(K%,9) = S * COS(BETAJ) - C * SIN(BETAJ)
5730   GOSUB 8230                                  ' Get K values
5740   '
```

```
5750   ' ... Assemble the upper triangle of the element stiffness matrix
5760   '
5770   L   = ELEM(K%,5)
5780   CI  = ELEM(K%,6)
5790   SI  = ELEM(K%,7)
5800   CJ  = ELEM(K%,8)
5810   SJ  = ELEM(K%,9)
5820   EAL = PROP(PN%,4)*PROP(PN%,2)/ELEM(K%,5)
5830   EIL = PROP(PN%,4)*PROP(PN%,3)/ELEM(K%,5)
5840   ML  = PROP(PN%,5)*ELEM(K%, 5)/840
5850   ESTIFF(1,1) =   EAL * CI * CI + K1 * EIL * SI * SI/L^2
5860   ESTIFF(1,2) =   EAL * SI * CI - K1 * EIL * SI * CI/L^2
5870   ESTIFF(1,3) = - K2 * EIL * SI/L
5880   ESTIFF(1,4) = - EAL * CI * CJ - K1 * EIL * SI * SJ/L^2
5890   ESTIFF(1,5) = - EAL * CI * SJ + K1 * EIL * SI * CJ/L^2
5900   ESTIFF(1,6) = - K3 * EIL * SI/L
5910   ESTIFF(2,2) =   EAL * SI * SI + K1 * EIL * CI * CI/L^2
5920   ESTIFF(2,3) =   K2 * EIL * CI/L
5930   ESTIFF(2,4) = - EAL * SI * CJ + K1 * EIL * CI * SJ/L^2
5940   ESTIFF(2,5) = - EAL * SI * SJ - K1 * EIL * CI * CJ/L^2
5950   ESTIFF(2,6) =   K3 * EIL * CI/L
5960   ESTIFF(3,3) =   K4 * EIL
5970   ESTIFF(3,4) = - K5 * EIL * SJ/L
5980   ESTIFF(3,5) =   K5 * EIL * CJ/L
5990   ESTIFF(3,6) =   K6 * EIL
6000   ESTIFF(4,4) =   EAL * CJ * CJ + K1 * EIL * SJ * SJ/L^2
6010   ESTIFF(4,5) =   EAL * SJ * CJ - K1 * EIL * SJ * CJ/L^2
6020   ESTIFF(4,6) = - K7 * EIL * SJ/L
6030   ESTIFF(5,5) =   EAL * SJ * SJ + K1 * EIL * CJ * CJ/L^2
6040   ESTIFF(5,6) =   K7 * EIL * CJ/L
6050   ESTIFF(6,6) =   K8 * EIL
6060   '
6070   ' ... Set up element mass matrix
6080   '
6090   EMASS(1,1) =   280 * ML * CI * CI + M1 * ML * SI * SI
6100   EMASS(1,2) =   280 * ML * SI * CI - M1 * ML * SI * CI
6110   EMASS(1,3) = - M2 * SI * L
6120   EMASS(1,4) =   140 * ML * CI * CJ + M3 * ML * SI * SJ
6130   EMASS(1,5) =   140 * ML * CI * SJ - M3 * ML * SI * CJ
6140   EMASS(1,6) = - M4 * ML * SI * L
6150   EMASS(2,2) =   280 * ML * SI * SI + M1 * ML * CI * CI
6160   EMASS(2,3) =   M2 * ML * CI * L
6170   EMASS(2,4) =   140 * ML * SI * CJ - M3 * ML * CI * SJ
6180   EMASS(2,5) =   140 * ML * SI * SJ + M3 * ML * CI * CJ
6190   EMASS(2,6) =   M4 * ML * CI * L
6200   EMASS(3,3) =   M5 * ML * L * L
6210   EMASS(3,4) = - M6 * ML * SJ * L
6220   EMASS(3,5) =   M6 * ML * CJ * L
6230   EMASS(3,6) =   M7 * ML * L * L
6240   EMASS(4,4) =   280 * ML * CJ * CJ + M8 * ML * SJ * SJ
6250   EMASS(4,5) =   280 * ML * SJ * CJ - M8 * ML * SJ * CJ
6260   EMASS(4,6) = - M9 * ML * SJ * L
6270   EMASS(5,5) =   280 * ML * SJ * SJ + M8 * ML * CJ * CJ
6280   EMASS(5,6) =   M9 * ML * CJ * L
6290   EMASS(6,6) =   M10 * ML * L * L
6300   '
6310   ' ... Set up the element code number
6320   '
6330   EFREE%(1) = FREE%(3*IN%-2)
6340   EFREE%(2) = FREE%(3*IN%-1)
6350   EFREE%(3) = FREE%(3*IN%)
6360   EFREE%(4) = FREE%(3*JN%-2)
6370   EFREE%(5) = FREE%(3*JN%-1)
6380   EFREE%(6) = FREE%(3*JN%)
6390   '
6400   ' ... Add the element stiffness to the structure stiffness matrix
6410   '
6420   FOR I% = 1 TO 6
6430     IF EFREE%(I%) = 0 THEN GOTO 6500 ELSE L% = EFREE%(I%)
6440     FOR J% = I% TO 6
6450       IF EFREE%(J%) = 0 THEN GOTO 6490
6460       M% = EFREE%(J%)
6470       K(L%,M%) = K(L%,M%) + ESTIFF(I%,J%)
6480       M(L%,M%) = M(L%,M%) + EMASS(I%,J%)
6490     NEXT J%
6500   NEXT I%
6510 NEXT K%
```

```
6520 '
6530 ' ... Added masses
6540 '
6550 FOR I% = 1 TO NAMASS%
6560   IN% = AMASS(I%,1) : J% = FREE%(3*IN%-2)
6570   IF J% <> 0 THEN M(J%,J%) = M(J%,J%) + AMASS(I%,2)
6580   J% = J% = FREE%(3*IN%-1)
6590   IF J% <> 0 THEN M(J%,J%) = M(J%,J%) + AMASS(I%,2)
6600 NEXT I%
6610 '
6620 ' ... Use symmetry to fill out the matrices and set [PHI] to the
6630 '     the identity matrix
6640 '
6650 FOR I% = 1 TO NDOF%
6660   FOR J% = 1 TO NDOF%
6670     IF I%<J% THEN K(J%,I%)   = K(I%,J%) : M(J%,I%) = M(I%,J%)
6680     IF I%=J% THEN PHI(I%,J%) = 1 ELSE PHI(I%,J%) = 0
6690   NEXT J%
6700 NEXT I%
6710 PRINT
6720 RETURN
6730 '
6740 '**************************     ANALYSE     ********************************
6750 '
6760 NN% = 0 : TOL=1000000! : ABSTOL = 1E-08 : DONE% = FALSE%
6770 PRINT : PRINT USING "Tolerance #.####^^^^     Iteration "; TOL;
6780 ROW%  = CSRLIN : COL%   = POS(0) : PRINT NN%;
6790 WHILE DONE% = FALSE%
6800   DONE% = TRUE%
6810   FOR I% = 1 TO NDOF%
6820     FOR J% = I%+1 TO NDOF%
6830     IF (ABS(K[I%,J%]) < TOL) AND (ABS(M[I%,J%]) < TOL) THEN GOTO 6880
6840       DONE = FALSE%
6850       GOSUB 6990                               '...  Calc alpha and beta
6860       NN% = NN%+1 : LOCATE ROW%, COL% : PRINT NN%;
6870       GOSUB 7100                               '...  Transform matrices
6880     NEXT J%
6890   NEXT I%
6900   IF DONE% = FALSE% THEN GOTO 6940
6910   IF TOL   < ABSTOL THEN GOTO 6940
6920   TOL = TOL*.001 : DONE% = FALSE%
6930   LOCATE ROW%, 11 : PRINT USING "#.####^^^^"; TOL;
6940 WEND
6950 GOSUB 9360                                 '...  Pause
6960 GOSUB 7490                                 '...  Calc eigenvalues
6970 RETURN
6980 '
6990 '************************     ALPHA & BETA     ****************************
7000 '
7010 KKII  = K[I%,I%]*M[I%,J%]  - M[I%,I%]*K[I%,J%]
7020 KKJJ  = K[J%,J%]*M[I%,J%]  - M[J%,J%]*K[I%,J%]
7030 KK    = K[I%,I%]*M[J%,J%]  - K[J%,J%]*M[I%,I%]
7040 IF(KK > 0) THEN MULT=1 ELSE MULT=-1
7050 X     = KK/2 + MULT*SQR(KK*KK/4 + KKII*KKJJ)
7060 ALPHA =  KKJJ/X
7070 BETA  = -KKII/X
7080 RETURN
7090 '
7100 '**********************     TRANSFORM MATRICES     ************************
7110 '
7120 ' ...  Transform the stiffness matrix
7130 '
7140 FOR K% = 1 TO NDOF%
7150   TEMP       = K[I%,K%]
7160   K[I%,K%] = TEMP + BETA*K[J%,K%]
7170   K[J%,K%] = ALPHA*TEMP + K[J%,K%]
7180 NEXT K%
7190 '
7200 FOR K% = 1 TO NDOF%
7210   TEMP       = K[K%,I%]
7220   K[K%,I%] = TEMP + BETA*K[K%,J%]
7230   K[K%,J%] = ALPHA*TEMP + K[K%,J%]
7240 NEXT K%
7250 '
7260 ' ...  Transform the mass matrix
7270 '
7280 FOR K% = 1 TO NDOF%
```

```
7290   TEMP      = M[I%,K%]
7300   M[I%,K%] = TEMP + BETA*M[J%,K%]
7310   M[J%,K%] = ALPHA*TEMP + M[J%,K%]
7320 NEXT K%
7330 '
7340 FOR K% = 1 TO NDOF%
7350   TEMP      = M[K%,I%]
7360   M[K%,I%] = TEMP + BETA*M[K%,J%]
7370   M[K%,J%] = ALPHA*TEMP + M[K%,J%]
7380 NEXT K%
7390 '
7400 ' ...  Transform PHI
7410 '
7420 FOR K% = 1 TO NDOF%
7430   TEMP        =  PHI[K%,I%]
7440   PHI[K%,I%]=  TEMP+BETA*PHI[K%,J%]
7450   PHI[K%,J%]= ALPHA*TEMP+PHI[K%,J%]
7460 NEXT K%
7470 RETURN
7480 '
7490 '**********************     CALCULATE EIGENVALUES     *********************
7500 '
7510 FOR K% = 1 TO NDOF%
7520   THETA[K%,1] = SQR(ABS(K[K%,K%]/M[K%,K%]))
7530   THETA[K%,2] = K%
7540 NEXT K%
7550 '
7560 ' ...  Normalise eigenvectors
7570 '
7580 FOR K% = 1 TO NDOF%
7590   TEMP = 0
7600   FOR L% = 1 TO NDOF%
7610     IF ABS(PHI[L%,K%])>ABS(TEMP) THEN TEMP=PHI[L%,K%]
7620   NEXT L%
7630   FOR L% = 1 TO NDOF%
7640     PHI[L%,K%] = PHI[L%,K%]/TEMP
7650   NEXT L%
7660 NEXT K%
7670 '
7680 ' ... Sort eigenvalues into order
7690 '
7700 DONE% = FALSE%
7710 WHILE DONE% = FALSE%
7720   DONE% = TRUE%
7730   FOR K% = 1 TO NDOF%-1
7740     IF THETA[K%,1]<=THETA[K%+1,1] THEN GOTO 7780
7750     DONE% = FALSE%
7760     TEMP  =THETA[K%,1] : THETA[K%,1]=THETA[K%+1,1] : THETA[K%+1,1]=TEMP
7770     TEMP  =THETA[K%,2] : THETA[K%,2]=THETA[K%+1,2] : THETA[K%+1,2]=TEMP
7780   NEXT K%
7790 WEND
7800 RETURN
7810 '
7820 '********************     OUTPUT THE EIGENPAIRS     **********************
7830 '
7840 CLS
7850 PRINT #2,
7860 PRINT #2, "* * * * * * * * * * * * * *"
7870 PRINT #2, "*   FREQUENCIES AND   *             * :-  RESTRAINT"
7880 PRINT #2, "*   MODE SHAPES        *"
7890 PRINT #2, "* * * * * * * * * * * * * *"
7900 LCOUNT% = 5
7910 K% = 1
7920 WHILE K% <= NDOF%
7930   IF (K%+4) > NDOF% THEN L% = NDOF% ELSE L% = K%+4
7940   GOSUB 9450 : PRINT #2, : PRINT #2, "NUMBER           ";
7950   FOR J% = K% TO L%
7960     PRINT #2, USING "     ###     "; J%;
7970   NEXT J%
7980   GOSUB 9450 : PRINT #2,
7990   PRINT #2, "FREQUENCY (rad/s)";
8000   FOR J% = K% TO L%
8010     PRINT #2, USING "  #.####^^^^"; THETA(J%,1);
8020   NEXT J%
```

```
8030   GOSUB 9450 : PRINT #2,
8040   PRINT #2, "FREQUENCY    (Hz)";
8050   FOR J% = K% TO L%
8060     PRINT #2, USING "  #.####^^^^"; THETA(J%,1)/(2*3.14159);
8070   NEXT J%
8080   GOSUB 9450 : PRINT #2, : GOSUB 9450 : PRINT #2,
8090   GOSUB 9450 : PRINT #2, "NODE  DIR"
8100   FOR I% = 1 TO 3*NNODE%
8110     PRINT #2, USING " ##    #          "; (I%+2)\3; I% - 3*((I%-1)\3);
8120     FOR J% = K% TO L%
8130       IF FREE%(I%) = 0 THEN PRINT #2, "       *    "; : GOTO 8150
8140       PRINT #2, USING "  #.####^^^^"; PHI(FREE%(I%),THETA(J%,2));
8150     NEXT J%
8160     GOSUB 9450 : PRINT #2,
8170   NEXT I%
8180   GOSUB 9360
8190   K% = L% + 1
8200 WEND
8210 RETURN
8220 '
8230 '********    STIFFNESS FACTORS FOR DIFFERENT END CONNECTIONS   *********
8240 '
8250 ' ... Zero stiffness factore (i.e. pinned/pinned element)
8260 '
8270 ON ELEM(K%,4) GOTO 8310, 8390, 8470, 8550
8280 '
8290 ' ... Rigid/rigid element
8300 '
8310 K1 =  12 : K2  =  6 : K3 =   6 : K4 =   4
8320 K5 =  -6 : K6  =  2 : K7 =  -6 : K8 =   4
8330 M1 = 312 : M2  = 44 : M3 = 108 : M4 = -26 : M5  =   8
8340 M6 =  26 : M7 =  -6 : M8 = 312 : M9 = -44 : M10 =   8
8350 GOTO 8590
8360 '
8370 ' ... Rigid/pinned element
8380 '
8390 K1 =   3 : K2 =   3 : K3 =   0 : K4 =   3
8400 K5 =  -3 : K6 =   0 : K7 =   0 : K8 =   0
8410 M1 = 408 : M2 =  72 : M3 = 117 : M4 =   0 : M5  =  16
8420 M6 =  33 : M7 =   0 : M8 = 198 : M9 =   0 : M10 =   0
8430 GOTO 8590
8440 '
8450 ' ... Pinned/rigid element
8460 '
8470 K1 =   3 : K2 =   0 : K3 =   3 : K4 =   0
8480 K5 =   0 : K6 =   0 : K7 =  -3 : K8 =   3
8490 M1 = 198 : M2  =  0 : M3 = 117 : M4 = -33 : M5  =   0
8500 M6 =   0 : M7 =   0 : M8 = 408 : M9 = -72 : M10 =  16
8510 GOTO 8590
8520 '
8530 ' ... Pinned/pinned element
8540 '
8550 K1 =   0 : K2 =   0 : K3 =   0 : K4 =   0
8560 K5 =   0 : K6 =   0 : K7 =   0 : K8 =   0
8570 M1 =   0 : M2 =   0 : M3 =   0 : M4 =   0 : M5  =   0
8580 M6 =   0 : M7 =   0 : M8 =   0 : M9 =   0 : M10 =   0
8590 RETURN
8600 '
8610 '********************    GET KEYBOARD RESPONSE    ********************
8620 '
8630 ROW% = CSRLIN: COL% = POS(0)
8640 REPLY% = 0
8650 WHILE REPLY% = 0
8660   LOCATE ROW%, COL%: INPUT "", WRKSTR$: REPLY% = VAL(WRKSTR$)
8670   IF (REPLY% < 1) OR (REPLY% > NOPTIONS%) THEN GOTO 8700
8680   IF (REPLY% < 10) AND (LEN(WRKSTR$) = 1) THEN GOTO 8710
8690   IF (REPLY% > 9)  AND (LEN(WRKSTR$) = 2) THEN GOTO 8710
8700   SOUND 700, 4: REPLY% = 0: LOCATE ROW%, COL%
8710 WEND
8720 PRINT
8730 RETURN
8740 '
8750 '***********************    RECORD NUMBERS    ****************************
8760 '
```

```
8770 NNOS% = 0 : DONE% = TRUE%
8780 IF EOF(1)<>0 THEN DATAEND% = TRUE% : GOTO 9090
8790 '
8800 ' ... Read next line and check if it is blank
8810 '
8820 LINE INPUT #1, WRKSTR$
8830 LCOUNT% = LCOUNT% + 1 : LOCATE ROW%, COL% : PRINT LCOUNT%;
8840 FOR I% = 1 TO LEN(WRKSTR$)
8850   IF MID$(WRKSTR$,I%,1)<> " " THEN DONE% = FALSE% : I% = LEN(WRKSTR$)
8860 NEXT I%
8870 IF DONE% = TRUE% THEN GOTO 9090
8880 FOR I% = 1 TO 8
8890   WKSP$(I%) = ""
8900 NEXT I%
8910 '
8920 ' ... Search for numbers
8930 '
8940 FOR I% = 1 TO LEN(WRKSTR$)
8950   WHILE (MID$(WRKSTR$, I%, 1) = " ") AND I% <= LEN(WRKSTR$)
8960     I% = I% + 1
8970   WEND
8980   IF (I% > LEN(WRKSTR$)) THEN GOTO 9070
8990   IF NNOS% < 8 THEN GOTO 9030
9000   PRINT
9010   PRINT "*** ERROR *** Too many numbers on line - run DCHECK"
9020   GOSUB 9360 : GOTO 9080
9030   NNOS% = NNOS% + 1: WKSP$(NNOS%) = ""
9040   WHILE (MID$(WRKSTR$, I%, 1) <> " ") AND I% <= LEN(WRKSTR$)
9050     WKSP$(NNOS%) = WKSP$(NNOS%) + MID$(WRKSTR$,I%,1): I% = I% + 1
9060   WEND
9070 NEXT I%
9080 WRKSTR$ = ""
9090 RETURN
9100 '
9110 '***************************    FIND STRING     ****************************
9120 '
9130 STRFOUND% = FALSE% : DONE% = FALSE%
9140 WHILE DONE% = FALSE%
9150   IF EOF(1)<>0 THEN DATAEND% = TRUE% : DONE% = TRUE% : GOTO 9240
9160   LINE INPUT #1, WRKSTR$
9170   LCOUNT% = LCOUNT% + 1 : LOCATE ROW%, COL% : PRINT LCOUNT%;
9180   IF INSTR(WRKSTR$,TARGET$)>0 THEN STRFOUND% = TRUE% : DONE% = TRUE%
9190   IF(TARGET$="PFDYNAM") AND LCOUNT% > 4 THEN DONE% = TRUE%
9200   '
9210   ' ... Check for new module
9220   '
9230   IF LEFT$(WRKSTR$,4) ="+   " THEN DONE% = TRUE%
9240 WEND
9250 RETURN
9260 '
9270 '***************************    READ LINES     *****************************
9280 '
9290 FOR I% = 1 TO NLINES%
9300   IF EOF(1)<>0 THEN DATAEND% = TRUE% : I% = NLINES% : GOTO 9330
9310   LINE INPUT #1, WRKSTR$
9320   LCOUNT% = LCOUNT% + 1 : LOCATE ROW%, COL% : PRINT LCOUNT%;
9330 NEXT I%
9340 RETURN
9350 '
9360 '*********************    ANY KEY TO CONTINUE    ***********************
9370 '
9380 IF DEVICE$<>"SCRN:" THEN GOTO 9430
9390 LOCATE PAGELEN%+1, 1
9400 PRINT SPACE$(40); "Press any key to continue";
9410 WHILE INKEY$ = "" : WEND
9420 CLS : LCOUNT%=0
9430 RETURN
9440 '
9450 '**********************    CHECK FOR PAGE END    *********************
9460 '
9470 IF LCOUNT%>=PAGELEN% THEN GOSUB 9360
9480 LCOUNT% = LCOUNT% + 1
9490 RETURN
9500 '
9510 ' =======================    PFDYNAM.BAS    ========================
```

Data Generation

The program PRE.BAS is used to generate data for PFDYNAM.BAS (PRE.BAS is described in Chapter 5). Once complete the data can be checked with DCHECK.BAS. Input from the keyboard is underlined.

```
===================================================================

 PROGRAM PRE                          Copyright (c) J.Balfour 1991

    This program preprocesses data for the analysis programs
    PTRUSS, STRUSS, PFRAME, PFDYNAM, GRID, and SFRAME

               For further information contact
 James A.D.Balfour, Heriot-Watt University, Riccarton, Edinburgh
              Tel 031-449-5111, Fax 031-451-3170

===================================================================


Do you wish to
  (1)  Create a new data file
  (2)  Modify an existing data file
1

Name of the data file  =  PFDYNAM1.DAT


Note that pressing 'return' sets the current value equal to the
corresponding value for the previously entered data item. For
example pressing the return key when prompted for, say, an
x-coordinate will set that coordinate equal to the x-coordinate
of the previous node.


Select problem type
  (1) Plane truss
  (2) Space truss
  (3) Plane frame - static  analysis
  (4) Plane frame - dynamic analysis
  (5) Grillage
  (6) Space frame
4


Maximum problem size for program PFDYNAM is as follows:

    maximum number of nodes          =  17
    maximum number of elements       =  25
    maximum number of properties     =  10
    maximum number of restraints     =  10
    maximum number of local axes     =  10
    maximum number of nodal loads    =  1
    maximum number of added masses   =  30
    maximum semi-bandwidth           =  51


The semi-bandwidth is determined by the element with the biggest
difference between the node numbers at its ends (MaxDiff) and
the number and location of the restraints.

An overestimate of the semi-bandwidth can be made from the
following expression
                       (MaxDiff+1) * 3


Input job title and units - job title is simply a string to
                            to identify the problem

                          - note that the units string is for
                            information only and is ignored
                            during calculation. Consistent units
                            must be used throughout.
```

```
                              - use units of Newtons, metres and
                                kilograms for dynamic analysis

Job title                         = CASF 2nd Edition - Example 11.4

Units (force, length and mass) = N m and kg


Input nodal coordinates

Nodal coordinates - set node number to
                        0 to stop
                        H for help
                        L to list nodal coordinates data
                  - node numbers must be consecutive integers
                    starting at 1 (can be entered in any order)
                  - set node number negative to cancel node

Node number       =  1
X-coordinate      =  0
Y-coordinate      =  0

Node number       =  2
X-coordinate      =  6
Y-coordinate      =  0

Node number       =  3
X-coordinate      =  10
Y-coordinate      =  0

Node number       =  0

+ + + + + + + + + + + + +
+   NODAL COORDINATES   +
+ + + + + + + + + + + + +

NODE           X             Y
  1            0             0
  2            6             0
  3            10            0

Edit nodal coordinates - set node number to H for help

Node number       =  0


Input element data

Element data - set lower node number to
                    0 to stop
                    H for help
                    L to list element data
             - set property no. to zero to cancel element
             - element types :  no pins        = 1,     pin at end 'j'      = 2
                                pin at end 'i' = 3,     pin at 'i' and 'j' = 4


Lower  node number (i)  =  1
Higher node number (j)  =  2
Property number         =  1
Element type            =  1

Lower  node number (i)  =  2
Higher node number (j)  =  3
Property number         =  1
Element type            =  1

Lower  node number (i)  =  0

+ + + + + + + + +
+   ELEMENTS    +
+ + + + + + + + +

ELEMENT    PROPERTY     TYPE
  1   2        1           1
  2   3        1           1
```

```
Edit element data - set lower node number to H for help

Lower  node number (i)  =  0

Input property data

Element properties - set property number to
                        0 to stop
                        H for help
                        L to list property data
                   - set cross sectional area to 0 to cancel property
                   - hit return to retain any value

Property number             =  1
Cross-sectional area  (A)   =  0.05
Second moment of area (I)   =  1.25E-6
Elastic constant      (E)   =  200E9
Mass per unit length  (m)   =  500

Property number             =  0

+ + + + + + + + + +
+   PROPERTIES    +
+ + + + + + + + + +

PROPERTY        A            I            E            m
    1     0.5000E-01   0.1250E-05   0.2000E+12   0.5000E+03

Edit property data - set property number to H for help

Property number             =  0

Input restraints

Restraints - set node number to
                0 to stop
                H for help
                L to list restraint data
           - set restraint direction(s) to 0 to cancel
           - restraint directions are
                x-translation = 1
                y-translation = 2
                rotation      = 3
           - enter restraints as a composite number e.g. if a
             node is restrained in directions 1 and 2 enter 12

Node to be restrained  =  1
Direction(s)           =  123

Node to be restrained  =  2
Direction(s)           =  12

Node to be restrained  =  3
Direction(s)           =  12

Node to be restrained  =  0

+ + + + + + + + + +
+   RESTRAINTS    +
+ + + + + + + + + +

NODE    DIRECTION(S)
  1        123
  2         12
  3         12

Edit restraints - set node number to H for help

Node to be restrained  =  0


Input local axes

Local axes - set node number to
                0 to stop
                H for help
                L to list local axes data
```

```
           - set rotation to zero to cancel local axes
           - note that anticlockwise rotation from the global
             axes is +ve

Node            =  0

Edit local axes - set node to H for help

Node            =  0


Input added masses

Added masses - set node number to
                    0 to stop
                    H for help
                    L to list added mass data
               - set mass to 0 to cancel
               - lumped masses are assumed to have no rotationalinertia

Node =  0

Edit added masses - set node to H for help

Node =  0


Do you wish to
   (1) Edit data
   (2) Check the data (data will be written to file first)
   (3) List all of the current data to the screen
   (4) List all of the current data to the printer
   (5) Analyse (write data to file first)
   (6) Create/edit a new file (data will be written to file first)
   (7) Stop (data will be written to file first)
7

Wait - writing data to file PFDYNAM1.DAT
```

Sample Run 1

The following example shows PFDYNAM.BAS being used to analyse the structure example 11.4. Note that the output shown here has been produced by selecting the "Analyse, output to file" option.

```
===========================================================================

    PROGRAM PFDYNAM                          Copyright(c) J.A.D.Balfour 1991

                         CASF 2nd Edition - Example 11.4

    File Name :- PFDYNAM1.DAT                                Date :- 05-03-92
    Units     :- N,m,kg                                      Time :- 20:07:38

===========================================================================


+ + + + + + + + + + + + +
+   NODAL COORDINATES   +
+ + + + + + + + + + + + +

NODE            X             Y
  1             0             0
  2             6             0

+ + + + + + + + +
+   ELEMENTS    +
+ + + + + + + + +

ELEMENT     PROPERTY      TYPE
  1   2         1           0
```

```
+ + + + + +  + + +
+   PROPERTIES   +
+ + + + + +  + + +

PROPERTY        A               I            E            m
    1      0.1000E+01      0.1250E-05   0.2000E+12   0.5000E+03

+ + + + + + + + + +
+   RESTRAINTS    +
+ + + + + + + + + +

NODE     DIRECTION(S)
  1          12
  2          12

* * * * * * * * * * * *
*   FREQUENCIES AND   *              * :-  RESTRAINT
*   MODE SHAPES       *
* * * * * * * * * * * *

NUMBER                      1           2
FREQUENCY (rad/s)  0.6804E+01  0.3118E+02
FREQUENCY    (Hz)  0.1083E+01  0.4963E+01

NODE  DIR
  1    1                *           *
  1    2                *           *
  1    3       0.1000E+01  0.1000E+01
  2    1                *           *
  2    2                *           *
  2    3       -.1000E+01  0.1000E+01
```

Sample Run 2

The following example shows PFDYNAM.BAS being used to analyse a structure taken from "Finite Element Methods in Structural Mechanics", C.T.F.Ross[18], page 244. Note that the output shown here has been produced by selecting the "Analyse, output to file" option.

```
==========================================================================

    PROGRAM PFDYNAM.BAS                      Copyright(c) J.A.D.Balfour

                         Ross F.E. Methods 7.4.7 p 244

    File Name :- DATA\PFD3                                Date :- 19-11-91
    Units     :- N m and kg                               Time :- 18:37:08

==============================================================================

+ + + + + + + + + + + + + +
+   NODAL COORDINATES     +
+ + + + + + + + + + + + + +

Node            X             Y
  1             0             0
  2             0             1.5
  3             0             3
  4             5             3
  5             10            3
  6             15            3
  7             20            3
  8             20            1.5
  9             20            0

+ + + + + + + + +
+   ELEMENTS    +
+ + + + + + + + +

ELEMENT     PROPERTY      TYPE
  1  2          1           1
```

```
  2  3          1          1
  3  4          2          1
  4  5          2          1
  5  6          2          1
  6  7          2          1
  7  8          1          1
  8  9          1          1

+ + + + + +  + + +
+   PROPERTIES   +
+ + + + + +  + + +

PROPERTY        A                 I           E            m
    1      0.2400E-02       0.4190E-05  0.6670E+11  0.6290E+01
    2      0.6800E-02       0.6491E-04  0.6670E+11  0.1782E+02

+ + + + + + + + + + +
+   RESTRAINTS      +
+ + + + + + + + + + +

NODE     DIRECTION(S)
  1          123
  9          123

* * * * * * * * * * * * 
*   FREQUENCIES AND    *                * :-  RESTRAINT
*      MODE SHAPES     *
* * * * * * * * * * * * 

NUMBER                      1            2            3            4            5
FREQUENCY (rad/s)  0.1537E+02   0.2380E+02   0.5256E+02   0.1135E+03   0.1560E+03
FREQUENCY    (Hz)  0.2447E+01   0.3787E+01   0.8365E+01   0.1807E+02   0.2483E+02

NODE  DIR
  1    1                *            *            *            *            *
  1    2                *            *            *            *            *
  1    3                *            *            *            *            *
  2    1       0.4649E-01   0.4680E+00   -.1462E+00   -.3256E+00   -.3191E-01
  2    2       0.2433E-03   0.5897E-04   -.1517E-02   -.4294E-02   -.1555E-03
  2    3       -.3003E-01   -.4836E+00   0.1167E+00   0.3064E+00   0.1400E-01
  3    1       -.7617E-02   0.9993E+00   -.5708E-01   -.7016E-02   -.1036E+00
  3    2       0.4866E-03   0.1179E-03   -.3034E-02   -.8582E-02   -.3113E-03
  3    3       0.1334E+00   -.1117E+00   -.2811E+00   -.4700E+00   0.1194E-01
  4    1       -.7360E-02   0.9999E+00   -.5736E-01   -.4703E-02   -.1055E+00
  4    2       0.6843E+00   -.2130E+00   -.9978E+00   -.6889E+00   0.3228E-02
  4    3       0.1173E+00   0.1752E-01   -.8817E-02   0.3421E+00   -.2433E-02
  5    1       -.7102E-02   0.1000E+01   -.5748E-01   -.2331E-02   -.1099E+00
  5    2       0.1000E+01   0.3879E-01   0.8035E-02   0.1000E+01   -.5919E-02
  5    3       -.5497E-03   0.6548E-01   0.3216E+00   -.2326E-01   -.4754E-03
  6    1       -.6841E-02   0.9995E+00   -.5744E-01   0.7032E-04   -.1170E+00
  6    2       0.6803E+00   0.2590E+00   0.1000E+01   -.7654E+00   0.4619E-01
  6    3       -.1175E+00   0.6674E-02   -.1276E-01   -.3134E+00   0.2773E-01
  7    1       -.6579E-02   0.9985E+00   -.5725E-01   0.2471E-02   -.1269E+00
  7    2       0.4902E-03   -.3632E-04   0.3071E-02   -.9406E-02   -.4309E-02
  7    3       -.1313E+00   -.1153E+00   -.2757E+00   0.4217E+00   -.1628E+00
  8    1       -.5231E-01   0.4487E+00   -.1260E+00   0.1609E+00   -.8481E+00
  8    2       0.2451E-03   -.1815E-04   0.1536E-02   -.4706E-02   -.2152E-02
  8    3       0.3558E-01   -.4595E+00   0.8250E-01   -.1651E-01   0.1000E+01
  9    1                *            *            *            *            *
  9    2                *            *            *            *            *
  9    3                *            *            *            *            *

NUMBER                      6            7            8            9           10
FREQUENCY (rad/s)  0.1581E+03   0.1610E+03   0.1661E+03   0.2260E+03   0.3604E+03
FREQUENCY    (Hz)  0.2516E+02   0.2563E+02   0.2644E+02   0.3597E+02   0.5735E+02

NODE  DIR
  1    1                *            *            *            *            *
  1    2                *            *            *            *            *
  1    3                *            *            *            *            *
  2    1       -.9225E-02   -.9143E+00   0.8465E+00   0.1720E+00   -.9355E-01
  2    2       0.6605E-03   0.3004E-02   -.8058E-03   -.1908E-01   0.3409E-01
  2    3       0.1309E-01   0.1000E+01   0.1000E+01   -.2376E+00   0.1402E+00
  3    1       -.2695E-01   0.1890E+00   0.2992E-02   0.1193E+00   0.2478E+00
  3    2       0.1320E-02   0.6004E-02   -.1614E-02   -.3807E-01   0.6778E-01
  3    3       0.4262E-01   -.2017E+00   0.5700E-01   -.8308E+00   0.1000E+01
  4    1       -.2634E-01   0.2055E+00   0.2671E-02   0.1070E+00   0.2432E+00
```

4	2	0.4075E-01	0.3949E+00	0.1391E-01	-.1050E+00	-.6195E+00
4	3	-.4373E-01	0.3737E-01	-.9167E-02	0.9968E+00	-.8785E+00
5	1	-.2508E-01	0.2167E+00	0.2422E-02	0.8934E-01	0.2082E+00
5	2	-.6659E-01	-.3166E+00	-.1879E-02	0.1166E-01	0.9470E+00
5	3	0.3373E-01	-.1100E+00	0.2979E-03	-.9995E+00	-.1706E-01
6	1	-.2321E-01	0.2225E+00	0.2239E-02	0.6726E-01	0.1471E+00
6	2	0.9155E-01	0.2400E+00	0.8552E-03	0.8086E-01	-.5976E+00
6	3	-.1621E-01	0.1607E+00	0.6485E-03	0.1000E+01	0.9011E+00
7	1	-.2077E-01	0.2226E+00	0.2117E-02	0.4183E-01	0.6773E-01
7	2	0.1268E-02	0.6540E-02	-.9661E-04	0.3777E-01	0.6887E-01
7	3	0.5502E-01	-.1162E+00	-.3453E-02	-.8529E+00	-.9735E+00
8	1	0.9110E+00	0.1745E-01	-.8207E-02	-.8849E-02	-.2512E-01
8	2	0.6345E-03	0.3274E-02	-.4824E-04	0.1893E-01	0.3463E-01
8	3	0.1000E+01	-.6429E-01	0.9659E-02	0.1788E+00	0.7975E-01
9	1	*	*	*	*	*
9	2	*	*	*	*	*
9	3	*	*	*	*	*

NUMBER		11	12	13	14	15
FREQUENCY (rad/s)		0.5082E+03	0.5677E+03	0.6638E+03	0.9068E+03	0.1157E+04
FREQUENCY (Hz)		0.8089E+02	0.9035E+02	0.1056E+03	0.1443E+03	0.1842E+03
NODE	DIR					
1	1	*	*	*	*	*
1	2	*	*	*	*	*
1	3	*	*	*	*	*
2	1	-.1716E-02	-.2074E-02	-.2603E-01	-.7153E-02	-.1227E-01
2	2	0.3595E-01	-.4816E-03	0.6103E-01	0.9131E-01	0.3190E-01
2	3	0.1289E-01	0.1665E-02	0.4574E-01	0.1538E-01	0.2300E-01
3	1	-.4020E+00	-.8124E-01	0.2654E+00	-.8238E-01	0.7094E+00
3	2	0.7109E-01	-.9771E-03	0.1197E+00	0.1761E+00	0.5992E-01
3	3	0.8648E+00	0.8517E-02	0.1000E+01	0.9193E+00	0.1828E+00
4	1	-.3482E+00	-.9669E-01	0.2155E+00	-.5004E-01	0.3033E+00
4	2	-.5440E+00	-.1269E-02	-.4335E+00	-.5626E-01	0.5391E-01
4	3	-.1349E+00	0.1114E-02	0.3539E+00	0.1000E+01	0.3247E+00
5	1	-.2097E+00	-.1445E+00	0.7857E-01	0.1791E-01	-.4300E+00
5	2	0.1421E+00	0.2416E-02	-.2938E+00	-.1328E+00	0.1071E+00
5	3	-.9019E+00	0.1641E-02	-.8601E+00	0.7100E+00	0.5604E+00
6	1	-.2012E-01	-.2405E+00	-.9002E-01	0.7312E-01	-.6995E+00
6	2	0.5163E+00	0.1361E+00	0.2152E+00	-.1702E+00	0.1851E+00
6	3	0.3474E+00	0.1466E+00	-.9837E+00	0.2449E+00	0.1000E+01
7	1	0.1743E+00	-.4169E+00	-.2223E+00	0.7631E-01	-.2144E+00
7	2	-.3781E-01	0.1146E+00	-.8524E-01	0.1052E+00	-.8827E+00
7	3	0.1000E+01	0.1000E+01	-.2943E+00	-.4889E-01	0.7671E+00
8	1	0.2803E-01	-.4788E-02	-.1076E-01	0.1078E-02	0.1726E-02
8	2	-.1912E-01	0.5648E-01	-.4346E-01	0.5457E-01	-.4685E+00
8	3	-.6415E-01	0.1774E-01	0.2183E-01	-.1900E-02	-.5413E-02
9	1	*	*	*	*	*
9	2	*	*	*	*	*
9	3	*	*	*	*	*

NUMBER		16	17	18	19	20
FREQUENCY (rad/s)		0.1392E+04	0.2093E+04	0.2250E+04	0.3229E+04	0.6172E+04
FREQUENCY (Hz)		0.2215E+03	0.3332E+03	0.3581E+03	0.5140E+03	0.9823E+03
NODE	DIR					
1	1	*	*	*	*	*
1	2	*	*	*	*	*
1	3	*	*	*	*	*
2	1	-.4344E-02	-.3930E-02	-.6616E-02	0.6993E-02	0.1930E-05
2	2	-.1606E-01	-.4350E+00	0.2787E+00	-.7800E-01	-.4940E-02
2	3	0.8583E-02	0.9664E-02	0.1488E-01	-.1733E-01	-.4296E-05
3	1	0.3494E+00	0.3353E-01	0.1000E+01	-.9703E+00	-.1130E-02
3	2	-.2945E-01	-.7114E+00	0.4413E+00	-.9386E-01	0.7754E-03
3	3	-.2062E-01	0.1000E+01	-.6744E+00	0.1788E+00	0.2416E-02
4	1	0.8046E-01	-.2729E-01	-.4476E+00	0.1000E+01	0.1917E-02
4	2	-.1600E-01	-.7164E-01	0.4922E-01	-.1926E-01	0.2223E-02
4	3	-.7018E-01	0.1860E+00	-.1476E+00	0.4136E-01	0.6233E-02
5	1	-.3047E+00	-.1967E-01	-.6782E+00	-.8135E+00	-.4950E-02
5	2	-.4022E-01	-.2220E-01	0.5006E-02	-.1010E-01	0.8079E-02
5	3	-.1508E+00	0.5877E-01	-.7423E-01	0.1043E-02	0.2041E-01
6	1	-.2497E+00	0.3718E-01	0.9352E+00	0.4510E+00	0.1378E-01
6	2	-.7858E-01	-.9286E-02	-.3106E-01	-.1798E-01	0.2456E-01
6	3	-.3319E+00	0.1488E-01	-.1090E+00	-.3752E-01	0.6715E-01
7	1	0.1659E+00	0.7486E-03	0.5877E-02	0.9019E-02	-.3873E-01
7	2	0.1000E+01	-.2062E-01	-.3032E+00	-.8640E-01	-.1569E+00
7	3	0.2703E+00	-.2018E-01	-.4667E+00	-.1644E+00	0.1449E-01

```
  8    1           0.3472E-02  -.6615E-04  -.1565E-02  -.4067E-03  -.2395E-03
  8    2           0.5454E+00  -.1260E-01  -.1915E+00  -.7195E-01  0.1000E+01
  8    3           -.8116E-02  0.1778E-03  0.4217E-02  0.1145E-02  0.6594E-03
  9    1                *           *           *           *           *
  9    2                *           *           *           *           *
  9    3                *           *           *           *           *

NUMBER                  21
FREQUENCY (rad/s)  0.6961E+04
FREQUENCY    (Hz)  0.1108E+04

NODE  DIR
  1    1                *
  1    2                *
  1    3                *
  2    1           -.1426E-02
  2    2           0.1000E+01
  2    3           0.3978E-02
  3    1           -.1797E-01
  3    2           -.5004E+00
  3    3           0.6421E+00
  4    1           0.5856E-02
  4    2           -.4037E-01
  4    3           0.9141E-01
  5    1           -.1898E-02
  5    2           -.1015E-01
  5    3           0.2485E-01
  6    1           0.6448E-03
  6    2           -.3099E-02
  6    3           0.6706E-02
  7    1           0.2638E-03
  7    2           -.2301E-02
  7    3           -.3293E-03
  8    1           0.2673E-04
  8    2           0.4851E-02
  8    3           0.2400E-04
  9    1                *
  9    2                *
  9    3                *
```

Chapter 12

Buying and Using Commercial Frame Analysis Programs

12.1 Introduction

This book is intended to give engineering students and practising engineers an understanding of the theory behind commercial frame analysis programs. It is also intended give some insight into how that theory is turned into useful computer programs.

The user of commercial frame analysis programs need know little about the program to obtain results. There is general agreement, however, that better and safer use of the program will come from the user having some understanding of the structural theory underlying the program and an appreciation of how that theory might have been implemented.

Not so many years ago, civil engineers made very limited use of computers for structural analysis. In those days few engineering concerns owned a computer, being forced to use large structural analysis programs supported by computer bureaux. These programs were sometimes unnecessarily complicated to use. Often the engineers using them had little experience of computing, and hence an underlying distrust of computers. On top of this computing costs were invariably high.

This situation has now dramatically changed. Today's engineers are much more computer-conscious: powerful, low-cost microcomputers are widely available, and using computers to analyse structures has become the norm rather than the exception. There are a number of reasons behind the growth in popularity of computer-based structural analysis. Firstly (and perhaps most importantly), computer-based structural analysis is cost-effective. Secondly, computer methods often yield "better" results because they demand less idealisation of the true structural behaviour. Finally it can be argued that computer analysis is less error-prone than traditional structural analysis because the arithmetic is done by the computer, rather than the engineer.

12.2 Buying a Frame Analysis Program

Program Language

Frame analysis programs are usually written in either BASIC or FORTRAN. BASIC is a compact, interpretive language that is widely used on microcomputers. The early versions of BASIC were heavily criticised for their lack of structure which tended to encourage poor programming habits. Later (extended) versions of BASIC are much better structured. The disadvantages of using extended versions of BASIC are that they require more memory and they may run more slowly. On some machines the BASIC interpreter is stored in read only memory (ROM) although the trend in now to load the interpreter into random access memory (RAM) from disk. The principal advantages of BASIC interpreters are their low cost, wide availability, and simplicity. Principal disadvantages are their slowness and the number of variants of the language. The increase in the power of desktop micros has led to BASIC compilers becoming increasingly readily available. These compilers allow an executable file to be produced from a BASIC program. A compiled BASIC program will typically run an order of magnitude faster than the same program running under an interpreter.

FORTRAN was one of the world's first high-level computing languages. It is a compiled language that is widely used by engineers and scientists. FORTRAN compilers require large amounts of memory which a few years ago restricted FORTRAN to mini and mainframe computers. Modern desktop computers are now so powerful that they have no difficulty in supporting FORTRAN compliers. Compiled code runs many times faster than code that has to be interpreted. This gives FORTRAN a distinct advantage over interpreted BASIC for the solution of large structural problems. Another advantage of FORTRAN is that it is a highly standardised language. This allows the same program to be run on different machines with little or no modification. The speed and portability of FORTRAN ensure that it will continue to be an important language in scientific computing.

C is currently the language of choice for the development of applications in business and commerce, and is gaining popularity amoung scientists and engineerings. The C language was originally design as a language for the development of operating systems (for instance UNIX is written in C). C is a more highly structured language than FORTRAN and it allows the programmer much better access to the computer hardware. A number of commercial structural engineering programs have recently been converted from FORTRAN to C, and it may be that C will be eventually supersede FORTRAN.

Matching Machine, Language, and Program

The engineer faced with the task of buying a frame analysis program and a computer system to run it on will find a bewildering variety of choice. One of his prime considerations will almost certainly be cost. At one end of the spectrum are general purpose finite element packages (which invariably support line elements) requiring powerful computers, where the cost of hardware plus software might run into tens of thousands of pounds. At the other end of the spectrum are simple frame analysis programs designed to run on inexpensive microcomputers, where the total cost of the system will be measured in hundreds of pounds. Many desktop computers are more powerful than mainframe computers of only a few years ago and there is an extensive range of structural analysis software that will run on them. Much of this software has been ported over from mainframe systems. Consequently desktop machines offer a cost-effective solution to all but the most demanding structural analysis tasks.

Commercial software is invariably sold only as compiled code. This is done for security reasons. It precludes the purchaser from modifying the code (e.g. the format of the output), and the program is machine-specific (i.e. compiled code has no portability). Most structural analysis software comes with some copy protection. This might a key disk or a dongle. A key disk cannot be copied using standard copying methods, and the disk must in the computer while the software is being used. A dongle is a hardware device that must be plugged into a computer interface (usually the parallel printer port, as almost all computers have one). The software will periodically check for the presence of the dongle and will abort if it is not present.

If many engineers are likely to need simultaneous access to the program then there are many ways to meet this requirement. These range from giving each engineer (or small group of engineers) a stand-alone system, through networked stand-alone systems, to terminals into a single, centralised mainframe computer.

There are always likely to be a number of alternatives and the program and machine purchased will depend upon the budget, the existing facilities, the hardware required by the program, the need for data sharing, the number of engineers involved, and the likely program usage. Each situation must be individually assessed because the computing requirements of each engineering concern are unique, and the cost and nature of the available hardware and software vary continuously.

The Scope of the Program

The facilities offered by different frame analysis programs vary widely. Not all programs deal with all types of structural frameworks. The maximum problem size that the program will deal with is usually related to the size of the computer's memory and available disk space. Modern frame analysis programs allow large problems to be solved by using equation solving techniques that require only part of the structure stiffness matrix to be held in the computer's memory at any one time, with the rest being stored on disk. Such equation solving techniques are quite complex and require a fair amount of coding. As they make heavy use of secondary storage they tend to be much slower than techniques that don't use secondary storage. Element loads, local axes, elastic supports, and elements with pins are facilities that are offered by some programs, but not others. The engineer should always check that any program considered offers the facilities that are required.

The User Interface

The only interaction that the engineer has with automatic frame analysis programs is in the preparation of the input data and in the interpretation of the results. Consequently the user's opinion of the program will be largely based upon the ease of data preparation, and the readability of the output. This being the case, no program should be purchased without first considering the way in which data is supplied to the program and the format of the output.

Graphical output is enormously valuable. If the program has facilities to draw the structure from the input data then input errors involving the structure's geometry are quickly detected. Even more useful is the facility to draw the deformed shape of the structure from the results. This gives the engineer a feel for the structure's behaviour, and may help detect erroneous input data. Input and output from frame analysis programs is usually available as arrays. This facilitates the implementation of graphical output.

The growth in the popularity of graphical user interfaces has led to the production of structural analysis programs that are predominately graphically driven. Such programs are usually easier to learn than text based programs and they give the user easy access to graphical representation of the structure.

12.3 Using Frame Analysis Programs

A scale drawing of the structure showing node numbers, dimensions, loads, and restraints aids the preparation of input data. Often data can be given directly to the computer from such a drawing.

Node Numbers

Most programs internally renumber the nodes to minimise the bandwidth of the structure stiffness matrix (some programs insist on renumbering the nodes). If the program does not the user should be careful to number the nodes so that the maximum difference between the node numbers at the ends of any element is kept to a minimum (or near minimum).

Initial Element Sizing

One of the problems facing the designer of a structural framework is the initial choice of element sizes. Once these have been chosen then the designer iterates around the loop shown in fig 12.1 until a satisfactory structure is found.

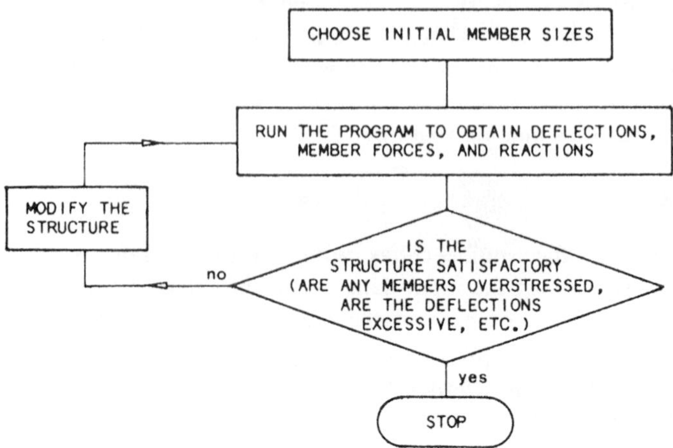

Figure 12.1 The design cycle

Initial element sizes can be selected in many ways. Using element sizes from a similar structure can prove efficient. Length to breadth (or span to depth) ratios can be used as the length to breadth of structural framework elements tends to be in the range 10 to 20. A preliminary analysis of the structure by hand can be used to determine initial element sizes - as can pure guesswork!

In the opinion of the author there are two good reasons in favour of spending some time over the initial sizing of elements. Firstly, the number of times the program is run before a satisfactory solution is found will usually be reduced. Secondly, and more importantly, the engineer will gain a feel for how the structure will perform. Should the structure behave differently from the engineer's expectations when it is subsequently analysed by the computer, then either the computer analysis is wrong due to erroneous input data, or the engineer has misunderstood how the structure will behave. In either case the source of the problem must be identified, understood, and corrected if necessary. To illustrate, consider the problem of choosing initial element sizes for the structure shown in fig 12.2.

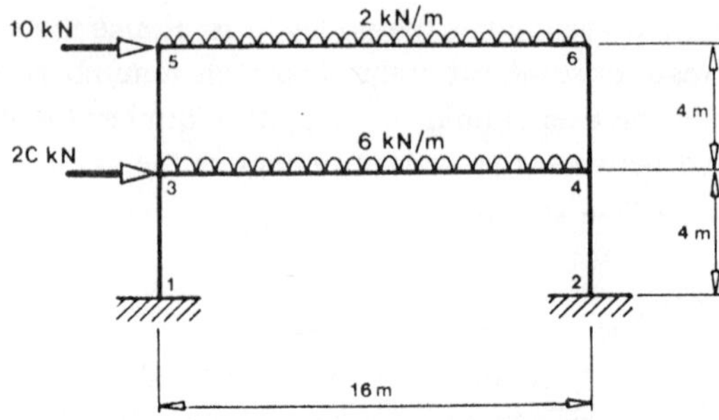

Figure 12.2 A two-storey portal frame

There follows an example of how the bending moments and sway deflections might be estimated (note that the initial element sizes can be calculated from the estimated bending moments). As the computer will later be used to analyse the structure, the preliminary calculations only serve to:

1. Enable reasonable initial element sizes to be selected.
2. Give an insight into the behaviour of the structure.
3. Provide a check that there are no gross errors in the input data.

The emphasis is therefore on simplicity, while still retaining the essential characteristics of the true structural behaviour. Provided that the calculations yield results that serve the objectives listed above then the method of calculation is immaterial. Hence the calculations that follow are only one example of how the problem might be tackled.

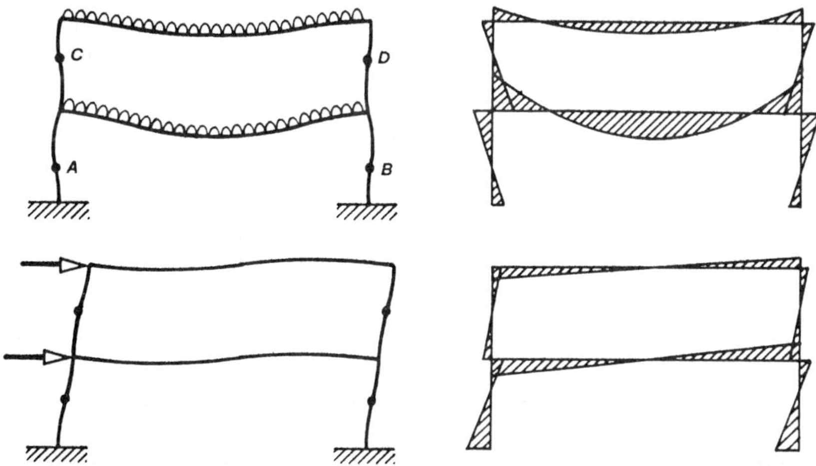

Figure 12.3 Deformed shapes and bending moment diagrams

For structures such as the portal shown in fig 12.2, axial deformations can be safely neglected. It is now helpful to decide the relative flexural stiffness (EI/L) of the elements. For simplicity all elements will be assumed to have the same flexural stiffness. If the effects of the vertical and the horizontal loads are separated then the general shape of the deformed structure and of the bending moment diagrams can be sketched as shown in fig 12.3.

Inspection of the deformations caused by the vertical loads indicates that there are points of contraflexure (zero curvature, therefore zero moment) at A, B, C, and D. The points of contraflexure at A and B occur at one-third of the column height. Consideration of the loading and the relative rotational stiffness of the joints indicates that joints 3 and 4 will rotate more than joints 5 and 6. It is, therefore, reasonable to assume that the points of contraflexure at C and D occur two-thirds of the way up the columns. Further, if the points of contraflexure are assumed to undergo no translation then the part of the structure bounded by points A, B, C, and D is as shown in fig 12.4.

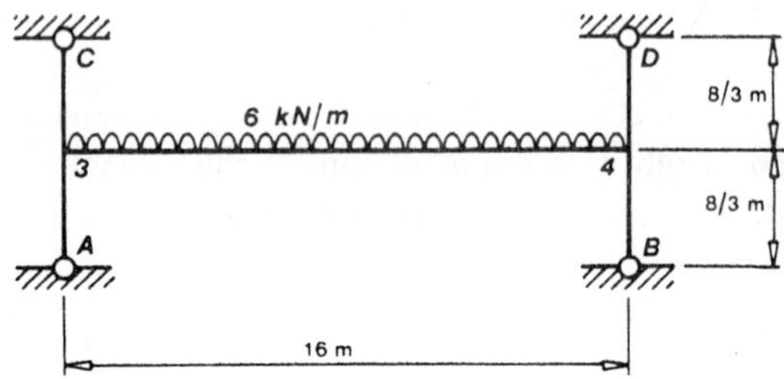

Figure 12.4 Substructure formed by A, B, C, D

The bending moments for the substructure shown in fig 12.4 are most easily calculated by moment distribution. As moment distribution is not within the scope of this book the bending moments will be evaluated using the stiffness method.

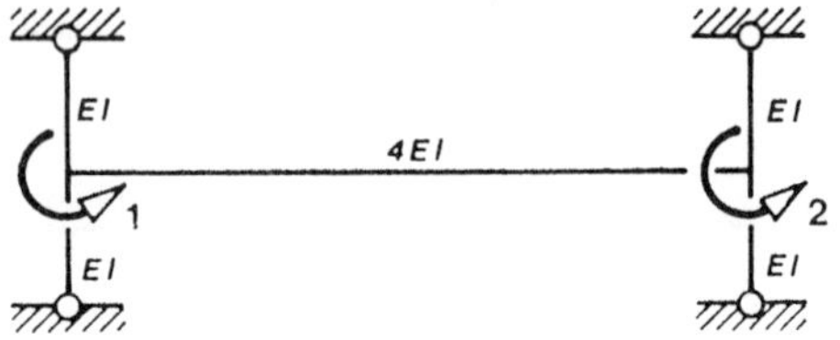

Figure 12.5 Freedoms and relative stiffness values

By treating the columns as elements with pins the substructure shown in fig 12.4 has two freedoms as illustrated in fig 12.5. This figure also shows the relative *EI* values of the elements (*EI/L* was previously assumed to be equal for all elements). The coefficients of the stiffness matrix can now be calculated as follows:

$$\begin{bmatrix} \frac{13EI}{4} & \frac{EI}{2} \\ \frac{EI}{2} & \frac{13EI}{4} \end{bmatrix} \begin{bmatrix} \theta_1 \\ \theta_2 \end{bmatrix} = \begin{bmatrix} 128 \\ -128 \end{bmatrix}$$

Noting that the rotations will be equal and opposite simplifies their evaluation, and θ_1 turns out to be $46.5/EI$.

From the rotations the bending moments at the joints are easily calculated, allowing the diagram of approximate bending moments to be drawn as shown in fig 12.6.

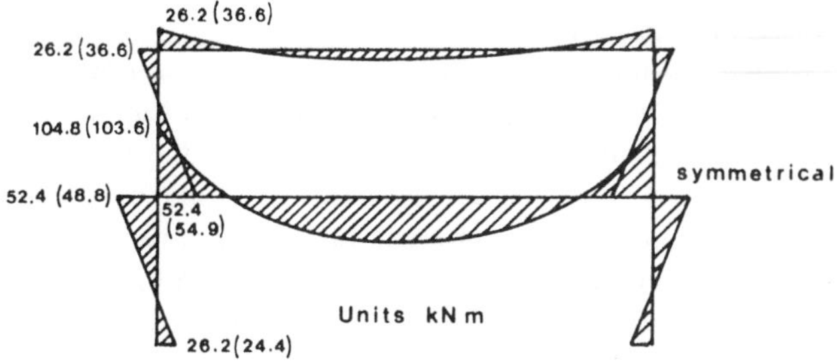

Figure 12.6 Approximate bending moments due to vertical load

The exact values (which are shown in brackets) indicate that, in general, the estimated bending moments are reasonable, and would certainly serve for the purposes of initial element sizing. It is interesting to note that the exact value of the rotation at joint 3 is $48.8/EI$.

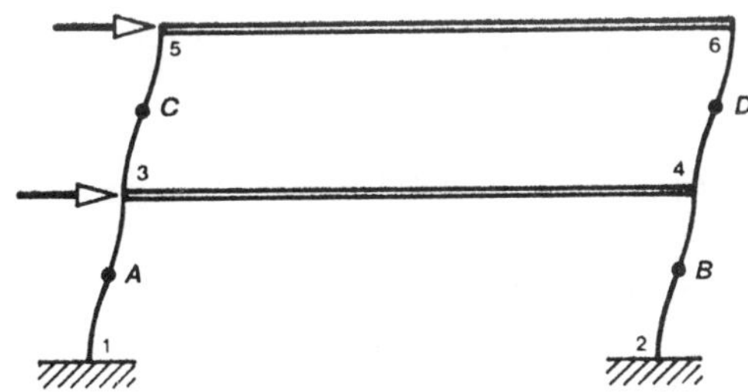

Figure 12.7 Deformed shape with no joint rotation

Attention will now be given to estimating the bending moments caused by the horizontal loads. If the beams where infinitely stiff compared to the columns then the joints would not rotate and the deformed shape would be as shown in fig 12.7.

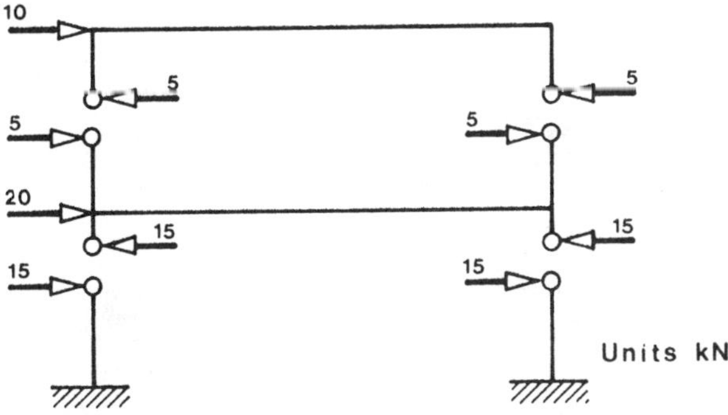

Figure 12.8 Freebody diagrams

There is a point of contraflexure on each column. If the stiffness of beam 3,4 is now reduced then the joints 3 and 4 will rotate causing points of contraflexure on the upper and lower columns to move together. If the stiffness of beam 5,6 is reduced (beam 3,4 remaining infinitely stiff) then the points of contraflexure on the upper columns will migrate towards the tops of these columns. The beams and columns have been assumed to have the same *EI/L* value for the purposes of these calculations. It is reasonable, therefore, to assume that the points of contraflexure at A and B occur three-quarters of the way up the lower columns, and the points of contraflexure at C and D occur half-way up the upper columns. This leads to the freebody diagrams shown in fig 12.8.
Hence the diagram of approximate bending moments for horizontal loads can be drawn as shown in fig 12.9. Again the exact values are given in brackets for the purposes of comparison.

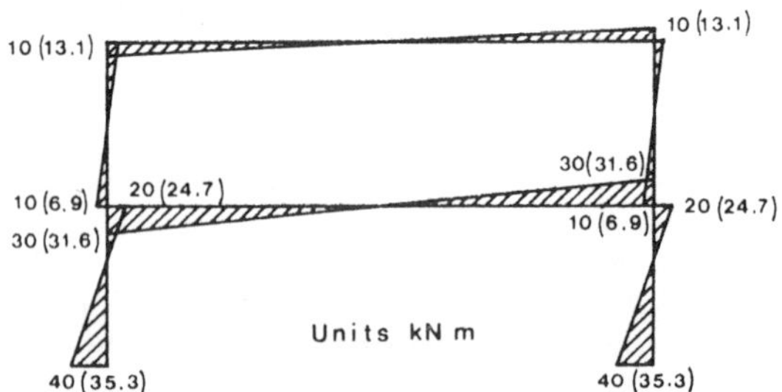

Figure 12.9 Approximate bending moment diagram for horizontal loads

In this instance the vertical loads cause no sidesway, and therefore do not appear in the calculation of sway displacements. The sway displacements associated with horizontal loads could be easily estimated from the approximate bending moment diagram for horizontal loads using techniques such as the moment-area method. However such methods are not within the scope of this book and here approximate values for the sway displacements will be found by assuming that the beams are infinitely stiff relative to the columns. The resulting deformed shape is shown in fig 12.7, and the sway displacements are:

At the level of the lower beam

$$\delta = \frac{FL^3}{12EI} = \frac{15(4^3)}{12EI} = \frac{80}{EI}$$

At the level of the upper beam

$$\delta = \frac{80}{EI} + \frac{FL^3}{12EI} = \frac{80}{EI} + \frac{5(4^3)}{12EI} = \frac{107}{EI}$$

These values form a lower bound to the sway displacements and the true displacements will be considerably bigger due to the rotation of the joints. The percentage error will be bigger at the level of the upper beam. A subsequent computer analysis gave deflections of $122/EI$ and $208/EI$ at the levels of the lower and upper beams respectively. Comparison with estimated displacements indicates that no gross error was present in the input data.

Element Properties

The initial element sizes are calculated from the estimated element forces. The initial element sizes allow the section properties of the elements to be evaluated.

For trusses and plane frames this process is usually straightforward. It is worth noting, however, that in the early stages of analysis it is not efficient to design the reinforcement for reinforced concrete elements as sufficiently accurate results can be obtained by assuming the elements to consist of plane concrete that won't crack in tension.

Grillages and space frames require the torsional constants for the elements. Often approximate formulae are employed. In some cases a parametric study of the sensitivity of the problem to the torsional stiffness of the elements is worthwhile. The engineer estimates maximum and minimum values of the torsional stiffness and compares results using these values.

In Chapter 10 the relationship between the element axes and the principal axes for a space frame element was seen to be at the discretion of the program writer. The program user should therefore consult the user manual to make sure that the data supplied to the program accurately describes the problem in hand.

Restraints

In section 2.6 boundary conditions were discussed in some detail. When preparing restraint data for a frame analysis program the engineer must consider how the boundary conditions that are input to the program will be achieved in practice. For instance, if a connection to a footing is assumed to be fully fixed then the engineer must ensure that the actual footing is big enough, and the connection is rigid enough for this assumption to be valid. Again parametric studies can be worthwhile and the facility to include elastic supports is particularly useful in this instance. If the analysis program caters for elements with pins the user should be careful not to over specify pinned supports by putting a pin at the support and a pin in the element attached to the support as this will result in node with rotational freedom, but no rotational stiffness (i.e. a mechanism).

Local Axes

Not all frame analysis programs allow the use of local axes. The way in which local axes are defined is at the discretion of the program writer. Hence the program user must pay careful attention to the user manual when preparing data for problems involving the use of local axes.

Element Loads

Most frame analysis programs allow the use of a variety of element loads. The user manual will describe the load types that are available and will define how they should be input to the computer.

Units

Any unit system can be used to prepare the input data for the analysis of frames. The only restriction is that the units must be consistent. For instance, if Newtons are used for force and centimetres for length then nodal coordinates must be given in cm, the elastic constants in N/cm^2, the cross sectional areas in cm^2, the torsional constants in cm^4, etc.

The user must be particularly careful about units for dynamic analysis. The units must again be consistent. In particular the user must ensure that one unit of force will produce one unit of acceleration on one unit of mass. To avoid error it is recommend that Newtons are used for force, metres for length, seconds for time, and kilograms for mass when conducting dynamic analysis.

Results

Results will be in the same units as the input data. For example, if the unit of force is the kgf and the unit of length is the metre, then displacements will be in m, rotations will be radians (always), forces will be in kgf, moments will be in kgf-m, etc. Nodal displacements will be output in the nodal axes systems (see section 3.6), element end forces (stress resultants) will be in the principal axes of the element, and reactions are normally output in nodal axes systems for the restrained nodes.

Appendix A

Definition of Program Variables

Note that % indicates an integer variable or an integer array, $ indicates a string variable or a string array, ( ) indicates a vector and (,) indicates a two dimensional array.

Variable	Description
A	Cross sectional area.
A(,)	GAUSSJ.BAS - the matrix A(,) contains the augmented matrix. Later overwritten by the decomposed matrix.
A2	Second moment of area.
ABSTOL	Limiting tolerance.
ACC	Acceleration factor for secant method in DSEARCH.BAS.
ALPHA	Anti-clockwise rotation of the element from the global *"x"* axis in degrees, or rotational factor in JACOBI.BAS.
AMASS(,)	Row *"i"* contains details of added mass *"i"*. (I%,1) Number of node where mass is located. (I%,2) Magnitude of the added mass.
AR	Value of ALPHA in radians.
B(,)	Matrix of right hand side vectors, stored column by column, and overwritten by the solution vectors.
BANNER$	Identification banner.
BAND%	Semi-bandwidth of the coefficient matrix.
BB%	Width of current row.
BETA	Rotational factor in JACOBI.BAS.
C	Cosine of ALPHA.
CI	Cosine of ALPHAI.
CJ	Cosine of ALPHAJ
C(,)	Used to preserve the coefficient matrix.
CO	Coefficient of thermal expansion.

COL%	Horizontal position of cursor on the screen.
COUNT%	Counter used when editing data.
CYZ	Cosine of the angle between the element "*y*" axis and the principal "*z*" axis for the element.
D(,)	Used to preserve the inverse of the coefficient matrix when calculating the inverse of the inverse.
DATAFILEOK$	Flag to indicate that data file has been successfully loaded.
DEN	Denominator.
DET	Current value of the determinant = DET * 2^POWER
DIRN%	Direction(s) of restraint.
DONE%	Completion flag.
E	Elastic constant (Young's modulus).
EAL	*E*A/L* for current element.
EIL	*E*I/L* for current element.
EJF( )	Vector of equivalent joint forces.
EDITING%	Flag to indicate that data is being edited.
EFREE%( )	Element code number.
ELEM(,)	Row "*i*" contains details of element "*i*". Note that * indicates a value that is calculated during analysis. See also PROP(,).

	PTRUSS.BAS	STRUSS.BAS	PFRAME.BAS
(I%,1)	I	I	I
(I%,2)	J	J	J
(I%,3)	PN	PN	PN
(I%,4)	L*	L*	TYPE
(I%,5)	C*	LX*	L*
(I%,6)	S*	MY*	CI*
(I%,7)		NX*	SI*
(I%,8)			CJ*
(I%,9)			SJ*

	PFDYNAM.BAS	GRID.BAS	SFRAME..BAS
(I%,1)	I	I	I
(I%,2)	J	J	J
(I%,3)	PN	PN	PN
(I%,4)	TYPE	L*	L*
(I%,5)	L*	C*	CI*
(I%,6)	CI*	S*	LX*
(I%,7)	SI*		MX
(I%,8)	CJ*		NX
(I%,9)	SJ*		

where

I	lower node.
J	higher node.

	PN	property number.
	TYPE	element type, depends upon the end connections and takes a value from 1 to 4 as follows 1 - rigid/rigid 2 - rigid/pinned 3 - pinned/rigid 4 - pinned/pinned
	L	length of element.
	C	cosine of ALPHA.
	S	sine of ALPHA.
	LX,MX,NX	direction cosines of element "*x*" axis.

ELOAD(,)	Row "*i*" contains details of element load "*i*". (I%,1) Lower node number of loaded element. (I%,2) Higher node number of loaded element. (I%,3) Load type, 1 for a a UDL, 2 for a point load. (I%,4) Magnitude of the load, note that a positive element load acts in the direction of the positive element "*y*" axis. (I%,5) Distance of point load from lower node.
EMASS(,)	Element mass matrix.
ERRORS%	Number of errors detected by DCHECK.BAS
ESTIFF(,)	Element stiffness matrix.
EVALUE(,)	Array of eigenvalues in DSEARCH.BAS (I%,1) eigenvalue. (I%,2) eigenvalue number.
EVECTOR(,)	Array of eigenvector in DSEARCH.BAS. Column "*i*" contains eigenvector "*i*".
FEF( )	Vector of fixed end forces.
FACT	Temporary floating point variable.
FALSE%	Boolean flag.
FORCE	Axial force in truss element.
FREE%()	Freedom vector.
GJL	*G*J/L* for current element.
I%	Loop counter or array index.
IN%	Number of node "*i*".
INPUTFILE$	Name of input file.
J%	Loop counter or array index.
JN%	Number of node "*j*".
K(,)	Structure stiffness matrix - initial or final.
K%	Loop counter or array index.
K1(,)	Stiffness matrix in JACOBI.BAS.
K2(,)	Transformed stiffness matrix in JACOBI.BAS.
K1-K8	Stiffness factors for plane frame elements with different end connections, and stiffness factors for space frames.

KK,KKII,KKJJ	Temporary values used when calculating ALPHA and BETA in JACOBI.BAS.
L	Length of element.
L%	Temporary integer variable.
LAXIS(,)	Row *"i"* contains details of local axes system *"i"*. (I%,1) Node where the local axes are located. (I%,2) Anti-clockwise rotation of the local axis system from the coordinate axes.
M%	Temporary integer variable.
M(,)	Structure mass matrix in PFDYNAM.BAS.
M1-M10	Mass coefficient to cater for different element end connections in PFDYNAM.BAS.
M1(,)	Mass matrix in JACOBI.BAS.
M2(,)	Transformed mass matrix in JACOBI.BAS.
MAXNEQ%	Maximum number of equations.
MAXBAND%	Maximum semi-bandwidth of coefficient or stiffness matrix.
MAXELEM%	Maximum number of elements.
MAXEVAL%	Maximum number of eigenvalues in DSEARCH.BAS.
MAXMBAND%	Maximum semi-bandwidth of the mass matrix in DSEARCH.BAS.
MAXNLOAD%	Maximum number of nodal loads.
MAXNODE%	Maximum number of nodes.
MAXPROP%	Maximum number of properties.
MAXREST%	Maximum number of restrained nodes.
MAXNRHS%	Maximum number of right-hand side vectors.
MAXSBAND%	Maximum semi-bandwidth of the stiffness matrix in DSEARCH.BAS.
MODSTR$	Name of the module being currently read.
ML	Mass of element.
NAMASS%	Number of added masses.
NAMASSNO%	Number of numbers when reading added masses.
NDOF%	Number of degrees of freedom.
NEIGEN%	Number of eigenvalues to be found in DSEARCH.BAS
NELEM%	Number of elements.
NELEMNO%	Number of numbers when reading elements.
NELOAD%	Number of element loads.
NELOADNO%	Number of numbers when reading element loads.
NFACT%	Number of factorisations in DSEARCH.BAS
NFREE%	Nodal freedom number.
NFOUND%	Number of eigenvalues evaluated in DSEARCH.BAS
NITER%	Iteration count.
NLAXIS%	Number of local axes.
NLESS%	Number of unknown eigenvalues less then the trial value.
NLOAD(,)	Row *"i"* contains details of nodal load *"i"*.

	(I%,1) number of loaded node.
	(I%,2) direction of the load.
	(I%,3) magnitude of the load
NNLOAD%	Number of nodal loads.
NNLOADNO%	Number of numbers when reading nodal loads.
NNODE%	Number of nodes.
NNODENO%	Number of numbers when reading nodal coordinates.
NNOS%	Number of numbers in input line.
NOPTIONS%	Number of options in the menu.
NPROP%	Number of element properties.
NPROPNO%	Number of numbers when reading properties.
NREST%	Number of restrained nodes.
NRESTNO%	Number of numbers when reading restraints.
NLINES%	Number lines to be skipped in data file.
NODE(,)	Row *"i"* contains nodal coordinates for node *"i"*.
	(I%,1) *"x"* coordinate.
	(I%,2) *"y"* coordinate.
	(I%,3) *"z"* coordinate (three dimensional structures only).
NNODE%	Number of nodes.
NRHS%	Number of right-hand side vectors.
NUM	Numerator.
OUTPUTFILE$	Name of output file.
P( )	Load vector.
PHI(,)	Matrix of eigenvectors in JACOBI.BAS.
PN%	Property number.
POWER	Current value of the determinant = DET * 2^POWER
PROGNAME$	String containing the name of the analysis program. This is a COMMON variable that allows PRE.BAS and DCHECK.BAS to CHAIN to the analysis program.
PROP(,)	Contains property data.

	PTRUSS.BAS	STRUSS.BAS	PFRAME.BAS
(I%,1)	N	N	N
(I%,2)	A	A	A
(I%,3)	E	E	A2
(I%,4)			E

	PFDYNAM.BAS	GRID.BAS	SFRAME..BAS
(I%,1)	N	N	N
(I%,2)	A	A	A
(I%,3)	A2	J2	IY
(I%,4)	M	G	IZ
(I%,5)		E	J2
(I%,6)			G
(I%,5)			E

	where
	N property number.
	A cross sectional area.
	A2 second moment of area.
	J2 torsional constant.
	IY second moment of area about principal *"y"* axis.
	IZ second moment of area about principal *"z"* axis.
	M mass per unit length.
	G modulus of rigidity.
	E elastic constant (Young's modulus).
REPLY%	Users response to a menu.
REST(,)	Row *"i"* contains details of restrained node *"i"*.
	(I%,1) number of restrained node.
	(I%,2) composite restraint components.
	(I%,3)-(I%,8) used to store reaction components.
RI	rotation at node *"i"*.
RJ	rotation at node *"j"*.
ROW%	Vertical position of cursor on screen.
S	Sine of ALPHA.
SI	Sine of ALPHAI.
SINGULAR%	Singular flag in DSEARCH.BAS.
SJ	Sine of ALPHAJ
STATUS%	Flag used to while reading input file.
STIFFMAT%	Boolean flag to indicate if the mass or the stiffness matrix is being processed.
T1-T9	Transformation factors for space frames.
TARGET$	Target string, used while reading data file.
TIME$	String containing time.
TEMP	Temporary floating point variable.
TEMP( )	Array of temporary floating point variables.
TEMP$	Temporary string variable.
THETA(,)	Matrix of eigenvalues.
	(I%,1) Square root of eigenvalue.
	(I%,2) Original position in matrix.
TITLE$	Job title.
TOL%	Current tolerance.
TRUE%	Boolean flag.
U( )	Used to store element end displacements and forces in SFRAME.BAS.
U1,U2,U3	Estimates of the next eigenvalue in DSEARCH.BAS.
UNITS$	String containing units - note that the units used must be consistent.
W	Point load on element.
W(,)	Workspace array in DSEARCH.BAS.

WARNINGS%	Number of warnings detected by DCHECK.BAS
WKSP$( )	Workspace array of strings.
X	Distance of element point load from node "*i*".
X( )	Workspace array in DSEARCH.BAS.
XXXX	Very small number - used for comparison purposes.
XI	"*x*" displacement at node "*i*".
XJ	"*x*" displacement at node "*j*".
Y	Distance of element point load from node "*j*".
Y( ),Y1( ),Y2( )	Workspace arrays in DSEARCH.BAS.
YI	"*y*" displacement at node "*i*".
YJ	"*y*" displacement at node "*j*".

Appendix B

Summary of Key Formulae and Matrices

B.1 Element Axes System

These are described in section 3.6 (see fig 3.12), and are shown in fig B.1

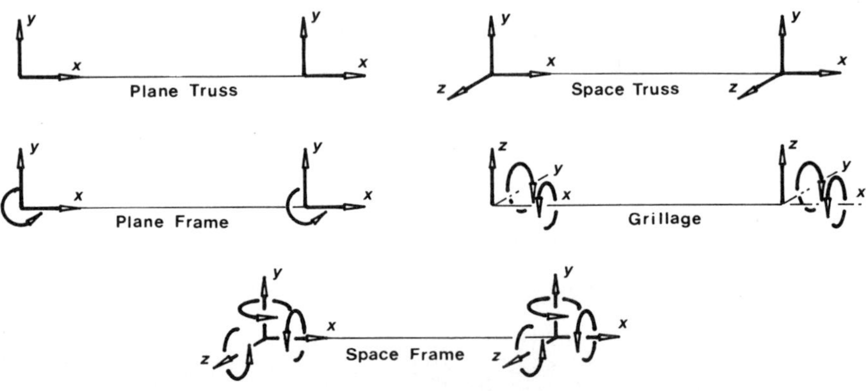

Figure B.1 Element Axes

B.2 Element Stiffness Matrices

The relationship between the element end displacements and the element end forces (expressed in the element axes system) is given by the following equation.

$$\begin{bmatrix} f_{ij} \\ f_{ji} \end{bmatrix} = \begin{bmatrix} k_{ii}^{j} & k_{ij} \\ k_{ji} & k_{jj}^{i} \end{bmatrix} \begin{bmatrix} \delta_{ij} \\ \delta_{ji} \end{bmatrix}$$

Plane Truss

$$\begin{bmatrix} f_{ijx} \\ f_{ijy} \\ f_{jix} \\ f_{jiy} \end{bmatrix} = \begin{bmatrix} \frac{EA}{L} & 0 & \frac{-EA}{L} & 0 \\ 0 & 0 & 0 & 0 \\ \frac{-EA}{L} & 0 & \frac{EA}{L} & 0 \\ 0 & 0 & 0 & 0 \end{bmatrix} \begin{bmatrix} \delta_{ijx} \\ \delta_{ijy} \\ \delta_{jix} \\ \delta_{jiy} \end{bmatrix}$$

Space Truss

$$\begin{bmatrix} f_{ijx} \\ f_{ijy} \\ f_{ijz} \\ f_{jix} \\ f_{jiy} \\ f_{jiz} \end{bmatrix} = \begin{bmatrix} \frac{EA}{L} & 0 & 0 & \frac{-EA}{L} & 0 & 0 \\ 0 & 0 & 0 & 0 & 0 & 0 \\ 0 & 0 & 0 & 0 & 0 & 0 \\ \frac{-EA}{L} & 0 & 0 & \frac{EA}{L} & 0 & 0 \\ 0 & 0 & 0 & 0 & 0 & 0 \\ 0 & 0 & 0 & 0 & 0 & 0 \end{bmatrix} \begin{bmatrix} \delta_{ijx} \\ \delta_{ijy} \\ \delta_{ijz} \\ \delta_{jix} \\ \delta_{jiy} \\ \delta_{jiz} \end{bmatrix}$$

Plane Frame - Various End Connections

Elements containing pins are dealt with by using the variables K_1 - K_8 in the element stiffness matrix as shown in the following equation. The values of the K factors are tabulated in Table B.1.

$$\begin{bmatrix} f_{ijx} \\ f_{ijy} \\ m_{ij} \\ f_{jix} \\ f_{jiy} \\ m_{ji} \end{bmatrix} = \begin{bmatrix} \frac{EA}{L} & 0 & 0 & -\frac{EA}{L} & 0 & 0 \\ & K_1\frac{EI}{L^3} & K_2\frac{EI}{L^2} & 0 & -K_1\frac{EI}{L^3} & K_3\frac{EI}{L^2} \\ & & K_4\frac{EI}{L} & 0 & K_5\frac{EI}{L^2} & K_6\frac{EI}{L} \\ & \text{symmetric} & & \frac{EA}{L} & 0 & 0 \\ & & & & K_1\frac{EI}{L^3} & K_7\frac{EI}{L^2} \\ & & & & & K_8\frac{EI}{L} \end{bmatrix} \begin{bmatrix} \delta_{ijx} \\ \delta_{ijy} \\ \theta_{ij} \\ \delta_{jix} \\ \delta_{jiy} \\ \theta_{ji} \end{bmatrix}$$

Table B.1 *K* Factors

ELEMENT TYPE	NODE "i"	NODE "j"	K_1	K_2	K_3	K_4	K_5	K_6	K_7	K_8
1	Rigid	Rigid	12	6	6	4	-6	2	-6	4
2	Rigid	Pinned	3	3	0	3	-3	0	0	0
3	Pinned	Rigid	3	0	3	0	0	0	-3	-3
4	Pinned	Pinned	0	0	0	0	0	0	0	0

Grillage

$$
\begin{bmatrix} f_{ijz} \\ m_{ijx} \\ m_{ijy} \\ f_{jiz} \\ m_{jix} \\ m_{jiy} \end{bmatrix} =
\begin{bmatrix}
\frac{12EI}{L^3} & 0 & -\frac{6EI}{L^2} & -\frac{12EI}{L^3} & 0 & -\frac{6EI}{L^2} \\
0 & \frac{GJ}{L} & 0 & 0 & -\frac{GJ}{L} & 0 \\
-\frac{6EI}{L^2} & 0 & \frac{4EI}{L} & \frac{6EI}{L^2} & 0 & \frac{2EI}{L} \\
-\frac{12EI}{L^3} & 0 & \frac{6EI}{L^2} & \frac{12EI}{L^3} & 0 & \frac{6EI}{L^2} \\
0 & -\frac{GJ}{L} & 0 & 0 & \frac{GJ}{L} & 0 \\
-\frac{6EI}{L^2} & 0 & \frac{2EI}{L} & \frac{6EI}{L^2} & 0 & \frac{4EI}{L}
\end{bmatrix}
\begin{bmatrix} \delta_{ijz} \\ \theta_{ijx} \\ \theta_{ijy} \\ \delta_{jiz} \\ \theta_{jix} \\ \theta_{jiy} \end{bmatrix}
$$

Space Frame

The element stiffness matrix for a plane frame is 12x12, and because of its size it is presented in terms of four element stiffness submatrices.

End "i"

$$
\begin{bmatrix} f_{ijx} \\ f_{ijy} \\ f_{ijz} \\ m_{ijx} \\ m_{ijy} \\ m_{ijz} \end{bmatrix} =
\begin{bmatrix}
\frac{EA}{L} & 0 & 0 & 0 & 0 & 0 \\
0 & \frac{12EI_z}{L^3} & 0 & 0 & 0 & \frac{6EI_z}{L^2} \\
0 & 0 & \frac{12EI_y}{L^3} & 0 & -\frac{6EI_y}{L^2} & 0 \\
0 & 0 & 0 & \frac{GJ}{L} & 0 & 0 \\
0 & 0 & -\frac{6EI_y}{L^2} & 0 & \frac{4EI_y}{L} & 0 \\
0 & \frac{6EI_z}{L^2} & 0 & 0 & 0 & \frac{4EI_z}{L}
\end{bmatrix}
\begin{bmatrix} \delta_{ijx} \\ \delta_{ijy} \\ \delta_{ijz} \\ \theta_{ijx} \\ \theta_{ijy} \\ \theta_{ijz} \end{bmatrix} +
$$

$$
\begin{bmatrix}
-\frac{EA}{L} & 0 & 0 & 0 & 0 & 0 \\
0 & -\frac{12EI_z}{L^3} & 0 & 0 & 0 & \frac{6EI_z}{L^2} \\
0 & 0 & -\frac{12EI_y}{L^3} & 0 & -\frac{6EI_y}{L^2} & 0 \\
0 & 0 & 0 & -\frac{GJ}{L} & 0 & 0 \\
0 & 0 & \frac{6EI_y}{L^2} & 0 & \frac{2EI_y}{L} & 0 \\
0 & -\frac{6EI_z}{L^2} & 0 & 0 & 0 & \frac{2EI_z}{L}
\end{bmatrix}
\begin{bmatrix} \delta_{jix} \\ \delta_{jiy} \\ \delta_{jiz} \\ \theta_{jix} \\ \theta_{jiy} \\ \theta_{jiz} \end{bmatrix}
$$

End "j"

$$
\begin{bmatrix} f_{jix} \\ f_{jiy} \\ f_{jiz} \\ m_{jix} \\ m_{jiy} \\ m_{jiz} \end{bmatrix} =
\begin{bmatrix}
-\frac{EA}{L} & 0 & 0 & 0 & 0 & 0 \\
0 & -\frac{12EI_z}{L^3} & 0 & 0 & 0 & -\frac{6EI_z}{L^2} \\
0 & 0 & -\frac{12EI_y}{L^3} & 0 & \frac{6EI_y}{L^2} & 0 \\
0 & 0 & 0 & -\frac{GJ}{L} & 0 & 0 \\
0 & 0 & -\frac{6EI_y}{L^2} & 0 & \frac{2EI_y}{L} & 0 \\
0 & \frac{6EI_z}{L^2} & 0 & 0 & 0 & \frac{2EI_z}{L}
\end{bmatrix}
\begin{bmatrix} \delta_{ijx} \\ \delta_{ijy} \\ \delta_{ijz} \\ \theta_{ijx} \\ \theta_{ijy} \\ \theta_{ijz} \end{bmatrix} +
$$

$$
\begin{bmatrix}
\frac{EA}{L} & 0 & 0 & 0 & 0 & 0 \\
0 & \frac{12EI_z}{L^3} & 0 & 0 & 0 & -\frac{6EI_z}{L^2} \\
0 & 0 & \frac{12EI_y}{L^3} & 0 & \frac{6EI_y}{L^2} & 0 \\
0 & 0 & 0 & \frac{GJ}{L} & 0 & 0 \\
0 & 0 & \frac{6EI_y}{L^2} & 0 & \frac{4EI_y}{L} & 0 \\
0 & -\frac{6EI_z}{L^2} & 0 & 0 & 0 & \frac{4EI_z}{L}
\end{bmatrix}
\begin{bmatrix} \delta_{jix} \\ \delta_{jiy} \\ \delta_{jiz} \\ \theta_{jix} \\ \theta_{jiy} \\ \theta_{jiz} \end{bmatrix}
$$

B.3 Element Mass Matrices

The relationships between the element end accelerations and the element end forces (expressed in the element axes system) for different types of structural frameworkds are given by the following equations.

Plane Truss

$$
\begin{bmatrix} f_{ijx} \\ f_{ijy} \\ f_{jix} \\ f_{jiy} \end{bmatrix} =
\begin{bmatrix}
\frac{mL}{3} & 0 & \frac{mL}{6} & 0 \\
0 & \frac{mL}{3} & 0 & \frac{mL}{6} \\
\frac{mL}{6} & 0 & \frac{mL}{3} & 0 \\
0 & \frac{mL}{6} & 0 & \frac{mL}{3}
\end{bmatrix}
\begin{bmatrix} \ddot{\delta}_{ijx} \\ \ddot{\delta}_{ijy} \\ \ddot{\delta}_{jix} \\ \ddot{\delta}_{jiy} \end{bmatrix}
$$

Plane Frame - Various End Connections

Elements containing pins are dealt with by using the variables M_1 - M_{10} in the element mass matrix as shown in the following equation. The values of the *"M"* factors are tabulated in Table B.2.

$$\begin{bmatrix} f_{ijx} \\ f_{ijy} \\ m_{ij} \\ f_{jix} \\ f_{jiy} \\ m_{ij} \end{bmatrix} = \frac{mL}{840} \begin{bmatrix} 280 & 0 & 0 & 140 & 0 & 0 \\ & M_1 & M_2L & 0 & M_3 & M_4L \\ & & M_5L^2 & 0 & M_6L & M_7L^2 \\ & \text{symmetric} & & 280 & 0 & 0 \\ & & & & -M_8 & M_9L \\ & & & & & M_{10}L^2 \end{bmatrix} \begin{bmatrix} \ddot{\delta}_{ijx} \\ \ddot{\delta}_{ijy} \\ \ddot{\theta}_{ij} \\ \ddot{\delta}_{jix} \\ \ddot{\delta}_{jiy} \\ \ddot{\theta}_{ji} \end{bmatrix}$$

Table B.2 Mass coefficients

Element Type	Node "i"	Node "j"	M_1	M_2	M_3	M_4	M_5	M_6	M_7	M_8	M_9	M_{10}
1	Rigid	Rigid	312	44	108	-26	8	26	-6	312	-44	8
2	Rigid	Pinned	408	72	117	0	16	33	0	198	0	0
3	Pinned	Rigid	198	0	117	-33	0	0	0	408	-72	16
4	Pinned	Pinned	0	0	0	0	0	0	0	0	0	0

B.4 Transformation Matrices

The following transformation matrices transform nodal load and displacement vectors from the nodal axes system to the element axes system. The transformation matrix T_{ij} transforms force and displacement vectors at nodes "*i*" and "*j*" from the *nodal axes system to the element axes system* (for element "*i,j*").

Plane Truss

$$T = \begin{bmatrix} \cos\alpha & \sin\alpha \\ -\sin\alpha & \cos\alpha \end{bmatrix}$$

Space Truss

$$T = \begin{bmatrix} l_x & m_x & n_x \\ l_y & m_y & n_y \\ l_z & m_z & n_z \end{bmatrix}$$

Plane Frame - Various End Connections

Note that as plane frame elements have been developed to cater for local axes the nodes "*i*" and "*j*" may have different angles associated with them.

$$T_{ij} = \begin{bmatrix} \cos\alpha_i & \sin\alpha_i & 0 \\ -\sin\alpha_i & \cos\alpha_i & 0 \\ 0 & 0 & 1 \end{bmatrix}$$

$$T_{ji} = \begin{bmatrix} \cos\alpha_j & \sin\alpha_j & 0 \\ -\sin\alpha_j & \cos\alpha_j & 0 \\ 0 & 0 & 1 \end{bmatrix}$$

Grillage

$$T = \begin{bmatrix} 1 & 0 & 0 \\ 0 & \cos\alpha & \sin\alpha \\ 0 & -\sin\alpha & \cos\alpha \end{bmatrix}$$

Space Frame

$$T = \begin{bmatrix} l_x & m_x & n_x & 0 & 0 & 0 \\ l_y & m_y & n_y & 0 & 0 & 0 \\ l_z & m_z & n_z & 0 & 0 & 0 \\ 0 & 0 & 0 & l_x & m_x & n_x \\ 0 & 0 & 0 & l_y & m_y & n_y \\ 0 & 0 & 0 & l_z & m_z & n_z \end{bmatrix}$$

B.5 Global Element Stiffness Matrices

The global element stiffness matrix relates the element end displacements and the element end forces (expressed in the nodal axes systems) is given by the following equation.

$$\begin{bmatrix} F_{ij} \\ F_{ji} \end{bmatrix} = \begin{bmatrix} K_{ii}^j & K_{ij} \\ K_{ji} & K_{jj}^i \end{bmatrix} \begin{bmatrix} \Delta_{ij} \\ \Delta_{ji} \end{bmatrix}$$

where

$$K_{ii}^j = T_{ij}^{-1}\, k_{ii}^j\, T_{ij}$$

$$K_{ij} = T_{ij}^{-1}\, k_{ij}\, T_{ji}$$

$$K_{ji} = K_{ij}^T$$

$$K_{jj}^i = T_{ji}^{-1}\, k_{ii}^j\, T_{ji}$$

Note that if local axes are not catered for $T_{ji} = T_{ij}$. In this text local axes have only been considered when dealing with plane frame elements .

Plane Truss

$$K_{ii}^{j} = \frac{EA}{L}\begin{bmatrix} cos^2\alpha & cos\alpha \, sin\alpha \\ cos\alpha \, sin\alpha & sin^2\alpha \end{bmatrix} = K_{jj}^{i}$$

Find K_{ij} and K_{ji} by multiplying corresponding elements of K_{ii}^{j} by

$$\begin{bmatrix} -1 & -1 \\ -1 & -1 \end{bmatrix}$$

Space Truss

$$K_{ii}^{j} = \frac{EA}{L}\begin{bmatrix} l_x^2 & l_x m_x & l_x n_x \\ l_x m_x & m_x^2 & m_x n_x \\ l_x n_x & m_x n_x & n_x^2 \end{bmatrix} = K_{jj}^{i}$$

Find K_{ij} and K_{ji} by multiplying corresponding elements of K_{ii}^{j} by

$$\begin{bmatrix} -1 & -1 & -1 \\ -1 & -1 & -1 \\ -1 & -1 & -1 \end{bmatrix}$$

Plane Frame - Various End Connections

$$K_{ii}^{j} = T_{ij}^{-1} \, k_{ii}^{j} \, T_{ij}$$

$$= \begin{bmatrix} \left(\frac{EA}{L} cos^2\alpha_i + K_1 \frac{EI}{L^3} sin^2\alpha_i\right) & \left(\frac{EA}{L} - K_1 \frac{EI}{L^3}\right) cos\alpha_i \, sin\alpha_i & -K_2 \frac{EI}{L^2} sin\alpha_i \\ \left(\frac{EA}{L} - K_1 \frac{EI}{L^3}\right) cos\alpha_i \, sin\alpha_i & \left(\frac{EA}{L} sin^2\alpha_i + K_1 \frac{EI}{L^3} cos^2\alpha_i\right) & K_2 \frac{EI}{L^2} cos\alpha_i \\ -K_2 \frac{EI}{L^2} sin\alpha_i & K_2 \frac{EI}{L^2} cos\alpha_i & K_4 \frac{EI}{L} \end{bmatrix}$$

$$K_{ij} = T_{ij}^{-1} \, k_{ij} \, T_{ji}$$

$$= \begin{bmatrix} \left(\frac{EA}{L} c\alpha_i c\alpha_j + K_1 \frac{EI}{L^3} s\alpha_i s\alpha_j\right) & \left(-\frac{EA}{L} c\alpha_i s\alpha_j + K_1 \frac{EI}{L^3} s\alpha_i c\alpha_j\right) & -K_3 \frac{EI}{L^2} s\alpha_i \\ \left(-\frac{EA}{L} s\alpha_i c\alpha_j + K_1 \frac{EI}{L^3} c\alpha_i s\alpha_j\right) & \left(-\frac{EA}{L} s\alpha_i s\alpha_j - K_1 \frac{EI}{L^3} c\alpha_i c\alpha_j\right) & K_3 \frac{EI}{L^2} c\alpha_i \\ -K_5 \frac{EI}{L^2} s\alpha_j & K_5 \frac{6EI}{L^2} c\alpha_j & K_6 \frac{EI}{L} \end{bmatrix}$$

where $c\alpha_i = cos\alpha_i$, $s\alpha_i = sin\alpha_i$, $c\alpha_j = cos\alpha_j$ and $s\alpha_j = sin\alpha_j$

$$\boldsymbol{K}_{jj}^{i} = \boldsymbol{T}_{ji}^{-1}\,\boldsymbol{k}_{jj}^{i}\,\boldsymbol{T}_{ji}$$

$$= \begin{bmatrix} \left(\frac{EA}{L}\cos^2\alpha_j + K_1\frac{EI}{L^3}\sin^2\alpha_j\right) & \left(\frac{EA}{L} - K_1\frac{EI}{L^3}\right)\cos\alpha_j\sin\alpha_j & -K_7\frac{EI}{L^2}\sin\alpha_j \\ \left(\frac{EA}{L} - K_1\frac{EI}{L^3}\right)\cos\alpha_j\sin\alpha_j & \left(\frac{EA}{L}\sin^2\alpha_j + K_1\frac{EI}{L^3}\cos^2\alpha_j\right) & K_7\frac{EI}{L^2}\cos\alpha_j \\ -\frac{EI}{L^2}\sin\alpha_j & K_7\frac{EI}{L^2}\cos\alpha_j & K_8\frac{EI}{L} \end{bmatrix}$$

Grillage

$$\boldsymbol{K}_{ii}^{j} = \boldsymbol{T}_{ij}^{-1}\,\boldsymbol{k}_{ii}^{j}\,\boldsymbol{T}_{ij}$$

$$= \begin{bmatrix} \frac{12EI}{L^3} & \frac{6EI}{L^2}\sin\alpha & -\frac{6EI}{L^2}\cos\alpha \\ \frac{6EI}{L^2}\sin\alpha & \left[\frac{GJ}{L}\cos^2\alpha + \frac{4EI}{L}\sin^2\alpha\right] & \left[\frac{GJ}{L} - \frac{4EI}{L}\right]\cos\alpha\sin\alpha \\ -\frac{6EI}{L^2}\cos\alpha & \left[\frac{GJ}{L} - \frac{4EI}{L}\right]\cos\alpha\sin\alpha & \left[\frac{GJ}{L}\sin^2\alpha + \frac{4EI}{L}\cos^2\alpha\right] \end{bmatrix}$$

$$\boldsymbol{K}_{ij} = \boldsymbol{T}_{ij}^{-1}\,\boldsymbol{k}_{ij}\,\boldsymbol{T}_{ij}$$

$$= \begin{bmatrix} -\frac{12EI}{L^3} & \frac{6EI}{L^2}\sin\alpha & -\frac{6EI}{L^2}\cos\alpha \\ -\frac{6EI}{L^2}\sin\alpha & \left[-\frac{GJ}{L}\cos^2\alpha + \frac{2EI}{L}\sin^2\alpha\right] & \left[-\frac{GJ}{L} - \frac{2EI}{L}\right]\cos\alpha\sin\alpha \\ \frac{6EI}{L^2}\cos\alpha & \left[-\frac{GJ}{L} - \frac{2EI}{L}\right]\cos\alpha\sin\alpha & \left[-\frac{GJ}{L}\sin^2\alpha + \frac{2EI}{L}\cos^2\alpha\right] \end{bmatrix}$$

Find $\boldsymbol{K}_{jj}^{i}$ by multiplying corresponding elements of $\boldsymbol{K}_{ii}^{j}$ by

$$\begin{bmatrix} 1 & -1 & -1 \\ -1 & 1 & 1 \\ -1 & 1 & 1 \end{bmatrix}$$

Find $\boldsymbol{K}_{ji}$ by multiplying corresponding elements of $\boldsymbol{K}_{ij}$ by

$$\begin{bmatrix} 1 & -1 & -1 \\ -1 & 1 & 1 \\ -1 & 1 & 1 \end{bmatrix}$$

Space Frame

The principal axes of plane frame and grillage elements are assumed to coincide with the element axes. For space frame elements this is not necessarily the case. Hence, the determination of the global element stiffness submatrices requires an additional transformation, as shown in the following equation.

$$K_{ii}^{j} = T_{ij}^{-1}\, T_{ij}^{*-1}\, k_{ii}^{j*}\, T_{ij}^{*}\, T_{ij}$$

$$K_{ij} = T_{ij}^{-1}\, T_{ij}^{*-1}\, k_{ij}^{*}\, T_{ij}^{*}\, T_{ij}$$

Using the transformation factors T_1 - T_9 and the stiffness factors K_1 - K_{10} as follows allows for the efficient calculation of the global element stiffness matrix.

$$T_{ij}^{*}\, T_{ij} = [\, T_{ij}^{-1}\, T_{ij}^{*-1}\,]^{T}$$

$$= \begin{bmatrix} T_1 & T_4 & T_7 & 0 & 0 & 0 \\ T_2 & T_5 & T_8 & 0 & 0 & 0 \\ T_3 & T_6 & T_9 & 0 & 0 & 0 \\ 0 & 0 & 0 & T_1 & T_4 & T_7 \\ 0 & 0 & 0 & T_2 & T_5 & T_8 \\ 0 & 0 & 0 & T_3 & T_6 & T_9 \end{bmatrix}$$

$$\begin{array}{lcllcl}
T_1 & = & l & \text{or} & = & 0 \\
T_2 & = & -\,(m \cos\beta \;+\; l\,n \sin\beta)\;/\;D & \text{or} & = & -n \sin\beta \\
T_3 & = & (m \sin\beta \;-\; l\,n \cos\beta)\;/\;D & \text{or} & = & -n \cos\beta \\
T_4 & = & m & \text{or} & = & 0 \\
T_5 & = & (l \cos\beta \;-\; m\,n \sin\beta)\;/\;D & \text{or} & = & c \\
T_6 & = & -\,(l \sin\beta \;+\; m\,n \cos\beta)\;/\;D & \text{or} & = & -s \\
T_7 & = & n & \text{or} & = & n \\
T_8 & = & D \sin\beta & \text{or} & = & 0 \\
T_3 & = & D \cos\beta & \text{or} & = & 0
\end{array}$$

(the right-hand expressions are those assumed for the special case where the element lies paallel to the coordinate "z" axis). Writing the k_{ii}^{j*}, k_{ij}^{*}, k_{ji}^{*} and k_{jj}^{i*} submatrices in terms of "K" factors allows the global stiffness submatrices to be evaluated in terms of the "T" and "K" factors as follows.

$$k_{ii}^{j*} = \begin{bmatrix} K_1 & 0 & 0 & 0 & 0 & 0 \\ 0 & K_2 & 0 & 0 & 0 & K_7 \\ 0 & 0 & K_3 & 0 & K_8 & 0 \\ 0 & 0 & 0 & K_4 & 0 & 0 \\ 0 & 0 & K_9 & 0 & K_5 & 0 \\ 0 & K_{10} & 0 & 0 & 0 & K_6 \end{bmatrix}$$

where

$$K_1 = EA/L \qquad K_2 = 12EI_z/L^3$$
$$K_3 = 12EI_y/L^3 \qquad K_4 = GJ/L$$
$$K_5 = 4EI_y/L \qquad K_6 = 4EI_z/L$$
$$K_7 = 6EI_z/L^2 \qquad K_8 = -6EI_y/L^2$$
$$K_9 = -6EI_y/L^2 \qquad K_{10} = 6EI_z/L^2$$

$$k_{ij}^* = \begin{bmatrix} -K_1 & 0 & 0 & 0 & 0 & 0 \\ 0 & -K_2 & 0 & 0 & 0 & K_7 \\ 0 & 0 & -K_3 & 0 & K_8 & 0 \\ 0 & 0 & 0 & -K_4 & 0 & 0 \\ 0 & 0 & -K_9 & 0 & -K_5/2 & 0 \\ 0 & -K_{10} & 0 & 0 & 0 & -K_6/2 \end{bmatrix}$$

$$K_{ii}^j = T_{ij}^{-1}\, T_{ij}^{*-1}\, k_{ii}^{j*}\, T_{ij}^*\, T_{ij}$$

$$= \begin{bmatrix} k_{11} & k_{12} & k_{13} & k_{14} & k_{15} & k_{16} \\ k_{21} & k_{22} & k_{23} & k_{24} & k_{25} & k_{26} \\ k_{31} & k_{32} & k_{33} & k_{34} & k_{35} & k_{36} \\ k_{41} & k_{42} & k_{43} & k_{44} & k_{45} & k_{46} \\ k_{51} & k_{52} & k_{53} & k_{54} & k_{55} & k_{56} \\ k_{61} & k_{62} & k_{63} & k_{64} & k_{65} & k_{66} \end{bmatrix}$$

where

$$k_{11} = K_1T_1T_1 + K_2T_2T_2 + K_3T_3T_3$$
$$k_{12} = K_1T_1T_4 + K_2T_2T_5 + K_3T_3T_6$$
$$k_{13} = K_1T_1T_7 + K_2T_2T_8 + K_3T_3T_9$$
$$k_{14} = K_7T_2T_3 + K_8T_3T_2$$
$$k_{15} = K_7T_2T_6 + K_8T_3T_5$$
$$k_{16} = K_7T_2T_9 + K_8T_3T_8$$
$$k_{21} = K_1T_4T_1 + K_2T_5T_2 + K_3T_6T_3$$
$$k_{22} = K_1T_4T_4 + K_2T_5T_5 + K_3T_6T_6$$
$$k_{23} = K_1T_4T_7 + K_2T_5T_8 + K_3T_6T_9$$
$$k_{24} = K_7T_5T_3 + K_8T_6T_2$$
$$k_{25} = K_7T_5T_6 + K_8T_6T_5$$
$$k_{26} = K_7T_5T_9 + K_8T_6T_8$$
$$k_{31} = K_1T_7T_1 + K_2T_8T_2 + K_3T_9T_3$$
$$k_{32} = K_1T_7T_4 + K_2T_8T_5 + K_3T_9T_6$$
$$k_{33} = K_1T_7T_7 + K_2T_8T_8 + K_3T_9T_9$$

$$
\begin{aligned}
k_{34} &= K_7T_8T_3 + K_8T_9T_2 \\
k_{35} &= K_7T_8T_6 + K_8T_9T_5 \\
k_{36} &= K_7T_8T_9 + K_8T_9T_8 \\
k_{41} &= K_9T_2T_3 + K_{10}T_3T_2 \\
k_{42} &= K_9T_2T_6 + K_{10}T_3T_5 \\
k_{43} &= K_9T_2T_9 + K_{10}T_3T_8 \\
k_{44} &= K_4T_1T_1 + K_5T_2T_2 + K_6T_3T_3 \\
k_{45} &= K_4T_1T_4 + K_5T_2T_5 + K_6T_3T_6 \\
k_{46} &= K_4T_1T_7 + K_5T_2T_8 + K_6T_3T_9 \\
k_{51} &= K_9T_5T_3 + K_{10}T_6T_2 \\
k_{52} &= K_9T_5T_6 + K_{10}T_6T_5 \\
k_{53} &= K_9T_5T_9 + K_{10}T_6T_8 \\
k_{54} &= K_4T_4T_1 + K_5T_5T_2 + K_6T_6T_3 \\
k_{55} &= K_4T_4T_4 + K_5T_5T_5 + K_6T_6T_6 \\
k_{56} &= K_4T_4T_7 + K_5T_5T_8 + K_6T_6T_9 \\
k_{61} &= K_9T_8T_3 + K_{10}T_9T_2 \\
k_{62} &= K_9T_8T_6 + K_{10}T_9T_5 \\
k_{63} &= K_9T_8T_9 + K_{10}T_9T_8 \\
k_{64} &= K_4T_7T_1 + K_5T_8T_2 + K_6T_9T_3 \\
k_{65} &= K_4T_7T_4 + K_5T_8T_5 + K_6T_9T_6 \\
k_{66} &= K_4T_7T_7 + K_5T_8T_8 + K_6T_9T_9
\end{aligned}
$$

The other three global element stiffness submatrices are evaluated in a similar fashion.

B.6 Element End Forces

If the element axes coincide with the principal axes of the section then the element stiffness submatrices in the element axes system can be used to calculate the element end forces from the element end displacements using the following equations.

$$
\begin{aligned}
f_{ij} &= k_{ii}^{j} \, T_{ij} \, \Delta_{ij} + k_{ij} \, T_{ij} \, \Delta_{ji} \\
f_{ji} &= k_{ji} \, T_{ij} \, \Delta_{ij} + k_{jj}^{i} \, T_{ij} \, \Delta_{ji}
\end{aligned}
$$

The element stiffness submatrices and the transformation matrices have already been presented. It is computationally advantageous to expand the above equations as follows.

Note also that if there are any fixed end forces these must be superimposed on the force system associated with nodal displacements.

Plane Truss

$$f_{ijx} = \frac{EA}{L}[\cos\alpha\,(\Delta_{ix} - \Delta_{jx}) + \sin\alpha\,(\Delta_{iy} - \Delta_{jy})]$$

$$f_{ijy} = 0$$

$$f_{jix} = -f_{ijx}$$

$$f_{jiy} = 0$$

Space Truss

$$f_{ijx} = \frac{EA}{L}[l_x(\Delta_{ix} - \Delta_{jx}) + m_x(\Delta_{iy} - \Delta_{jy}) + n_x(\Delta_{iz} - \Delta_{jz})]$$

$$f_{ijy} = 0$$

$$f_{ijz} = 0$$

$$f_{jix} = -f_{ijx}$$

$$f_{jiy} = 0$$

$$f_{jiz} = 0$$

Plane Frame - Various End Connections

Note that the "K" factors for different end connections are given in Table B.1.

$$f_{ijx} = \frac{EA}{L}(\Delta_{ix}\cos\alpha_i - \Delta_{jx}\cos\alpha_j + \Delta_{iy}\sin\alpha_i - \Delta_{jy}\sin\alpha_j)$$

$$f_{ijy} = K_1\frac{EI}{L^3}(-\Delta_{ix}\sin\alpha_i + \Delta_{jx}\sin\alpha_j + \Delta_{iy}\cos\alpha_i - \Delta_{jy}\cos\alpha_j) + \frac{EI}{L^2}(K_2\,\Theta_i + K_3\,\Theta_j)$$

$$m_{ij} = \frac{EI}{L^2}(-K_2\,\Delta_{ix}\sin\alpha_i - K_5\,\Delta_{jx}\sin\alpha_j + K_2\,\Delta_{iy}\cos\alpha_i - K_5\Delta_{jy}\cos\alpha_j) + \frac{EI}{L}(K_4\,\Theta_i + K_6\,\Theta_j))$$

$$f_{jix} = -f_{ijx}$$

$$f_{jiy} = -f_{ijy}$$

$$m_{ji} = -m_{ij} + L f_{ijy}$$

Grillage

$$f_{ijz} = \frac{12EI}{L^3}(\Delta_{iz} - \Delta_{jz}) + \frac{6EI}{L^2}[(\Theta_{ix} + \Theta_{jx})sin\alpha - (\Theta_{iy} + \Theta_{jy})cos\alpha]$$

$$m_{ijx} = \frac{GJ}{L}[(\Theta_{ix} - \Theta_{jx})cos\alpha + (\Theta_{iy} - \Theta_{jy})sin\alpha]$$

$$m_{ijy} = \frac{6EI}{L^2}[-\Delta_{iz} + \Delta_{jz}] - \frac{2EI}{L}[(2\Theta_{ix} + \Theta_{jx})sin\alpha - (2\Theta_{iy} + \Theta_{jy})cos\alpha]$$

$$f_{jiz} = -f_{ijz}$$

$$m_{jix} = -m_{ijx}$$

$$m_{jiy} = -m_{ijy} - Lf_{ijz}$$

Space Frame

To find the element end forces for a space frame element requires an additional transformation as shown in the following equation.

$$f_{ij}^{*} = k_{ii}^{j*} T_{ij}^{*} T_{ij} \Delta_i + k_{ij}^{*} T_{ij}^{*} T_{ij} \Delta_j$$

Using the "T" and "K" factors described previously.

$$f_{ijx}^{*} = K_1 (T_1\Delta_{ix} + T_4\Delta_{iy} + T_7\Delta_{iz} - T_1\Delta_{jx} - T_4\Delta_{jy} - T_7\Delta_{jz})$$

$$f_{ijy}^{*} = K_2 (T_2\Delta_{ix} + T_5\Delta_{iy} + T_8\Delta_{iz} - T_2\Delta_{jx} - T_5\Delta_{jy} - T_8\Delta_{jz})$$

$$f_{ijz}^{*} = K_3 (T_3\Delta_{ix} + T_6\Delta_{iy} + T_9\Delta_{iz} - T_3\Delta_{jx} - T_6\Delta_{jy} - T_9\Delta_{jz})$$

$$m_{ijx}^{*} = K_4 (T_1\Delta_{ix} + {}_4\Delta_{iy} + T_7\Delta_{iz} - T_1\Delta_{jx} - T_4\Delta_{jy} - T_7\Delta_{jz})$$

$$m_{ijy}^{*} = K_8 (T_3\Delta_{ix} + T_6\Delta_{iy} + T_9\Delta_{iz} - T_3\Delta_{jx} - T_6\Delta_{jy} - T_9\Delta_{jz})$$

$$m_{ijz}^{*} = K_7 (T_2\Delta_{ix} + T_5\Delta_{iy} + T_8\Delta_{iz} - T_2\Delta_{jx} - T_5\Delta_{jy} - T_8\Delta_{jz})$$

$$f_{jix}^{*} = -f_{ijx}^{*}$$

$$f_{jiy}^{*} = -f_{ijy}^{*}$$

$$f_{jiz}^{*} = -f_{ijz}^{*}$$

$$m_{jix}^{*} = -m_{ijx}^{*}$$

$$m^*_{jiy} = -m^*_{ijy} - L f^*_{ijy}$$

$$m^*_{jiz} = -m^*_{ijz} + L f^*_{ijz}$$

B.7 Reactions

The reactions at the restraints can be calculated from the following equation (note that the summation is for all elements framing into node "i").

$$\boldsymbol{R}_i = \sum_{j=1}^{n} \boldsymbol{F}_{ij} - \boldsymbol{P}_i$$

The element end force equations produce forces in the element axes system. These forces must be transformed into the global axes prior to summation, i.e.

$$\boldsymbol{R}_i = \sum_{j=1}^{n} \boldsymbol{T}^{-1}_{ij} \boldsymbol{f}_{ij} - \boldsymbol{P}_i$$

In this equation the principal axes of the element are assummed to coincide with the element axes. In the case of space frames this is not necessarily the case and an additional transformation is necessary as shown in the following equation.

$$\boldsymbol{R}_i = \sum_{j=1}^{n} \boldsymbol{T}^{-1}_{ij} \boldsymbol{T}^{*-1}_{ij} \boldsymbol{f}_{ij} - \boldsymbol{P}_i$$

Bibliography

1 Bathe, K.J., and Wilson, E.L., *Numerical Methods in Finite Element Analysis,* Prentice-Hall, 1976.

2 Bathe, K.J., and Wilson, E.L., *Solution Methods for Eigenvalue Problems in Structural Mechanics,* Int. Journal for Numerical Methods in Engineering, Vol 6, pp213-226.

3 Bathe, K.J., and Wilson, E.L., *Eigensolution of Large Structural Systems with Small Bandwidth,* A.S.C.E. Journal of the Engineering Mechanics Division, Vol 99, pp467-479.

4 Bhat, P., *Solution of Problems in the Matrix Analysis of Skeletal Structures,*Construction Press, 1981.

5 Coates, R.C., Coutie, M.G., Kong, F.K., *Structural Analysis* (3rd ed.), Chapman and Hall, 1990.

6 Clough, R.W. and Penzien, J., *Dynamics of Structures,* McGraw-Hill, 1975.

7 Gere, J.M., Timoshenko, S.P., *Mechanics of Materials* (2nd ed.), Van Nostrand Reinhold, 1987.

8 Ghali and Neville, *Structural Analysis - A Unified Classical and Matrix Approach* (3rd ed.), Chapman and Hall, 1989.

9 Gordon, J.E., *Structures or Why Things Don't Fall Down,* Penguin 1978.

10 Hurty, W.C., and Rubinstein, M.F., *Dynamics of Structures,* Prentice-Hall, 1964.

11 Jenkins, W.M., *Structural Analysis using Computers,* Longman Scientific and Technical, 1990.

12 Livesley, R.K., *Matrix Methods of Structural Analysis,* Pergamon (2nd ed.), 1964 .

13 Majid. K.I., *Theory of Structures with Matrix Notation,* Newnes-Butterworth, 1978.

14 Marshal and Nelson, *Structures,* Longman Scientific and Technical, 1990

15 Neal, B.G., *Structural Theorems and Applications*, Pergamon, 1964.

16 Pilkey, W.D., and Pilkey, O.H., *Mechanics of Solids*, Quantum, 1974.

17 Przemieniecki, J.S., *Theory of Matrix Structural Analysis*, McGraw-Hill, 1968.

18 Ross, C.T.F., *Finite Element Methods in Structural Mechanics*, Ellis Horwood, 1985.

19 Ross, C.T.F., and Johns, T., *Computer Analysis of Skeletal Structures*, SPON, 1981.

20 Rubenstein, M.F., *Computer Matrix Analysis of Structures*, Prentice-Hall, 1966.

21 Smith, I.M., *Programming the Finite Element Method*, Wiley, 1982.

22 Tauchert, T.R., *Energy Principles in Structural Mechanics*, McGraw-Hill, 1974.

23 Timoshenko, S.P., and Young, D.H., *Theory of Structures*,McGraw-Hill, 1965.

24 Timoshenko, S.P., and Goodier, J.N., *Theory of Elasticity*, (3rd ed.) McGraw-Hill, 1970.

25 Warburton, G.B., *The Dynamical Behaviour of Structures*, (2nd ed.), Pergamon, 1976.

26 Weaver, W., Gere, J.M., *Matrix Analysis of Framed Structures* (2nd ed.), Van Nostrand, 1980.

27 West, Harry.H., *Analysis of Structures* (2nd ed.), Wiley, 1989.

28 Wilkinson, J.H., *The Algebraic Eigenvalue Problem*, Clarendon Press, 1965.

Index

Software Order Form

The programs described in the book are available on disk either as ASCII source code files or in compiled form. Compiled programs can be run from the DOS command by simply typing the name of the program, but they have the disadvantage that they cannot be modified by the user. Source code programs can, of course, be altered/extended by the user. They have the disadvantage that they will only run under a BASIC compiler or an interpreter. The programs will run without modification on any IBM Personal Computer, or 100% compatible IBM PC-compatible computer with a minimum of 640K of memory and a floppy disk drive (either 5.25" 1.2 megabyte, or 3.5", 1.4 megabyte).

Note that a basic compiler (QBASIC, QuickBASIC, Microsoft PDS) or interpreter (e.g. GW-BASIC) is required to run the programs in source code form. The disk of programs is accompanied by a user manual. The site licence includes source and compiled code, and two copies of the manual.

Please copy and complete the form below, enclose a sterling cheque for the total amount due, and mail to:

Qtek Systems
5 Hermitage Terrace
Edinburgh EH10 4RP

Description			Price	No.	Total
Demonstration disk			Free		
Disk of source code			£45		
Disk of compiled code			£45		
Site licence (5 users)			£120		
Site licence (10 users)			£180		
Disk Size	3.5"	5.25"		TOTAL	

(Note that the price includes VAT, postage and packing)

Name: . . . .

Address: . . . .

. . . .

. . . .

. . . .

Postcode . . . . Date: . . . .

Note: The software is supplied without any warranty.